Visual Dysfunction in Diabetes

Ophthalmology Research

Joyce Tombran-Tink, PhD, and Colin J. Barnstable, DPhil
Series Editors

For further volumes
http://www.springer.com/series/7660

Visual Dysfunction in Diabetes

The Science of Patient Impairment and Health Care

Edited by

Joyce Tombran-Tink, PhD
Department of Ophthalmology
Department of Neural and Behavioral Sciences
Milton S. Hershey Medical Center
Penn State University College of Medicine, Hershey, PA, USA

Colin J. Barnstable, DPhil
Department of Neural and Behavioral Sciences
Milton S. Hershey Medical Center
Penn State University College of Medicine, Hershey, PA, USA

Thomas W. Gardner
Department of Ophthalmology and Visual Sciences, Kellogg Eye Center
University of Michigan Medical School, Ann Arbor, MI, USA

Humana Press

Editors
Joyce Tombran-Tink, PhD
Department of Ophthalmology
Department of Neural
and Behavioral Sciences
Milton S. Hershey Medical Center
Penn State University College of Medicine
Hershey, PA, USA
jttink@aol.com

Colin J. Barnstable, DPhil
Department of Neural
and Behavioral Sciences
Milton S. Hershey Medical Center
Penn State University College of Medicine
Hershey, PA, USA
cbarnstable@hmc.psu.edu

Thomas W. Gardner
Department of Ophthalmology
and Visual Sciences
Kellogg Eye Center
University of Michigan Medical School
Ann Arbor, MI, USA
tomwgard@med.umich.edu

http://extras.springer.com

ISBN 978-1-4939-6168-9   ISBN 978-1-60761-150-9 (eBook)
DOI 10.1007/978-1-60761-150-9
Springer New York Dordrecht Heidelberg London

Printed on acid-free paper

Humana Press is part of Springer Science+Business Media (www.springer.com)

Preface

INTRODUCTION

This monograph is intended to serve two functions: first, to help readers understand the impact of vision impairment in people living daily with diabetes rather than considering diabetic retinopathy solely as a medical problem; second, to explore what we know and what we do not know about the ways diabetes affect the eye. Even with the plethora of new information being generated, there are still a series of fundamental questions that must be addressed if we are to develop effective treatments for diabetic retinopathy.

In the first chapter of this volume, Stuckey relates her experiences with proliferative diabetic retinopathy (PDR) and associated laser treatment. She provides a perspective on the visual and emotional component of vision loss that can be explained only by someone who has experienced it firsthand. She describes not only the loss of vision from the vitreous hemorrhage, the pain of the laser treatments, but also the permanent consequence of reduced peripheral vision and ability to adapt to dark conditions and from dark to light. Thus, it is clear that ophthalmologists do not "cure" diabetic retinopathy with retinal photocoagulation, but merely keep people from really becoming blind. Stuckey provides powerful incentives for us to do a better job to understand the nature of the problems she and other people with diabetes face, or at least dread. She also provokes us to prevent diabetic retinopathy or at least maintain vision without the need for destructive treatment.

HOW IS DIABETIC RETINOPATHY DETECTED?

For the detection and diagnosis of diabetic retinopathy in standard clinical practice, each patient is assessed individually with standard clinical tools including indirect ophthalmoscopy and slit lamp biomicroscopy following pupillary dilation. These methods of physical examination not only provide structural information about the ocular media and the status of the retinal blood vessels and optic nerve, but also provide little information regarding the structure or function of the neural retina, the part that is key to vision. So, the evaluation of large populations for the presence of retinopathy is usually done by photographic methods; the analysis of the resulting images has dramatically reduced vision impairment in communities of countries such as Iceland and Norway. However, the protocols for capturing and assessing the images continue to evolve because they require manual interpretation and are not quantitative.

Scanlon summarizes the progress in screening for diabetic retinopathy based on his extensive experience in the United Kingdom. Clearly, screening in European countries is much more widely implemented and successful than in the United States or elsewhere, revealing the distinct cultural and economic differences in response to a common problem across the oceans. Thus, there is no single solution to population screenings for diabetic retinopathy and multiple approaches may be needed to achieve optimal specificity and sensitivity.

Adams and Bearse detail their extensive cross-sectional and longitudinal studies of patients with diabetes and no or mild nonproliferative retinopathy using multifocal ERGs and visual field tests. They find that prolonged implicit time on the mfERG, an indicator of bipolar cell and outer plexiform layer integrity, predicts the development of vascular lesions, with topographical correspondence. This technique has the advantage of being independent of patient responses and can assess nearly the entire retina. Their data clearly show the early impact of diabetes on the neurosensory retina prior to the loss of visual acuity, and illustrate the potential to diagnose retinal impairment early so that it can be slowed if treatments can be developed.

HOW DOES DIABETES AFFECT THE EYE?

The clinical impact of diabetes on the eye is generally discussed in terms of diabetic retinopathy, but Midena reinforces the importance of corneal neuropathy which predisposes patients to epithelium breakdown, and is reflected by changes in the corneal structure as seen with confocal microscopy and by reduced corneal sensation. Diabetic corneal neuropathy has little direct impact on visual function but is further evidence of the widespread impact of diabetes in the eye. Furthermore, diabetes often frequently causes dysfunction of the autonomic nerves that regulate pupil size. Taken together with the impact of diabetes on sensory neurons in the retina, it is now evident that diabetes causes widespread neuropathic changes in the eye.

Cunha-Vaz and colleagues point out that there may be variable phenotypes of diabetic retinopathy based on clinical findings of microaneurysm turnover, vascular leakage, and macular thickening. In several longitudinal studies, they have quantified microaneurysm turnover on fundus photographs as well as vascular leakage and macular thickening to form a composite multimodal retinal analysis system that provides a more comprehensive assessment of retinopathy grade than any measure alone.

The clinical phenotype of diabetic retinopathy has generally been descriptive with little effort to provide quantitative parameters that predict the progress of diabetic retinopathy. The composite scoring system developed by Cunha-Vaz et al. is one of the first endeavors to account for consequences of increased vascular leakage and capillary closure. They found a greater rate of microaneurysm formation turnover in patients with more severe diabetes and worse visual acuity. This careful analysis of various patterns of vascular damage is an important step toward an improved understanding of diabetic retinopathy.

Medina and Vujosevic address the fundamental issue of the impact of diabetes on various aspects of vision. They trace a series of investigation into this question over the past 3 decades in which increasingly sensitive tests have been used to quantify defects in the inner vs. outer retina, and macular vs. mid-peripheral retinal in patients with various stages of diabetes. Most studies have evaluated a limited number of parameters in small cohorts of patients, so it remains difficult to have a comprehensive assessment of the impact of the range of diabetic retinopathy on vision over time. However, the net knowledge at this point that there is evidence of ganglion cell and inner retinal defects, as well as defects in the photoreceptor/pigmented epithelium with increased retinopathy grade, macular edema, and proliferative retinopathy. However, it remains uncertain which cellular defects primarily give rise to loss of visual acuity or the relationship of functional defects to alterations in retinal structure.

Two chapters examine various aspects of blood–retinal barrier break down in diabetic retinopathy. First, Hafezi-Moghadam discusses the normal role of the blood–retinal barrier to protect the neural retina and the role of inflammation and BRB permeability in diabetic retinopathy. In particular, he summarizes the role of inflammatory leukocyte recruitment to capillary endothelium by adhesion molecules such as ICAM-1, integrins, and other molecules that allow leukocytes to migrate through extracellular matrix. One of the mechanisms by which leukocytes increase permeability is through the release of azurocidin, a protease that attracts other inflammatory cells and increases vascular permeability. The actions of azurocidin can be blocked by a protease inhibitor such as aprotinin in experimental models of diabetic retinopathy, and he points out that aprotinin is used clinically in patients undergoing cardiothoracic and orthopedic surgery to reduce vascular leakage. In sum, this model suggests that leukocyte recruitment and activation may play a critical role in retinal vascular leakage particularly media through azurocidin release and this strategy may provide a therapeutic target.

Runkle, Titchenell, and Antonetti detail the known cellular and molecular regulation of the blood–retinal barrier and its compromise by diabetes, notably VEGF. VEGF induces phosphorylation and ubiquitination of occludin, leading to its internalization and movement away from the plasma membrane, and increased endothelial cell permeability, as mediated by activation of protein kinase C (PKC) isoforms. Several of these steps may be targets for therapeutic regulation.

In addition to a change in the barrier function of the retinal vasculature, the vessels themselves undergo pathological changes. Kern describes the capillary nonperfusion and degeneration that are early hallmarks of diabetic retinopathy. These changes can lead to preretinal neovascularization, and many of the current therapeutic approaches are based on the premise that blocking the early vascular pathology will prevent this subsequent pathology.

Extracellular serine proteinases include urokinase plasminogen activator (uPA) and members of the family of zinc-dependent endopeptidases called matrix metalloproteinases (MMPs). These proteinases participate in the degradation of interstitial extracellular matrices and basement membranes, and help in the recruitment of progenitor cells into the extracellular matrix during tissue remodeling. Proteinases are expressed by normal cells in tissue remodeling events and also during pathological events such as tumor angiogenesis and metastasis. The roles of these proteinases in diabetic retinopathy are summarized in the chapter by Rangasamy, McGuire, and Das.

Urokinase activates its cognitive receptor, a member of the lymphocyte antigen receptor superfamily, and leads to MAPK activation. MMPs release extracellular matrix from angiogenic growth factors such as VEGF and bFGF. They are expressed in multiple retinal cell types and are potential targets for therapeutic manipulation, either directly or via tissue inhibitors of matrix proteinases (TIMPs). To date most of the work in the eye relates to the control of abnormal vascular leakage and macular edema or neovascularization.

One of the ways of gaining insight into the biochemical changes occurring in diabetic retinopathy is to examine the proteins in the vitreous. Feener describes the identification of several hundred proteins in the human vitreous and the changes that occur in diabetes. Though many of the changes seen can be attributed a breakdown in the blood–retinal barrier, other may represent proteins secreted from the retina or attempts by the retina to counteract the deleterious effects of diabetes. As well as providing insights into the pathogenesis of the disease, these proteomic studies may give us sensitive biomarkers to indicate the stage and prognosis for patients.

Diabetic retinopathy is much more than a vascular disease and Barber, Robinson, and Jackson summarize the current knowledge of neurodegeneration in diabetic retinopathy. There are close similarities in structure in alterations and structure and function of the retina in animal models of diabetic retinopathy and humans. That is, there is delayed oscillatory potentials and reduction of the b-wave amplitude that corresponds with, but is not necessarily the direct result of increased death of retinal ganglion cells, amacrine neurons, bipolar neurons, and photoreceptors and/or reduced neurotransmission. Together, this extensive evidence clearly shows that there is neurodegeneration in early stages of diabetic retinopathy concomitant with the early detection of vascular changes. These findings are fundamental to our understanding of the nature of diabetic retinopathy and have a great impact on future efforts in diagnosis, prevention, and treatment.

Khan and Chakrabarti summarize the mechanisms by which hyperglycemia depresses the viability and function of retinal endothelial cells such that they have an increased rate of apoptosis, alters their participation in autoregulation, damages basement membranes matrix constituents, and contributes to neovascularization. Multiple biochemical changes have been described in animal models of diabetes and endothelial cells and cultural but the understanding of their roles in human diabetic retinopathy remains limited.

Stahl and coworkers discuss regarding insulin-like growth factor binding protein-3 (IGFBP-3) as a regulator of the growth hormone/insulin-like growth factor pathway in proliferative retinopathies. They summarize the relationship between VEGF-induced angiogenesis in retinopathy of prematurity (ROP) and PDR. Both conditions are characterized by peripheral retinal capillary closure, followed by peripheral retinal neovascularization, and treatments for both conditions are currently limited to growth factor inhibition and/or laser photocoagulation after the development of neovascularization. Their previous work in experimental models of ROP suggests that there are reduced insulin-like growth factor-1 (IGF-1) levels in the serum of premature infants associated with a loss of peripheral retinal vessels, and that systemic IGF-1 administration increases the risk of neovascularization. Likewise, patients with type 1 diabetes have reduced serum IGF-1 levels in the preproliferative stage, and systemic IGF treatment can accelerate the development of ocular neovascularization. Elevated serum IGF-1 levels are associated with accelerated proliferative retinopathy in pregnant diabetic women.

The authors describe the role of (IGFBP-3) which forms a molecular complex with insulin-like growth factors in the serum and retards their degradation. They propose that IGFBP-3 could be used as an adjunct to IGF-1 supplementation during the nonproliferative phase of retinopathy. In the proliferative phase IGF-1 may accelerate the involvement of neovascularization. Thus, titration of the levels of IGF and binding proteins may allow for improved regulation of proliferative retinopathies.

Murray and Ma summarize the panoply of proteins that exert prosurvival and differentiation features in retinal vascular and neuronal cells. They emphasize that despite laboratory-based studies of the biological roles of these factors, most of them have not been studied sufficiently to enable clinical trials. Moreover, most of them are studied as single factors whereas they function in combination with others in vivo. Nevertheless, these naturally derived biological products have potential for clinical application.

The most severe forms of diabetic retinopathy occur due to vitroretinal traction leading to epiretinal membranes with tingental or anterior traction, frequently resulting in retinal detachment and blindness.

For the past 15 years, the major emphasis in diabetic retinopathy research has been VEGF-induced neovascularization but the cause of fibrosis following treatment of neovascularization has remained unclear. van Geest et al. have pioneered the concept that connect tissue growth factor (CTGF) is increased during the fibrotic stage of diabetic retinopathy, or at least is expressed without the opposition of VEGF. In fact, they also show in strong evidence that CTGF expression increases in the blood vessels of diabetic rats shortly after diabetes induction suggesting that the fibrotic process actually starts in the preclinical stage of diabetic retinopathy, concomitant with basement lamina thickening, gloss of pericytes, and capitulary occlusion. Further studies will help to determine if CTGF inhibition can prevent fibrosis within the retina and the risk of tractional retinal detachment.

HOW CAN VISION LOSS BE LIMITED: EXPERIMENTAL THERAPIES

The ultimate test of a proposed disease mechanism lies in its relevance as a therapeutic target. Since the initial discovery of increased VEGF levels in human diabetic retinopathy in 1994, numerous studies have demonstrated a relationship with DME and increasing severity of retinopathy. Kim, Do, and Nguyen review the literature on the effects of intravitreally administered VEGF antagonists on DME. The positive effects of repeated treatments have now been shown in several clinical trials, but the authors remind us that the mechanisms by which vision improves after VEGF inhibition remain uncertain. As they also point out, it is unknown precisely why and how vision is impaired by DME in the first place. The growing evidence of a key role of VEGF and its inhibition will stimulate further investigations into these important questions.

Simo and colleagues point out that the metabolic pathways leading to retinal neurodegeneration are poorly understood, but there is likely an imbalance of neuroprotective factors vs. neurotoxic metabolites such as glutamate. The authors also emphasize the use of the *db/db* mouse with a leptin receptor mutation as a model to study retinal neurodegeneration in diabetes because it eliminates any potential for confounding effects of streptozotocin on the findings.

The range of neuropeptides in the retina is extensive and includes pigment epithelial-derived factor (PEDF), somatostatin (SST), erythropoietin (Epo), neuroprotectin D1 (NPD1), brain-derived neurotrophic factor (BDNF), glial cell line-derived neurotrophic factor (GDNF), ciliary neurotrophic factor (CNTF), and adrenomedullin (AM). SST is potentially interesting in diabetes since its general function in the peripheral tissues is to mediate the effects of growth hormone and IGF-1. In the retina, SST is expressed by amacrine cells and pigmented epithelium, and is reduced in diabetic rats and in diabetic human vitreous. Retinal lipids are also important because docosahexaenoic acid is a precursor to NPD1.

One group of cells that serve as an important source of active peptides in the retina are the glial cells. Sawada and colleagues document the effects that cytokines released from glial cells can have on the blood–retinal barrier and discuss treatments that may show some benefit by altering the pattern of expression of these cytokines.

Begg and colleagues thoroughly reviewed the effects of improved diabetes control on the development and progression of diabetic retinopathy, detailing the results of the DCCT and EDIC studies. They also cite less known findings, such as the improved outcome in patients undergoing panretinal photocoagulation who have HBA1c<8% at the time of treatment than those whose control is worse.

In addition, they summarize the studies that confirm strong beneficial effects of pancreas transplantation and islet cell transplantation, although the ocular benefits arise at the cost of more hypoglycemia and side effects of immunosuppression. In short, the prognosis for vision is markedly better with better metabolic control, irrespective of the means by which it is achieved.

From the chapters in this volume, it will be apparent that we have an overview of the timing and pathology of vascular lesions in the retinas of patients with diabetes. We also know that macular edema is a major factor in the loss of visual acuity and that laser photocoagulation and anti-VEGF therapies convey substantial benefit to many patients.

The list of what we do not know is much longer. We need to know whether metabolic factors beyond glucose contribute to vision-threatening diabetic retinopathy and how these lead to vision impairment. Is diabetic retinopathy a response to systemic metabolic abnormalities or are there unique ocular problems related to insulin resistance? Perhaps, the most fundamental gap in our knowledge is the relationship between the neural, vascular, and inflammatory abnormalities in diabetic retinopathy. Do they represent a pathological cascade induced sequentially or simultaneous responses to one or more metabolic perturbations? If we do not address these questions, it is possible that the long process of developing new therapeutics will target only one arm of the pathology and leave the retina open to damaging consequences of the others. Although we think of the changes detected in diabetes as being pathological, many of them may be an attempt by the tissue to restore normal function. This is certainly true in inflammatory responses, and we need to distinguish protective from damaging inflammatory responses.

Although there is much about the biology of the normal and diabetic eye that still needs to be learned, we also have an urgent need to develop tools that will help in the testing and application of new therapeutics. We clearly need to define optimal indices of retinal structure and function that predict development of diabetic retinopathy and vision impairment; indices that can be used as dynamic parameters for clinical trials of therapeutics.

While the list of outstanding questions is long, the tools to address them are now available and we can look forward to rapid progress in knowledge and, more importantly, new scientific approaches that lessen the vision impairment associated with diabetes.

Joyce Tombran-Tink
Colin J. Barnstable
Thomas W. Gardner

Contents

Contributors

ANTHONY J. ADAMS • *School of Optometry, University of California, Berkeley, CA, USA*
DAVID A. ANTONETTI • *Departments of Cellular and Molecular Physiology and Ophthalmology, Penn State College of Medicine, Hershey, PA, USA*
ALISTAIR J. BARBER • *Departments of Ophthalmology and Cellular and Molecular Physiology, Penn State College of Medicine, Hershey, PA, USA*
MARCUS A. BEARSE JR. • *School of Optometry, University of California, Berkeley, CA, USA*
IAIN S. BEGG • *Department of Ophthalmology and Visual Sciences, University of British Columbia, Vancouver, BC, Canada*
RUI BERNARDES • *AIBILI, Azinhaga Santa Comba, Celas, Coimbra, Portugal*
SUBRATA CHAKRABARTI • *Department of Pathology, University of Western Ontario, London, ON, Canada*
JOSÉ CUNHA-VAZ • *AIBILI, Azinhaga Santa Comba, Celas, Coimbra, Portugal*
ARUP DAS • *Division of Ophthalmology, University of New Mexico School of Medicine, Albuquerque, NM, USA*
DIANA V. DO • *Wilmer Eye Institute, Johns Hopkins University, Baltimore, MD, USA*
EDWARD P. FEENER • *Joslin Diabetes Center, Boston, MA, USA*
THOMAS W. GARDNER • *Department of Ophthalmology and Visual Sciences, Kellogg Eye Center, University of Michigan Medical School, Ann Arbor, MI, USA*
ALI HAFEZI-MOGHADAM • *Department of Radiology, Harvard Medical School, Center for Excellence in Functional and Molecular Imaging Brigham and Women's Hospital, Boston, MA, USA*
ANN HELLSTROM • *Department of Ophthalmology, Harvard Medical School, Children's Hospital Boston, Boston, MA, USA*
CRISTINA HERNÁNDEZ • *Diabetes Research Unit, Institut de Recerca Hospital Universitari Vall d'Hebron, Barcelona, Spain*
SHUICHIRO INATOMI • *Department of Ophthalmology, Sapporo Medical University School of Medicine, Sapporo, Japan*
GREGORY R. JACKSON • *Departments of Ophthalmology and Neural and Behavioral Sciences, Penn State College of Medicine, Hershey, PA, USA*
TIMOTHY S. KERN • *Departments of Medicine and Ophthalmology, Case Western Reserve University, Cleveland, OH, USA*
ZIA A. KHAN • *Department of Pathology, University of Western Ontario, London, ON, Canada*
BEN J. KIM • *Wilmer Eye Institute, Johns Hopkins University, Baltimore, MD, USA*
I. KLAASSEN • *Department of Ophthalmology, Ocular Angiogenesis Group, Academic Medical Center, University of Amsterdam, Amsterdam, The Netherlands*

E.J. KUIPER • *Department of Ophthalmology, Ocular Angiogenesis Group, Academic Medical Center, University of Amsterdam, Amsterdam, The Netherlands*
CONCEIÇÃO LOBO • *AIBILI, Azinhaga Santa Comba, Celas, Coimbra, Portugal*
CHATARINA LOFQVIST • *Department of Ophthalmology, Harvard Medical School, Children's Hospital Boston, Boston, MA, USA*
JIAN-XING MA • *Department of Physiology, University of Oklahoma Health Sciences Center, Oklahoma City, OK, USA*
PAUL MCGUIRE • *Department of Cell Biology and Physiology, University of New Mexico School of Medicine, Albuquerque, NM, USA*
EDOARDO MIDENA • *Department of Ophthalmology, University of Padova, Padova, Italy and Fondazione GB Bietti per l'Oftalmologia IRCSS, Rome, Italy*
ANNE R. MURRAY • *Department of Physiology, University of Oklahoma Health Sciences Center, Oklahoma City, OK, USA*
QUAN DONG NGUYEN • *Wilmer Eye Institute, Johns Hopkins University, Baltimore, MD, USA*
NAMI NISHIKIORI • *Department of Ophthalmology, Sapporo Medical University School of Medicine, Sapporo, Japan*
HIROSHI OHGURO • *Department of Ophthalmology, Sapporo Medical University School of Medicine, Sapporo, Japan*
SAMPATHKUMAR RANGASAMY • *Department of Cell Biology and Physiology, University of New Mexico School of Medicine, Albuquerque, NM, USA*
WILLIAM F. ROBINSON • *Departments of Ophthalmology, Penn State College of Medicine, Hershey, PA, USA*
E. AARON RUNKLE • *Department of Pathology,, Penn State College of Medicine, Hershey, PA, USA*
NORIMASA SAWADA • *Department Pathology, Sapporo Medical University School of Medicine, Sapporo, Japan*
PETER SCANLON • *Harris Manchester College, University of Oxford, Oxford, UK*
R.O. SCHLINGEMANN • *Department of Ophthalmology, Ocular Angiogenesis Group, Academic Medical Center, University of Amsterdam, Amsterdam, The Netherlands*
RAFAEL SIMÓ • *Diabetes Research Unit, Institut de Recerca Hospital Universitari Vall d'Hebron, Barcelona, Spain*
LOIS SMITH • *Department of Ophthalmology, Harvard Medical School, Children's Hospital Boston, Boston, MA, USA*
ANDREAS STAHL • *Department of Ophthalmology, Harvard Medical School, Children's Hospital Boston, Boston, MA, USA*
HEATHER STUCKEY • *Department of Medicine, Penn State University College of Medicine, Hershey, PA, USA*
DAVID M. THOMPSON • *Department of Medicine, University of British Columbia, Vancouver, BC, Canada*
PAUL M. TITCHENELL • *Department of Cellular & Molecular Physiology,, Penn State College of Medicine, Hershey, PA, USA*

R.J. van Geest • *Department of Ophthalmology, Ocular Angiogenesis Group, Academic Medical Center, University of Amsterdam, Amsterdam, The Netherlands*
C.J.F. van Noorden • *Department Cell Biology and Histology, Ocular Angiogenesis Group, Academic Medical Center, University of Amsterdam, Amsterdam, The Netherlands*
Marta Villarroel • *Diabetes Research Unit, Institut de Recerca Hospital Universitari Vall d'Hebron, Barcelona, Spain*
Stela Vujosevic • *Department of Ophthalmology, University of Padova, Padova, ItalyFondazione GB Bietti per l'Oftalmologia IRCSS, Rome, Italy*
Garth L. Warnock • *Department of Surgery, University of British Columbia, Vancouver, BC, Canada*

Part I
Living with Diabetic Retinopathy

1
Living with Diabetic Retinopathy: The Patient's View

Heather Stuckey

CONTENTS

Keywords Dark adaptation • Floaters • Insulin-dependent diabetes • Laser treatment • Micro aneurysm • Quality of life

The men of experiment are like the ant, they only collect and use; the reasoners resemble spiders, who make cobwebs out of their own substance. But the bee takes the middle course: it gathers its material from the flowers of the garden and field, but transforms and digests it by a power of its own. Not unlike this is the true business of philosophy (science); for it neither relies solely or chiefly on the powers of the mind, nor does it take the matter which it gathers from natural history and mechanical experiments and lay up in the memory whole, as it finds it, but lays it up in the understanding altered and digested. Therefore, from a closer and purer league between these two faculties, the experimental and the rational, much may be hoped.

—Francis Bacon

Although many of us can understand diabetic retinopathy from a scientific, rational view, this chapter takes us deeper into the personal experience of having diabetic retinopathy. It explores some of the fears, uncertainties, and hope from people who have diabetes, including my own. Like some of you reading this chapter, I am a researcher motivated by improving diabetes. Not unlike the bee, I am also in the unique position of having insulin-dependent diabetes myself since the age of 12. This dual role of researcher and patient gives me the opportunity to narrate the complex relationship of living a life with diabetes and a complication of diabetic retinopathy, while maintaining an active research agenda with diabetes.

From: *Ophthalmology Research: Visual Dysfunction in Diabetes*
Edited by: J. Tombran-Tink et al. (eds.), DOI 10.1007/978-1-60761-150-9_1

From this insider patient perspective, diabetes is different than when it is viewed as only a science. It takes audacity to inject a needle under the skin four or five times a day or to start an insulin pump. It requires persistence to handle a disease that is relentless. It takes understanding to put yourself in the place of a patient who crawls on the kitchen floor while trying to get a cup of juice, trembling in sweat and fuzziness. It takes courage to accept the news that you have diabetic retinopathy, and you need immediate surgery to prevent blindness. From a distance, the decisions about medical care and diabetes treatment look different than when they are happening to you.

Until there is a cure for diabetes and retinopathy, we need to continue to search for the best advances in medical care, and how our actions are affecting those we serve. We need to listen to the experiences of our patients to balance our scientific knowledge about the disease. Rita Charon, a general internist and literary scholar, focuses on the outcomes of documenting the experiences and narratives of patients, and how these narratives function in the construction of knowledge [1–3]. Charon [4] said she "came to understand that I had accrued deep knowledge about my patients that remained unavailable" because she had not written down the stories of the patients (p. 404). Sharing what she has learned with her patients is therapeutic, often deepening their mutual commitment and investment. She went on to say, "I feel privileged to have discovered how to fortify my medicine with the narrative gifts of perception, imagination, curiosity, and the indebtedness we listeners accrue toward those we hear."

The chapter begins with my personal experience of having diabetes and diabetic retinopathy. Toward the end of the chapter, there are stories included from other individuals who've mentioned their experiences with diabetic retinopathy. Within the narratives, there is a common thread of fear of the unknown in the foreground, yet a promise of hopefulness. There is hope that we will find a cure for diabetes and that we can make the treatment for retinopathy less destructive.

MY PATIENT EXPERIENCE

It is difficult to imagine a life without eyesight or world without shape and color. When much younger, I used my eyes to draw, to write, and to see the world through the imagination. To stare at the clouds and dream of dragons, ships, and explorers across the blue vastness was one of my favorite hobbies. During my kindergarten years, my eyesight began to blur—very slowly—until I could no longer see the blackboard clearly in my classroom, and the teacher moved my seat to the front of the class. Signs looked fuzzy, and trees no longer looked like they had leaves, but were morphed lumps of green, yellow, and orange colors. This was my first experience with myopia, corrected with glasses, and the world was restored. If only all problems in the 1970s could have been solved with a glass lens and a plastic frame! From that young age, I've been wearing some sort of corrective eyewear and have always respected the power of the eyes.

At the age of 12, I was diagnosed with insulin-dependent diabetes. My mother noticed the symptoms of diabetes—constant thirst, with my drinking nearly a gallon of milk at a time, and frequent urination, every hour on the hour. She knew the symptoms because her mother had lived with type 2 for a number of years before being diagnosed. The time in the hospital was fuzzy, but friends and teachers would ask what it was like to give myself shots and what foods I was "allowed" to eat. At that time, I didn't want to talk about my diabetes. My disease was something I would have rather ignored. I always gave myself my shots, but didn't frequently check my blood sugar. It wasn't something that seemed that imperative. Certainly, I understood that one of the primary complications of diabetes was blindness, but I didn't want to think that it could happen to me. I was young and felt indestructible, but had no realistic grasp of what the elevated blood sugars were doing to the tiny vessels in my eyes. I had no idea at all—until my first visit to the office of ophthalmology in 1995 after my left eye had hemorrhaged.

I had been taking a shower when I first noticed a spider web off to my left. The black swirl appeared ominous against the white porcelain. Although I tried to whisk it away, I couldn't seem to reach the shadowy web. Terrified, I realized it was *inside* my eye, not an external web. Hundreds of thoughts burst into my mind. What is it? What's happening? Is this a complication of diabetes? Am I going blind? The ophthalmologist, Dr. Gardner, assured me that he would do his best to prevent blindness, to stop the progression of the disease. But, that would mean immediate surgery.

At first, it was difficult to understand what having proliferative diabetic retinopathy meant. Maybe it was the suddenness of the onset or the startled reaction of the diagnosis, but my memory is somewhat cloudy. In my recollection, it was explained that my blood vessels were trying to get oxygen, and to maintain adequate oxygen levels, they started to form smaller blood vessels. Unfortunately, these vessels were much more tenuous and fragile than the original. They broke easily, and what I was seeing was some of the blood leaking into the retina and vitreous, causing floaters. It looked like a shadow moving across my eye, rather than something definitive. It was shapeless, and I watched the kaleidoscope of blood start as a large woven mass, then slowly break into little parts over the next few hours, eventually forming a fog which hindered my sight for several months. At that time, I didn't understand that the technical name was neovascularization. I simply knew that things were not as they should be, and that my eyes were calling for help.

On the day of my appointment, I entered a small room with bright cinder block walls. Humming sounds and drips were ominous, as I waited for the unknown. Dr. Gardner asked if I had any questions before beginning the hour-long procedure. "No," I told him. "But please be careful. I know you've done this a 1,000 times before, but I'm scared." Clasping my hand in his, he silently communicated trust. He encouraged me to be strong as he glued the round stabilizer to my eyelid. I tried to blink, but the surrounding metal resisted motion. He turned his back to prepare a syringe of relaxant solution. "You might feel a pinch," he said, as what felt like a 6-in. needle penetrated my bottom-left eyelid. Wincing, I adjusted the Sony headphones over my ears so I could hear the music of Enya rather than the chilling drip, drip, drip around me.

With my chin and forehead trapped against steel, Dr. Gardner skillfully aimed the first laser shot. At first, I didn't feel pain. Two, three, still nothing. Twenty, thirty, forty, the back of my eye pinched. Two hundred, three hundred. My eye ached from the sharpness. As the doctor consoled me with, "You're doing fine" and "Hang in there," one strong emotion surfaced: anger; anger at my eyes for being imperfect, anger at myself for not keeping my diabetes in control, and anger at my diabetes for being so cruel.

For a day or two, I wore a patch over my eye and slept. As the patch was peeled away, things appeared brighter than before, but not unbearable. The room felt full of sunbeams, even on the somewhat cloudy day. The white-painted walls mingled with the space in front of me, and it took a moment to find the dimensions of both, where one started while the other began. After the adjustment, I could see the shapes of my lamp, the bedposts, the pillows, all of my personal books, and items within the bedroom. This familiar sight reassured me that the surgery was successful, and I felt the tension leave my body. The whiteness and disorientation faded over the next few hours, but the sensitivity to light and reduced peripheral vision remains.

What has helped the most in getting through this complication is the attention of the ophthalmologist himself, Dr. Gardner. My experience of having a physician who is soft-spoken and compassionate has soothed my fears and communicated trust. His ability to give undivided attention, and remembering to ask questions about my family or a personal situation, has connected me with him. He is attentive and gently touches my shoulder when he walks in the room to ask how I am doing. His personalized interactions have made the difference in my optimism about the future of my eyesight and improved quality of life. When my eyes don't seem quite right, or I am experiencing a new symptom, such as flashes or unusual coloring, I can call or e-mail him to ask him whether it is necessary for me to come for a visit, or whether these side effects are "normal" in patients with diabetic proliferative retinopathy. He is responsive and respects my value as a patient and as a colleague. These are qualities that have helped me both physically with my retinopathy as well as psychologically with the anxiety associated with the complications. I am indebted to his skill as a physician, his vision as a researcher, and his personal mission to help all patients see to the best of their ability. These are qualities which help physicians continue to excel in their practice.

The complications of retinal surgery are difficult to adjust to, and it requires a supportive physician and patient interaction to be successful. Even after 15 years of living with the disease, I'm not used to the difficulty of seeing at night and in bright lights. This was a complication that I knew would be a probability, but it is very different when actually going through the experience. One spring, I took a trip to Washington, DC, with four of my childhood friends. We were amazed at the marble steps and pillars of the Lincoln Memorial, commemorating the 16th president of the USA. All of us walked the low steps that led to the central hall, where the solitary figure of Abraham Lincoln sat. Along the side walls were carved inscriptions of the Inaugural and the Gettysburg Address, sending us the message of equality and a new birth of freedom. After viewing the monument, my friends started to walk down the stairs, as we were planning to walk around the National Mall. I was still looking at the marble Lincoln, and as I turned around, I realized I was alone. I walked out to the front of the monument and shaded my eyes from the glaring sun. As I looked down, all I could see was a white slate, instead of distinguishable steps. I knew there were steps there—I'd walked up them and my friends walked down—but where was the next step? My eyes had not adjusted, and I began to get anxious. I called out to one of my friends, "Tammy," but she didn't hear me. I sensed there were many other people around me, but the world was just so sparkling white that I couldn't really see anything. For a moment, I was paralyzed, standing at the top of the steps, staring blankly. A wave of panic rolled through my forehead. I scrunched down and walked on four limbs like a crab down the stairs. My friends were laughing at the bottom of the steps, "What are you doing?" because they thought I was trying to be funny. I told them I couldn't see, but I'm sure they didn't quite understand. Honestly, I didn't understand. Now, I'm aware that I need to be careful in places where there is a shift from dark to bright light. Something simple like walking out onto the patio of my house on a sunny day requires me to tap the space in front of me to find the concrete step below. It's a reminder that I need to be cautious and that my eyes need time to adjust.

This also happens when I go from light to dark areas. I used to be one of those people who would sneak into a movie theater while the previews were playing, just in time for the feature presentation. Now, I'm one of the first to sit down while there are still dim lights in the theater. My 12-year-old son and I were going to the movies, and we were a few minutes late. He stopped and asked if I was OK. With popcorn in my right hand and a soda in the other, it was difficult to find another hand to grab onto his coat to make my way through the aisles. Coming into a poorly lit room makes it impossible for me to move forward until my eyes adjust. It takes me at least 5 min to begin to see silhouettes of images or people in the room. I can no longer trust my sense of sight because my eyes have been damaged by laser surgery and years of high blood sugars; instead, I intently rely on the sense of feel and memory.

Another simple event that causes difficulty is heading out to see the fireworks at dusk. I had an experience of following a friend up a road that led to a grassy path. My friend went ahead, but I wasn't sure where the road stopped and the grass began. It appeared as though the terrain had changed, but the road in front of me looked like a dark lake, and I wasn't sure I could trust what it was seeing. I could tell that other people were moving around me, quite quickly, as I stepped quietly, one toe at a time to find my way. My friend turned around and took my arm, leading me with her across the grass. It's times like these that I am keenly aware of my altered vision.

An enjoyment of mine is going to amusement parks, but having reduced vision makes seeing through the indoor queue lines quite difficult because of the sudden shift from light to dark. Recently, we were in Disneyland, California, ready to ride "Indiana Jones Adventure." The entryway halls were dark for effect, with a strange-looking hologram on the wall. I squinted, but still couldn't quite make out the image. It was all I could do to navigate the left-to-right line to keep up. I held onto my son's shirt so that I didn't lose my way, but I heard the people in back of me grow impatient. They stepped on the back of my shoes and said, "move forward." They could see fine, so what was my problem? After all, I didn't look blind, and my healthy, strong body shouldn't have needed assistance.

My vision issues don't just stop with transitions from dark to light. I'm concerned about when I'm going to have my next episode of severe floaters in my right eye. I've been bothered by these floaters ever since my surgery. I'm never sure if my sudden loss of vision is going to be permanent. At the most unfortunate time, when I was trying to conduct my dissertation work, I developed a large floater in my right eye, making it impossible to see. The reason and timing for the appearance of floaters seem to be unpredictable—I was watching television and noticed the fireworks explosion of fluid filling my eye. As if writing a dissertation isn't stressful enough, I was trying to meet the deadlines with only one functioning eye. I tried to look around the web by moving my head, having to rely on my left eye to read. I think about these floaters often, and wonder when the next one might hit. The rational I knows it will be a few weeks, or months, until the cloud dissipates, but a side of me also wonders whether the obstruction will be permanent. As it's been well over a decade since my last surgery, the floaters are becoming more sporadic, and my eyes are more stable. I'm also getting used to the signs and symptoms of a floater, and no longer am surprised by having limited vision. However, I'm still never certain that they will go away.

The effects of the laser treatment also restrict my driving in unknown places. I am reluctant to drive at night because I am afraid that I won't be able to see properly. It's difficult to see the transition in the road from highway to ramps, especially in rural areas that are dimply lit at night. Rainstorms in the dark magnify the problem. Driving on a snowy, sunny day can be worse because the intense whiteness is simply blinding. It is the same situation as the fireworks path, where things appear to be a continuous row without distinction between one terrain and the other. I lose the ability to distinguish depth, distance, and shading. Now I limit my driving at night to places that are familiar to me or allow someone else to drive me. My driving record is safe, but it is better to take a precaution to not drive than find myself in an unknown situation. Because of the eye damage, I think twice about whether I can go into our local caverns with my son because of the darkness, or any kind of fun house, haunted house, or darkened museum. It's not like being in a dark room, where you can still see shapes and patterns. This is complete black, like being blindfolded. There's no depth to anything, so it's a matter of feeling my way around the room.

Having had several laser treatments, my peripheral vision is also limited. It hasn't affected much of my life, but it is funny when I go for the yearly eye exam, and I realize how much I really can't see. The technician checking my vision is holding out his fingers to the right saying, "How many do I have up?" and I'm thinking, "Man, I really can't see anything." It isn't a real problem, except that I need to remember to look down, especially in the kitchen where I typically run into the corner of the side table or the cat dishes on the floor. It's also common for me to trip over the open dishwasher. Part of this comes from the fact that I was never considered graceful, but I'm sure having limited peripheral vision doesn't help. My experience with having diabetic retinopathy has been filled with both laughter at my inadequacies and fear at the uncertainties.

OTHERS' EXPERIENCES

These kinds of uncertainties have also been the experience of others with diabetic retinopathy. In a qualitative study of ten people with diabetes, we examined how this group coped, or made meaning of their diabetes. The purpose of the pilot study was to understand more about the experience of diabetes and its complications, in order to help adults live more harmoniously with their chronic disease [5,6]. The average age of the participant was 42, with an age at diagnosis between the years of 4 and 25 (average = 10.8). They had type 1 diabetes from a minimum of 12 years to a maximum of 52 years (average = 31), with 311 cumulative years of experience with diabetes.

The study began by asking the participants to tell me about their diagnosis of diabetes, which was difficult for most to do as they had not thought about how that diagnosis may have affected the way that they are currently caring for their disease. My work did not specifically include the transcripts of the participants' fears of retinopathy and other complications. But because the patient's experience of retinopathy is an important point to be made for this chapter, I have included their comments (with pseudonyms used) below.

Six out of the ten participants had at least one retinal surgery, and they found it to be a difficult experience. In one participant's story of retinopathy in 2003, Karla said a floater happened where she least expected it—St. John, US Virgin Islands. She woke up around 3:00 a.m. in her camp cottage and began to violently dry heave and vomit. Approximately 30 min later, she woke up, looked around, and realized her vision had something obstructing it. She tells of her experience in this way:

> I blinked to see if I was dreaming, but knew immediately that it was a dreaded "floater." I had to turn my head to the side so I could see out of that eye. It was as if I constantly had a bug flying into my line of vision. Being that it was 3:30 in the morning and not much healthcare available on the island, I waited until the sun rose to tell my friends I needed to go to the clinic.

She told them her suspicions about a microaneurysm bursting from the force of the dry heaves, but there was nothing they could do for her at St. John, so she left for the island of St. Thomas via ferry ride. She arrived at the ER, where the on-call physician examined her eye and said there was nothing he could do for her, either. He called the local ophthalmologist to see if she was available, but was not hopeful since it was a Saturday. Luckily, the ophthalmologist was still in her office, which was only a block away. She told Karla that she did have a bleed in her eye and that she should avoid scuba diving, sneezing, coughing, or anything that would put pressure on her eye. Karla was "so afraid to even fly home to the states." She was scheduled for laser surgery about a week later, and says:

> I was given the option of having a numbing medicine injected for the procedure, but decided the needle might be worse than how the doctor described the surgery. Instead, I just took two Advil an hour prior to surgery. I was led into a pitch dark room and had something placed in my eye to keep it open. Then I proceeded to see bright green flashes of light and heard sounds like a video game (like Asteroids, if you are old enough to remember Atari). My doctor warned me when he got closer to a nerve, because that did cause more discomfort than other areas. It was like a twinge or someone hitting your funny bone, only in your eyes.

She said her eye felt sore for an hour or so after the procedure, but overall was not "as bad as I had psyched up myself to expect. The worst part of the whole thing was having your eye held open when you had an extreme urge to blink." She is still frightened of the end results if a full retinal detachment were to occur, because she loves photography and sightseeing, but is no longer afraid of the laser surgery procedure. She had only one surgery, and so far, it has been successful. She thanks God every day for the gift of her sight. Having the surgery has been a reminder not to take her sight for granted. The pictures below are the microaneurysm that bled in her left eye (Figs. 1 and 2).

As another participant described her surgery for diabetic retinopathy, she explained how it hurt, but also that she was fortunate to have not gone blind. She understands that the "flip side" of dealing with diabetes is that she could have lost a limb already, or been blind, and she could have had "so much happened to me that hasn't." She could get through the retinal surgery, knowing that she would be able to watch the sunset, or look in her garden, and see her children grow up to graduate or to get married. Knowing that

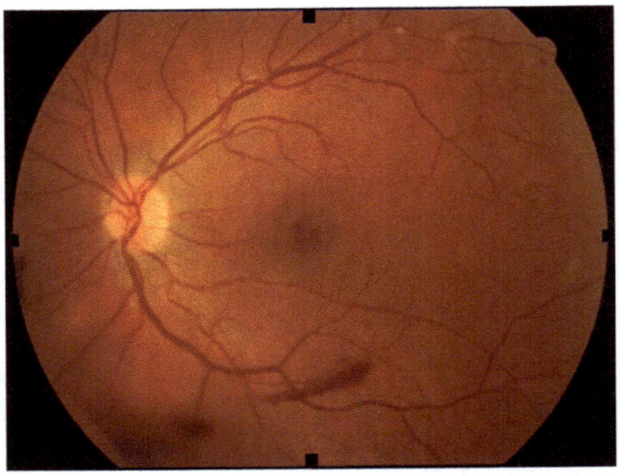

Fig. 1. Left eye microaneurysm.

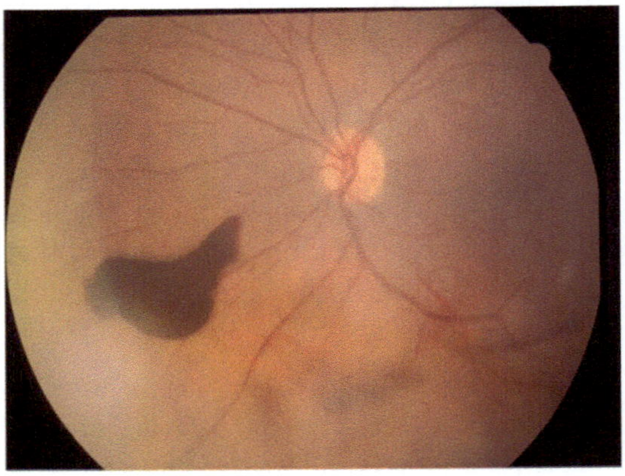

Fig. 2. Left eye subhyaloid hemorrhage.

she is able to see, having the retinal surgery was not as bad as the alternative. Camilla summarized her gratitude in this way:

> When it comes down to it, I count myself truly blessed because I could have had things so much worse. I just learned to deal with what I've been given, and just think it could be worse. Just be grateful that this is all you have to deal with.

Because of her retinopathy, Camilla also relies on her husband to do most of the driving, especially at night and in the rain. Her husband was supportive of her when she developed retinopathy and had to go to the eye doctor. She called him at work because she was seeing something in front of her eye. She explained to him,

> 'I have this claw-looking thing,' and he's like, 'Can you see it?' And I say, 'Yeah, I can see it,' not thinking he thinks that it's something that's protruding out of my eye. So he rushes over to meet me at the eye doctor, and he says, 'Well, you look OK. I was thinking I was going to see this monster.' [He thought the "claw" was outside, not inside, her eye.]

One of the more ominous thoughts about diabetes for these participants was the possibility of going blind. Before going into laser surgery for the first time, Camilla spent some time with her children, and she vividly described her feelings as she spent the day with them:

> The whole time, it was a dreary day, and I was just taking in everything. What the clouds looked like. They're so gray, in the dark over here, and trying to keep everything pictured in my mind. What the trees looked like. What the Dairy Queen sign looked like. My husband's profile. I just kept looking at him and the children. I gave the kids a hug, and I tried to remember.

For one participant, it was difficult for her to help other people understand what it is like to get laser surgery for diabetic retinopathy. She said, "They have no idea," but she was grateful to be able to talk to the group, who could relate to her complications on some level. All she can do is try to "stay ahead of it" on a day-to-day basis and make the best of the difficult days. Amber was used to dealing with diabetes, the way that she was "used to" dealing with the blood she has in both her eyes from retinopathy. She said that the bleeds in her eyes have become a part of her vision, and she tells herself to keep going. "You know," she said, "You've got to deal with what you have."

Like the others in the group, I generally take a positive spin on diabetes. Sometimes you need to laugh a little. One woman told her daughter, "If I ever go blind, don't put me in a polka-dotted shirt." We sometimes make light of our disease. After several years, it still requires creativity to figure out where to put an insulin pump on a swimsuit. The pump does make my life easier and better, especially at night. Before the pump, I would wake up with multiple low blood sugars while sleeping because the NPH insulin was peaking. These days, it's less common to have a low blood sugar at night. I also think that things could be worse, whether I'm talking about the insulin pump or talking about my complications. Having diabetes is not as bad as being—and I could finish the sentence a thousand ways—in the intensive care unit, diagnosed with MS or some forms of cancer, or dead. And yet, we may have some of the same fears and feelings as those who have a terminal illness.

Marie shared the story of being diagnosed with diabetes in 1984, which serves as an example of the fears. She has not had retinopathy surgery, but faces the prospect of blindness as a complication of diabetes:

> As I went to get my insulin and syringes from the pharmacy, I cried all the way there. Not only did I fear shots, but I've always been petrified of going blind and here I had a disease that actually had blindness as a possibility. I never did like anyone messing with my eyes. As a child, I would 'flip out' when I got an eyelash in my eye and had to work it out. Just thinking about having any kind of eye surgery or people invading my eyes is totally stressful. I am also somewhat claustrophobic, and blindness is very black, dark and confining… the ultimate in being locked in a car trunk or trapped in an elevator. My yearly eye exam is always tense, and I breathe a big sigh of relief when I hear that all is well. I am hoping that my eyes remain healthy because facing retinopathy is not anything I could easily deal with (and I've been through a lot… breast cancer with chemotherapy, major reconstructive surgery, carpel tunnel surgery, two broken wrists). None of these comes close to the fear I have of going blind.

Having diabetes is frightening and confusing, and the fear of going blind is pervasive, like the humidity of summer. My purpose is to help myself and others make meaning of diabetes and see how we can find greater strength and wellness with the opportunity for healing, even if not a cure. Even if we don't understand all the root causes of diabetes or retinopathy, as patients, we can reflect on what we do know and how we can help others live more fully with the disease. As medical professionals, researchers, and scientists, that fear is something we can seek to eliminate.

PHOTOS OF THE MEANING OF DIABETES

To put these thoughts of the diagnosis and the meaning of diabetes in visual form, the photo below represents the day of my diabetes diagnosis (Fig. 3). It is labeled "unnatural" because having diabetes meant I would need to take some form of insulin injection every day for the rest of my life and should avoid sugar. I might go blind when I grow older or lose my kidney function. These things are unnatural, especially as a young child, represented by the bright orange slash. The slash appears among the ground and the grass of the earth, meaning growth and natural life. Although originally, the photo was about the diagnosis of diabetes, it also relates to its complications, such as retinopathy. Having retinal surgery is unnatural, as some blood vessels are sacrificed in order to save others and to preserve the site for long term. Although some eye procedures can be expected at an older age, it is unnatural, and frightening, to have surgery at age 25.

This next photo (Fig. 4) of a cell block also represents my thoughts of having diabetes and diabetic retinopathy. I took this picture at the Eastern State Penitentiary in Philadelphia, Pennsylvania. As the website states (http://www.easternstate.org/), the Penitentiary was once the most famous and expensive prison in the world, but stands today as a world of crumbling cellblocks and empty guard towers. My eyes used to be unscathed by disease, but have slowly deteriorated, like the plaster on the floor of the cell and the table that has fallen down from the weight of gravity over the years. My eyes show

Fig. 3. Unnatural.

Fig. 4. Hydrant.

evident signs of damage in the pin-points of burning laser that penetrated my retina, and my lack of peripheral vision.

The room (and my sight) is not gone, however, because the building has not collapsed. The structure remains intact. Although my eyes may be ragged and somewhat worn out, they still perform the job that they were intended to do. I can see. I realize the room will not be restored to complete newness, but it can be cleaned and maintained. Keeping my diabetes under control and my body healthy, there's hope that I will be able to see for my lifetime.

It is a wonderful thing to have vision, to experience life in color, to read, to watch the clouds move mysteriously on an overcast day, and to be able to turn my head and see my son when he was younger, yelling, "Watch this, mom," from the playground. As he gets older, my eyes soak in the shape of his face and the curl of his hair and study the speckles of light in his eyes. I can see, and my prognosis for continued vision is very good. Each year, I schedule an appointment with Dr. Gardner, and my eyesight remains stable.

Rather than destroying the retina and damaging vision, we need to find easier, gentler ways to treat diabetic retinopathy to detect ways of catching the disease earlier so the fear of blindness is much less. That is what is important to us who have retinopathy. But scientific research to find a less destructive treatment is only part of the story. Behind every project or procedure, there's a human element—a *person* who is frightened, wondering whether he's going to go blind. He's giving his eyes, one of his most valuable possessions to you, the clinician. Besides vessels and fluid, what do you see? Do you see the way they are looking at you for hope? Do you see how they are afraid that they might go blind? They don't want to go through laser treatment. They are afraid there will be complications with the surgery, and they will go blind. They won't remember the hue of the sky or the color of the cornfield. What did snow really look like? And what did the shadow of my toddler's head look like at night? This person with diabetic retinopathy might go blind. And they are looking to you for *hope.* Regardless of your relationship to research, there is a *patient*, not a retina, who needs hope. What do you see? How can you give them that hope? How can you communicate trust to them? The best advice I can give is to look them with a soft face and tell them that you are going to do whatever it takes to preserve their sight. Their probability for continued eyesight is going to be very good. There are other promising methods for treatment, and you will make sure that they are getting the best treatment possible. This is *really* seeing. How can you improve your eyesight, your communication of hope to the patient? If you give me laser surgery treatment, you're treating maybe half of my disease. But if you give me hope that I won't go blind, you treat the other half.

Perhaps some of you have diabetes, or have loved ones and friends who have a chronic illness, or have diabetic retinopathy. This personal connection is what stirred you. Maybe your interest also comes from a deep desire to improve the lives of so many who suffer with diabetes and its complications or the science of discovering a cure or a breakthrough in treatment. For me, understanding the experience of diabetes is not only a research interest, but a personal quest. My hope is that you will see what having diabetes, and diabetic retinopathy, means to someone with diabetes, and you will understand how very important your work is to those of us who have this chronic illness.

The research in this book is groundbreaking and exciting. Research like this has preserved the eyesight of myself and many others and improved our quality of life. Over the past 20 years, I have seen many outstanding medical achievements in diabetes care: blood glucose machines, which achieve accurate results in 5 s, short-acting human insulin, needles which come in ultrathin shapes and sizes, and the insulin pump, continuous glucose monitoring and new advances in knowledge, medication, and technology that have made it possible for people with *diabetes* to live long, productive lives.

Ultimately, I hope we will be able to find a cure for diabetes. Diabetes is a demanding, frightening, exasperating disease. I fully support research that finds ways to make it easier to live with the complications of diabetes. As a fellow researcher, a patient, and as a friend, I thank all of you reading this chapter who have worked to preserve our eyesight, in whatever way. I encourage you to continue to find research to improve the lives of those with diabetic retinopathy, not only to restore sight but also to give hope.

REFERENCES

1. Charon R, Spiegel M (2006) Reflexivity and responsiveness: the expansive orbit of knowledge. Lit Med 51:vi–xi
2. Charon R (2004) Narrative and medicine. New Engl J Med 350(9):862–865
3. Charon R (2001) Narrative medicine: a model for empathy, reflection and trust. J Am Med Assoc 286(15):1897–1902
4. Charon R (2004) Physician writers: Rita Charon. Lancet 363(9406):404
5. Stuckey H, Tisdell E (2010) The role of creative expression in diabetes: an exploration into the meaning-making process. Qual Health Res 20:42–56
6. Stuckey H (2009) Creative expression as a way of knowing in diabetes adult health education: an action research study. Adult Educ Q 60:46–64

Part II
How Is Diabetic Retinopathy Detected?

2
Diabetic Retinopathy Screening: Progress or Lack of Progress

Peter Scanlon

CONTENTS

Keywords Screening • Diabetic retinopathy • Visual Impairment • Blindness • Diabetes control and complications trial • United Kingdom prospective diabetes study • Early treatment diabetic retinopathy study • St. Vincent Declaration

DEFINITIONS OF SCREENING FOR DIABETIC RETINOPATHY

The definition of screening that was adapted by the WHO [1] in 1968 was "the presumptive identification of unrecognized disease or defect by the application of tests, examinations or other procedures which can be applied rapidly. Screening tests sort out apparently well persons who probably have a disease from those who probably do not. A screening test is not intended to be diagnostic. Persons with positive or suspicious findings must be referred to their physicians for diagnosis and necessary treatment."

Applying the principles for screening for human disease that were derived from the public health papers produced by the WHO [1] in 1968 to sight-threatening diabetic retinopathy raises the following questions [2]:

1. Is there evidence that sight-threatening diabetic retinopathy is an important public health problem?
2. Is there evidence that the incidence of sight-threatening diabetic retinopathy is going to remain the same or become an even greater public health problem?
3. Is there evidence that sight-threatening diabetic retinopathy has a recognizable latent or early symptomatic stage?
4. Is there evidence that treatment for sight-threatening diabetic retinopathy is effective and agreed universally?

From: *Ophthalmology Research: Visual Dysfunction in Diabetes*
Edited by: J. Tombran-Tink et al. (eds.), DOI 10.1007/978-1-60761-150-9_2

5. Is a suitable and reliable screening test available, acceptable to both health-care professionals and (more importantly) to the public?
6. Are the costs of screening and effective treatment of sight-threatening diabetic retinopathy balanced economically in relation to total expenditure on health care – including the consequences of leaving the disease untreated?

Is There Evidence That Sight-Threatening Diabetic Retinopathy Is an Important Public Health Problem?

Studies Reporting the Prevalence of Diabetic Retinopathy

Reports from North America have shown that diabetic retinopathy continues to be prevalent in the **USA**:

1. In 2008–2009, Klein [3] reported the 25-year progression of retinopathy and of macular edema [4] in persons with type 1 diabetes from the Wisconsin Epidemiological Study of Diabetic Retinopathy (WESDR study). The 25-year cumulative rate of progression of DR was 83%, progression to proliferative DR (PDR) was 42%, and improvement of DR was 18%. The 25-year cumulative incidence was 29% for macular edema and 17% for clinically significant macular edema.
2. In 1995, Klein [5] reported the incidence of macular edema over a 10-year period. This was 20.1% in the younger-onset group, 25.4% in the older-onset group taking insulin, and 13.9% in the older-onset group not taking insulin.
3. In 2004, Kempen [6] reported that, among an estimated 10.2 million US adults 40 years and older known to have DM, the estimated crude prevalence rates for retinopathy and vision-threatening retinopathy were 40.3 and 8.2%, respectively.

Worldwide reports have shown that sight-threatening diabetic retinopathy is prevalent in both type 1 and type 2 diabetes in the UK [7], India [8], Germany [9], Ethiopia [10], Australia [11], Denmark [12], Singapore [13], and China [14].

Reports on Blindness and Visual Impairment

In 1994, Moss [15] reported on the 10-year incidence of blindness in the WESDR study. 1.8, 4.0, and 4.8% in the younger-onset, older-onset taking insulin, and older-onset not taking insulin groups, respectively. Respective 10-year rates of visual impairment were 9.4, 37.2, and 23.9%.

In 1995, Evans [16] reported on the causes of blindness and partial sight in England and Wales from an analysis of all BD8 forms for the year April 1990 to March 1991. Among people of working age (ages 16–64), diabetes was the most important cause (13.8%) with 11.9% due to diabetic retinopathy. This study was repeated 10 years later and reported by Bunce [17] in 2006, and diabetic retinopathy was still the commonest cause of visual loss in the working age group.

In 2001, Cunningham [18] reported that 45 million people worldwide fulfill the World Health Organization's criterion for blindness and the cause of one-quarter of all blindness, which affects people in both developed and developing nations, includes diabetic retinopathy and macular degeneration. In 2002, Kocur [19] reported that in people of working age in Europe, diabetic retinopathy is the most frequently reported causes of serious visual loss.

Zhang [20] reported results from the national health and nutrition examination survey in the USA. People with diabetes were more likely to have uncorrectable VI than those without diabetes.

Is There Evidence That the Incidence of Sight-Threatening Diabetic Retinopathy Is Going to Remain the Same or Become an Even Greater Public Health Problem?

Numerous studies have shown that there is a rising incidence of diabetes and its complications in all age groups, both in the UK and worldwide.

In 1997, Amos [21] estimated that 124 million people worldwide have diabetes, 97% NIDDM, and that by 2010, the total number with diabetes is projected to reach 221 million.

In 2000, Sorensen [22] reported that the World Health Organization has recognized that there is a "global epidemic of obesity," and the prevalence of type 2 diabetes is rising in parallel.

In 2001, Boyle [23] estimated the number of Americans with diagnosed diabetes is projected to increase from prevalence of 4.0% in 2000 to a prevalence of 7.2% in 2050.

The International Diabetes Federation estimated the prevalence of diabetes in 2003 in 20–79 age groups and projected this to an estimate in 2025. They predicted rises in numbers of people with diabetes of 7.07–15.04 million in Africa, of 19.24–39.41 million in Eastern Mediterranean and Middle East Region, of 48.38–58.64 million in Europe, of 23.02–36.18 million in America, of 14.16–26.16 million in South and Central American Region, of 39.3–81.57 million in Southeast Asian Region, and of 43.02–75.76 million in Western Pacific Region.

Is There Evidence That Sight-Threatening Diabetic Retinopathy Has a Recognizable Latent or Early Symptomatic Stage?

Numerous reports from the Wisconsin Epidemiological Study [24, 25] have shown that sight-threatening diabetic retinopathy in both type 1 and type 2 diabetes has a recognizable latent or early symptomatic stage. In patients with type 1 diabetes, Klein [3] reported that the 25-year cumulative rate of progression of DR was 83%, progression to PDR was 42%, and improvement of DR was 18%.

The Early Treatment Diabetic Retinopathy [26] documented all the photographic lesions of diabetic retinopathy and the risks of progression of DR relating to those lesions.

The United Kingdom Prospective Diabetes Study [27] documented the incidence and progression of diabetic retinopathy over 6 years from diagnosis of type 2 (non-insulin-dependent) diabetes.

Is There Evidence That Treatment for Sight-Threatening Diabetic Retinopathy Is Effective and Agreed Universally?

The Evidence That Diabetic Retinopathy Can Be Prevented or the Rate of Deterioration Reduced by Improved Control of Blood Glucose, Blood Pressure and Lipid Levels, and by Giving Up Smoking

Evidence for the link between poor glucose control and greater progression of diabetic retinopathy (DR) was provided by numerous early studies [28, 29]. The study that confirmed that intensive blood glucose control reduces the risk of new-onset DR and slows the progression of existing DR for patients with IDDM was the Diabetes Control and Complications Trial (DCCT) [30].

Similarly, for type 2 diabetes, the United Kingdom Prospective Diabetes Study (UKPDS) [31] demonstrated that intensive blood glucose control reduces the risk of new-onset DR and slows the progression of existing DR for patients with type 2 diabetes.

Control of systemic hypertension has been shown [32, 33] to reduce the risk of new-onset DR and slow the progression of existing DR.

There is evidence [34, 35] that elevated serum lipids are associated with macular exudates and moderate visual loss, and partial regression of hard exudates may be possible by reducing elevated lipid levels.

There is some evidence that smoking may be a risk factor in progression of diabetic retinopathy in type 1 diabetes as described by Muhlhauser [36] and Karamanos [37]. However, in type 2 diabetes, the evidence is controversial [27].

The Evidence that Laser Treatment Is Effective

Evidence for the efficacy of laser treatment for diabetic eye disease has been shown from the Diabetic Retinopathy Study [38] and the Early Treatment Diabetic Retinopathy Study [39]. In 1976, the organizers of the Diabetic Retinopathy Study [40] modified the trial protocol and recommend treatment for control eyes with "high-risk characteristics." In 1981, they reported [41] that photocoagulation, as used in the study, reduced the 2-year risk of severe visual loss by 50% or more.

In 1985, a report [42] from the Early Treatment Diabetic Retinopathy Study showed that focal photocoagulation of "clinically significant" diabetic macular edema (CSMO) substantially reduced the risk of visual loss.

Further studies that have shown evidence for the longer-term efficacy of laser treatment for diabetic eye disease have been reported by Blankenship [43] and Chew [44].

The Evidence That Vitrectomy for More Advanced Disease Is Effective

Smiddy [45], he noted that, according to the Early Treatment Diabetic Retinopathy Study, at least 5% of eyes receiving optimal medical treatment will still have progressive retinopathy that requires laser treatment and pars plana vitrectomy. He also noted that, although vitrectomy improves the prognosis for a favorable visual outcome, preventive measures, such as improved control of glucose levels and timely application of pan retinal photocoagulation, are equally important in the management.

There have been reports of improving visual results during the last 20 years following vitrectomy, the most recent being from Yorston [46].

Is a Suitable and Reliable Screening Test Available, Acceptable to Both Health-Care Professionals and (More Importantly) to the Public?

There is an increasing acceptance that, in population-based screening programs, digital photography offers the best method of screening for sight-threatening diabetic retinopathy. Digital photography has been shown to provide higher sensitivities and specificities across large numbers of operators than examination techniques such as direct ophthalmoscopy [47, 48], or slit lamp biomicroscopy [49, 50]. Digital photography also has the advantage that a percentage of images can be reexamined for quality assurance purposes.

The acceptance of digital photography for population-based screening does not imply that this replaces the comprehensive eye examination as pointed out by Chew [51].

In screening studies, far more controversial than the use of digital photography has been the use of mydriasis or nonmydriasis and the number of fields photographed.

There have been strong proponents [52] of nonmydriatic photography for many years. However, it has been recognized in more recent years that ungradable image rates for nonmydriatic digital photography in a predominantly white Caucasian population [53, 54] are of the order of 19–26%. Scotland has developed a national screening program based on one-field nonmydriatic photography following a report [55] from the Health Technology Board for Scotland. Other proponents of nonmydriatic digital photography have attempted to capture three-fields [56], five-fields [57], and remarkably Shiba [58] excluded the over 70 years age group and attempted 9× overlapping nonmydriatic 45° fields.

Mydriatic digital photography studies [49, 53] have shown that consistently good results can be achieved, with sensitivities of >80% and high levels of specificity. In these studies, specificity does vary depending on whether ungradable images are regarded as test positive, but levels of >85% are consistently achieved. England has developed a national screening program [7] based on two-field mydriatic photography.

In 2004, Williams produced a report [59] for the American Academy of Ophthalmology summarizing the use of single-field fundus photography for diabetic retinopathy screening.

In 2007–2008, reports of diabetic retinopathy screening were published from France [60], Spain [61], the Canary Islands [62], Western Cape [63], the USA [64], and England [7].

The debate over whether mydriasis should be used for screening and the number of fields used has continued around the world with two of the recent studies coming to very different conclusions [60, 61].

Are the Costs of Screening and Effective Treatment of Sight-Threatening Diabetic Retinopathy Balanced Economically in Relation to Total Expenditure on Health Care – Including the Consequences of Leaving the Disease Untreated?

In 1982, Savolainen [65] reported on the cost-effectiveness of photocoagulation for sight-threatening diabetic retinopathy in the UK. There have been reports of computer simulation models of diabetic retinopathy screening by Javitt [66, 67], Dasbach [68], Caro [69], and Fendrick [70], based on the health systems in the USA and Sweden, that concluded that screening for sight-threatening diabetic retinopathy was cost-effective.

James et al [71]. reported results for an organized screening program in the UK using 35-mm retinal photography and demonstrated this to be more cost-effective than the previous system of opportunistic screening.

Meads [72] reviewed published studies of the costs of blindness and compared Fould's 1983 estimate [73] inflated to £7,433 in 2002 costs, Dasbach's 1991 estimate [68] inflated to £5,391 in 2002 costs, and Wright's 2000 estimate [74] inflated to £7,452 (4,070–£11,250) in 2002 costs. He concluded that much of the uncertainty in any sensitivity analysis of the cost of blindness in older people is associated with the cost of residential care and that the excess admission to care homes caused by poor vision is impossible to quantify at the present time.

Only four studies have been published that assess the costs of screening using digital photography. The first was from a telemedicine program in Norway [75] where, at higher workloads, telemedicine was cheaper. The second compared an optometry model with a digital photographic model in the UK [76]. However, in this study, there were poor compliance rates in the newly introduced screening program in both models. A cost-effectiveness analysis [77] of use of a telemedicine screening program in a prison population in Texas concluded that teleophthalmology holds great promise to reduce the cost of inmate care and reduce blindness caused by diabetic retinopathy in type 2 diabetic patients. Tung [78] concluded that screening for DR in Chinese with type 2 diabetes is both medically and economically worthwhile and recommended annual screening.

PROGRESS OF LACK OF PROGRESS IN SCREENING FOR DIABETIC RETINOPATHY IN DIFFERENT PARTS OF THE WORLD

In 1990, the St. Vincent Declaration [79] recognized diabetes and diabetic retinopathy to be a major and growing European health problem, a problem at all ages and in all countries. The first of the five-year targets that were unanimously agreed by government health departments and patient's organizations from all European countries was to reduce new blindness due to diabetes by one-third or more. In 2005 in Liverpool UK, a conference took place to review progress in the prevention of visual impairment due to diabetic retinopathy since the publication of the St. Vincent Declaration. Delegates attended as representatives from 29 European countries, and there were invited experts from Europe and the US. It was clear from this meeting that the health-care systems in Europe were at very different stages of development, and the funding of those health-care systems varied considerably. For example, if the population did not have access to adequate treatment facilities, there was little point in concentrating on screening for diabetic retinopathy until adequate treatment facilities were established.

Hence, the conference recommended the following steps in the development of systematic screening programs for sight-threatening DR:

Step 1

Access to effective treatment

- Minimum number of lasers per 100,000 population
- Equal access for all patient groups
- Maximum time to treatment from diagnosis, 3 months

Step 2

Establish opportunistic screening

- Dilated fundoscopy at time of attendance for routine care
- Annual review
- National guidelines on referral to an ophthalmologist

Step 3

Establish systematic screening

- Establish and maintain disease registers
- Systematic call and recall for all people with diabetes
- Annual screening
- Test used has sensitivity of ≥80% and specificity of ≥90%
- Coverage ≥80%

Step 4

Establish systematic screening with full quality assurance and full coverage

- Digital photographic screening
- All personnel involved in screening will be certified as competent
- 100% coverage
- Quality assurance at all stages
- Central/regional data collection for monitoring and measurement of effectiveness

The European countries that were most advanced in development of national screening programs were those that had nationalized health systems that facilitated the development of public health screening programs. Iceland, England, Scotland, Wales, and Northern Ireland had all developed national screening programs, whereas Denmark, Finland, and Sweden had regional programs, all with good coverage. At that time, these countries had an estimated overall prevalence of diabetes in Europe approximating 4%.

The wealthier European countries that had private health-care systems (e.g., Eire, France, Germany, Greece, Israel, Italy, Luxembourg, the Netherlands, Portugal, Spain) had developed local screening programs, many of which are based upon the initiatives of individual persons. However, there was a lack of uniformity between different centers on screening methodology and classification of diabetic retinopathy. More recently, there have been attempts within some of these countries to standardize [80] their screening systems and to develop a framework [81] for the development of a national screening program.

With respect to Eastern Europe (Czech Republic, Turkey, Hungary, Romania, and Serbia and Montenegro), the Czech Republic introduced diabetic retinopathy screening and treatment guidelines published in 2002; Hungary, Romania, and Turkey have local or regional screening programs. Turkey reported that 7.2% of their population was known to have diabetes. Serbia and Montenegro reported that they did not have a formalized screening program, but had taken steps to introduce protocols. In parts of Serbia, there was a lack of available lasers.

Posters were also presented from the following countries—Albania, Bulgaria, Georgia, Kazakhstan, Lithuania, Uzbekistan, and St. Petersburg. Bulgaria has 17 lasers, but there are insufficient in the other countries: Uzbekistan appears to have none and Kazakhstan only one or two. Lasers are available for the "general" population in Lithuania, with one in Albania, one in St. Petersburg, and some in Bulgaria. Other lasers are in private offices.

In Australia, there are local screening programs that have developed to serve individual populations such as the aboriginal [82] population and rural Victoria [83].

Similarly, localized screening programs have developed in the Western Cape [63], India [8], Japan [58], and China [14].

A recent study [84] by Boucher from Canada attempted to increase uptake of diabetic retinopathy screening by locating mobile screening imaging units within pharmacies. This produced further communication within the same journal to which Boucher replied [85], "Despite efforts to educate both patients and physicians about the importance of routine diabetic screening and despite the publication of Canadian screening guidelines, a large percentage of the diabetic population continues to receive inadequate retinopathy screening. This has led to the search for strategies to better detect vision-threatening retinopathy and reduce the incidence of complications and blindness from diabetic retinopathy."

In America, health-care delivery is chiefly driven by market forces, and the key to any new preventive health program is reimbursement. Provision of medical care is based on private insurance for those who can pay for it and a patchwork of Federal programs for the indigent and the elderly. It is estimated that there are more than 43 million Americans who have no health-care insurance whatsoever.

The Center for Medicare and Medicaid Services (CMS) sets reimbursement standards for Federal programs and also influences private insurers' reimbursement policies. Currently, CMS does not offer reimbursement for image-based diabetic retinopathy screening, and only a few private insurers do so.

Hence, screening programs in America have usually been developed by enthusiasts such as the Vine Hill program [64] where digital retinal imaging is undertaken in an inner-city primary care clinic, in the Joslin Diabetes Center [56], or in a Veterans Affairs Medical Center [86].

REFERENCES

1. Wilson J, Jungner G. The principles and practice of screening for disease. Public Health Papers 34. Geneva: WHO; 1968.
2. Scanlon P. An evaluation of the effectiveness and cost-effectiveness of screening for diabetic retinopathy by digital imaging photography & technician ophthalmoscopy & the subsequent change in activity, workload and costs of new diabetic ophthalmology referrals. [M.D.]. London; 2005.
3. Klein R, Knudtson MD, Lee KE, Gangnon R, Klein BE. The Wisconsin Epidemiologic Study of Diabetic Retinopathy: XXII the twenty-five-year progression of retinopathy in persons with type 1 diabetes. Ophthalmology. 2008;115(11):1859–68.
4. Klein R, Knudtson MD, Lee KE, Gangnon R, Klein BE. The Wisconsin Epidemiologic Study of Diabetic Retinopathy XXIII: the twenty-five-year incidence of macular edema in persons with type 1 diabetes. Ophthalmology. 2009;116(3):497–503.
5. Klein R, Klein BE, Moss SE, Cruickshanks KJ. The Wisconsin Epidemiologic Study of Diabetic Retinopathy. XV. The long-term incidence of macular edema. Ophthalmology. 1995;102(1):7–16.
6. Kempen JH, O'Colmain BJ, Leske MC, Haffner SM, Klein R, Moss SE, et al. The prevalence of diabetic retinopathy among adults in the United States. Arch Ophthalmol. 2004;122(4): 552–63.
7. Scanlon PH. The English national screening programme for sight-threatening diabetic retinopathy. J Med Screen. 2008;15(1):1–4.
8. Raman R, Rani PK, Reddi Rachepalle S, Gnanamoorthy P, Uthra S, Kumaramanickavel G, et al. Prevalence of diabetic retinopathy in India: Sankara Nethralaya Diabetic Retinopathy Epidemiology and Molecular Genetics Study report 2. Ophthalmology. 2009;116(2):311–8.

9. Hesse L, Grusser M, Hoffstadt K, Jorgens V, Hartmann P, Kroll P. Population-based study of diabetic retinopathy in Wolfsburg. Ophthalmologe. 2001;98(11):1065–8.
10. Seyoum B, Mengistu Z, Berhanu P, Abdulkadir J, Feleke Y, Worku Y, et al. Retinopathy in patients of Tikur Anbessa Hospital diabetic clinic. Ethiop Med J. 2001;39(2):123–31.
11. Tapp RJ, Shaw JE, Harper CA, de Courten MP, Balkau B, McCarty DJ, et al. The prevalence of and factors associated with diabetic retinopathy in the Australian population. Diabetes Care. 2003;26(6):1731–7.
12. Knudsen LL, Lervang HH, Lundbye-Christensen S, Gorst-Rasmussen A. The North Jutland County Diabetic Retinopathy Study: population characteristics. Br J Ophthalmol. 2006;90(11):1404–9.
13. Wong TY, Cheung N, Tay WT, Wang JJ, Aung T, Saw SM, et al. Prevalence and risk factors for diabetic retinopathy: the Singapore Malay Eye Study. Ophthalmology. 2008;115(11): 1869–75.
14. Wang FH, Liang YB, Zhang F, Wang JJ, Wei WB, Tao QS, et al. Prevalence of diabetic retinopathy in rural China: the Handan Eye Study. Ophthalmology. 2009;116(3):461–7.
15. Moss SE, Klein R, Klein BE. Ten-year incidence of visual loss in a diabetic population. Ophthalmology. 1994;101(6):1061–70.
16. Evans J. Causes of blindness and partial sight in England and Wales 1990–1991. London: OPCS; 1995. p. 1–29.
17. Bunce C, Wormald R. Leading causes of certification for blindness and partial sight in England & Wales. BMC Public Health. 2006;6:58.
18. Cunningham Jr ET. World blindness–no end in sight. Br J Ophthalmol. 2001;85(3):253.
19. Kocur I, Resnikoff S. Visual impairment and blindness in Europe and their prevention. Br J Ophthalmol. 2002;86(7):716–22.
20. Zhang X, Gregg EW, Cheng YJ, Thompson TJ, Geiss LS, Duenas MR, et al. Diabetes mellitus and visual impairment: national health and nutrition examination survey, 1999-2004. Arch Ophthalmol. 2008;126(10):1421–7.
21. Amos AF, McCarty DJ, Zimmet P. The rising global burden of diabetes and its complications: estimates and projections to the year 2010. Diabet Med. 1997;14 Suppl 5:S1–85.
22. Sorensen TI. The changing lifestyle in the world. Body weight and what else? Diabetes Care. 2000;23 Suppl 2:B1–4.
23. Boyle JP, Honeycutt AA, Narayan KM, Hoerger TJ, Geiss LS, Chen H, et al. Projection of diabetes burden through 2050: impact of changing demography and disease prevalence in the U.S. Diabetes Care. 2001;24(11):1936–40.
24. Klein R, Klein BE, Moss SE, Davis MD, DeMets DL. The Wisconsin Epidemiologic Study of Diabetic Retinopathy. IX. Four-year incidence and progression of diabetic retinopathy when age at diagnosis is less than 30 years. Arch Ophthalmol. 1989;107(2):237–43.
25. Klein R, Klein BE, Moss SE, Davis MD, DeMets DL. The Wisconsin Epidemiologic Study of Diabetic Retinopathy. X. Four-year incidence and progression of diabetic retinopathy when age at diagnosis is 30 years or more. Arch Ophthalmol. 1989;107(2):244–9.
26. Early Treatment Diabetic Retinopathy Study Research Group. Fundus photographic risk factors for progression of diabetic retinopathy. ETDRS report number 12. Ophthalmology. 1991;98(5 Suppl):823–33.
27. Stratton IM, Kohner EM, Aldington SJ, Turner RC, Holman RR, Manley SE, et al. UKPDS 50: risk factors for incidence and progression of retinopathy in Type II diabetes over 6 years from diagnosis. Diabetologia. 2001;44(2):156–63.
28. Brinchmann-Hansen O, Dahl-Jorgensen K, Sandvik L, Hanssen KF. Blood glucose concentrations and progression of diabetic retinopathy: the seven year results of the Oslo study. BMJ. 1992;304(6818):19–22.

29. Danne T, Weber B, Hartmann R, Enders I, Burger W, Hovener G. Long-term glycemic control has a nonlinear association to the frequency of background retinopathy in adolescents with diabetes. Follow-up of the Berlin Retinopathy Study. Diabetes Care. 1994;17(12):1390–6.
30. The Diabetes Control and Complications Trial Research Group. The effect of intensive treatment of diabetes on the development and progression of long-term complications in insulin-dependent diabetes mellitus. N Engl J Med. 1993;329(14):977–86.
31. UK Prospective Diabetes Study (UKPDS) Group. Intensive blood-glucose control with sulphonylureas or insulin compared with conventional treatment and risk of complications in patients with type 2 diabetes (UKPDS 33). Lancet. 1998;352(9131):837–53.
32. Chase HP, Garg SK, Jackson WE, Thomas MA, Harris S, Marshall G, et al. Blood pressure and retinopathy in type I diabetes. Ophthalmology. 1990;97(2):155–9.
33. Matthews DR, Stratton IM, Aldington SJ, Holman RR, Kohner EM. Risks of progression of retinopathy and vision loss related to tight blood pressure control in type 2 diabetes mellitus: UKPDS 69. Arch Ophthalmol. 2004;122(11):1631–40.
34. Chew EY, Klein ML, Ferris FL, Remaley NA, Murphy RP, Chantry K, et al. Association of elevated serum lipid levels with retinal hard exudate in diabetic retinopathy. Early Treatment Diabetic Retinopathy Study (ETDRS) report 22. Arch Ophthalmol. 1996;114(9):1079–84.
35. Cusick M, Chew EY, Chan CC, Kruth HS, Murphy RP, Ferris 3rd FL. Histopathology and regression of retinal hard exudates in diabetic retinopathy after reduction of elevated serum lipid levels. Ophthalmology. 2003;110(11):2126–33.
36. Muhlhauser I, Bender R, Bott U, Jorgens V, Grusser M, Wagener W, et al. Cigarette smoking and progression of retinopathy and nephropathy in type 1 diabetes. Diabet Med. 1996;13(6):536–43.
37. Karamanos B, Porta M, Songini M, Metelko Z, Kerenyi Z, Tamas G, et al. Different risk factors of microangiopathy in patients with type I diabetes mellitus of short versus long duration. The EURODIAB IDDM complications study. Diabetologia. 2000;43(3):348–55.
38. The Diabetic Retinopathy Study Research Group. Indications for photocoagulation treatment of diabetic retinopathy: Diabetic Retinopathy Study Report no. 14. Int Ophthalmol Clin. 1987;27(4):239–53.
39. Treatment techniques and clinical guidelines for photocoagulation of diabetic macular edema. Early treatment Diabetic Retinopathy Study Report Number 2. Early treatment Diabetic Retinopathy Study Research Group. Ophthalmol. 1987;94(7):761–74.
40. Spalter HF. Photocoagulation of circinate maculopathy in diabetic retinopathy. Am J Ophthalmol. 1971;1(1 Part 2):242–50.
41. The Diabetic Retinopathy Study Research Group. Photocoagulation treatment of proliferative diabetic retinopathy. Clinical application of Diabetic Retinopathy Study (DRS) findings, DRS Report Number 8. Ophthalmology. 1981;88(7):583–600.
42. Early Treatment Diabetic Retinopathy Study research group. Photocoagulation for diabetic macular edema. Early Treatment Diabetic Retinopathy Study report number 1. Arch Ophthalmol. 1985;103(12):1796–806.
43. Blankenship GW. Fifteen-year argon laser and xenon photocoagulation results of Bascom Palmer eye institute's patients participating in the diabetic retinopathy study. Ophthalmology. 1991;98(2):125–8.
44. Chew EY, Ferris 3rd FL, Csaky KG, Murphy RP, Agron E, Thompson DJ, et al. The long-term effects of laser photocoagulation treatment in patients with diabetic retinopathy: the early treatment diabetic retinopathy follow-up study. Ophthalmology. 2003;110(9):1683–9.
45. Smiddy WE, Flynn Jr HW. Vitrectomy in the management of diabetic retinopathy. Surv Ophthalmol. 1999;43(6):491–507.

46. Yorston D, Wickham L, Benson S, Bunce C, Sheard R, Charteris D. Predictive clinical features and outcomes of vitrectomy for proliferative diabetic retinopathy. Br J Ophthalmol. 2008;92(3):365–8.
47. Moss SE, Klein R, Kessler SD, Richie KA. Comparison between ophthalmoscopy and fundus photography in determining severity of diabetic retinopathy. Ophthalmology. 1985;92(1): 62–7.
48. Harding SP, Broadbent DM, Neoh C, White MC, Vora J. Sensitivity and specificity of photography and direct ophthalmoscopy in screening for sight threatening eye disease: the Liverpool Diabetic Eye Study. BMJ. 1995;311(7013):1131–5.
49. Olson JA, Strachan FM, Hipwell JH, Goatman KA, McHardy KC, Forrester JV, et al. A comparative evaluation of digital imaging, retinal photography and optometrist examination in screening for diabetic retinopathy. Diabet Med. 2003;20(7):528–34.
50. Warburton TJ, Hale PJ, Dewhurst JA. Evaluation of a local optometric diabetic retinopathy screening service. Diabet Med. 2004;21(6):632–5.
51. Chew EY. Screening options for diabetic retinopathy. Curr Opin Ophthalmol. 2006;17(6): 519–22.
52. Leese GP, Ahmed S, Newton RW, Jung RT, Ellingford A, Baines P, et al. Use of mobile screening unit for diabetic retinopathy in rural and urban areas. BMJ. 1993;306(6871): 187–9.
53. Scanlon PH, Malhotra R, Thomas G, Foy C, Kirkpatrick JN, Lewis-Barned N, et al. The effectiveness of screening for diabetic retinopathy by digital imaging photography and technician ophthalmoscopy. Diabet Med. 2003;20(6):467–74.
54. Murgatroyd H, Ellingford A, Cox A, Binnie M, Ellis JD, MacEwen CJ, et al. Effect of mydriasis and different field strategies on digital image screening of diabetic eye disease. Br J Ophthalmol. 2004;88(7):920–4.
55. Facey K, Cummins E, Macpherson K, Morris A, Reay L, Slattery J. Organisation of Services for Diabetic Retinopathy Screening. Glasgow: Health Technology Board for Scotland; 2002. p. 1–224.
56. Bursell SE, Cavallerano JD, Cavallerano AA, Clermont AC, Birkmire-Peters D, Aiello LP, et al. Stereo nonmydriatic digital-video color retinal imaging compared with early treatment diabetic retinopathy study seven standard field 35-mm stereo color photos for determining level of diabetic retinopathy. Ophthalmology. 2001;108(3):572–85.
57. Massin P, Erginay A, Ben Mehidi A, Vicaut E, Quentel G, Victor Z, et al. Evaluation of a new non-mydriatic digital camera for detection of diabetic retinopathy. Diabet Med. 2003;20(8):635–41.
58. Shiba T, Yamamoto T, Seki U, Utsugi N, Fujita K, Sato Y, et al. Screening and follow-up of diabetic retinopathy using a new mosaic 9-field fundus photography system. Diabetes Res Clin Pract. 2002;55(1):49–59.
59. Williams GA, Scott IU, Haller JA, Maguire AM, Marcus D, McDonald HR. Single-field fundus photography for diabetic retinopathy screening: a report by the american academy of ophthalmology. Ophthalmology. 2004;111(5):1055–62.
60. Aptel F, Denis P, Rouberol F, Thivolet C. Screening of diabetic retinopathy: Effect of field number and mydriasis on sensitivity and specificity of digital fundus photography. Diabetes Metab. 2008;34(3):290–3.
61. Baeza M, Orozco-Beltran D, Gil-Guillen VF, Pedrera V, Ribera MC, Pertusa S, et al. Screening for sight threatening diabetic retinopathy using non-mydriatic retinal camera in a primary care setting: to dilate or not to dilate? Int J Clin Pract. 2009;63(3):433–8.
62. Lopez-Bastida J, Cabrera-Lopez F, Serrano-Aguilar P. Sensitivity and specificity of digital retinal imaging for screening diabetic retinopathy. Diabet Med. 2007;24(4):403–7.

63. Mash B, Powell D, du Plessis F, van Vuuren U, Michalowska M, Levitt N. Screening for diabetic retinopathy in primary care with a mobile fundal camera–evaluation of a South African pilot project. S Afr Med J. 2007;97(12):1284–8.
64. Taylor CR, Merin LM, Salunga AM, Hepworth JT, Crutcher TD, O'Day DM, et al. Improving diabetic retinopathy screening ratios using telemedicine-based digital retinal imaging technology: the Vine Hill study. Diabetes Care. 2007;30(3):574–8.
65. Savolainen EA, Lee QP. Diabetic retinopathy - need and demand for photocoagulation and its cost-effectiveness: evaluation based on services in the United Kingdom. Diabetologia. 1982;23(2):138–40.
66. Javitt JC, Aiello LP, Chiang Y, Ferris 3rd FL, Canner JK, Greenfield S. Preventive eye care in people with diabetes is cost-saving to the federal government. Implications for health-care reform. Diabetes Care. 1994;17(8):909–17.
67. Javitt JC, Aiello LP. Cost-effectiveness of detecting and treating diabetic retinopathy. Ann Intern Med. 1996;124(1 Pt 2):164–9.
68. Dasbach EJ, Fryback DG, Newcomb PA, Klein R, Klein BE. Cost-effectiveness of strategies for detecting diabetic retinopathy. Med Care. 1991;29(1):20–39.
69. Caro JJ, Ward AJ, O'Brien JA. Lifetime costs of complications resulting from type 2 diabetes in the U.S. Diabetes Care. 2002;25(3):476–81.
70. Fendrick AM, Javitt JC, Chiang YP. Cost-effectiveness of the screening and treatment of diabetic retinopathy. What are the costs of underutilization? Int J Technol Assess Health Care. 1992;8(4):694–707.
71. James M, Turner DA, Broadbent DM, Vora J, Harding SP. Cost effectiveness analysis of screening for sight threatening diabetic eye disease. BMJ. 2000;320(7250):1627–31.
72. Meads C, Hyde C. What is the cost of blindness? Br J Ophthalmol. 2003;87(10):1201–4.
73. Foulds WS, MacCuish A, Barrie T. Diabetic retinopathy in the West of Scotland: its detection and prevalence, and the cost-effectiveness of a proposed screening programme. Health Bull. 1983;41(6):318–26.
74. Wright SE, Keeffe JE, Thies LS. Direct costs of blindness in Australia. Clin Experiment Ophthalmol. 2000;28(3):140–2.
75. Bjorvig S, Johansen MA, Fossen K. An economic analysis of screening for diabetic retinopathy. J Telemed Telecare. 2002;8(1):32–5.
76. Tu KL, Palimar P, Sen S, Mathew P, Khaleeli A. Comparison of optometry vs digital photography screening for diabetic retinopathy in a single district. Eye. 2004;18(1):3–8.
77. Aoki N, Dunn K, Fukui T, Beck JR, Schull WJ, Li HK. Cost effectiveness analysis of telemedicine to evaluate diabetic retinopathy in a prison population. Am J Ophthalmol. 2005;139(2):399.
78. Tung TH, Shih HC, Chen SJ, Chou P, Liu CM, Liu JH. Economic evaluation of screening for diabetic retinopathy among Chinese type 2 diabetics: a community-based study in Kinmen, Taiwan. J Epidemiol. 2008;18(5):225–33.
79. Diabetes care and research in Europe: the Saint Vincent declaration. Diabet Med. 1990;7(4):360.
80. Massin P, Chabouis A, Erginay A, Viens-Bitker C, Lecleire-Collet A, Meas T, et al. OPHDIAT: a telemedical network screening system for diabetic retinopathy in the Ile-de-France. Diabetes Metab. 2008;34(3):227–34.
81. HSE. Framework for the development of a diabetic retinopathy screening programme for Ireland. Dublin, 2008:1–96.
82. Jaross N, Ryan P, Newland H. Incidence and progression of diabetic retinopathy in an Aboriginal Australian population: results from the Katherine Region Diabetic Retinopathy Study (KRDRS). Report no. 2. Clin Experiment Ophthalmol. 2005;33(1):26–33.

83. Harper CA, Livingston PM, Wood C, Jin C, Lee SJ, Keeffe JE, et al. Screening for diabetic retinopathy using a non-mydriatic retinal camera in rural victoria. Aust N Z J Ophthalmol. 1998;26(2):117–21.
84. Boucher MC, Desroches G, Garcia-Salinas R, Kherani A, Maberley D, Olivier S, et al. Teleophthalmology screening for diabetic retinopathy through mobile imaging units within Canada. Can J Ophthalmol. 2008;43(6):658–68.
85. Boucher MC, Desroches G, Garcia-Salinas R, Kherani A, Maberley D, Olivier S, et al. Diabetic retinopathy screening. Can J Ophthalmol. 2009;44(1):100–1.
86. Cavallerano AA, Cavallerano JD, Katalinic P, Blake B, Rynne M, Conlin PR, et al. A telemedicine programme for diabetic retinopathy in a Veterans Affairs Medical Center–the Joslin Vision Network Eye Health Care Model. Am J Ophthalmol. 2005;139(4):597–604.

3
Functional/Neural Mapping Discoveries in the Diabetic Retina: Advancing Clinical Care with the Multifocal ERG

Anthony J. Adams and Marcus A. Bearse Jr.

CONTENTS

Keywords Multifocal electroretinogram • Non proliferative diabetic retinopathy • Neuropathy • Microvascular disease

INTRODUCTION

Diabetes, now an epidemic, has devastating effects on the eye and vision. The treatments of the eye complications are currently limited to relatively advanced stages and primarily to slow down the progressive retinal vasculopathy (diabetic retinopathy). New, nonfoveal measures of early retinal function abnormalities, including neural abnormalities, could change the focus of patient research and management to a more preventative agenda. We have found that multifocal electroretinogram implicit time (mfERG IT) delays are spatially associated in the retina with sites containing nonproliferative diabetic retinopathy (NPDR) and edema. These delays also occur, albeit to a lesser extent, in the retinas of patients with diabetes and no retinopathy. More important, we have shown that the mfERG IT, in combination with other risk factors such as blood glucose concentration and duration of diabetes, combines to provide remarkably accurate predictors of new retinopathy development at specific locations within the central 45° of the retina. Very recently, we showed that these mfERG IT delays are also predictive of the onset (initial appearance) of NPDR in adults. The importance and value of these local measures of neural retina function and health seems obvious. Understanding their relationship to

From: *Ophthalmology Research: Visual Dysfunction in Diabetes*
Edited by: J. Tombran-Tink et al. (eds.), DOI 10.1007/978-1-60761-150-9_3

systemic factors that are known to be associated with type 2 diabetes before and after the appearance of NPDR and using other known risk factors to further increase an already excellent predictive model, are the next logical research steps. Both offer promise of improved patient care and more personal patient management options.

DIABETES AND AN UNRESOLVED DIABETIC EYE MANAGEMENT PROBLEM

The Diabetes Epidemic

In the United States, 17.9 million people, 5.9% of the population, have diabetes [1]. There are also an estimated 5.7 million who have undiagnosed diabetes and 57 million who are prediabetic [1]. Diabetic retinopathy, the vascular eye complication, is the leading cause of blindness in the US among adults aged 20–74 years [1].

Current Treatment Focus

Treatments of the potentially devastating retinal complications are currently aimed at slowing the progression of vision loss after vascular-related structural damage within the retina which is funduscopically obvious. Laser photocoagulation, an invasive treatment that destroys retinal tissue, is used in cases of clinically significant macular edema (CSME). In cases of advanced retinopathy, panretinal laser treatment is applied to as many as thousands, or more, of retinal locations to destroy tissue and consequently reduce the retina's demand for oxygen, thereby slowing progression of neovascularization. Although these gold-standard treatments significantly reduce the loss of visual acuity, they have side effects, including loss of paracentral vision (important for reading and other tasks) and peripheral and night vision, and they are also associated with many adverse events [2]. Furthermore, despite these treatments, vision loss still continues at a disturbing rate [3–5]. Additional treatments are emerging, including intraocular and retrobulbar injection of steroids, anti-VEGF agents, PKC inhibitors, PEDF (pigment endothelium-derived factor) inducers, and several indirect growth factor modulators. These therapies are targeted at reducing macular edema, treating advanced disease, or reducing the risks of neovascularization. These important treatment improvements remain focused on the relatively advanced stages of vision loss produced by diabetes complications.

Vasculopathy and Neuropathy of the Retina

Increasing attention is being paid to the fact that there are both neural and vascular components involved in very early stages of diabetic retinopathy. The concept that diabetes directly affects the neurosensory retina, independent of clinically observed vascular changes, has been proposed for decades [6]. Bresnick proposed, almost 25 years ago, to redefine diabetic retinopathy as a neurosensory disorder resulting from both metabolic and systemic insults to the retina, in addition to the clinically apparent vascular changes [7]. Many sensitive human electrophysiological measurements of retinal neural function and psychophysical measurements of visual function now indicate that there are early abnormalities that appear before the clinical signs of diabetic retinopathy (vasculopathy) [8–10]. Consistent with this, results obtained in animal models of diabetes show that there are increased inflammatory factors, structural changes of the glia, and ganglion cell apoptosis in the retina before there are overt vascular changes associated with clinical retinopathy [11].

THE NEED TO GO BEYOND VISUAL ACUITY AND BEYOND FOVEAL FUNCTION

The Early Efforts

For almost three decades, research in our laboratory has involved the pursuit of retinal function and vision markers early in, or preceding, the diabetes complications of the retina. Quite clearly, visual acuity and visual fields are poor candidates, being quite late consequences of retinal vascular complications. Indeed, visual acuity is reduced only with edema in the foveal area of the macula, or as a result of fairly obstructive retinal/vitreous hemorrhages. For more than a century, there had been clinical reports of blue–yellow color vision changes in diabetes, even with foveal testing with fairly traditional clinical tests. Based on this, we began our first studies trying to isolate the vision function underlying specific cone photoreceptor types using a suprathreshold variation of the "two-color threshold" technique known to allow individual populations of cone receptor activity to be manifest in vision measures. In the early 1980s, we found quite dramatic reductions in the blue cone (S cone) sensitivity when deep violet patches of light were detected only by S cones against a bright yellow background [12, 13]. These losses of blue cone system sensitivity were even present prior to the clinically observable onset of the vascular retinopathy of diabetes. Later, we developed a method to make these same measurements across the retina and found losses in localized areas across the central 50° of the retina [14, 15]. [Parenthetically, our work on this followed on with Chris Johnson, then at UC Davis and led to the development of "blue-cone" (S cone) automated perimetry [16], which later was referred to as SWAP perimetry [17] with many applications in glaucoma patient management.]

In patients with diabetes, we much later reported that blue-cone perimetry revealed about 40% of central visual field zones as abnormal in the patients who had mild to moderate retinopathy and even 20% abnormal in the retinas of those with diabetes and no retinopathy [18]. However, disappointingly, we found little correlation of these field abnormalities with the locations of visible retinopathy.

Some Breakthroughs

By marked contrast, our first efforts with measuring local neural function across the retina with a newly emerging tool, the multifocal electroretinogram (mfERG), provided clear association of abnormal neural function (observed as delays in the local mfERG responses) with visible retinopathy [19]. This encouraged us to pursue the measures further with both cross-sectional and longitudinal studies. With evidence of association of neural dysfunction and visible retinopathy, the correlation between abnormality and retinopathy severity and the observation that many patches of retina without retinopathy had abnormal mfERG responses [19], we enrolled patients without retinopathy and with minimal retinopathy. Our goal was to see if the abnormal mfERG delays were present in eyes that had not yet shown clinical retinopathy and to explore whether abnormal neural function measures might be predictive of new retinopathy development.

Our studies over the 4–5 years that followed confirmed the initial promise of this measure, and we now know that the neural latency abnormalities (mfERG delays) observed in the earlier studies are not only present prior to retinopathy onset [19, 20] and correlated with the severity of the retinopathy at the local site of the retinal vascular signs of retinopathy [19–21], but are also predictive (precedes) retinopathy onset at locations in eyes that already have some retinopathy [20, 22–24]. Our longitudinal studies over 1, 2, and 3 years have shown that predictive models based on mfERG delays revealed remarkable potential clinical and research tools with high sensitivity and specificity [20, 23, 24]. In confirmation studies, the high sensitivity (prediction of retinopathy onset in a specific location) and specificity (prediction of normal retina at specific locations) remain high [24]. One of our recent publications also reveals that the mfERG measures are predictive of the onset of retinopathy in eyes that had no prior retinopathy [25].

These research results with early stage emergence of neural dysfunction measures in the retina are in striking contrast to the natural history of change in visual acuity. Visual acuity loss occurs many years after retinopathy appears, and then only with severe retinopathic events or edema that impact the fovea. By that time, the vascular events are very apparent to the clinician. So, as a functional outcome measure, visual acuity is primarily useful as a measure of success in slowing late-stage retinopathy, or for assessing the impact of treatments applied at that late stage. It is not useful to signal imminent retinal problems, early retinopathy progression, or the efficacy of any preventative treatments. In contrast, the implicit time (delay) measure of the mfERG has emerged as an exciting future clinical tool in the management of patients at earlier stages and for the exploration of new candidate treatments and interventions. With it, we have produced formal predictive models. It is the critical component of predictive models of retinopathy onset over a relatively short time frame and, as such, is an obvious candidate as an outcome measure for relatively brief clinical trials of proposed pharmaceutical preventatives at the earliest stages of diabetic retinal complications.

Predictive Models of Visible Retinopathy Onset at Specific Locations

Using multivariate logistic regression techniques, we formulated models incorporating mfERG IT and risk factors such as duration of diabetes and blood glucose control that predict the development of retinopathy in new retinal locations with high sensitivity and specificity (approx. 80–90%) [20, 23, 24]. Recently, we formulated another multifactor model, based on mfERG IT, that predicts the initial clinical onset of diabetic retinopathy [25].

HOW IS THE MFERG MEASURED AND WHAT IS IT MEASURING?

So, what is the mfERG and how is it actually measured? The mfERG is a noninvasive technique for measuring neural function in up to hundreds of contiguous retinal areas within the central retina [26, 27]. The implicit time (IT) of the P1 component of the local mfERG response waveform is a highly reproducible and sensitive indicator of neural function in the retina. Figure 1 provides a brief overview of the stimulus and response outputs across the central 45° of human retina [18–30].

Briefly, 103 local retinal responses to 200 cd/m^2 flash stimulation (actually, first-order response kernels) are recorded from the central ~45° of the retina during an ~8 min session using a 75 Hz frame rate and 10–100 Hz filtering. The responses are recorded using a bipolar contact lens electrode, and a ground electrode is clipped to the right earlobe. Fixation is monitored using an in-line infrared video camera. The session is broken into 16 segments for subject comfort.

The first prominent positive peak (P1) of the mfERG response (Fig. 1D) that our group has investigated is the easiest to measure, and the implicit time measure of P1 is far less variable than the amplitude measure (one-tenth of the coefficient of variation of the amplitude measure in healthy control subjects) [20].

Where Are These Neural Signals Generated in the Retina?

It is generally believed that mfERG IT delay, in the absence of reduced response amplitude, reflects abnormality of the outer plexiform layer and bipolar cells, as it does for the conventional full-field flash ERG. The P1 component of the mfERG waveform, from which we measure mfERG IT, is generated primarily by the opposing electrical polarities of the ON and OFF bipolar cell responses in the middle layers of the retina [31–33]. The retina is particularly susceptible to the early pathological vascular changes associated with type 2 diabetes because of its high metabolic demand, minimal retinal vascular supply, and low oxygen tension of the inner retinal layers [8]. It has been proposed that mfERG IT delays in the absence of mfERG response amplitude reductions represent the effects of reduced perfusion and resulting hypoxia/ischemia [19, 20, 22, 23, 30, 34]. Recently, more direct evidence supporting this view has been reported. In diabetic patients with enlarged foveal avascular zones, the area of the vascular-free zone has been shown to be correlated with increasing mfERG IT delay, but not mfERG amplitude reduction, in and adjacent to the fovea [34].

Some Key Results

Before highlighting the evolution of our predictive models, since 2004, it is illustrative to look at a single patient example of the way in which the local mfERG implicit time delay predicted subsequent retinopathy in a patient (Fig. 2).

In one of our first publications, we reported the sensitivity and specificity of the mfERG implicit time as part of a "one-year" predictive model. It certainly included what we later learned were both retinopathy that was transient and retinopathy that was likely to be persistent. Based on our data then, primarily for study patients with mild diabetic retinopathy, we found relatively high sensitivity (86%) and specificity (84%) [23].

This quantitative model was the first to make predictions of diabetic retinopathy lesions in discrete retinal areas. The study involved only one follow-up visit and thus could not examine whether the lesions that were evidenced were transient or sustained in nature. More recently, our review included new data that extended the study by Han et al. [23] for another year [20].

Two years later, we reported on a 3-year prediction model with similarly high sensitivity and specificity for patients who already had some retinopathy [24]. Eighteen

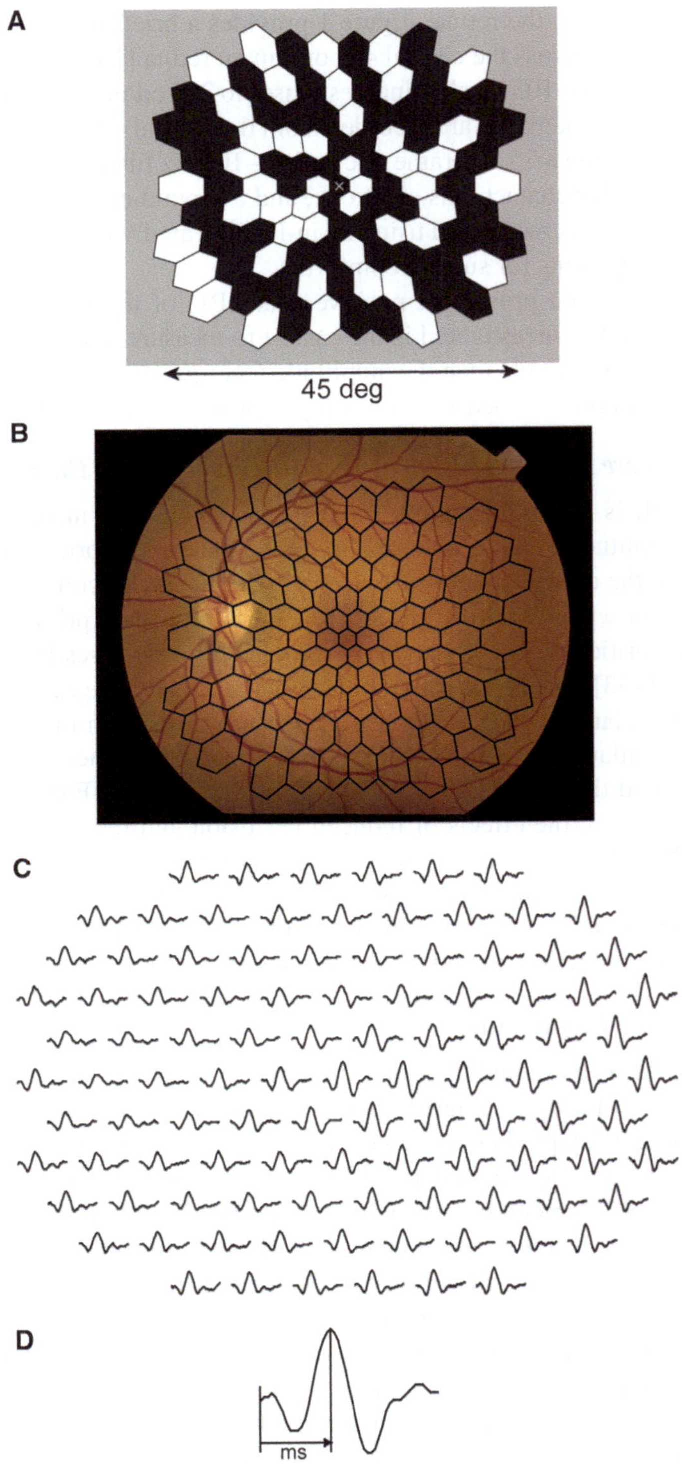

Fig. 1. Stimulus array of 103 scaled hexagons (**A**), its relationship with the retinal area tested (**B**), an example array of the 103 local mfERGs (**C**), and the mfERG implicit time (IT) measure from the stimulus flash to the P1 peak (**D**).

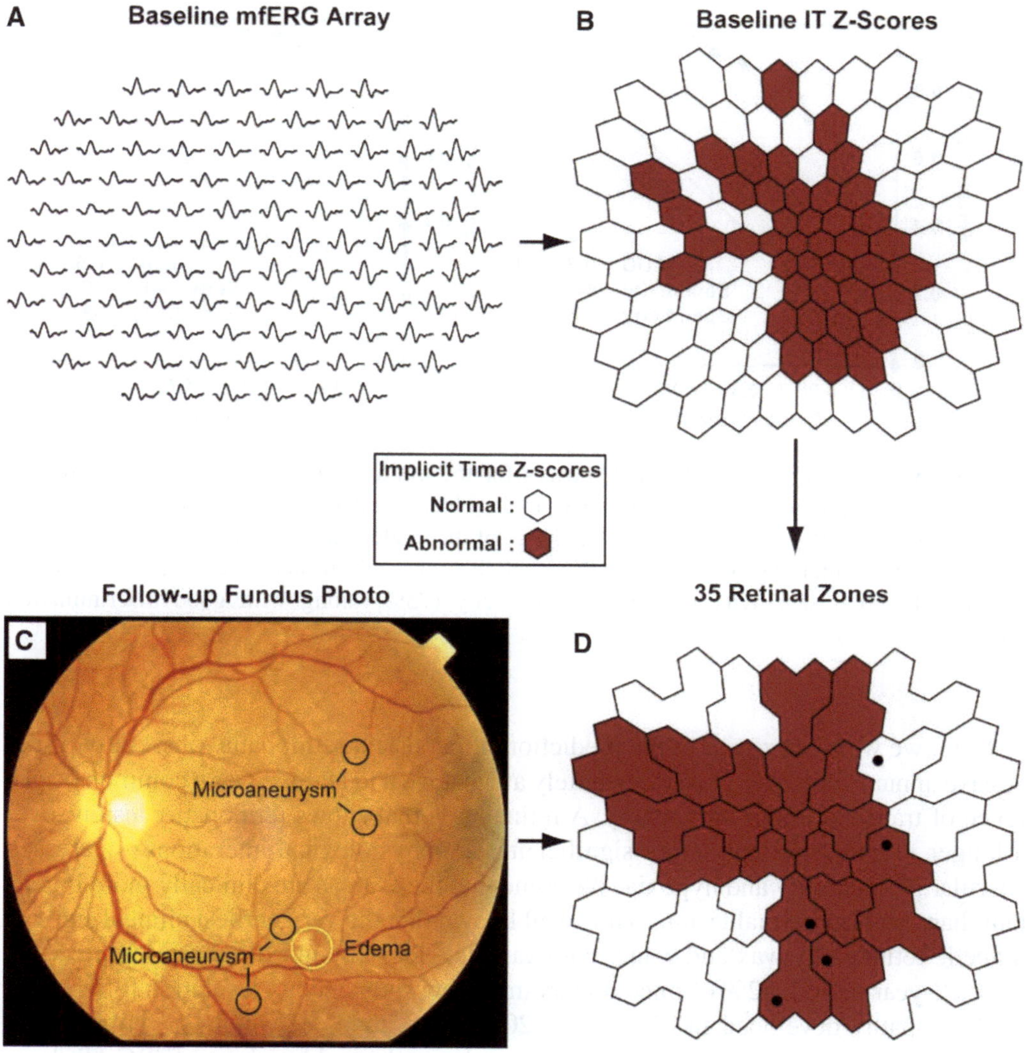

Fig. 2. Shows an example of the predictive power of the mfERG IT. The baseline mfERGs are shown in (**A**). At baseline, this patient had no retinopathy. The mfERG implicit time was, however, abnormally delayed ($P < 0.023$) in many of the 103 locations (*red hexagons* in (**B**)) and many of the 35 retinal zones used for analysis (*red patches* in (**D**)). On follow-up 1 year later, new patches of retinopathy and edema had developed, as indicated in the fundus photograph (**C**) and as *black dots* (**D**). As can be seen in (**D**), four of the five new lesions are associated with significantly delayed mfERG IT one year earlier, and the fifth lesion is very close to a delayed zone. (Fig. 2 from Bearse et al. [20]).

diabetic patients were examined at baseline and at three annual follow-ups. Again, 35 retinal zones were constructed from the 103-element stimulus array, and each zone was assigned the maximum IT *z*-score within it based on 30 age-similar control subjects. Logistic regression was used to investigate the development of retinopathy in relation to baseline mfERG IT delays and additional diabetic health variables. Again, receiver operating characteristic (ROC) curves were used to evaluate the models.

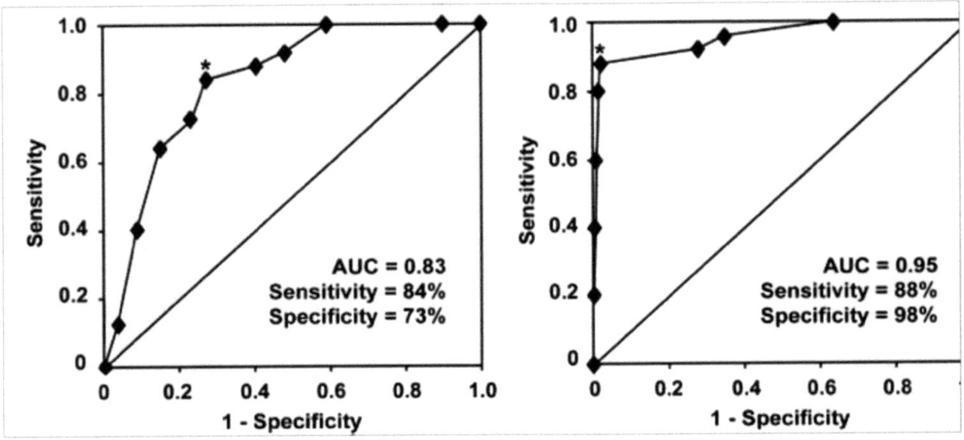

Fig. 3. (from Ng et al. [24]) ROC curves for the multivariate model (*right*) that predicts recurring retinopathy over the course of 3 years in a local retinal area. The area under the ROC curve (AUC) of 0.95 provides an overall measure of the model's discrimination accuracy (95%). Even a model containing only the mfERG implicit time, and no other factors (*left*), provided surprisingly good sensitivity (84%) and specificity (73%) along with good discrimination (AUC=0.83) [24].

Here, we were interested in the prediction of persistent retinopathy onset at two successive annual visits. We looked separately at what we had learned was a common occurrence of transient initial retinopathy. A retinal area that shows retinopathy lesions over a longer period represents more significant pathophysiological alterations—increased vascular permeability and hypoxia. We argue that these areas are clinically more important than transient retinal lesions. (It is well known that the very earliest clinical signs of diabetic retinopathy wax and wane. For example, Hellstedt and Immonen reported that over a 2-year period, 52% of microaneurysms show spontaneous resolution [35].)

Retinopathy developed in 77 of the 1,208 retinal zones of which 25 retinal zones had recurring retinopathy. The multivariate analyses showed baseline mfERG IT, duration of diabetes, and blood glucose concentration as the most important predictors of recurring retinopathy but were not at all predictive of transient retinopathy. ROC curves revealed sensitivity of 88% and specificity of 98% for the recurring retinopathy we were interested in (see Fig. 3). A tenfold cross-validation confirmed the high sensitivity and specificity of the model.

In a recent publication, we report on the onset of diabetic retinopathy in a study group of patients with diabetes but no clinically visible signs of retinopathy [25]. Again, the predictive multivariate models incorporating mfERG IT delay and other risk variables showed excellent ability to predict the onset of retinopathy with high sensitivity and specificity. Seventy-eight eyes from 41 diabetes patients were tested annually for several years. The presence or absence of DR at the last study visit was the outcome measure, and measurements of risk factors from the previous visit were used for prediction. Nearly 40% of the 78 eyes developed retinopathy for a total of 80 of 2,730 retinal zones. In short, mfERG IT was again a good predictor of diabetic retinopathy onset,

1 year later, even in patients without any prior retinopathy. It can be utilized to assess the risk of DR development in these patients and may be a valuable outcome measure in evaluation of novel prophylactic therapeutics directed at impeding DR.

Adolescents and Adult Diabetes

Are the mfERG abnormalities we see in adult diabetic subjects also present in adolescent patients with diabetes?

In 2005, the Center for Disease Control (CDC) in the US estimated that there are 206,000 people under the age of 20 that have diabetes, and approximately one in six overweight adolescents have prediabetes (CDC, 2005). Type 2 diabetes now accounts for up to 20% of all newly diagnosed adolescent cases [36].

In 2008, we reported that indeed, adolescents with type 2 diabetes do have abnormal neural function in the retina [37]. We also noted early indications of abnormal dilation of venules and abnormal thinning of the retina. Adolescents with type 2 diabetes often present with comorbidities such as obesity, hyperinsulinemia, hypertension, and hyperlipidemia. All of these conditions can impact both the vascular and neurologic health of the patient. Our study was the first of its kind to examine the neural retinal function, structure, and retinal vascular health in adolescents with type 2 diabetes.

Fifteen adolescents diagnosed with type 2 diabetes, aged 13–21 years with a mean diabetes duration of 2.1 ± 1.3 years, were examined. Twenty-six age-matched control subjects were also tested. The mfERGs of the type 2 diabetic patients were significantly delayed ($p = 0.03$). The diabetic group also showed significant retinal thinning and significant venular dilation.

Type 1 vs. Type 2: Differences in Retinal Function

In a recent paper, we noted differences in type 1 and type 2 adults with diabetes [25]. Neural function in the retina was distinctly poorer in the type 2 patients. We have noted this same difference when comparing adolescents with type 1 and type 2 diabetes [38]. This raises questions about possible underlying differences in pathophysiology of the retina (and beyond). Type 2 diabetes patients typically have more numerous cardiovascular risk factors and comorbidity factors than type 1 patients. Our current research is looking at this more carefully.

THE HORIZON FOR PATIENT CARE OF DIABETES RETINA AND RESEARCH AGENDA

The early neural changes in the retina of eye, produced by diabetes well before clinical signs of vascular retinopathy, have quite significant implications for patient care and management of eye complications as we look to the horizon. The mfERG implicit time, measured with clinical instrumentation, clearly identifies almost 20% of the central retina of patients with diabetes as functioning abnormally prior to visible retinopathy. This "neuropathy" is consistent with the changing view of the retinal complications of diabetes that has previously had almost entirely a "vascular" label; it still does with most clinicians. Regardless of the perspective of neural preceding vascular or vice versa—the debate will likely hinge on whichever new technical assessment tools are most sensitive—it is clear that the identification of functional deficits, early in the disease complication process of the eye, provides new opportunities for the development of new therapies and assessment tools for the staging of retinal changes.

Clinicians have primarily been limited to assessment of visual acuity at one central and tiny location of the retina, and to visual fields with relatively insensitive markers in diabetes. In fact, both visual acuity and visual fields by conventional perimetry are characteristic of fairly late stage vasculopathy of the retina—well after any prophylactic treatments could be applied. The early "warning signals" of the mfERG, coupled with an apparently powerful predictive ability for future retinopathy within a year or two, are an exciting advance in the potential management of the diabetic complications of the retina. New candidate interventions, aimed at preventing or slowing the path of retinopathy progression at early stages, may now be contemplated with biological and objective markers of functional improvement. With visual acuity loss typically occurring only after many years, it becomes a most unattractive outcome measure for any early intervention efficacy studies.

In management, it is conceivable that patient monitoring, based on the progression of neural abnormality, will be a valuable tool in the hands of eye care practitioners. Ophthalmologists and optometrists could have the ability to gauge both the severity of neural dysfunction and the likelihood of incipient local retinopathy and use this information to stage an appropriate and timely intervention.

Looking even further ahead, it is conceivable that as other functional measures of the retina, known to be altered at early stages of the diabetic complications (e.g., alterations in the retinal pigment epithelium function, or systemic serum markers or indices known to be risk factors) that might make the predictive models of incipient damage in the retina even more powerful than they already are. It is important to examine the potential relationships between the mfERG IT delays in diabetes and to look at systemic markers of glycemic control, diabetes-related inflammation, microvascular damage, and dyslipidemia (abnormal concentrations of lipids in the blood). These systemic markers are associated with diabetes and microvascular disease including diabetic retinopathy.

Taken a step further, as research links systemic serum risk factors to particular retinal structure changes, whether neural or vascular, it is conceivable that personalized treatment and management options will evolve for diabetic retinal health. Certainly, the opportunities for research to unveil those relationships and the underlying mechanisms provide an exciting opportunity in clinical research.

REFERENCES

1. American diabetes association web site. Diabetes statistics; 2010.
2. Aiello LM. Perspectives on diabetic retinopathy. Am J Ophthalmol. 2003;136:122–35.
3. Early Treatment Diabetic Retinopathy Study research group. Photocoagulation for diabetic macular edema. Early treatment diabetic retinopathy study report number 1. Arch Ophthalmol. 1985;103:1796–806.
4. Hansson-Lundblad C, Holm K, Agardh CD, Agardh E. A small number of older type 2 diabetic patients end up visually impaired despite regular photographic screening and laser treatment for diabetic retinopathy. Acta Ophthalmol Scand. 2002;80:310–5.
5. Vine AK. The efficacy of additional argon laser photocoagulation for persistent, severe proliferative diabetic retinopathy. Ophthalmology. 1985;92:1532–7.

6. Wolter JR. Diabetic retinopathy. Am J Ophthalmol. 1961;51:1123–41.
7. Bresnick GH. Diabetic retinopathy viewed as a neurosensory disorder. Arch Ophthalmol. 1986;104:989–90.
8. Antonetti DA, Barber AJ, Bronson SK, Freeman WM, Gardner TW, Jefferson LS, Kester M, Kimball SR, Krady JK, LaNoue KF, Norbury CC, Quinn PG, Sandirasegarane L, Simpson IA. Diabetic retinopathy: seeing beyond glucose-induced microvascular disease. Diabetes. 2006;55:2401–11.
9. Jackson GR, Barber AJ. Visual dysfunction associated with diabetic retinopathy. Curr Diab Rep. 2010;10:380–4.
10. Tzekov R, Arden GB. The electroretinogram in diabetic retinopathy. Surv Ophthalmol. 1999;44:53–60.
11. Kern TS. Contributions of inflammatory processes to the development of the early stages of diabetic retinopathy. Exp Diabetes Res. 2007;2007:95103.
12. Adams AJ. Chromatic and luminosity processing in retinal disease. Am J Optom Physiol Opt. 1982;59:954–60.
13. Adams AJ, Zisman F, Rodic R, Cavender JC. Chromaticity and luminosity changes in glaucoma and diabetes. Doc Ophthalmol Proc Series. 1982;33:413–8.
14. Heron G, Adams AJ, Husted R. Foveal and non-foveal measures of short wavelength sensitive pathways in glaucoma and ocular hypertension. Ophthalmic Physiol Opt. 1987;7:403–4.
15. Heron G, Adams AJ, Husted R. Central visual fields for short wavelength sensitive pathways in glaucoma and ocular hypertension. Invest Ophthalmol Vis Sci. 1988;29:64–72.
16. Johnson CA, Adams AJ, Lewis RA. Automated perimetry of short-wavelength-sensitive mechanisms in glaucoma and ocular hypertension; preliminary findings. In: Heijl A, editor. Proceedings of the VIIIth international perimetric society meeting. Amsterdam: Kuglrer & Ghedini Publications; 1989. p. 31–7.
17. Sample PA, Johnson CA, Haegerstrom-Portnoy G, Adams AJ. Optimum parameters for short-wavelength automated perimetry. J Glaucoma. 1996;5:375–83.
18. Han Y, Adams AJ, Bearse Jr MA, Schneck ME. Multifocal electroretinogram and short-wavelength automated perimetry measures in diabetic eyes with little or no retinopathy. Arch Ophthalmol. 2004;122:1809–15.
19. Fortune B, Schneck ME, Adams AJ. Multifocal electroretinogram delays reveal local retinal dysfunction in early diabetic retinopathy. Invest Ophthalmol Vis Sci. 1999;40:2638–51.
20. Bearse Jr MA, Adams AJ, Han Y, Schneck ME, Ng J, Bronson-Castain K, Barez S. A multifocal electroretinogram model predicting the development of diabetic retinopathy. Prog Retin Eye Res. 2006;25:425–48.
21. Schneck ME, Bearse Jr MA, Han Y, Barez S, Jacobsen C, Adams AJ. Comparison of mfERG waveform components and implicit time measurement techniques for detecting functional change in early diabetic eye disease. Doc Ophthalmol. 2004;108:223–30.
22. Han Y, Bearse Jr MA, Schneck ME, Barez S, Jacobsen CH, Adams AJ. Multifocal electroretinogram delays predict sites of subsequent diabetic retinopathy. Invest Ophthalmol Vis Sci. 2004;45:948–54.
23. Han Y, Schneck ME, Bearse Jr MA, Barez S, Jacobsen CH, Jewell NP, Adams AJ. Formulation and evaluation of a predictive model to identify the sites of future diabetic retinopathy. Invest Ophthalmol Vis Sci. 2004;45:4106–12.
24. Ng JS, Bearse Jr MA, Schneck ME, Barez S, Adams AJ. Local diabetic retinopathy prediction by multifocal ERG delays over 3 years. Invest Ophthalmol Vis Sci. 2008;49:1622–8.
25. Harrison WW, Bearse MA, Jr., Ng JS, Jewell N, Barez S, Burger D, Schneck ME, Adams AJ. Multifocal electroretinograms predict onset of diabetic retinopathy in adult patients with diabetes. Invest Ophthalmol Vis Sci. 2011;52:6825–31.

26. Bearse Jr MA, Han Y, Schneck ME, Adams AJ. Retinal function in normal and diabetic eyes mapped with the slow flash multifocal electroretinogram. Invest Ophthalmol Vis Sci. 2004;45:296–304.
27. Bearse Jr MA, Sutter EE. Imaging localized retinal dysfunction with the multifocal electroretinogram. J Opt Soc Am A Opt Image Sci Vis. 1996;13:634–40.
28. Han Y, Bearse Jr MA, Schneck ME, Barez S, Jacobsen C, Adams AJ. Towards optimal filtering of "standard" multifocal electroretinogram (mfERG) recordings: findings in normal and diabetic subjects. Br J Ophthalmol. 2004;88:543–50.
29. Harrison WW, Bearse Jr MA, Ng JS, Barez S, Schneck ME, Adams AJ. Reproducibility of the mfERG between instruments. Doc Ophthalmol. 2009;119:67–78.
30. Palmowski AM, Sutter EE, Bearse Jr MA, Fung W. Mapping of retinal function in diabetic retinopathy using the multifocal electroretinogram. Invest Ophthalmol Vis Sci. 1997;38: 2586–96.
31. Hare WA, Ton H. Effects of APB, PDA, and TTX on ERG responses recorded using both multifocal and conventional methods in monkey. Effects of APB, PDA, and TTX on monkey ERG responses. Doc Ophthalmol. 2002;105:189–222.
32. Hood DC. Assessing retinal function with the multifocal technique. Prog Retin Eye Res. 2000;19:607–46.
33. Horiguchi M, Suzuki S, Kondo M, Tanikawa A, Miyake Y. Effect of glutamate analogues and inhibitory neurotransmitters on the electroretinograms elicited by random sequence stimuli in rabbits. Invest Ophthalmol Vis Sci. 1998;39:2171–6.
34. Tyrberg M, Ponjavic V, Lovestam-Adrian M. Multifocal electroretinogram (mfERG) in patients with diabetes mellitus and an enlarged foveal avascular zone (FAZ). Doc Ophthalmol. 2008;117:185–9.
35. Hellstedt T, Immonen I. Disappearance and formation rates of microaneurysms in early diabetic retinopathy. Br J Ophthalmol. 1996;80:135–9.
36. CDC. National diabetes fact sheet: General information and national estimates on diabetes in the United States. US Department of health and human services, centers for disease control and prevention, Atlanta, GA; 2005.
37. Bronson-Castain KW, Bearse Jr MA, Neuville J, Jonasdottir S, King-Hooper B, Barez S, Schneck ME, Adams AJ. Adolescents with Type 2 diabetes: early indications of focal retinal neuropathy, retinal thinning, and venular dilation. Retina. 2009;29:618–26.
38. Bronson-Castain KW, Bearse Jr MA, Neuville J, Jonasdottir S, King-Hooper B, Barez S, Schneck ME, Adams AJ. Early neural and vascular changes in the adolescent type 1 and type 2 diabetic retina. Retina. 2011; Aug 30. [Epub ahead of print.]

Part III

How Does Diabetes Affect the Eye?

4
Corneal Diabetic Neuropathy

Edoardo Midena

CONTENTS

Keywords Sub-basal corneal nerve plexus • Corneal nerve fibers • Corneal confocal microscopy • Peripheral diabetic neuropathy

INTRODUCTION

The prevalence of diabetes mellitus is dramatically increasing worldwide, and consequently, the prevalence of chronic complications due to diabetes will increase in the near future [1]. The most common cause of chronic disability in diabetic patients is diabetic neuropathy, mainly, peripheral diabetic neuropathy. Peripheral diabetic neuropathy affects 50% of diabetic patients within 25 years of diagnosis [2]. Long-term effects of undetected and untreated peripheral diabetic neuropathy can lead to foot infections that do not heal, as well as foot ulcers. Patients may require subsequent amputation of the foot and digits, which can lead to a decreased quality of life and increased mortality [3].

The effective and reliable diagnosis and quantification of peripheral diabetic neuropathy are relevant in defining at risk patients, decreasing patient morbidity, and assessing new therapies [4, 5]. The clinical diagnosis of peripheral diabetic neuropathy is often missed or peripheral neuropathy is lately diagnosed, mainly because a simple noninvasive method for early detection of peripheral diabetic neuropathy is not yet available [6]. Clinical diagnosis is commonly made only when patients with peripheral diabetic neuropathy become symptomatic. Early diagnosis is currently based on electrophysiological tests or on skin biopsy, probably the gold standard in identifying small fiber peripheral diabetic neuropathy. Electrophysiological tests cannot detect the minute fiber nerve fiber damage in patients with diabetes [7]. Although skin biopsy may detect the minute damage in small peripheral nerve fibers, it has a major limitation because skin biopsy is an invasive test [8, 9].

From: *Ophthalmology Research: Visual Dysfunction in Diabetes*
Edited by: J. Tombran-Tink et al. (eds.), DOI 10.1007/978-1-60761-150-9_4

Recently, a new approach to the detection of very early small fiber peripheral diabetic neuropathy has been proposed and validated. It involves the detection and quantification of the alteration of corneal nerves in diabetes, mainly the subbasal corneal nerve plexus [10]. This is a monolayer of nerve fibers located at the border between corneal epithelium and stroma, which may be detected in vivo even in a noninvasive way (see below) and probably represents the best model to have clinical information on diabetic peripheral neuropathy.

CORNEAL CONFOCAL MICROSCOPY

Corneal confocal microscopy (CCM) is a diagnostic test used to investigate at a microscopic level the different layers of the cornea. It is based on the same physical principle of any confocal microscope, allowing to have in focus just one layer of the examined tissue. Light reflected by any layer out of focus is eliminated allowing to have a high magnification, sharp image of the layer under investigation. Using corneal confocal microscope, the individual structures of any corneal layer may be easily documented: from the endothelium through the stroma (containing keratocytes, nerve fibers, and sometimes Langherans cells) up to the epithelium (with each layer) and tear film. The procedure may be a contact or noncontact one. The noncontact procedure allows to repeat CCM in a safe way, as much as necessary and with high reliability [10]. In our studies, CCM was performed by using Confoscan 4.0 (Nidek, Gamogori, Japan) equipped with an Achroplan nonapplanating ×40 immersion objective lens (Zeiss, Oberkochen, Germany) and with a Z-ring adapter system. Each examination is performed according to a standard procedure, as previously described [11]. Briefly, before the examination, a drop of topical anesthetic (0.4% oxybuprocaine chlorohydrate) is instilled in the lower conjunctival fornix of the eye. One drop of 0.2% polyacrylic gel is applied onto the objective tip to serve as an immersion fluid. The patient is positioned in the chin and forehead rest, and when an image of stroma appears on the monitor of the confocal microscope, the recording button is pressed and a micrometric motor-driven system automatically completes the alignment. The focal plane is automatically moved to reach the anterior chamber and begins recording several scans of the entire depth of the cornea. The Z-ring device is used for all examinations, and only the central cornea is analyzed. Illumination intensity is kept constant in all cases. The images collected using this procedure are analyzed in a qualitative and/or quantitative way. The endothelium is automatically analyzed using a dedicated software available with the machine. Both stromal and epithelial cells may be quantified in a semiautomatic way. The analysis of corneal sub-basal nerve plexus (CSNP) has been recently validated in a large group of normal and pathological eyes (Figs. 1 and 2) [10].

The assessment of CSNP was performed according to the following standardized procedure. The standard dimension of each image produced was 340×255 mm (area = 0.132 mm^2) with an optical section thickness of 5.5 mm. For each examined cornea, the best sharp focus frame of CSNP was chosen. For each frame of the CSNP images, five parameters were analyzed: nerve fiber length (NFL), number of fibers (NF), number of branching (NBr), number of beadings (NBe), and fiber tortuosity (FT) (Fig. 3). NFL was calculated using an image processing computer tool, the Neuron J© program to

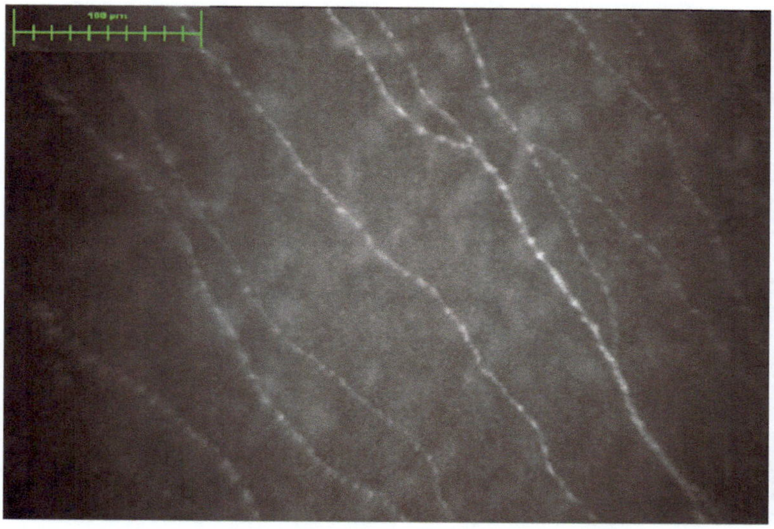

Fig. 1. Corneal subbasal nerve plexus (CSNP) in a normal subject, as shown by corneal confocal microscopy (CCM). It appears as a monolayer of straight nerve fibers with hyperreflective spots along the nerve (nerve beadings).

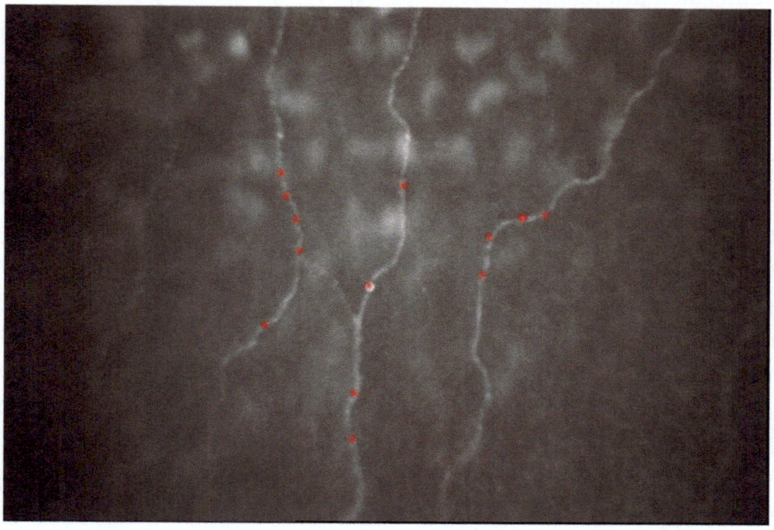

Fig. 2. CSNP in diabetes, examined with CCM. The most evident aspect is the reduction of nerve beadings (colored in *red*) along the nerve fibers.

outline nerve fibers from each CSNP frame. NFL for each image was calculated as the total length of the nerves (micrometers) divided by the area of the image (0.132 mm^2) and expressed as micrometers per square millimeters (mm/mm^2). NF was manually calculated and defined as the total number of principal nerve trunks and their branches per image. NBr was manually calculated and defined as the total NBr per image. NBe was defined as the number of hyperreflective points manually calculated considering 100 mm

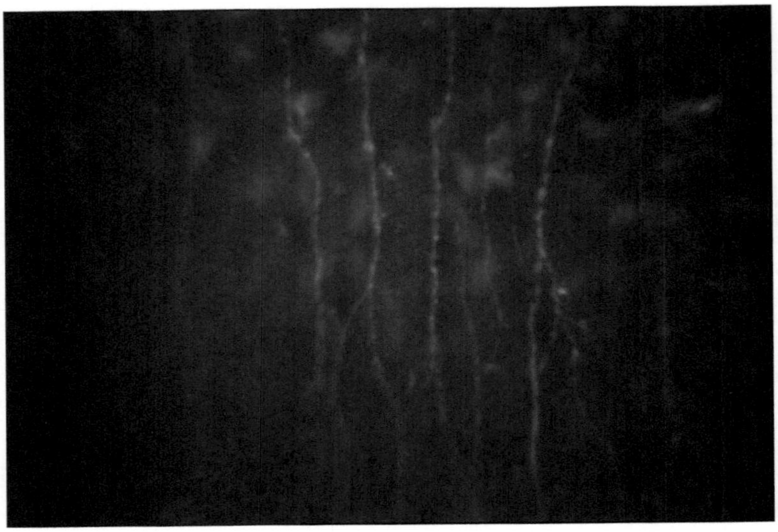

Fig. 3. Normal nerve tortuosity in the corneal subbasal nerve layer.

of one fiber. The fiber was randomly chosen by the operator between all the best focused fibers. The same standard magnification was kept for all the images during the counting. The score system proposed by Oliveira-Soto and Efron [12] was used to analyze FT.

CORNEAL NERVES AND DIABETES

The cornea is the most densely innervated tissue in the body and is richly supplied by sensory and autonomic nerve fibers [13, 14]. Nerve bundles enter the cornea at the periphery in a radial manner parallel to the corneal surface. The nerve bundles lose their perineurium and myelin sheaths approximately 1 mm from the limbus and continue into the cornea surrounded by Schwann cell sheaths, and then subdivide several times into smaller branches. Stromal nerve trunks move from the periphery toward the corneal center and eventually turn 90°, proceeding toward the corneal surface and penetrating Bowman's layer. After penetrating Bowman's layer, the large nerve bundles divide into several smaller bundles, which turn another 90° and continue parallel to the corneal surface between Bowman's layer and the basal epithelial cell layer, creating the subbasal corneal nerve plexus. The CSNP is characterized by local axon enlargements, or beading, which are accumulations of mitochondria and glycogen particles located at the periphery of the bundle. Corneal nerve fibers exert important trophic influences on the corneal epithelium and contribute to the maintenance of a healthy ocular surface [13]. Corneal abnormalities caused by diabetes include superficial punctuate keratopathy, recurrent epithelial defects, neurotrophic keratopathy, and corneal ulcer [15–19]. These abnormalities have been reported to occur in 50–74% of patients with diabetes who never underwent surgery, and many of these patients are asymptomatic [18, 19]. Corneal sensation is reduced in diabetic patients and progresses with the severity of neuropathy, suggesting that corneal nerve fiber damage accompanies diabetic somatic nerve fiber damage [20–22], one of the most important and invalidating diabetic chronic complica-

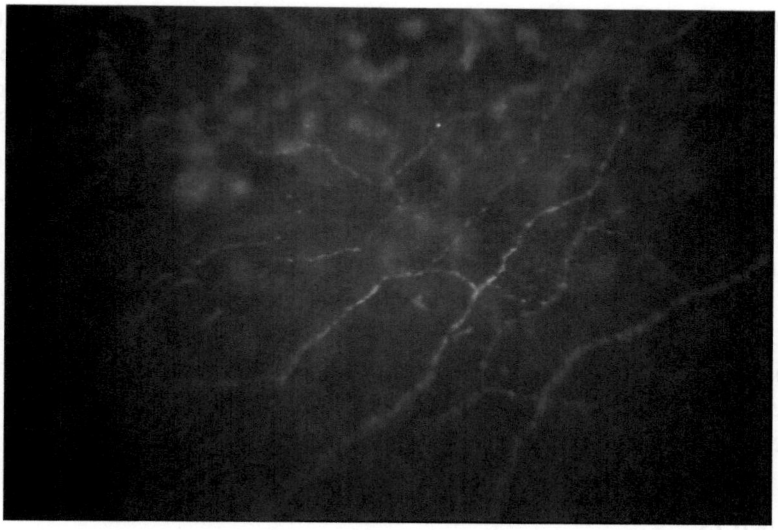

Fig. 4. Altered (increased) tortuosity of the subbasal nerve plexus in diabetes. This image is classified as stage 4 tortuosity.

tions [23]. A growing interest in corneal morphology in diabetic patients, especially in CSNP, is documented [21, 24–27]. Corneal nerve changes secondary to diabetes mellitus have been recently analyzed with CCM using a multiparametric approach and termed corneal diabetic neuropathy (CDN) [21].

CDN, as defined using CCM, is characterized by relevant modifications (vs. control subjects) of CSNP parameters which may be summarized as follows: decrease in the number of fibers, branching pattern and number of beadings, and increase in nerve tortuosity in diabetic patients (Fig. 4) [21]. Rosenberg et al. [22] found a reduction in long nerve fiber bundle in patients with mild to moderate neuropathy, and a reduction in corneal mechanical sensitivity only in patients with severe neuropathy, suggesting that decrease in nerve fiber bundle counts precede impairment of corneal sensitivity and that reduction in neurotrophic stimuli in severe neuropathy may induce a thin epithelium that may lead to recurrent erosions. Chang et al. [24] defined diabetic alterations in the corneal innervations using CCM, finding a decrease in nerve fiber density and nerve branch density and an increase of tortuosity, demonstrating that reduced density in basal epithelial cell is correlated with changes in innervations. Malik et al. [26] showed a progressive reduction in the number of corneal nerve fibers in diabetes, suggesting enhanced degeneration, and showed reduction in the number of corneal nerve branches, suggesting a reduction in regenerative capacity, with a progression of neuropathic severity. Quattrini et al. [27] quantified nerve fiber pathological changes using CCM and found a progressive reduction in corneal nerve fiber and branch density, but the latter was significantly reduced even in diabetic patients without neuropathy. Kallinikos et al. [25] demonstrated that tortuosity coefficient of nerve fibers was significantly greater in the severe diabetic neuropathic group than in control subjects and in the mild and moderate neuropathic groups, suggesting that this morphologic abnormality relates to the severity of somatic neuropathy and may reflect an alteration in the degree of degeneration in diabetes. Moreover, Mehra et al. [28] demonstrated, using CCM, a significant improvement in corneal nerve fiber density and nerve fiber length within 6 months after pancreas transplantation in patients with type 1 diabetes, indicating an early repair process with the restoration of euglycemia. Regeneration of CSNP was demonstrated after refractive surgery [29, 30]. In diabetes, nerve fiber damage is caused by hyperglycemia and oxidative stress [31–33] and not by fiber axotomy, as in refractive surgery [29, 30]. Neurons are obligate glucose users, and whereas some neurons express glucose transporters, glucose may enter the cell solely based on concentration gradient [32]. This makes neurons of the peripheral nervous system particularly vulnerable to hyperglycemia [32]. Vincent et al. [31] reviewed the evidence that indicates that glucose-mediated oxidative stress is an inciting event in the development of diabetic neuropathy. In a pilot study on CSNP regeneration in diabetic patients under topical antioxidant therapy, we observed an increase in NF, NFL, NBe, and nerve sprouting.

CONCLUSION

CCM is currently the key diagnostic technique in evaluating and monitoring CSNP and CDN in vivo. Quantification of CSNP parameters allows a correct, reproducible, and objective in vivo, noninvasive approach to CDN, allowing to characterize peripheral diabetic neuropathy, a potentially highly disabling complication of diabetes, and CCM may represent a valid tool in monitoring CSNP regeneration, which may have important implications for corneal healing and health.

REFERENCES

1. Mokad A, Ford ES, Bowman BA, Nelson DE, Englegau MM, Vinicor F, et al. Diabetes trends in the US: 1990–1998. Diabetes Care. 2000;23:1278–83.
2. Gooch C, Podwall D. The diabetic neuropathies. Neurologist. 2004;10:311–22.
3. Partanen J, Niskanen L, Lehtinen J, Mervaala E, Siitonen O, Uusitupa M. Natural history of peripheral diabetic neuropathy in patients with non-insulin-dependent diabetes mellitus. N Engl J Med. 1995;333:89–94.
4. Park TS, Park JH, Beak HS. Can diabetic neuropathy be prevented? Diabetes Res Clin Pract. 2004;66:S53–8.
5. Boucek P. Advanced diabetic neuropathy: a point of no return. Rev Diabet Stud. 2006;3: 143–50.
6. Rahman M, Griffin SJ, Rahtmann W, Wareham NJ. How should peripheral neuropathy be assessed in people with diabetes in primary care? A population based comparison of four measures. Diabet Med. 2003;20:368–74.
7. Daube JR. Electrophysiologic testing in diabetic neuropathy. In: Dyck P, Thomas P, editors. Diabetic neuropathy. Philadelphia: WB Saunders; 1999. p. 213–5.
8. Smith AG, Howard JR, Kroll R, Ramachandaran P, Hauer P, Singleton JR, et al. The reliability of skin biopsy with measurement of intraepidermal nerve fiber density. J Neurol Sci. 2005;228:65–9.
9. Umapathi T, Tan WL, Cheong Loke S, Cheow Soon P, Tavintharan S, Huak Chan Y. Intraepidermal nerve fiber density as a marker of early diabetic retinopathy. Muscle Nerve. 2007;35:591–8.
10. Midena E, Cortese M, Miotto S, Gambato C, Cavarzeran F, Ghirlando A. Confocal microscopy of corneal sub-basal nerve plexus: a quantitative and qualitative analysis in healthy and pathologic eyes. J Refract Surg. 2009;25:S125–9.

11. Brugin E, Ghirlando A, Gambato C, Midena E. Central corneal thickness. Z-ring corneal confocal microscopy versus ultrasound pachimetry. Cornea. 2007;26:303–7.
12. Oliveira-Soto L, Efron N. Morphology of corneal nerves using confocal microscopy. Cornea. 2001;20:374–84.
13. Muller LJ, Pels L, Vrensen GFJM. Ultrastructural organization of human corneal nerves. Invest Ophthalmol Vis Sci. 1996;37:476–88.
14. Muller LJ, Marfurt CF, Kruse F, Tervo TMT. Corneal nerves: structure, contents and function. Exp Eye Res. 2003;76:521–42.
15. Ohashi Y. Diabetic keratopathy. Nippon Ganka Gakkai Zasshi. 1997;101:105–10.
16. Creuzot-Garcher C, Lafontaine PO, Gualino O, D'Athis P, Petit JM, Bron A. Study of ocular surface involvement in diabetic patients. J Fr Ophtalmol. 2005;28:583–8.
17. Rao GN. DR P Siva Reddy Oration. Diabetic keratopathy. Indian J Ophthalmol. 1987;35:16–36.
18. Schultz RO, Peters MA, Sobocinski K, et al. Diabetic corneal neuropathy. Trans Am Ophthalmol Soc. 1983;81:107–24.
19. Didenko TN, Smoliakova GP, Sorokin EL, et al. Clinical and pathogenetic features of neurotrophic corneal disorders in diabetes. Vestn Oftalmol. 1999;115:7–11.
20. Tavakoli M, Kallinikos PA, Efron E, Boulton AJM, Malik RAM. Corneal sensitivity is reduced and relates to the severity of neuropathy in patients with diabetes. Diabetes Care. 2007;30:1895–7.
21. Midena E, Brugin E, Ghirlando A, Sommavilla M, Avogaro A. Corneal diabetic neuropathy: a confocal microscopy study. J Refract Surg. 2006;22:S1047–52.
22. Rosenberg ME, Tervo TMT, Immonen IJ, Muller LJ, Gronhagen-Riska C, Vesaluoma H. Corneal structure and sensitivity in type 1 diabetes mellitus. Invest Ophthalmol Vis Sci. 2000;41:2915–21.
23. Boulton AJ, Vinik AI, Arezzo JC, et al. Diabetic neuropathies: a statement by the American Diabetes Association. Diabetes Care. 2005;28:956–62.
24. Chang PY, Carrel H, Huang JS, et al. Decreased density of corneal basal epithelium and subbasal corneal nerve bundle changes in patients with diabetic retinopathy. Am J Ophthalmol. 2006;142:488–90.
25. Kallinikos P, Berhanu M, O'Donnel C, Boulton AJM, Efron N, Malik RA. Corneal nerve tortuosity in diabetic patients with neuropathy. Invest Ophthalmol Vis Sci. 2004;45:418–22.
26. Malik RA, Kallinikos P, Abbott CA, et al. Corneal confocal microscopy: a non-invasive surrogate of nerve fibre damage and repair in diabetic patients. Diabetologia. 2003;46:683–8.
27. Quattrini C, Tavakoli M, Jeziorska M, et al. Surrogate markers of small fiber damage in human diabetic neuropathy. Diabetes. 2007;56:2148–54.
28. Mehra S, Tavakoli M, Kallinikos PA, et al. Corneal confocal microscopy detects early nerve regeneration after pancreas transplantation in patients with type 1 diabetes. Diabetes Care. 2007;30:2608–12.
29. Cavillo MP, McLaren JW, Hodge DO, Bourne WM. Corneal reinnervation after LASIK. Prospective 3-year longitudinal study. Invest Ophthalmol Vis Sci. 2004;45:3991–6.
30. Erie J, McLaren JW, Hodge DO, Bourne WM. Recovery of corneal subbasal nerve density after PRK and LASIK. Am J Ophthalmol. 2005;140:1059–64.
31. Vincent AM, Russel JW, Low P, Feldman EL. Oxidative stress in the pathogenesis of diabetic neuropathy. Endocr Rev. 2003;25:612–8.
32. Sullivan KA, Feldman EL. New developments in diabetic neuropathy. Curr Opin Neurol. 2005;18:586–90.
33. McHugh JM, McHugh WB. Diabetes and peripheral sensory neurons: what we don't know and how it can hurt us. AACN Clin Issues. 2004;15:136–49.

5
Clinical Phenotypes of Diabetic Retinopathy

José Cunha-Vaz, Rui Bernardes, and Conceição Lobo

CONTENTS

Keywords Blood–retinal barrier • Retinal vascular endothelium • Macular edema • Retinal leakage analyzer • Multimodal macula mapping • Microaneurysm turnover • Retinopathy progression

Diabetic retinopathy is characterized by gradually progressive alterations in the retinal microvasculature and is the leading cause of new cases of legal blindness among Americans between the ages of 20 and 74 years [1].

Diabetic retinopathy occurs in both type 1 (also known as juvenile-onset or insulin-dependent diabetes) and type 2 (also known as adult-onset or noninsulin-dependent diabetes) diabetes. All the features of diabetic retinopathy may be found in both types of diabetes, but characteristically the incidence of its major complications and main causes of vision loss, macular edema, and retinal neovascularization is quite different for each type of diabetes [1]. Diabetic retinopathy in type 1 diabetes induces vision loss mainly due to the formation of new vessels in the eye fundus and development of proliferative retinopathy, whereas in type 2 diabetes, vision loss is most commonly due to macular edema and proliferative retinopathy is relatively rare.

From: *Ophthalmology Research: Visual Dysfunction in Diabetes*
Edited by: J. Tombran-Tink et al. (eds.), DOI 10.1007/978-1-60761-150-9_5

It is apparent, from the data available from a variety of large longitudinal studies, that the evolution and progression of diabetic retinopathy vary between the two types of diabetes involved and between different patients even when belonging to the same type of diabetes, and does not necessarily progress to clinically significant macular edema (CSME) or proliferative retinopathy leading to vision loss.

NATURAL HISTORY

Diabetic retinopathy shows initially minimal fundus abnormalities and progresses over time to more and more advanced microvascular changes. The main alterations occurring in the diabetic retina are: breakdown of the blood–retinal barrier (BRB), evidenced by abnormal vascular leakage and capillary closure leading to progressive tissue ischemia. These two main alterations lead, as they progress, to the two major complications of diabetic retinal disease which are associated with vision loss: CSME and proliferative retinopathy.

The retinal changes that result from diabetes before the development of the two main complications referred above are conventionally described under the name of nonproliferative diabetic retinopathy (NPDR).

The initial stages of NPDR are, therefore, characterized by the presence of microaneurysms (MA), hemorrhages, hard exudates or cotton-wool spots, indirect signs of vascular hyperpermeability, and capillary closure.

These are the alterations that dominate the initial stages of NPDR, and it is crucial to analyze their development and progression, in order to clarify their relative importance in the progression of diabetic retinopathy [2]. They are not present in every patient in the same way nor at the same rate.

It is fundamental to realize that the course and rates of progression of the retinopathy vary between patients. MA, for example, may come and go. Once you get an MA, you do not necessarily continue to have that MA. MA may disappear due to vessel closure, which is an indication of worsening of the retinopathy because of progressive vascular closure [3]. Hemorrhages will obviously come and go as the body heals them. Frequently, there is apparent clinical improvement with fewer lesions visible, but in reality, it masks worsening of the disease.

A prominent feature of diabetic retinopathy, focal edema, can spontaneously resolve itself. Indeed, it is resolved in approximately a third of patients over a period of 6 months, without any intervention [4].

The initial pathological changes occurring in the diabetic retina are characteristically located in the small retinal vessels of the posterior pole of the retina, i.e., in the macular area. The structural changes in the small vessels include endothelial cell and pericyte damage and thickening of basement membrane [2, 5].

Retinal vascular endothelium is a fundamental part of the BRB, which has many similarities with the blood–brain barrier. It functions as a selective barrier which has shown to be altered in experimental and human diabetes [6]. It is altered in the early stages of diabetic retinal disease.

Pericyte damage has also been reported as one of the earliest findings in diabetic retinal disease since the introduction of retinal digest studies [7]. However, pericyte damage may be more prominent just because it is more readily detectable than endothelial cell damage, because the pericytes are encased in basement membrane and thus less accessible to the clearing effect of blood flow, whereas dying endothelial cells slough off into the capillary lumen and are rapidly cleared by the blood stream.

The simplest paradigm that explains the initial retinal microvascular changes in diabetes, capillary hyperpermeability, and capillary closure is damage to the vascular endothelium. In the retina, endothelial cells are the site of the BRB, a specific blood–tissue barrier, and, as in all vessels, provide a nonthrombogenic surface for blood flow. Both these properties are compromised by diabetes from the initial stages of diabetic retinal disease.

In addition, diabetes also affects the neural and glial cells of the retina. Consequently, we have an initial pathological picture characterized by endothelial and pericyte alterations associated with basement membrane thickening and MA formation, together with retinal tissue changes.

These alterations when seen as a whole are characteristic for NPDR, particularly the alteration of the BRB, the pericyte damage, and the MA formation, but they also occur in a variety of retinal diseases unrelated to diabetes. There is clear site specificity, not disease specificity [2].

Which are then the features of the retinal circulation which are specific to the retina and may be responsible for the site specificity of diabetic retinopathy? They are the BRB and the autoregulation of retinal blood flow. Both serve the needs of the neuronal and glial cells of the retina.

An abnormality of the BRB, demonstrated both by vitreous fluorometry and fluorescein angiography, has repeatedly been demonstrated to be an early finding both in human and in experimental diabetes [6, 8, 9]. Loss of retinal blood flow autoregulation contributes to capillary closure that ultimately leads to retinal ischemia and to one of the two major complications of diabetic retinal disease, proliferative retinopathy, which causes the most tragic outcomes: vitreous hemorrhage, rubeosis iridis, retinal detachment, etc.

It is becoming apparent that at least three processes can contribute to retinal capillary occlusion and obliteration in diabetes: proinflammatory changes, microthrombosis, and apoptosis [10].

MA FORMATION AND DISAPPEARANCE RATES

MA and hemorrhages are the initial changes seen on ophthalmoscopic examination and fundus photography (FP). MA counting has been suggested as an appropriate marker of retinopathy progression [11, 12].

It must be realized that MA formation and disappearance are dynamic processes. During a 2-year follow-up of 24 type 1 diabetics with mild background diabetic retinopathy using fluorescein angiography, Hellstedt and Immonen [13] observed 395 new MA and the disappearance of 258 previously identified.

Generally, the disappearance of an MA is not a reversible process and indicates vessel closure and progressive vascular damage. Therefore, to assess progression of retinopathy, MA counting should take into account every newly developed MA identified in a new location.

We have developed software for MA counting in fundus-digitized images where the location of each MA is taken into account and registered [14]. In a follow-up study with repeated fundus images obtained at regular intervals, all MAs in the fundus were counted and added as they became visible in new locations in the retina. The results of MA counting using this method, in a 2-year follow-up study of a series of eyes with

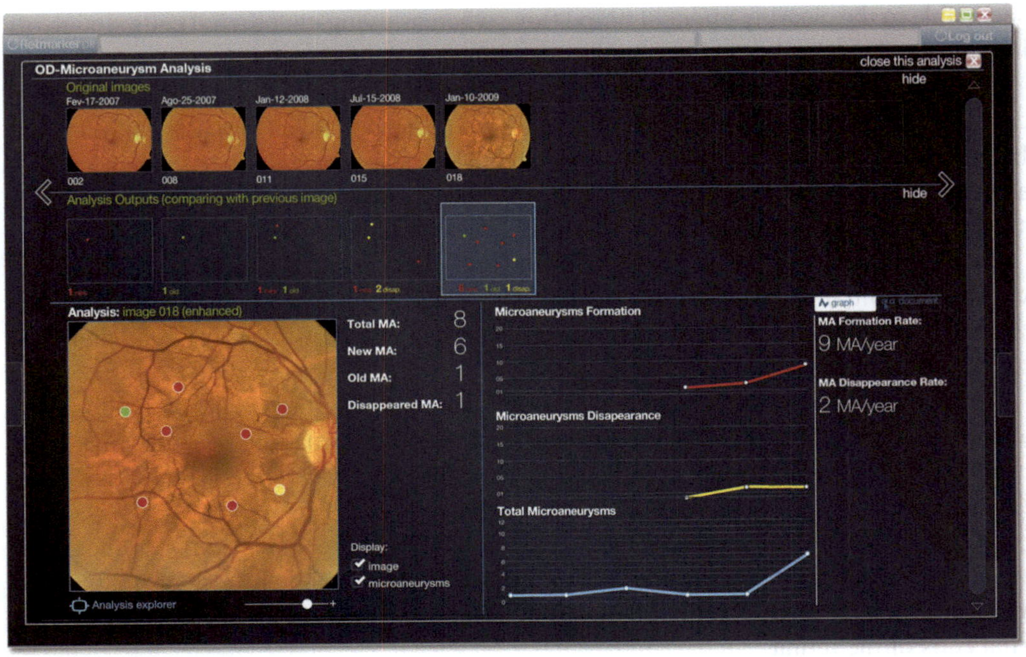

Fig. 1. Microaneurysm analysis.

mild nonproliferative retinopathy in subjects with type 2 diabetes, maintaining a stable metabolic control during the period of the study, suggested that MA counting may be a good marker of disease progression in the initial stages of NPDR [15].

In order to improve the identification and counting of MA on color fundus images, the software included algorithms for eye movement compensation, color correction, and identification of each MA by its coordinates.

Using the software's ability to identify each MA as a single entity, in a specific location with identifiable coordinates, the following parameters were assessed: cumulative number of MA, MA formation rate, and MA disappearance rate.

In a study involving 50 eyes/patients over a period of 2 years, with examinations performed every 6 months, using the traditional procedure, the total amount of MA detected at every visit remained stable. However, using the software to identify MA location, the cumulative number of MA rose from 115 at the first visit to 505 at the last visit, showing a marked increase in new MA. It is now obvious that there were many more new MA in the fundus in this 2-year time period than expected using data for each examination separately.

One of the advantages of the method used is the ability to count the number of real new MA appearing at every visit (MA formation rate) (Fig. 1). The rate of formation (MA/year) ranged from 0 to 22. The results showed that eyes in the same retinopathy stage from different patients show very different MA formation rates. Values for MA formation rate higher than 3 MAs/year correlated well with increased fluorescein leakage measured by vitreous fluorometry and capillary closure identified by a damaged foveal avascular zone (FAZ), demonstrating a direct correlation with faster retinopathy progression [16].

The MA disappearance rate ranged from 0 to 16 MAs/year. MA disappearance rates also varied quite markedly in eyes from different patients and showed similar correlations.

MA formation represents particularly well diabetic retinopathy because MAs are associated with localized proliferation of endothelial cells, loss of pericytes, and alterations of the capillary basement membrane, alterations that occur in the initial stages of diabetic retinal disease and have been considered to be directly involved in its pathophysiology [2, 17, 18].

MA closure and their disappearance are most probably due to thrombotic phenomena leading to subsequent rerouting of capillary blood flow and progressive remodeling of the retinal vasculature in diabetes [19]. These thrombotic changes are probably enhanced by changes in the red and white cells occurring as a result of diabetes.

MA counting on fundus photographs and MA counting on fluorescein angiography have been proposed as predictive indicators of progression of diabetic retinopathy [20, 21]. The software developed by our research group allows the identification of the exact location of each MA in successive fundus photographs performed in each eye. The identification of the exact location of an individual MA is considered particularly important because a new MA is considered to develop only once in a specific location, its disappearance being generally associated with capillary closure, leaving in its place mainly remnants of basement membrane [2, 18].

Our studies demonstrated a steady turnover of MAs in the diabetic retina, even in the initial stages of retinopathy. In fact, most MAs show a lifetime of less than 1 year, with new ones being formed and disappearing at rates which vary between different patients, confirming previous reports [22].

Most interestingly, however, is the observation that some patients show much higher rates of MA formation and disappearance, suggesting that they may represent specific phenotypes of diabetic retinopathy. These eyes showed also faster progression in other retinal lesions, with increased fluorescein leakage, i.e., alterations of BRB, and progression in capillary closure.

Using this new methodology, we have recently analyzed data from a group of 113 type 2 diabetic patients with mild-to-moderate NPDR, followed up for 2 years as controls in diabetic retinopathy clinical trials, and thereafter, by usual care at the same institution for a period of 10 years [23].

MA turnover from the initial 2 years was correlated with the occurrence of CSME during the following 8 years.

Patients were maintained under acceptable metabolic control during this period, and underwent ophthalmological examinations (including color fundus photography) every month.

At baseline, all patients showed mild-to-moderate retinopathy and were classified as levels 20 (MA only) or 35 (MA/hemorrhages and/or hard exudates) according to the Early Treatment of Diabetic Retinopathy Study (ETDRS) grading scale.

At the end of the 10-year follow-up period, 17 out of the 113 patients developed CSME needing photocoagulation.

When counting the total number of MA over the first 2 years of the follow-up, a significant increase in the number of MA was found for the CSME eyes ($p=0.002$), while for the non-CSME eyes, the number of MA remained relatively constant ($p=0.647$).

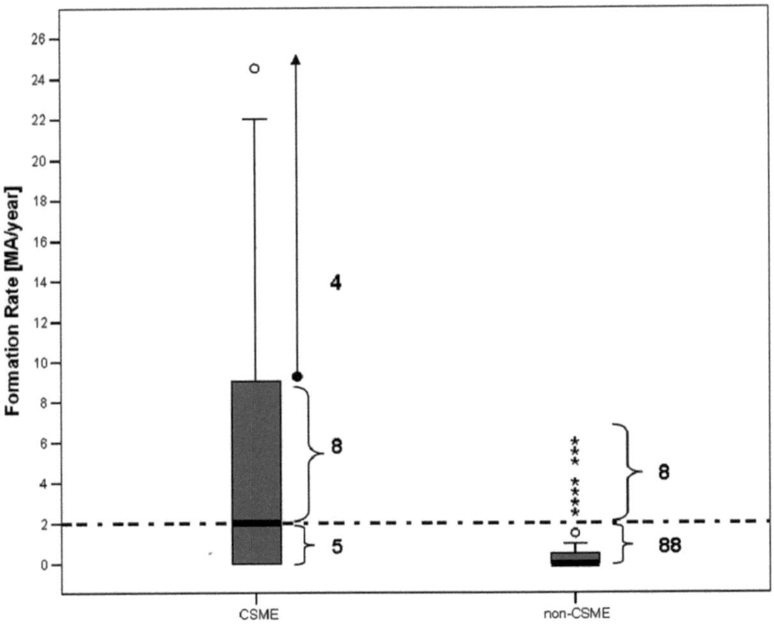

Fig. 2. Boxplot for the microaneurysm formation rate for clinically significant macular edema (CSME) and non-CSME eyes, and number of eyes for the different values of the microaneurysm formation rate.

When computing the MA turnover for the same period of time, a higher MA turnover was found in the group of patients/eyes that developed CSME (higher MA formation and disappearance rates). Formation and disappearance rates of 9.2 ± 18.2 and 7.5 ± 16.6 MAs/year, respectively, were found for the eyes that developed CSME, while rates of 0.5 ± 1.2 and 0.5 ± 1.2 MAs/year were found for the non-CSME eyes ($p < 0.001$).

A MA turnover of at least 2 MAs/year was found in 12 of the 17 eyes that developed CSME (70.6%), whereas this was only found in 8 of the 96 eyes that did not develop CSME during the 10-year follow-up period (8.3%) (Fig. 2).

This study shows that in the initial stages of diabetic retinopathy, higher MA counts and MA turnover obtained from color fundus photography are good indicators of retinopathy progression and development of CSME needing photocoagulation.

We also found that MA turnover is more reliable than simple MA counts and that there was much better agreement between graders when determining MA turnover than MA counts.

Recently, Sharp et al. [24] found that the MA turnover varied widely between eyes of the same retinopathy level. This is also consistent with our findings. MA turnover has been shown in this study to vary between patients that were classified with the same retinopathy level. Particularly relevant is the finding that the patients who have higher MA turnover values are the ones that go on to develop CSME within a period of 10 years and show a more rapid retinopathy progression, particularly in association with poor metabolic control demonstrated by higher HbA_{1C} values.

It appears that it is possible to use MA turnover computed from noninvasive color fundus photographs as a biomarker to identify eyes/patients at risk of progression to CSME.

ALTERATION OF THE BLOOD–RETINAL BARRIER

Fluorescein angiography confirmed most of what was known of the initial pathological picture of diabetic retinopathy and showed in the initial stages of the disease focal leaks of fluorescein, demonstrating, in a clinical setting, the existence of focal breakdowns of the BRB.

In 1975, vitreous fluorometry, a clinical quantitative method for the study of the BRB, was introduced by our group [6], showing that an alteration of the BRB could be detected and measured in some diabetic eyes presenting clinically normal fundi. These results were confirmed by Waltman et al. [9] and demonstrated that breakdown of the BRB plays an important initiating role in the development of the diabetic retinopathy.

One major limitation of the available commercial instrumentation for vitreous fluorometry was associated with the fact that the permeability of the BRB is measured as an average over the posterior pole. Accurate mapping of localized changes in the permeability of the BRB would be beneficial for early diagnosis, to explain the natural history of retinal disease, and to predict its effect on visual acuity.

We have recently developed a new method of retinal leakage mapping, the retinal leakage analyzer (RLA), which is capable of measuring localized changes in fluorescein leakage across the BRB while simultaneously imaging the retina (Fig. 3). The instrument is based on a confocal scanning laser ophthalmoscope that was modified into a confocal scanning laser fluorometer [25]. Two types of information are obtained simultaneously, distribution of fluorescein concentration (retina and vitreous) and

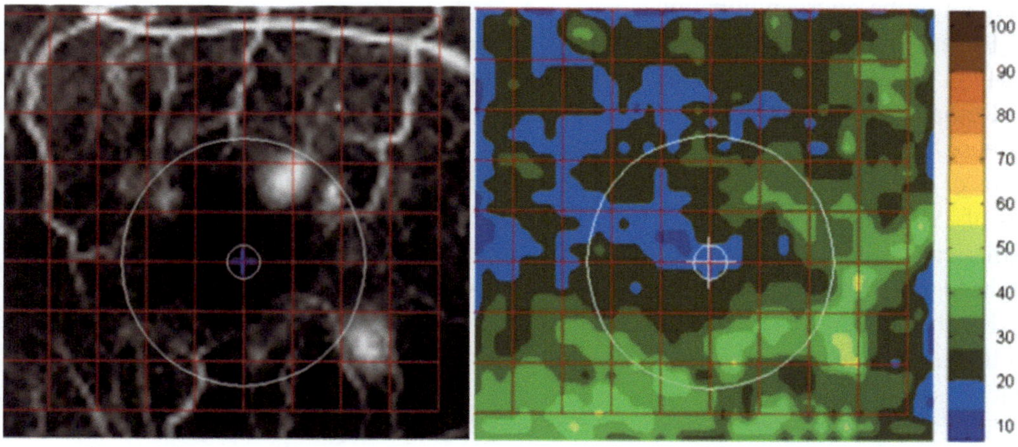

Fig. 3. Macula from a patient with diabetes type 2. Fluorescein angiography obtained by scanning laser ophthalmoscope (*left*). Retinal leakage analyzer (RLA) blood–retinal barrier (BRB) permeability map (RLmap) in a false color-code map (*right*).

morphology of the eye fundus. This simultaneous acquisition is crucial because it allows a direct correlation to be established between the maps of permeability and the morphological information.

RETINAL CAPILLARY CLOSURE

Retinal ischemia due to vascular closure develops relatively early in the course of diabetic retinopathy and is attributed to changes in vascular autoregulation and microthrombosis formation. Retinal blood flow changes are considered to lead to the development of poor perfusion facilitating microthrombosis formation [19].

Alterations in retinal blood flow have been identified in the different stages of the progression of retinopathy, but a major problem associated with these measurements is their technical complexity and variability. Our observations indicate that in some diabetic eyes, even before the development of visible retinopathy, there is (probably due to local factors) a marked increase in retinal capillary blood flow with maximal utilization of the retinal capillary net, whereas other eyes do not show this circulatory response, suggesting individual variations in the response to the altered metabolic status. This increase in retinal blood flow may contribute to localize endothelial damage and establish the appropriate conditions for microthrombosis formation.

NEURONAL AND GLIAL CELL CHANGES: RETINAL THICKNESS MEASUREMENTS

We have stated previously that the simplest paradigm to explain increased capillary permeability and the advent of capillary closure centers on vascular endothelial damage. There are, however, a number of reports showing changes in the neuronal and glial cells of the retina in diabetes very early in the course of the disease [26]. This is clearly of major potential importance, and it may indicate at least a contributory role in the development of the microangiopathy.

Recent evidence suggests that retinal glial and Muller cells, in particular, are affected early in the course of both experimental and human diabetes.

Retinal edema is a frequent alteration occurring in the initial stages of diabetic retinal disease. As the disease progresses, it may cause CSME, one of the two major complications of disease associated with loss of vision. Based on WESDR data, it was estimated (as of 1993) that of approximately 7,800,000 people with diabetes, about 84,000 North Americans would develop proliferative retinopathy and about 95,000 would develop sight loss from macular edema over a 10-year period [11, 12].

Edema of the retina is any increase of water of the retinal tissue resulting in an increase in its volume, i.e., because of the structural organization of the retina, an increase in its thickness.

This increase in water content of the retinal tissue may be initially intracellular or extracellular. In the first case, also called cytotoxic edema, there is an alteration of the cellular ionic exchanges with an excess of Na^+ inside the cell. In the second case, also called vasogenic edema, there is predominantly extracellular accumulation of fluid directly associated with the alteration of the BRB [27].

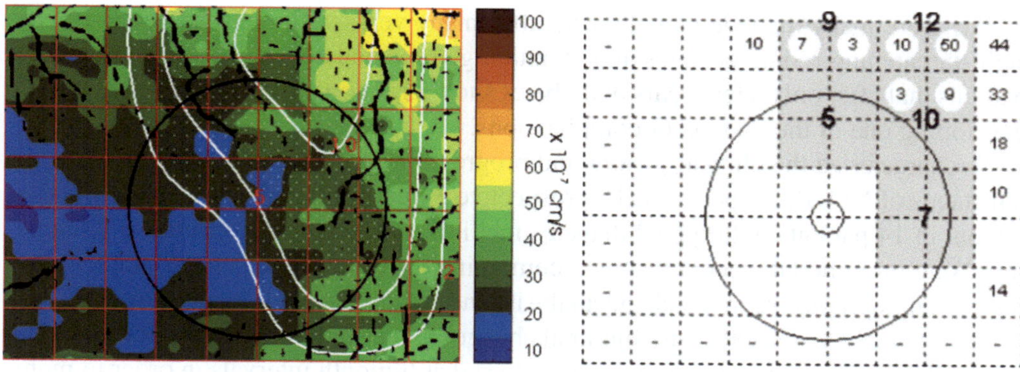

Fig. 4. Multimodal macula mapping of an eye with mild NPDR showing localized increases in leakage and retinal thickness. The *background* represents the leakage using a false color code. *Units* are $\times 10^{-7}$ cm/s (*left*). The *gray areas* represent increased retinal thickness (shown in *white dots on the left image*) (*right*).

It is now possible to objectively measure retinal thickness. Optical coherence tomography (OCT, Carl Zeiss Meditec, Dublin, CA, USA) is a powerful tool for the objective assessment of macular edema.

Measurements of retinal thickness show that localized areas of retinal edema are a frequent finding in the diabetic retina in the initial stages of NPDR in subjects with diabetes type 2 and allow to follow its progression to CSME.

MULTIMODAL MACULA MAPPING

The initial changes occurring in the diabetic retina involve the macula, and an alteration of the macula will, sooner or later, affect visual acuity.

There are a variety of diagnostic tools and techniques to examine the macular region and to obtain information on its structure and function. The different methods available offer different perspectives and fragmentary information. It has been our objective, in recent years, to combine different methodologies and to obtain maps of the alterations occurring in the macular region of the retina (Fig. 4).

Our research group has been developing methods to combine and integrate data from fundus photography, angiographic images (scanning laser ophthalmoscope–fluorescein angiography), maps of fluorescein leakage into the vitreous (scanning laser ophthalmoscope–retinal leakage analyzer), and maps of retinal thickness of the macular area to achieve multimodal macula mapping [25, 28, 29].

CLINICAL RETINOPATHY PHENOTYPES

It is well recognized that duration of diabetes and level of metabolic control are important risk factors for development of diabetic retinopathy.

However, these risk factors do not explain the great variability that characterizes the evolution and rate of progression of the retinopathy in different diabetic individuals. There is clearly great individual variation in the course of diabetic retinopathy.

There are many diabetic patients who after many years with diabetes never develop sight-threatening retinal changes and maintain good visual acuity. There are also other patients that after only a few years of diabetes show a retinopathy that progresses rapidly developing one of the two major complications.

To characterize the clinical picture and progression of the retinal changes in the initial stages of NPDR, we performed a prospective 3-year follow-up study of the macular region, in 14 patients with type 2 diabetes mellitus and mild nonproliferative retinopathy, using multimodal macula mapping combining data from fundus photography, fluorescein angiography, retinal leakage analysis, and retinal thickness measurements [30].

In a span of 3 years, eyes with minimal changes at the start of the study (levels 20 and 35 of ETDRS-Wisconsin grading) were followed at 6-month intervals in order to monitor progression of the retinal changes.

The most frequent alterations observed were, by decreasing order of frequency MA, leaking sites on the RLA and areas of increased retinal thickness.

Increased rates of MA formation were found in eyes that showed more MA at baseline and higher values of BRB permeability during the study.

RLA-leaking sites were a very frequent finding and reached very high BRB permeability values in some eyes. These sites of alteration of the BRB, well identified in RLA maps, maintained, in most cases, the same location on successive examinations, but their BRB permeability values fluctuated greatly between examinations, indicating reversibility of this alteration.

There was, in general, a correlation between the BRB permeability values and the changes in HbA_{1C} levels occurring in each patient. This correlation was particularly clear when looking at eyes that showed, at some time during the follow-up period, BRB permeability values within the normal range. A return to normal levels of BRB permeability was, in this study and in each patient, always associated with a stabilization or decrease in HbA_{1C} values.

Areas of increased retinal thickness were another frequent finding in these eyes. They were present in every eye at some time during the follow-up and were absent, at baseline, in only 2 of the 14 eyes. This confirms previous observations by our group [25] and by others [31]. However, the areas of increased retinal thickness varied in their location over subsequent examinations and did not correlate with changes in HbA_{1C} levels. They may represent a delayed response in time to other changes occurring in the retina, such as increased leakage [25]. They certainly represent in most cases zones of extracellular edema, an interpretation supported by the frequent shift observed in their location in subsequent examinations.

Multimodal imaging of the macula made apparent three major evolving patterns occurring during the follow-up period of 3 years: Pattern A includes eyes with a slow rate of MA formation, relatively little abnormal fluorescein leakage, and a normal FAZ. This group appears to represent eyes presenting slowly progressing retinal disease. Pattern B includes eyes with persistently high leakage values, indicating an important alteration of the BRB, high rates of MA formation, and a normal FAZ. All these features suggest a more rapid and progressive form of the disease. This group appears to identify a "wet" form of diabetic retinopathy. Pattern C includes eyes with high rates of MA formation and disappearance, variable leakage, and an abnormal FAZ. This group is less

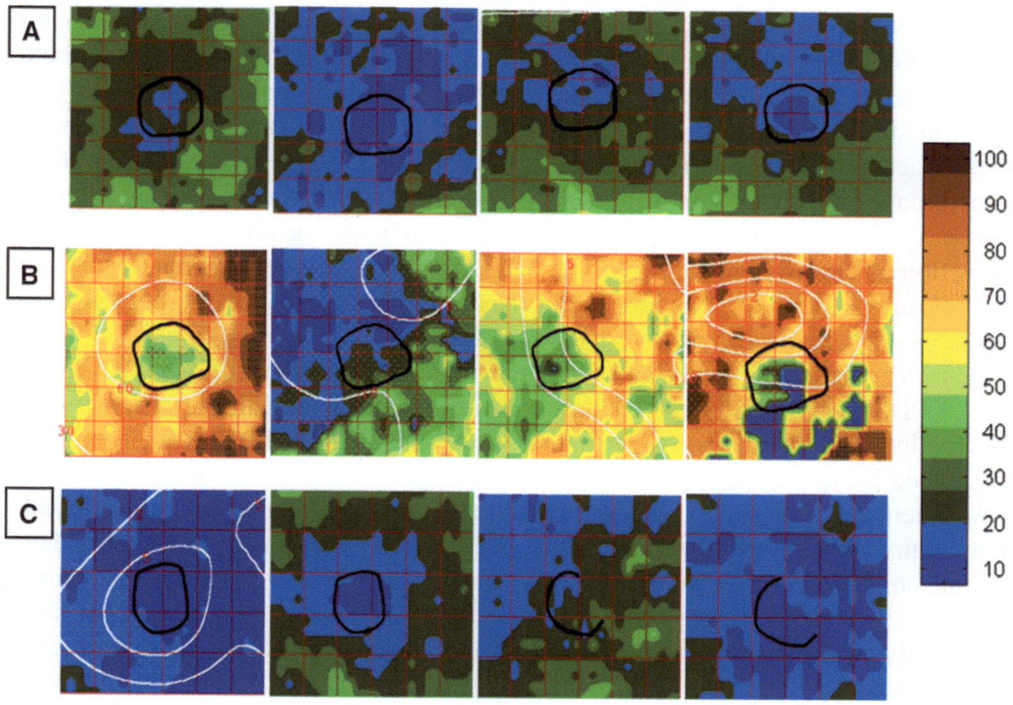

Fig. 5. Multimodal images from three different patients, at yearly intervals, showing for each visit the foveal avascular zone (FAZ) contour, RLA results, and retinal thickness analyzer results. (**A**) Pattern A. Note the little amount of retinal leakage over the four represented visits and the normal FAZ contour. This patient showed a slow rate of microaneurysm formation. (**B**) Pattern B. Note the high retinal leakage showing a certain degree of reversibility and the normal FAZ contour. This patient showed a high rate of microaneurysm accumulation over the 3-year follow-up period. (**C**) Pattern C. Note the reversible retinal leakage and the development of an abnormal FAZ contour. This patient showed a high rate of microaneurysm formation.

well characterized considering the small number of eyes that showed an abnormal FAZ. It may be that abnormalities of the FAZ may occur as a late development of groups A and B or progress rapidly as a specific "ischemic" form (Fig. 5).

We have now extended our observations after following for 2 years 59 patients with diabetes type 2 and mild NPDR. In this larger study, these three different phenotypes were again clearly identified. The discriminative markers of these phenotypes were MA formation and disappearance rates, degree of fluorescein leakage, and signs of capillary closure in the capillaries surrounding the FAZ [23].

It must be realized that levels of hyperglycemia and duration of diabetes, i.e., exposure to hyperglycemia, are expected to influence the evolution and rate of progression tentatively classified in these three major patterns.

If diabetic retinopathy is a multifactorial disease—in the sense that different factors or different pathways may predominate in different groups of cases with diabetic retinopathy—then it is crucial that these differences and the different phenotypes are identified.

Diabetes mellitus is a familial metabolic disorder with strong genetic and environmental etiology. Familial aggregation is more common in type 2 than in type 1 diabetes. Rema et al. [32] reported that familial clustering of diabetic retinopathy was 3 times higher in siblings of type 2 subjects with diabetic retinopathy. Presence or absence of genetic factors may play a fundamental role in determining specific pathways of vascular disease and, as a consequence, different progression patterns of diabetic retinal disease. It could be that certain polymorphisms would make the retinal circulation more susceptible to an early breakdown of the BRB (pattern B) or microthrombosis and capillary closure (pattern C). The absence of these specific genetic polymorphisms would lead to an evolving pattern of pattern A.

It is clear from this study and from previous large studies such as DCCT [33] and UKPDS [34] that hyperglycemia plays a determinant role in the progression of retinopathy. It is interesting to note that HbA_{1C} levels are also largely genetically determined [35].

An interesting perspective of our observations, analyzed under the light of available literature, depicts diabetic retinopathy as a microvascular complication of diabetes mellitus conditioned in its progression and prognosis by a variety of different genetic polymorphisms and modulated in its evolution by HbA_{1C} levels, partly genetically determined and partly dependent on individual diabetes management. The interplay of these multiple factors and the duration of this interplay would finally characterize different clinical pictures or phenotypes of diabetic retinopathy.

RELEVANCE FOR CLINICAL TRIAL DESIGN

It is crucial in order to design an appropriate clinical trial to test the efficacy of a drug, to identify not only the meaningful clinical endpoints but also the surrogate endpoints that may demonstrate efficacy of a drug in a realistic and feasible period of time [36].

It is clear that such process implies the validation of surrogate endpoints by the associated occurrence of hard clinical outcomes such as significant visual loss. It is here that the problem lies. Diabetic retinopathy progresses to irreversible stages of the disease with relatively little visual loss, and when macular edema or proliferative retinopathy is present, it becomes ethically mandatory to perform photocoagulation treatment.

The development of an effective drug must take into account the need to demonstrate efficacy on the earliest and reversible stages of diabetic retinal disease by demonstrating its effect on surrogate endpoints which can be followed for shorter periods of time. The assumption would be that those surrogate endpoints would ultimately be validated by association with more hard clinical outcomes.

It is therefore an urgent priority to identify endpoints which can be accepted as surrogates and be validated in longer natural history studies.

The candidates for surrogate endpoints in the initial stages of the retinal disease are not many: mean differences on the ETDRS retinopathy scale, reduction in fluorescein leakage, reduction in macular thickening, and microaneurysm turnover.

The problem of using the ETDRS retinopathy scale lies in the fact that in the initial stages of the retinopathy, even a two-step eye change means that we have to wait for an important worsening of the retinopathy and the presence of irreversible changes.

The second possibility, reduction in fluorescein leakage, evaluates one of the two main factors in the progression of the retinopathy, the alteration of the BRB permeability. It has, however, a major drawback; it involves intravenous fluorescein administration and is technically difficult.

The third candidate, reduction in macular thickening by measuring the mean change with dedicated instrumentation, has been shown to correlate poorly with visual acuity loss, and changes in retinal thickness do not necessarily correlate with progression of the retinopathy [29].

Finally, the fourth possibility, calculation of MA turnover from fundus photographs, taking into account every new MA according to their specific location in the eye fundus is noninvasive and has the potential to become an extremely informative marker of the overall progression of diabetic retinal vascular disease. By calculating MA turnover on digital fundus images, using appropriate software to identify the specific location of each MA, we may be able to measure the rate of progression of diabetic vascular retinal disease [23]. Our studies suggest that MA turnover may contribute decisively to design feasible clinical trials to test the efficacy of new drugs.

Another fundamental step in this procedure is the characterization of the different phenotypes of diabetic retinal disease.

The design of future clinical trials should consider only groups of patients characterized by their homogeneity: patients presenting a specific retinopathy phenotype (wet/leaky or ischemic), with similar duration of diabetes and at similar levels of blood pressure and metabolic control (HbA_{1C} values). Patients that have retinopathy characterized by low progression with low values of MA turnover, which are the majority, should not be included in relatively short-term clinical trials.

RELEVANCE FOR CLINICAL MANAGEMENT

It is accepted that in the initial stages of diabetic retinopathy when the fundus alterations detected by ophthalmoscopy or slit-lamp examination are limited to MA, hemorrhages and hard or soft exudates, i.e., mild or NPDR, an annual examination is indicated to every patient with 5 or more years of duration of their diabetes. This is the recommendation of the American Academy of Ophthalmology Guidelines for Diabetic Retinopathy [37].

Our observations and the identification of different diabetic retinopathy phenotypes in the initial stages of diabetic retinopathy, i.e., mild or moderate NPDR, characterized by different rates of progression of the retinopathy suggest that specific approaches should be used when managing these different retinopathy phenotypes.

A patient with mild or moderate NPDR, presenting retinopathy phenotype B (wet/leaky), characterized by marked breakdown of the BRB, identified by high MA formation rates and increased values of fluorescein leakage into the vitreous, registered during a period of 1–2 years of follow-up, and indicating rapid retinopathy progression, should be watched more closely and examined at least at 6-month intervals. Furthermore, blood pressure values and metabolic control should be closely monitored at least at 3-month intervals and paying close attention to HbA_{1C} levels. Communication channels should be rapidly established between ophthalmologist and their diabetologist, internist, or general health care provider. Information should be given indicating that the chances of rapid retinopathy progression to more advanced stages of disease are in these patients relatively high, calling for immediate tighter control of both glycemia and blood pressure.

A patient with mild or moderate NPDR presenting retinopathy phenotype C, ischemic, characterized by high MA formation and disappearance rates and signs of capillary closure would similarly indicate the need for shorter observation intervals than 1 year with particular attention for other systemic signs of microthrombosis. Here, however, control of hyperglycemia and blood pressure must be addressed with some degree of caution. Improved metabolic and blood pressure control must be progressive and less aggressive than with phenotype B. It is realized that the ischemia that characterizes phenotype C may become even more apparent in eyes submitted to rapid changes in metabolic control, and lowering rapidly the blood pressure may increase the retinal damage associated with ischemia.

Finally, a patient with mild or moderate NPDR, presenting phenotype A, identified by low levels of fluorescein leakage, no signs of capillary closure, low MA formation rates, and with a diabetes duration of more than 10 years, all signs indicating a slowly progression subtype of diabetic retinopathy, may be followed at intervals longer than 1 year. If the examination performed at 2-year intervals confirms the initial phenotype characterization, the patient and his diabetologist, internist, or general health care provider should be informed of the good prognosis associated with this retinopathy phenotype.

TARGETED TREATMENTS

It would be of great benefit to have a drug available which would prevent the need for destructive photocoagulation of the retina. Furthermore, many diabetic patients are not well controlled, they do not come to the doctor often, and they become blind because they do not get medical attention in time for photocoagulation.

The major large clinical trials have shown that tight glycemic control slows the development and progression of diabetic retinopathy. But the constantly increasing incidence of type 2 diabetes and the evidence that retinal damage begins early on underscore the need for a medical treatment that is targeted to the initial retinal alterations.

Several key pathways have been incriminated in the process of triggering diabetic retinal disease, and they may play specific roles in the development of specific retinopathy phenotypes. Four candidates, the polyol pathway, nonenzymatic glycosylation, growth factors, and protein kinase C, may be playing leading roles in the development of diabetic retinal disease.

It is possible that all these different mechanisms of disease play complementary roles in the progression of diabetic retinal disease.

The identification of different retinopathy phenotypes characterized by different rates of progression and different dominant retinal alterations may indicate that different disease processes predominate in specific retinopathy phenotypes.

Identification of well-defined retinopathy phenotypes may be an essential step in the quest for a successful treatment of diabetic retinopathy. After the characterization of specific retinopathy phenotypes, the predominant disease mechanisms involved may be identified, and drugs directly targeted at the correction of these disease mechanisms may be used with greater chances of success.

REFERENCES

1. Aiello LP, Gardner TW, King GL, Blankenship G, Cavallerano JD, Ferris III FL, Kein R. Diabetic retinopathy. Diabetes Care. 1998;21:143–56.
2. Cunha-Vaz JG. Pathophysiology of diabetic retinopathy. Br J Ophthalmol. 1978;62:351–5.
3. Cunha-Vaz JG. Perspectives in the treatment of diabetic retinopathy. Diabetes Metabol Rev. 1992;8:105–16.
4. Ferris F, Davis M. Treating 20/20 eyes with diabetic macula edema. Arch Ophthalmol. 1990;117:675–6.
5. Garner A. Pathogenesis of diabetic retinopathy. Semin Ophthalmol. 1987;2:4–11.
6. Cunha-Vaz JG, Faria de Abreu JR, Campos AJ, Figo GM. Early breakdown of the blood–retinal barrier in diabetes. Br J Ophthalmol. 1975;59:649–56.
7. Cogan DG, Kwabara T. Capillary shunts in the pathogenesis of diabetic retinopathy. Diabetes. 1963;12:293–300.
8. Waltman SR, Krupin T, Hanish S, Oestrich C, Becker B. Alteration of the blood–retinal barrier in experimental diabetes mellitus. Arch Ophthalmol. 1978;96:878–9.
9. Waltman SR, Oestrich C, Krupin T, Hanish S, Ratzan S, Santiago J, Kilo C. Quantitative vitreous fluorophotometry: a sensitive technique for measuring early breakdown of the blood–retinal barrier in young diabetic patients. Diabetes. 1978;27:85–7.
10. Gardner TW, Aiello LP. Pathogenesis of diabetic retinopathy. In: Flynn Jr HW, Smiddy WE, editors. Diabetes and ocular disease: past, present, and future therapies, AAO monograph no. 14. San Francisco: The Foundation of the American Academy of Ophthalmology; 2000. p. 1–17.
11. Klein R, Klein BEK, Moss SE, Cruikschanks KJ. The Wisconsin epidemiologic study of diabetic retinopathy. XV the long-term incidence of macular edema. Ophthalmology. 1995;102:7–16.
12. Klein R, Meuer SM, Moss SE, Klein BEK. Retinal microaneurysms counts and 10-year progression of diabetic retinopathy. Arch Ophthalmol. 1995;113:1386–91.
13. Hellstedt T, Immonen I. Disappearance and formation rates of microaneurysms in early diabetic retinopathy. Br J Ophthalmol. 1996;80:135–9.
14. Bernardes R, Nunes S, Pereira I, Torrent T, Rosa A, Coelho D, Cunha-Vaz J. Computer-assisted microaneurysm turnover in the early stages of diabetic retinopathy. Ophthalmologica. 2009;223:284–91.
15. Torrent-Solans T, Duarte L, Monteiro R, Almeida E, Bernardes R, Cunha-Vaz J. Red-dots counting on digitalized fundus images of mild nonproliferative retinopathy in diabetes type 2. Invest Ophthalmol Vis Sci. 2004:2985 (Abstract number 2985/B620).
16. Nunes S, Pires I, Rosa A, Duarte L, Bernardes R, Cunha-Vaz J. Microaneurysm turnover is a biomarker for diabetic retinopathy progression to clinically significant macular edema: findings for type 2 diabetics with nonproliferative retinopathy. Ophthalmologica. 2009;223: 292–7.
17. Ashton N. Studies of retinal capillaries in relation to diabetic and others retinopathies. Br J Ophthalmol. 1963;47:521–38.
18. Ashton N. Vascular basement membrane changes in diabetic retinopathy. Montgomery lecture, 1973. Br J Ophthalmol. 1974;58:344–7.
19. Boeri D, Maiello M, Lorenzi M. Increased prevalence of microthromboses in retinal capillaries of diabetic individuals. Diabetes. 2001;50:1432–9.

20. Kohner EM, Sleightholm M. Does microaneurysm count reflect severity of early diabetic retinopathy? Ophthalmology. 1986;93:586–9.
21. Klein R, Klein BE, Moss SE. How many steps of progression of diabetic retinopathy are meaningful? The Wisconsin epidemiologic study of diabetic retinopathy. Arch Ophthalmol. 2001;119:547–53.
22. Kohner EM, Dollery CT. The rate of formation and disappearance of microaneurysms in diabetic retinopathy. Trans Ophthalmol Soc U K. 1970;90:369–74.
23. Nunes S, Bernardes RC, Duarte L, Cunha-Vaz J. Identification of different phenotypes of mild non proliferative retinopathy of type 2 diabetes using cluster and discriminant mathematical analysis. Invest Ophthalmol Vis Sci. 2006;47:E-Abstract 1018.
24. Sharp PF, Olson J, Strachan F, Hipwell J, O'Donnell M, Wallace S, Goatman K, Grant A, Waugh N, McHardy K, Forrester JV. The value of digital imaging in diabetic retinopathy. Health Technol Assess. 2003;7(30):iii–x.
25. Lobo CL, Bernardes RC, Santos FJ, Cunha-Vaz JG. Mapping retinal fluorescein leakage with confocal scanning laser fluorometry of the human vitreous. Arch Ophthalmol. 1999;117: 631–7.
26. Lorenzi M, Gerhardinger C. Early cellular and molecular changes induced by diabetes in the retina. Diabetologia. 2001;44:791–804.
27. Cunha-Vaz JG, Travassos A. Breakdown of the blood–retinal barriers and cystoid macular edema. Surv Ophthalmol. 1984;28:485–92.
28. Lobo CL, Bernardes RC, Cunha-Vaz JG. Alterations of the blood–retinal barriers and retinal thickness in preclinical retinopathy in subjects with type 2 diabetes. Arch Ophthalmol. 2000;118:1664–9.
29. Bernardes R, Lobo C, Cunha-Vaz JG. Multimodal macula mapping: a new approach to study diseases of the macula. Surv Ophthalmol. 2002;47:580–9.
30. Lobo CL, Bernardes RC, Figueira JP, Faria de Abreu JR, Cunha-Vaz JG. Three-year follow-up of blood–retinal barrier and retinal thickness alterations in patients with type 2 diabetes mellitus and mild nonproliferative diabetic retinopathy. Arch Ophthalmol. 2004;122:211–7.
31. Fritsche P, VanderHeijde R, Suttorp-schulten MSA, Pollack BC. Retinal thickness analysis (RTA). An objective method to assess and quantify the retinal thickness in healthy controls and diabetics without diabetic retinopathy. Retina. 2002;22:768–71.
32. Rema M, Saravan G, Deepa R, Mohan V. Familial clustering of diabetic in South Indian Type diabetic patients. Diabet Med. 2002;19(11):910–6.
33. The Diabetes Control and Complication Trial/Epidemiology of Diabetes Interventions and Complications Research Group. Retinopathy and nephropathy in patients with type 1 diabetes four years after a trial of intensive insulin therapy. N Engl J Med. 2000;342:381–9.
34. Stratton IM, Kohner EM, Aldington SJ, Turner RC, Holman RR, Manley SE, Matthews DR; for the UKPDS Group. UKPDS 50: risk factors for incidence and progression of retinopathy in type II diabetes over 6 years from diagnosis. Diabetologia. 2001;44:156–63.
35. Snieder H, Sawtell PA, Ross L, Walker J, Spector TD, Leslie RDG. HbA_{1C} levels are genetically determined even in type 1 diabetes. Evidence from healthy and diabetic twins. Diabetes. 2001;50:2858–63.
36. Cunha-Vaz JG. Diabetic retinopathy. Surrogate outcomes for drug development for diabetic retinopathy. Ophthalmologica. 2000;214:377–80.
37. Fong DS, Aiello L, Gardner TW, King GL, Blankenship G, Cavallerano JD, Ferris FL, Klein R. Diabetic retinopathy. Diabetes Care. 2003;26:226–9.

6
Visual Psychophysics in Diabetic Retinopathy

Edoardo Midena and Stela Vujosevic

CONTENTS

Keywords Visual acuity • Snellen chart • Color vision dysfunction • Contrast sensitivity • Macular recovery function • Perimetry

INTRODUCTION

Irreversible and severe visual loss may represent the end of long lasting diabetic retinopathy. The progression of visual impairment and the quantification of final residual visual function are currently determined by means of diagnostic tests which rely on the physiological and mathematical principles of psychophysics. The best known among these tests is the quantification of visual acuity: a classic visual function psychophysical test. Visual psychophysical tests are the cornerstone of visual function investigation, and any physical or pharmacological therapy for the treatment of diabetic retinopathy still has the maintenance (or improvement) of visual function as primary endpoint. More recently, subtle and precocious neurosensory visual abnormalities have been quantified in diabetic patients in order to detect early visual dysfunction, even before the onset of clinically detectable retinopathy. The aim of these investigations is to try to identify among diabetic subjects a population at higher risk of developing vision-threatening retinopathy [1].

Psychophysics is a science which developed as a way to measure the internal sensory and perceptual responses to external stimuli [2]. Psychophysical visual function testing

From: *Ophthalmology Research: Visual Dysfunction in Diabetes*
Edited by: J. Tombran-Tink et al. (eds.), DOI 10.1007/978-1-60761-150-9_6

may reflect the neural activity of the whole visual pathway, but it is known that these tests are valuable clinical indicators of retinal function derangements induced by the metabolic changes secondary to diabetes mellitus. In fact, in diabetic patients, impaired vision in dim light and difficulties in recognizing the contour of objects in low-contrast conditions are common complaints even with good visual acuity and full visual fields [3]. Moreover, health-related quality of life can become affected in diabetics even prior to vision loss due to anxiety about the future and emotional reaction to diagnosis and treatment of retinopathy [4].

Visual acuity is still considered the gold standard in clinical practice of vision testing, but it does not entirely reflect functional vision. Functional vision describes the impact of sight on quality of life that represents *the patient's point of view* [5, 6]. This approach is better quantified using available psychophysical tests (visual acuity, color vision, contrast sensitivity, macular recovery function, perimetry, and microperimetry).

VISUAL ACUITY

The quantification of visual acuity (VA) is the best known and most widely used test for assessing the integrity of the visual function in clinical settings. It represents the ability to discriminate, at high contrast (black symbols/letters on a white background), two separated stimuli. The Snellen chart is the most widely used tool for VA assessment, and it is routinely used in any clinical setting worldwide. The prototype of this chart was developed in 1862 by the Dutch ophthalmologist Hermann Snellen. He defined "standard vision" as the ability to recognize one of his optotypes at a visual angle of 1 min of arc. Later, the original chart was modified and became what is now known as a standard Snellen chart. This chart has well-documented limits owing to design flaws, such as inconsistent progression of letter size from one line to another, unequal legibility of letters used, unequal and unrelated spacing between letters and rows, and large gaps between acuity levels at the lower end of the chart [7–10]. Variability in background ambient illumination and contrast and poor reliability during test–retest evaluation make, in some cases, Snellen measurements clinically inadequate and prevent reliable evaluation of data obtained from different studies [11–13].

Therefore, new and standardized charts with logMAR (logarithm of the minimal angle of resolution) progression have been developed and introduced into clinical practice, based on design suggested by Bailey and Lovie in 1976, lately described in detail by Ferris et al., and adopted for the Early Treatment Diabetic Retinopathy Study (ETDRS chart) [14, 15]. The major advantages of this chart are regular geometric progression of the size and spacing of the letters, following a logarithmic scale with 0.1 log units steps, equal number of letters in each row, five Sloan optotypes, comparable legibility of the *sans serif* letters, high accuracy, and reliability for both high and low levels of VA [14–17]. Thus, the ETDRS chart has become the gold standard for measuring VA at least in clinical trials.

In diabetic patients, the full functional impact of macular edema (diabetic macular edema, DME) and the functional effects of its treatment on visual function are still poorly documented and understood [18]. Ang et al. found that VA was a poor predictor of presentation and type of DME and that its usefulness as a sole screening tool is limited [19]. On the contrary, Sakata et al. [20] reported a correlation of VA with macular microcirculation characteristics (perifoveal capillary blood flow velocity and severity of perifoveal capillary occlusion) and central foveal retinal thickness in diabetics.

Since the ETDRS study demonstrated that focal macular laser photocoagulation prevents moderate vision loss in approximately 50% of cases, visual acuity has been considered the primary endpoint in all clinical trials evaluating both the natural history as well as the efficacy of any treatment strategy in clinically significant diabetic macular edema (CSME) [21–26]. But in clinical practice, DME is currently assessed not only with VA but also with optical coherence tomography (OCT), a retinal structure test. Therefore, the correlation between these two investigations, one functional and one structural, has been widely, even if not definitively, investigated. Recently, the Diabetic Retinopathy Clinical Research Network reported only modest correlation between VA and OCT-measured center point retinal thickness with a possible wide range of VA for a given degree of retinal edema. These authors also found modest correlation of changes in retinal thickening and VA after focal laser treatment for DME [27]. Browning et al. [28] found no correlation between the extension of DME by OCT and changes of VA after laser photocoagulation, during 12 months follow-up. These results suggest that OCT measurement alone may not be a good surrogate for VA as a primary outcome in studies of DME. Moreover, VA data needs to be integrated with more comprehensive visual function information.

COLOR VISION

As a predominantly macular function, color discrimination may be impaired by any degenerative process affecting the central retina [29]. In diabetes, the underlying mechanism of color dysfunction is uncertain and may relate to metabolic derangement in the neural retina other than to microvascular disease [30]. Several hypotheses have been proposed such as (a) osmotic distortion of the retina caused by the fluid shifts inside the retina, followed by distortion and dysfunction of the neural cells and (b) disorders of metabolisms of neural cells caused by direct diabetes damage or mediated by the alterations of the retinal microcirculation [31–35]. Dean et al. [36] suggested a major role of retinal hypoxia showing that color vision deficits in diabetics with retinopathy can be partially reversed by inhalation of pure oxygen. Different tests are available to assess color vision; unfortunately, most of them are negatively affected by lens opacities [37]. Moreover, approximately 10% of male population and 0.5% of female population show varying degree of congenital color deficiency. Therefore, studies evaluating color vision in diabetics should account for all these factors. One of the most widely known and reported test is the Farnsworth–Munsell 100-Hue Test (FM 100 Hue Test); this is also the most time-consuming diagnostic procedure [38].

Since the first report (in 1905) describing the association between abnormalities in color vision and diabetes mellitus, many researchers have reported the relationship between diabetic retinopathy and color vision dysfunction [39–43]. The first controlled study of color vision in diabetics was reported by Kinnear et al. [44] and Lakowski et al. [29] who showed in a large group of subjects that blue-yellow and blue-green color vision losses were found significantly more among diabetic patients with retinopathy than in normal controls. Other studies confirmed that the blue-yellow axis (the short-wavelength-sensitive cone system) is more vulnerable to diabetes than the green and the red axes [45, 46]. But this conclusion is not unanimously accepted. Hue discrimination in diabetics without retinopathy or with only microaneurysms has been reported not to significantly differ from controls, whereas other studies concluded that diabetics show abnormal results in color vision tests and a tritanopic reduction in a chromatic-contrast threshold when compared with normal controls [47–50] (Table 1). Different studies showed deficits in blue-yellow color discrimination in both adults and adolescents with type 1 diabetes mellitus who had no evidence of retinopathy [41, 44, 51–60]. Hardy et al. [61] found in young patients with insulin-dependent diabetes mellitus (IDDM) that FM 100 Hue Test was more sensitive and specific in detecting dysfunction of the visual pathway than both flash and pattern electroretinogram, and proposed this test for the early visual dysfunction evaluation without success. In the ETDRS, the FM 100 Hue Test was performed in 2,701 patients and showed abnormal hue discrimination in approximately 50% of cases when compared with published data on normal subjects [62]. Macular edema severity, age, and the presence of new vessels were the factors most strongly associated with impaired color discrimination, especially the tritan-like defect [62].

Green et al. [63] examined the FM 100 Hue Test as a screening device for sight-threatening diabetic retinopathy and reported sensitivity of 73% and specificity of 66%, concluding that the test was not sensitive enough for screening of sight-threatening diabetic retinopathy. In a similar study, Bresnick et al. [41] reported sensitivity of 65% and specificity of 59%. Therefore, new color vision tests have been proposed and evaluated. The Mollon–Reffin "Minimalist" test showed sensitivity of 88.9% and specificity of 93.3% in detecting DME [64]. An automated tritan contrast threshold showed 94% sensitivity and 95% specificity in screening for sight-threatening diabetic retinopathy, mainly for DME before the onset of visual loss [65, 66]. Although more advanced stages of retinopathy and DME show greater effect on color vision, subtle specific spectral losses, especially related to blue-yellow discrimination, seem widespread in patients with diabetes, irrespective of the presence of retinopathy and duration of diabetes. Moreover, decreased hue discrimination is present after successful panretinal laser photocoagulation for proliferative DR [67]. These data are also confirmed by studies on contrast sensitivity, and they should be considered in the evaluation and counseling of patients with diabetic retinopathy.

CONTRAST SENSITIVITY

Perhaps the chief merit of the human contrast sensitivity function is that it provides considerably more information than visual acuity: The contrast sensitivity function is a description of the visual system's sensitivity to course-scale detail and medium-scale detail as well to fine detail, while visual acuity quantifies sensitivity to fine detail only. For any given spatial frequency, contrast sensitivity is the reciprocal of contrast detection threshold. The contrast sensitivity function is a plot of the reciprocal of the contrast detection threshold for a grating vs. the spatial frequency of that grating. Contrast sensitivity (CS) function may be quantified using different laboratory and clinical tests [68]. CS determines the person's contrast detection threshold, the lowest contrast at which

Table 1. Studies which have investigated color vision in patients with diabetic retinopathy

Principal investigator/ year of publication	Types of study	Sample size	Age in years: mean/range	DR status and VA	Nature of stimulus	Conclusions
Roy et al. [54]	Case-control	12 Pts (23 eyes)	45.33 (36–56)	7-Mild 5-Moderate retinopathy More than 25 years of diabetes VA: 20/20	Farnsworth–Munsell 100-Hue Test (FM 100 Hue Test)	There was significant difference between mild and moderate group in CV defects; but there was not significant difference from normal subjects' CV
Bresnick et al. [41]	Case-control	Cases-90 pts (and eyes) Controls-published age norms data	Median: 36 (19–68)	12-No/mild/ moderate DR 29-Severe DR 49-PDR VA: –	FM 100 Hue Test	Tritanlike axis was comparable with scores of normal population; yellow-blue hue discrimination defect correlated significantly with severity of retinopathy and maculopathy, and with fluorescein leakage in the macula
Green et al. [63]	Case-control (small number of controls)	Cases-126 pts (232 eyes) Controls-16 subjects (18 eyes)	–	115 (eyes)-No DR 55-bDR 42-PDR 20-Exudative maculopathy VA: –	FM 100 Hue Test	CV deteriorated with increasing severity of diabetic retinopathy
Roy et al. [37]	Case-control	Cases-51 pts (95 eyes) Controls-41 pts (81 eyes)	Cases: 37.0 ± 10.5 Controls: 33.9 ± 11.8	Mild retinopathy (only five or fewer microaneurysms) VA: 20/20	Lanthony desaturated D-15 test FM 100 Hue Test Gunkel chromograph test	Diabetic pts showed significantly more CV defects than controls on all three tests. Among diabetic pts no significant differences were found correlating to age, duration of diabetes or glycosylated hemoglobin

(continued)

Table 1. (continued)

Principal investigator/ year of publication	Types of study	Sample size	Age in years: mean/range	DR status and VA	Nature of stimulus	Conclusions
Greenstein et al. [95]	Case-control	Cases-24 pts and eyes Controls-age-similar normal data from Verriest et al. [124]	Cases: 45.8 (24–68)	From no DR to severe NPDR; from no macular edema to center involving edema VA: 20/30 or better	FM 100 Hue Test + Two-color increment threshold test	No correlation was found between Farnsworth's result and levels of DR; S-cone pathway, measured by Two-Color Increment Threshold Test showed significant correlation with level of both retinopathy and maculopathy
Hardy et al. [55]	Case-control	Cases-38 (pts) Controls-36	Cases: 26.1 (16–40)	No DR VA: 6/9 or better	FM 100 Hue Test	Diabetic pts had significant abnormal results compared with normal subjects; no significant correlation was found between CV abnormalities and diabetes duration or glycosylated hemoglobin values
Maár et al. [64]	Case-control	Cases-10 (pts) with CSME Controls-29 without CSME	Cases: 33.7 ± 7.75 Controls: 28.07 ± 5.67	Cases + controls: 12-No DR 18-Mild DR 4-Moderate DR 3-Severe DR 2-PDR Cases-VA: 0.07 ± 2.01 logMAR Controls-VA: –0.06 ± 0.17 logMAR	Lanthony desaturated D-15 test Mollon–Reffin Minimalist test version 6.0	Highly significant correlation was found between the tritan value of the Mollon test and the presence of CSME; Lanthony test did not show a significant correlation with presence/absence of CSME

Giusti [60]	Case-control	Cases-39 pts Controls-39 pts	17.14 ± 8.2 18.1 ± 3.1	Cases-No DR; VA: 1.08 ± 0.15 logMAR Controls-VA: 1.07 ± 0.24 logMAR	Standard Pseudoisochromatic Plates (SPP2) Roth 28-Hue test FM 100 Hue Test Lanthony D-15 Hue test	SPP2 and Roth tests did not show differences between cases and controls; Farnsworth and Lanthony tests showed significant difference between diabetic pts and normal subjects
Ong et al. [65]	Cross-sectional study	510 pts: 493-NSTDR 17-STDR	NSTDR: 60.9 ± 13.9 STDR: 60.4 ± 11.3	NSTDR: VA: 0.06 ± 0.09 383 no DR 110 bDR STDR: VA: 0.1 ± 0.11 3 Pre-proliferative DR 2 PDR 12 Maculopathy	Automated Tritan Contrast Threshold (TCT)	Sensibility of 94% and specificity of 95% were found in detecting STDR; no association was found between abnormal values of TCT and clinical parameters (HbA_{1c}, duration of diabetes, micro-albuminuria)
Wong et al. [125]	Case-control	Cases-35 (pts and eyes) Controls-115	60 (median)	CSME (cases)-35; VA: 0.20 (median) NPDR (controls)-115; VA: 0.20 (median)	ChromaTest	Statistically significant results were found between NPDR group and CSME group for both tritan and protan color contrast threshold; sensitivity and specificity of ChromaTest were respectively of 71 and 70% in detecting CSME in diabetic pts

Pts patients; *VA* visual acuity; *DR* diabetic retinopathy; *NPDR* non proliferative diabetic retinopathy; *bDR* background diabetic retinopathy; *PDR* proliferative diabetic retinopathy; *CV* color vision; *STDR* sight-threatening diabetic retinopathy; *NSTDR* non sight-threatening diabetic retinopathy; *CSME* clinically significant diabetic macular edema a certain pattern can be seen. An assumption which often underlies the clinical use of the CS function is that it predicts whether a patient is likely to have difficulty in seeing visual targets typical of everyday life. A contrast sensitivity assessment procedure consists of presenting the observer with a sine-wave grating target of a given spatial frequency (i.e., the number of sinusoidal luminance cycles per degree of visual angle). The contrast of the target grating is then varied while the observer's contrast detection threshold is determined. Typically, contrast thresholds of this sort are collected using vertically oriented sine-wave gratings varying in spatial frequency from 0.5 (very wide) to 32 (very narrow) cycles per degree of visual angle.

Whereas standard visual acuity testing is a high-contrast test by definition and it measures only size, it does not provide full information about visual function in the everyday life activities. Contrast sensitivity measures the two major variables: size and contrast, offering a more realistic quantification of visual impairment. There are different types of chart tests to capture the different aspects of the CS function (charts with white and black bars of decreasing contrast, charts with letters). Among them, the Pelli–Robson chart is the most commonly used chart in clinical trials. It consists of letters of a single (large) size (low spatial frequency). The chart is arranged by triplets of letters and each triplet is 0.15 log units higher in contrast than the preceding triplet.

Both hue discrimination and contrast sensitivity may reflect (if the lens is clear) macular function, but their exact physiological relationship has not yet been fully explained. Some data suggest that the CS function more significantly correlates to DR grading than color vision and macular recovery function [69, 70]. Unfortunately, data about CS function in diabetics are still controversial. This difference in clinical results may be, at least methodologically, explained by the different methods used to quantify CS, as well as the lack of homogeneity in the examined groups (type of diabetes, age, criteria, and methods for DR evaluation). This fact points to the importance of developing a standardized test to accurately and reliably quantify contrast sensitivity function in both clinical practice and clinical trials. Diabetic patients with retinopathy and good visual acuity frequently show spatial resolution defects, which can be detected measuring CS function. The reductions in CS involve mainly the intermediate and medium-high spatial frequencies in relation to the severity of retinopathy and previous laser photocoagulation; nevertheless, some patients show losses at the medium-low spatial frequencies [71–74]. In DME, Arend et al. [75] found that loss of CS correlates with the enlargement of the foveal vascular zone. Midena et al. [76] studied the effect of both focal and grid laser photocoagulation on CS of patients with DME and found that CS function improved after treatment, but it never normalized. The same finding was reported by Talwar et al. [77] who found improved CS and stabilization of visual acuity after focal argon laser photocoagulation for CSME.

Farahvash et al. described the early improvement of CS at midfrequencies after macular laser photocoagulation. This benefit appeared only in patients with resolved CSME, suggesting that CS is probably a more sensitive parameter than visual acuity for early monitoring of CSME after laser photocoagulation [78]. The significant reduction in CS function documented in diabetics with retinopathy is not confirmed when a subject has no retinopathy: There is still not strong evidence of significant difference in CS between diabetics without retinopathy and normal controls. According to Arend et al., there was no difference in CS function between diabetics without retinopathy and controls, whereas Ghafour et al. [71], using the same test, found that diabetics without retinopathy were abnormal at 3.2 and 6.3 c/deg. Using the Vision Contrast Test System in patients with little or no retinopathy, Trick et al. [69] found reduced mean CS at each spatial frequencies when compared to controls; however, a post hoc analysis yielded no statistical difference between the groups. Sokol et al. studied separately insulin- and non-insulin-dependent diabetes mellitus (IDDM and NIDDM) patients and found that patients with IDDM and no DR had normal CS function, whereas patients with NIDDM, normal VA, and no DR had abnormal CS at high spatial frequencies. If background retinopathy was present, abnormal CS at all spatial frequencies was found [73]. Della Sala et al. [72], using the Cambridge low-contrast sensitivity charts, showed abnormal CS in 9 of 22 patients without diabetic retinopathy and in only 6 of 20 patients with background retinopathy (Table 2). Therefore, the contrast sensitivity losses in IDDM and NIDDM patients may not be similar, and further studies are needed to substantiate this hypothesis. Contrast sensitivity testing, as color vision testing, shows significant changes in diabetics and there is some correlation with glycemic control, although prospective studies are required to assess this relationship over a longer time period. Although both tests show similar patterns in diabetics, direct comparisons of the two tests seem to indicate the CS function test as more sensitive and specific.

MACULAR RECOVERY FUNCTION (NYCTOMETRY)

Macular recovery function (nyctometry) is a dynamic measure of the initial 2-min course of macular recovery function following preadaptation to a strong uniform illumination of a large area of the retina. It is a standardized technique, which lasts only 6.5 min. It quantifies not only the dark adaptation of the cone system but also the macular sensitivity to glare [79]. Gliem and Schulze reported a progressive reduction in macular recovery related to deterioration of DR [80]. Midena et al. [79] showed, in a well-defined series of patients, that reduced nyctometry is directly and strongly related to the progression of retinal (functional and anatomical) derangement due to diabetes mellitus. Different authors suggested, but never definitively proved, that nyctometry can be used to predict the progression of background DR to proliferative DR. They suggested the use of nyctometry as a screening method in selecting patients at high risk for proliferative DR [81–83]. Verrotti et al. [84] found altered nyctometry in microalbuminuric diabetic children vs. normoalbuminuric and normal controls. Reported values were independent of both the level and the fluctuations of glycemia. However, Lauritzen et al. [85] found improved performance of nyctometry in the first year in patients on a intensive insulin regimen. In two separate studies, Andersen et al. [86] and Frost-Larsen et al. [87] found significant improvement in macular recovery function in newly diagnosed juvenile diabetics after a 10-day period of superregulation in the biostator. This indicates that in metabolic dysregulation, the results of nyctometry are reversible to a certain extent provided the reduced values of nyctometry are mainly due to functional changes in the retina [83].

In CSME, 1 week after macular laser photocoagulation, nyctometry was shown to decrease significantly, followed by slow improvement toward the initial value [76].

Table 2. Studies which have investigated contrast sensitivity in patients with diabetic retinopathy

Principal investigator/ year of publication	Types of study	Sample size	Age in years: mean/range	DR status and VA	Nature of stimulus	Conclusions
Ghafour et al. [71]	Case-control	Cases-93 Controls-80	Cases-47 (27–70) Controls-46 (24–68)	42-No DR 22-Background DR 29-PDR VA: 6/5 to 6/36	Arden grating test	Diabetic pts without DR have increased thresholds at the higher spatial frequencies. Significative difference CS threshold found between each group (controls-no DR-background-PDR)
Hyvärinen et al. [3]	Case-control	Cases-19 Controls-from Virsu et al. [127]	Cases-32 (19–59)	5-Micro±hemorrhages but normal vision (20/20) 5-bDR 6-PDR 3-PDR + central cataract	Sinusoidal grating (cathode-ray-display)	CS seems to correlate better with DR status then with VA. Diabetic pts without DR has no significant reduction of CS in comparison to normal subjects. In the third group (cataract) CS was better than expected
Regan and Neiman [128]	Case-control	Cases-15 Controls-40	Cases-49 (24–75)	6-VA ³6/7.5 or better 9-VA <6/7.5	Regan chart	Diabetic pts had significative CS loss
Sokol et al. [73]	Case-control	Cases-64 Controls-117	–	31-IDDM and no DR; VA: 20/25 or better 33-NIDDM and no (n = 16) or background (n = 17) DR; VA: 20/30 or better	Sinusoidal grating (microprocessor-controlled video system)	Pts with NIDDM, normal VA and no DR had abnormal CS at high spatial frequencies. Pts with NIDDM and bDR had abnormal CS at all spatial frequencies. Pts with IDDM and no DR had normal CS

Della Sala et al. [72]	Case-control	Cases-42 Controls-84	Cases-12–75 Controls-14–68	22-No DR 19-bDR 1-PDR VA: 1.0 or better	Cambridge low-contrast sensitivity charts	Diabetic pts showed decreased CS
Trick et al. [69]	Case-control	Cases-57 Controls-35	No DR-36.9 ± 11.1 bDR-37.9 ± 8.6 Controls-33.3 ± 9.3	37-No DR 20-bDR 18-NIDDM 39-IDDM VA: 20/30 or better	Vistech VCTS 6500 distance chart	All diabetic pts showed decreased CS, particularly with mid-spatial frequency gratings. No difference between IDDM and NIDDM pts
Khosla et al. [129]	Case-control	Cases-38 eyes (22 pts) Controls-20 eyes (10 pts)	No DR-50.0 ± 11.8 bDR-47.1 ± 10.3 Controls-47.2 ± 13.5	22 (eyes)-No DR 16-bDR VA: 6/6	Cambridge low-contrast sensitivity charts	Significant decreased CS in the retinopathy group. No significant difference in CS between non retinopathy group and normal subjects
Midena et al. [76]	Prospective non comparative study	30 Diabetic pts	55 (40–70)	Minimal- to mild-bDR CSME VA: 1.0	Arden grating test	CS improved after photocoagulation, with a significant difference after 3 months, but did not reach normal values
Bangstad et al. [130]	Case-control	Cases-30 pts with micro-albuminuria Controls-27 pts with normo-albuminuria	Micro-albuminuria group: 19 (14–29) Normo-albuminuria group: 19 (14–24)	Micro-albuminuria group: 12 pts-No DR 18 pts-bDR Normo-albuminuria group: 12 pts-no DR 15 pts-bDR	Vistech VCTS 6500 distance chart	Micro-albuminuric pts showed worse CS at middle and high spatial frequencies, but significantly only for 18 cpd

(continued)

Table 2. (continued)

Principal investigator/ year of publication	Types of study	Sample size	Age in years: mean/range	DR status and VA	Nature of stimulus	Conclusions
Arend et al. [75]	Case-control	Cases-20 pts Controls-Normal subjects from database (arch ophth 75;610)	Cases-42 ± 12	6-No DR 8-Only microaneurisms 4-Mild retinopathy 1-Severe retinopathy 1-PDR VA: 20/25 or better	CSV-1000 (Vector vision; Dayton, OH)	Diabetic pts had significantly lower CS at 6 and 12 c/deg than control subjects. Foveal avascular zone and perifoveal intercapillary area correlated significantly with CS at 12 cpd
De Marco [131]	Case-control	Cases-66 IDDM Controls-66	Cases: 30–10 ± 1.22 years 36–16.07 ± 2.3 years Controls: 30–9.93 ± 2.02 years 36–17 ± 3.16 years	No DR VA: 1.0 or better	Conel CST automatic (Roma, ITA) (sinusoidal gratings)	No difference was found between diabetic aretinopatic pts and controls. CS was not correlated with sexual maturity or duration of diabetes
Verrotti et al. [132]	Case-control	Cases-40 + 20 Controls-20	Cases-16.9 ± 4.9 Controls-16.9 ± 4.6	20-No DR 30-bDR 10-Preproliferative/PDR	CSV-1000 (Vector vision; Dayton, OH)	Diabetic pts showed a significative decrease of CS at 12 and 18 c/deg; preproliferative and PDR pts had decreased CS at all frequencies, and significative lower than CS of aretinopathic pts

Mackie et al. [133]	Case-control	Cases-90 Controls-50	Cases: Young pts-33.2 ± 7.9 Older pts-65.5 ± 8.2 Controls: Young pts-30.8 ± 7.9 Older pts-66.6 ± 10.1	VA: 0.3 or better	Pelli–Robson chart	There was a progressive reduction of CS threshold through the five groups of diabetic pts (no DR, bDR, PPR, treated retinopathy, treated maculopathy), significative between groups in which there was a difference of at least two adjacent degrees of retinopathy. No significative difference was found between controls and aretinopathic pts
Lövestam-Adrian et al. [134]	Case control	Cases-20 Controls-19	Cases-32 (15–27) Controls-30 (20–42)	Cases-Treated PDR or severe NPDR; VA: 0.9 (0.4–1.0) Controls-No DR or non treated mild bDR; VA: 1.0 (0.5–1.0)	Precision Vision chart (Preisler instrument AB, Illinois, USA)	Pts treated with panretinal photocoagulation had significative higher contrast threshold than untreated diabetic pts
Talwar et al. [77]	Prospective noncomparative study	14 Eyes with untreated CSME	47–60 years	CSME VA: 0.49 (1.0–0.1)	Cambridge low-contrast sensitivity charts	The CS improved significantly after photocoagulation treatment
Stavrou et al. [135]	Case-control	Cases-20 pts Controls-24 pts	Cases-62.67 ± 11.21 Controls-67.36 ± 7.35	12-No/minimum DR 4-Mild DR 4-Moderate/severe DR VA: 6/9.5 or better 8 Pts had macular edema	Pelli–Robson chart	CS was lower in diabetic pts than CS of controls, but significative difference was observed only between no/minimum DR group and controls.

(continued)

Table 2. (continued)

Principal investigator/ year of publication	Types of study	Sample size	Age in years: mean/range	DR status and VA	Nature of stimulus	Conclusions
						Presence/absence of macular edema was correlated to decreased CS, but there was no significative difference between the two groups
Farahvash et al. [78]	Prospective noncomparative study	17 Diabetic pts (34 eyes)	–	26 Eyes-diffuse macular edema 8 Eyes-focal macular edema VA pre-treatment: 0.21 VA post-treatment: 0.24	Metrovision with a high resolution cathodic ray tube stimulator	CS had a significative improved after photocoagulation treatment only in the frequency of 6.4 cpd

Pts patients; *VA* visual acuity; *CS* contrast sensitivity; *DR* diabetic retinopathy; *bDR* background diabetic retinopathy; *PDR* proliferative diabetic retinopathy; *CSME* clinically significant diabetic macular edema; *IDDM* insulin-dependent diabetes mellitus; *NIDDM* non insulin-dependent diabetes mellitus; *cpd* cycles per degree

Frost-Larsen et al. [83] demonstrated a close correlation of the oscillatory potential and nyctometry in IDDM patients, suggesting a common retinal mechanism responsible for the changes of both parameters in DR. Macular recovery function is a complex phenomenon consisting of photochemical, neural receptor, and network adaptation, the resultant achievement being an optimized interaction of all three mechanisms [88]. Although the mechanisms responsible for the increased recovery time in the initial phase of this test are unknown, the phenomenon appears related to disturbances primarily in the neural network adaptation. The site of the neuronal mechanisms of this test is likewise believed to be located to the inner nuclear layer, and it might be influenced by the same functional disturbances which suppress the generation of the oscillatory potential [83, 89]. Unfortunately, the technology to perform this test is no more available and a new electronic version is under investigation.

PERIMETRY

Perimetry represents a systematic measurement of visual field sensitivity function. It encompasses the assessment of differential light threshold of retinal locations from the fovea to the preplanned periphery. The two most commonly used types of perimetry are Goldmann kinetic perimetry and (threshold) static automated perimetry. Kinetic perimetry is particularly useful for obtaining the outline of extensive defects and identifying major scotomas. Static perimetry is particularly useful for detailed probing in carefully selected areas and represents the current cornerstone of visual field testing. Standard threshold static automated perimetry quantifies the differential light threshold required to detect a static white light stimulus in the visual field. Since standard threshold perimetry uses a static achromatic stimulus, it is thought to nonselectively evoke both major groups of retinal ganglion cells: (1) the parasol ganglion cells of the magnocellular visual pathway subserving motion perception, low spatial resolution, high contrast sensitivity, and stereopsis and (2) the midget ganglion cells of the parvocellular visual pathway subserving central visual acuity, color perception, low contrast sensitivity, high spatial resolution, static stereopsis, pattern recognition, and shape. There is considerable overlap in the receptive fields of these cell types; therefore, a nonselective, white-on-white stimulus cannot detect the earliest loss of retinal ganglion cells, and standard threshold perimetry therefore may not detect visual field loss until the whole population of retinal ganglion cells is significantly damaged. In addition to new algorithms, visual field testing is becoming more sophisticated with the development of new perimetric technologies. New technologies are aimed at earlier detection of subtle deficits and enhancing diagnostic accuracy. The sensitivity to short-wavelength stimuli can be measured in different regions of the visual field by blue-on-yellow perimetry (short-wavelength automated perimetry, SWAP). It is accomplished by determining the sensitivity to blue stimuli (thus stimulating the short-wavelength cone system) on a bright yellow background. In this way, long- and medium-wavelength cone system sensitivity is reduced and rods are saturated.

In DME, visual acuity loss is quite relevant and irreversible when long lasting edema involves the center of the macula; in these cases, the outcome of laser treatment is poor. But before the loss of visual acuity is reported by patients, they may suffer from other disturbances of visual function such as waviness, blurring, relative scotoma, and decrease of contrast sensitivity which are not assessed and quantified in routine examination. Therefore, a visual function test aimed at identifying vision-threatening retinopathy before visual acuity is affected would be of great value. One possible approach may be to identify decreased sensitivity in paracentral areas using perimetry.

It has been reported that patients with diabetic retinopathy show sensitivity loss in the midperipheral field by white-on-white perimetry (WWP) and that this sensitivity loss is correlated with the retinal areas of nonperfusion [90–92]. The sensitivity loss was closely associated with microangiopathy and was greater in the midperipheral area than in the paracentral area. Bek and Lund-Andersen evaluated with Humphrey Field Analyzer retinal sensitivity over cotton wool spots in patients with diabetic retinopathy and reported localized nonarcuate scotomata in the visual field, which may persist even when the funduscopic lesions resolve [93]. A selective loss of short-wavelength sensitive pathway has been demonstrated in diabetic patients with minimal or no diabetic retinopathy [94–97]. SWAP has been suggested as a useful tool for defining visual function loss in diabetic patients with early ischemic damage of the macula or clinically significant macular edema [98, 99]. Decreased blue-on-yellow sensitivity has also been demonstrated in diabetic children without clinically detectable retinopathy [100] (Table 3).

When comparing SWAP and WWP in diabetics, SWAP seems superior for macular localized field loss determination and early ischemic macular damage evaluation. Uncertainty remains about its use in macular edema. Moreover, SWAP showed to be highly lens opacity–dependent [98, 99, 101]. On the other hand, WWP correlates better with the ETDRS severity scale than SWAP or visual acuity determination, and it might be better in separating groups with different levels of retinopathy [102]. As elegantly stated by Sunness et al. [103], conventional visual field examination is inadequate for the accurate functional evaluation of macular diseases and detection of small scotoma, particularly when foveal function is compromised and the patient may have unstable and extrafoveal fixation. Accuracy of the conventional visual field rests on the assumption that fixation is foveal and stable. Moreover, the detection of the site and stability of retinal fixation (foveal or extrafoveal) and the quantification of retinal threshold over small and discrete retinal lesions are beyond the possibilities of conventional, automatic, and nonautomatic perimetry [2].

MICROPERIMETRY (FUNDUS-RELATED PERIMETRY)

The integration of retinal details with function has been achieved by fundus-related perimetry, more widely known as microperimetry. Microperimetry allows for the exact topographic correlation between fundus abnormalities and corresponding functional alterations by integration, with different methods, of differential light threshold (more commonly known as retinal sensitivity) and fundus imaging. It also allows to quantify fixation characteristics, by exactly defining location and stability of any foveal or extrafoveal (PRL, preferred retinal locus) fixation site, as well as determination of size, site, and shape of scotoma. Moreover, the possibility of an automatic follow-up examination (using the microperimeter MP-1, Nidek Co, Japan) which allows the evaluation of exactly the same retinal points tested at baseline, regardless of any change in fixation

Table 3. Studies which have investigated perimetry in patients with diabetic retinopathy

Principal investigator/ year of publication	Types of study	Sample size	Age in years: mean/range	DR status and VA	Nature of stimulus	Conclusions
Bek et al. [136]	Cross-sectional	20 Pts	–	Hard exudates and/or localized leakage of fluorescein VA: 6/18 or better	Humphrey field analyzer	No topographical correlation was found between barrier leakage and decreased light sensitivity
Lutze et al. [137]	Case-control	Cases-31 pts Controls-50 pts	30 (Median) (19–59)	No DR-6 Mild retinopathy-10 Moderate retinopathy-4 Severe retinopathy-4 PDR-7 VA: 20/80 or better	Humphrey field analyzer	S-cone sensitivity and achromatic sensitivity were not significantly reduced in diabetic pts, but they showed localized sensitivity losses in visual fields in diabetic pts. Localized sensitivity losses of SWAP were significantly correlated to the level of DR
Hudson et al. [99]	Case-control	Cases-24 pts and eyes Controls-400 pts	59.75 (45–75) 48 (18–84)	CSME (Early Treatment Diabetic Retinopathy Study (ETDRS)) VA: 0.25 or better	Humphrey field analyzer	SWAP test showed greater sensitivity than WWP test in detecting visual field defects. The position of localized field loss assessed by SWAP corresponded with clinical mapping of the area of DME
Nomura et al. [138]	Case-control	Cases-31 pts Controls-11 pts	Cases: No DR 50.9 (40–59) bDR 51.3 (40–59) Controls: 51.7 (40–59)	No DR-21 bDR-10 VA: 20/20	Humphrey field analyzer 750	No significant correlation was found between level of DR and FM 100 Hue Test. The SWAP sensitivity of the upper half of the central 20–30° area was significantly reduced in bDR group; no significant sensitivity loss was detected with WWP

(continued)

Table 3. (continued)

Principal investigator/ year of publication	Types of study	Sample size	Age in years: mean/range	DR status and VA	Nature of stimulus	Conclusions
Remky et al. [98]	Case-control	Cases-31 pts and eyes Controls-31	35 ± 12	No DR-9 Only microaneurysms-5 Mild retinopathy-13 Moderate retinopathy-1 Severe retinopathy-2 VA: 20/25 or better	Humphrey field analyzer 750	SWAP thresholds were significantly correlated with increasing size of FAZ and PIA; WWP thresholds and VA were not correlated with diabetic changes of the perifoveal capillary area
Verrotti [139]	Prospective study	Cases-60 pts	15.9 (14–18)	No DR VA: 1.0 or better	Humphrey field analyzer 640	The probability of retinopathy development after 8 years of follow-up was significantly higher in subgroups of patients with mean sensitivity in areas 2 and 3 below cut-off
Afrashi et al. [140]	Case-control	Cases-43 pts Controls-30 pts	31.03 (16–38) 30.13 (21–35)	No DR VA: 20/20	Humphrey field analyzer 750	There was no difference in sensitivity between the diabetic and the control group. The values of mean deviation by blue-on-yellow perimetry in diabetic pts were significantly higher than in the control group. WWP did not show this difference
Remky et al. [141]	Case-control	Cases-45 pts and eyes Controls-58 pts	37.2 ± 10.4 37.2 ± 14.1	No/mild macular changes (not edema) No DR-13 Only microaneurysms-11	Humphrey field analyzer 750	SWAP thresholds were significantly more reduced in pts with advanced DR than those of WWP. In pts with no DR sensitivity was not affected

				Advanced DR-21 Cases-VA: 0.015 ± 0.042 Controls-VA: 0.013 ± 0.034		
Han et al. [142]	Case-control	Cases-22 pts and eyes Controls-18 pts	52.4 (32–59) 43.5 (26–64)	Cases-mild (20) or moderate (2) DR Controls-no DR VA: 20/25	Humphrey field analyzer	Both groups showed reduced sensitivity at SWAP test. Also mfERG showed similar number of significant abnormalities. In diabetic pts with DR SWAP and mfERG also showed some spatial agreement
Bengtsson et al. [102]	Cross-sectional	59 Pts and eyes	50.6 (20–69)	–	Humphrey field analyzer 750	WWP was correlated with degree of peripheral DR better than VA or SWAP test. SWAP was superior to both WWP and VA in measuring effects caused by enlarged FAZ and PIAs
Agardh et al. [101]	Cross-sectional	59 Pts and eyes	50.6 (20–69)	DME-20 No DME-39 VA: –0.04 (median) (–0.22 to +0.82)	Humphrey field analyzer 750	VA was correlated to the thickness of macula when edema involved the center of the macula. SWAP was able to detect macular edema, WWP was not. SWAP and WWP were correlated to FAZ and PIA. Visual field defects reflected ischemic damage of the macula rather than macular edema per se

(continued)

Table 3. (continued)

Principal investigator/ year of publication	Types of study	Sample size	Age in years: mean/range	DR status and VA	Nature of stimulus	Conclusions
Nitta et al. [143]	Case-control	Cases-33 pts and eyes Controls-33	41.7 ± 6.8 41.2 ± 6.3	No DR VA: 20/20 or better	Humphrey field analyzer 750	There was a correlation between decreasing of mean deviation and increasing clinical data (duration of diabetes, fructosamine concentration, glycatet hemoglobin) with SWAP test, with not in WWP test
Lobefalo et al. [100]	Case-control	Cases-50 pts (100 eyes) Controls: 50 pts	13.3 (10.1–16.3)	No DR VA: 0.8 or better	Humphrey field analyzer 640	Mean perimetric sensitivity of SWAP showed significant lower values in micro-albuminuric group than values of normo-albuminuric group. Mean perimetric sensitivity of WWP did not show significant differences between micro-albuminuric and normo-albuminuric diabetic pts, either between diabetic pts and controls
Pahor [144]	Case-control	Cases-32 eyes (25 pts) Controls-30 eyes	51.2 (22–71) 48.3 (17–64)	Moderate DR-17 eyes Severe DR-15 VA: 6/9 or better	Humphrey field analyzer	There was a significant correlation between visual field defects and areas of reduced retinal perfusion

Pts patients; *DR* diabetic retinopathy; *VA* visual acuity; *bDR* background diabetic retinopathy; *PDR* proliferative diabetic retinopathy; *CSME* clinically significant diabetic macular edema; *DME* diabetic macular edema; *FAZ* foveal avascular zone; *PIA* Perifoveal Intercapillary Area; *WWP* white-on-white perimetry; *SWAP* short-wavelength automated perimetry; *mfERG* multifocal electroretinogram characteristics, is a valuable tool of this technique, mainly in the evaluation of treatment outcome. Microperimetry offers several advantages vs. standard perimetry in the quantification of macular sensitivity, such as direct real-time fundus control, direct correlation between sensitivity and fundus details, detection of central microscotomata, and continuous monitoring of fixation.

The original Scanning Laser Ophthalmoscope (SLO, Rodenstock, Germany) was the first instrument combining static perimetric testing and simultaneous observation of the fundus. SLO allowed a real-time examination by an infrared (IR) source of the retina and allowed the manual projection of visual stimuli of different shapes, sizes, and intensities over selected retinal areas. The sensitivity map, obtained according to the stimulation pattern (in dB or pseudocolors), was available at the end of the examination. This map contained the fixation area, the fixation target, and the threshold data. This instrument is no more commercially available.

With the introduction of a new microperimeter, a liquid crystal display (LCD) microperimeter (MP-1) coupled with a color fundus camera, visualization of color fundus details allows to directly report functional data onto clinical fundus image, and automatic tests are also obtained. MP-1 microperimeter has both an infrared and a color fundus camera, as well as an automatic real-time tracking system that allows for a full automatic retinal fixation and threshold determination as well as automatic follow-up and differential maps determination, independently from fixation characteristics. The main technical characteristics of this instrument have been previously described in detail [104–106]. Rohrschneider et al. compared MP-1 and SLO microperimeters and found that both instruments analyzed retinal sensitivity and fixation characteristics, and the results obtained from both instruments were directly comparable. However, MP-1 is superior to SLO due to the automatic real-time alignment system, a larger field of (fundus) view (44°×36° MP-1 vs. 33°×2° SLO) and color image [107].

The most relevant characteristics of advanced microperimetry performed with the MP-1 microperimeter may be briefly summarized as follows:

- Exact fundus-related stimulation
- Automatic eye-tracking system
- Automatic static and kinetic stimulation (with standardized or customized grids and centration)
- Normative age-related database [108]
- Age-related differential maps (local defect determination, shallow defects determination, etc.)
- Automatic follow-up and differential maps
- Screening tests (short test duration: <5 min)
- Morpho/functional relationship investigation (overlapping of sensitivity maps over different types of fundus images)

MP-1 microperimetry is a mesopic test that requires a 5–10-min dark light adaptation before starting the examination.

In the last 15 years, microperimetry has been successfully used in the diagnosis and follow-up of different macular disorders, including: age-related macular degeneration, myopic maculopathy, macular dystrophies, and diabetic macular edema [105, 109–117].

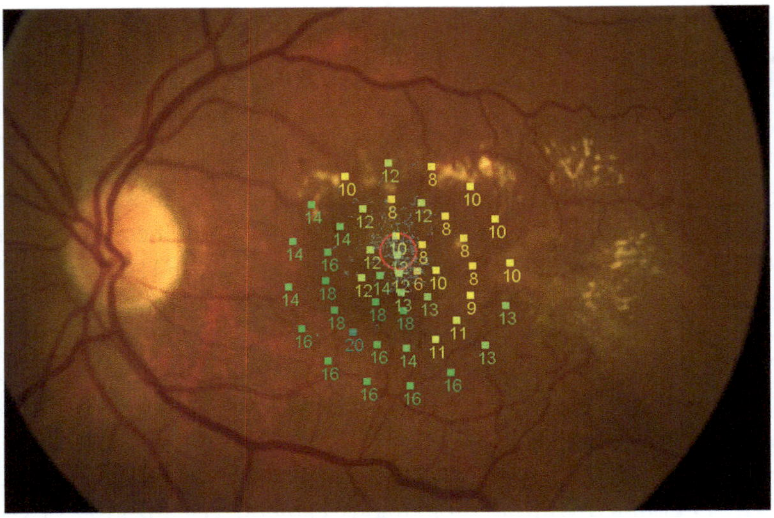

Fig. 1. Microperimetry map (in decibels) superimposed onto the color fundus image in a case of clinically significant diabetic macular edema (CSME). Decrease of retinal sensitivity is shown on the temporal side of the macular region.

In DME, microperimetry has been used for the quantification of macular sensitivity; the correlation of macular sensitivity to macular thickness, visual acuity, and fundus autofluorescence data; and the fixation patterns determination in different stages and types of edema.

Different studies report the correlation between retinal sensitivity, determined with microperimetry, and VA in patients with CSME [102, 108, 118]. Moreover, reduced retinal sensitivity is related to increasing retinal thickness [102, 114, 118] (Table 4). In a study published by Vujosevic et al. [104], a significant inverse relationship was found in patients with CSME, between retinal sensitivity and normalized retinal thickness values obtained with OCT, with a decay of 0.83 dB ($p<0.0001$) for every 10% of deviation of retinal thickness from the normal values (Fig. 1). This means that normalized macular thickness better copes with macular function than any absolute value [104]. Microperimetry seems to represent a better functional testing than BCVA for quantifying visual function in diabetic patients, because it incorporates a functional measure that may potentially supplement the predictive value of OCT and visual acuity [104, 118, 119].

Besides retinal sensitivity, microperimetry allows to quantify retinal fixation characteristics. Fixation characteristics (location and stability) are relevant parameters for understanding patient's quality of vision, especially reading ability, and its knowledge may be important in planning laser treatment [110, 119–121]. Reading ability better correlates with subjective quality of vision rather than distant visual acuity [110]. Whereas different studies agree that macular sensitivity deteriorates in patients with DME, data about fixation characteristics are quite contrasting [104, 109, 110, 114, 118, 122] (Table 4). Kube et al. [114] found decreased fixation stability in patients with DME using SLO microperimetry. Carpineto et al. [122] found that all eyes with eccentric or unstable fixation had cystoid DME. Vujosevic et al. [119] found that fixation patterns are not significantly

Table 4. Studies which have investigated microperimetry in patients with diabetic retinopathy

Principal investigator/ year of publication	Types of study	Sample size	Age in years: mean/range	DR status and VA	Nature of stimulus	Conclusions
Rohrschneider et al. [110]	Prospective	30 Pts and eyes	63 (37–81)	CSME VA: From 20/200 to 20/20	SLO 101 Rodenstock	In ten eyes VA significantly improved after laser photocoagulation, in nine eyes it decreased. Fifteen eyes showed improving in mean light sensitivity after treatment, seven showed decreasing. Nine eyes improved in fixation stability, five eyes demonstrated a deterioration. There was no significant correlation between stability of fixation and visual acuity or subjective patient changes
Mori et al. [111]	Cross-sectional	19 Pts and eyes	63 (45–78)	CSME with: Dense scotoma-4 Relative scotoma-10 No scotoma-5 VA: 0.7 (–0.2 to 2) logMAR	SLO 101 Rodenstock	Significant difference was found between the three groups VA. There were significant differences in the prevalence of cystoid changes, diffuse edema, unstable fixation among the three groups. Group with dense scotoma showed a great association with all these three clinical characteristics, group with no scotoma did not show any of these characteristics

(continued)

Table 4. (continued)

Principal investigator/ year of publication	Types of study	Sample size	Age in years: mean/range	DR status and VA	Nature of stimulus	Conclusions
Moller and Bek [145]	Prospective	24 Pts and eyes	66.9 (38–85)	CSME treated with standard argon laser treatment (ETDRS protocol) VA: I group: −0.05 to 0.2 II group: 0.21–0.4 III group: 0.41–0.6 IV group: 0.61–0.8	SLO 101 Rodenstock	A significant negative correlation was found between the changes in VA and the changes in the retinal areas covered by hard exudates. In four pts hard exudates covered fovea at baseline, and the site of fixation was at the border of the exudate. After laser treatment, in two eyes hard exudates reduced, resulting in an increased VA and a shift of the site of fixation, in one eye hard exudates increased, followed by a VA impairment and a more peripheral site of fixation
Kube et al. [114]	Case-control	Cases-27 pts Controls-61	54 (17–81) 45 (18–85)	Presence of diabetic maculopathy Cases-VA: 0.6 ± 0.32 Controls-VA 1.0 ± 0.1	SLO 101 Rodenstock	Fixation stability was significantly decreased in diabetic pts in comparison to controls. Macular light sensitivity was worse in diabetic pts than in controls, and temporal parts of the macula were the most affected. No correlation was found between VA and foveal light sensitivity nor foveal fixation

Vujosevic et al. [104]	Cross-sectional	61 Eyes (32 pts)	56.1 ± 12.5	Non edema (NE)-16; VA: –0.07 ± 0.18 logMAR NCSME-30; VA: 0.12 ± 0.48 CSME-15; VA: 0..33 ± 0.36	MP-1 Nidek	VA and central macular sensitivity correlated significantly in the NCSME group, but not in the NE or in the CSME group. There was a significant correlation between retinal sensitivity and normalized macular thickness detected by OCT scans
Okada et al. [118]	Retrospective case-control	Cases-32 eyes (25 pts) Controls-17 pts	Cases-58.8 (25–76) Controls-42–76	CSME VA: 0.7 (0.1–0.7) Controls: –0.1 (–0.2 to –0.1)	MP-1 Nidek	Mean sensitivities in diabetic pts were lower than in healthy controls. VA and macular sensitivities were significantly correlated. A significant negative correlation was also found between foveal thickness (by OCT) and the mean retinal sensitivities at central 2° and 10°
Carpineto et al. [122]	Cross-sectional	Cases: 84 pts and eyes	66.35 (45–81)	CSME (67% cystoid) VA: 0.60 ± 0.29 logMAR	MP-1 Nidek	VA, central retinal sensitivity, foveal thickness, duration of symptoms, HbA_{1c} levels and the presence of cystoid macular edema were significantly associated with fixation impairment. The three groups (stable vs. unstable and central vs. eccentric fixation) showed statistically differences in VA, central retinal sensitivity, and foveal thickness. Cystoid macular edema was significantly more frequent in the eccentric and unstable group

(continued)

Table 4. (continued)

Principal investigator/ year of publication	Types of study	Sample size	Age in years: mean/range	DR status and VA	Nature of stimulus	Conclusions
Unoki et al. [146]	Prospective cross-sectional	20 Eyes (17 pts)	62.9 (43–78)	Severe NPDR-11 PDR-9 All showed a nonperfused area in the temporal macula VA: 0.28 ± 0.30 logMAR	MP-1 Nidek	Areas of capillary nonperfusion detected by FA were associated with the loss of retinal sensitivity. The average sensitivity of the next nearest points from the area of capillary nonperfusion was significantly reduced compared with that of the other areas. OCT scans showed morphological changes of the nonperfused areas
Grenga et al. [147]	Prospective	20 Eyes	65.7 ± 13.3	Chronic diffuse macular edema VA: 0.13 ± 0.09 decimal units	MP-1 Nidek	Three months after injection of intravitreal triamcinolone, VA, macular thickness and mean retinal sensitivity improved significantly. At 6 months after injection follow-up of the data were similar to those at baseline
Vujosevic et al. [119]	Prospective	179 Eyes (98 pts)	58.4 ± 11.2	NCSME-32 CSME-147 VA: from worse than 20/200 to 20/25 or better	MP-1 Nidek	Site and stability of fixation were associated. A significant association was found between fixation characteristics and visual acuity, but they were not influenced by edema characteristics (diffuse, focal, cystoid, spongelike edema, with or without neuroretinal detachment). Subfoveal hard exudates were significantly associated with eccentric and unstable fixation, juxtafoveal or no exudates were not

Pts patients; *VA* visual acuity; *DR* diabetic retinopathy; *PDR* proliferative diabetic retinopathy; *NPDR* non-proliferative diabetic retinopathy; *CSME* clinically significant diabetic macular edema; *SLO* scanning laser ophthalmoscope; *MP-1* Microperimeter MP-1

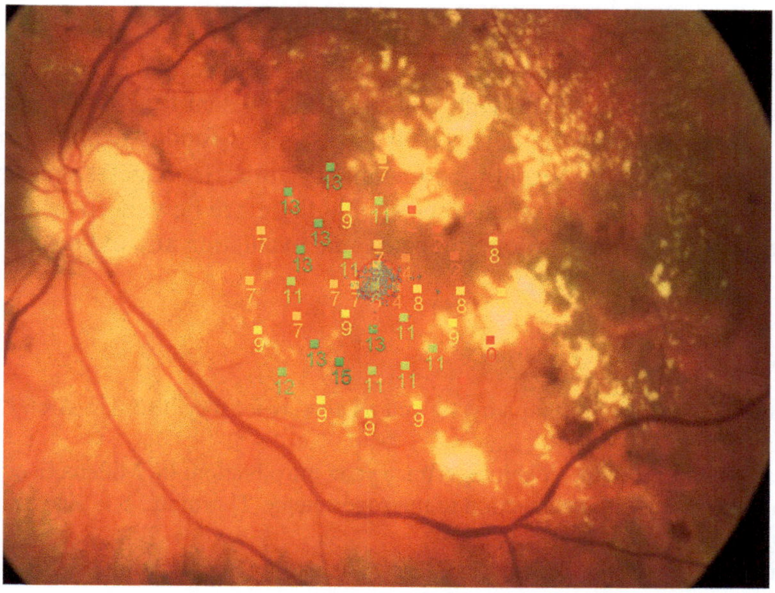

Fig. 2. Microperimetry map (in decibels) superimposed onto the color fundus image in a case of severe CSME with large hard exudates. Over hard exudates the retina shows some dense scotomatous zones. Fixation (*tiny light blue spots centred onto the fovea*) is stable and central.

influenced by either topographical extension of edema (focal or diffuse) or by the OCT classification of edema. Moreover, fixation pattern was not significantly influenced by the presence of subfoveal serous neuroretinal detachment, showing a different fixation behavior compared to age-related macular degeneration [105, 119]. The only parameter influencing fixation was the presence of subfoveal hard exudates. In these cases, the knowledge of fixation location and stability is fundamental in order to avoid complications due to the photocoagulation of newly developed fixation area (Fig. 2).

The duration of DME, which cannot be exactly quantified in a cross-sectional study, might have a relevant impact on the survival and/or functional reserve of macular cells undergoing mechanical and toxic stress induced by edema, and this may explain the difference in fixation results described above. It seems that in patients with DME, the damage to photoreceptor occurs as a late phenomenon and probably is not related to intraretinal cysts formation. In diabetic retinopathy, retinal neurodegeneration may precede photoreceptor loss, as previously reported [123].

Therefore, microperimetry may be of value in predicting the functional outcome of DME after interventions that seem equally effective in restoring normal foveal thickness. This hypothesis has been recently confirmed by a randomized and prospective study conducted by Vujosevic et al. [124]. These authors have demonstrated that subthreshold micropulse diode laser is as effective as modified ETDRS photocoagulation in reducing central retinal thickness. But with subthreshold treatment, retinal macular sensitivity stabilizes or improves, whereas with standard photocoagulation, it significantly deteriorates, manifesting as progressive microscotomata.

CONCLUSION

Diabetes has a relevant impact on visual function, up to permanent visual acuity loss when retinopathy is clinically evident, but changes in visual function may occur long before any structural change is detected by experienced fundus examination or even by fluorescein angiography. Visual function abnormalities in diabetes, mainly detected and quantified using psychophysical tests, should therefore be viewed as a new way of detecting and quantifying diabetic retinopathy and evaluating any old or new treatment approach.

REFERENCES

1. Bresnick GH. Diabetic retinopathy viewed as a neurosensory disorder. Arch Ophthalmol. 1986;104:989–90.
2. Midena E. Fundus perimetry-microperimetry: an introduction. In: Midena E, editor. Perimetry and the fundus: an introduction to microperimetry. Thorofare, NJ: Slack Incorporated; 2006. p. 1–7.
3. Hyvärinen L, Laurinen P, Rovamo J. Contrast sensitivity in evaluation of visual impairment due to diabetes. Acta Ophthalmol. 1983;61:94–101.
4. Sharma S, Oliver-Fernandez A, Liu W, et al. The impact of diabetic retinopathy on health-related quality of life. Curr Opin Ophthalmol. 2005;16:155–9.
5. Owsley C, Sloane ME. Contrast sensitivity, acuity, and the perception of 'real-world' targets. Br J Ophthalmol. 1987;71:791–6.
6. Midena E, editor. Perimetry and the fundus: an introduction to microperimetry. Thorofare, NJ: Slack Incorporated; 2006.
7. Gibson RA, Sanderson HF. Observer variation in ophthalmology. Br J Ophthalmol. 1980;64:457–60.
8. Wick B, Schor CM. A comparison of the Snellen chart and the S-chart for visual acuity assessment in amblyopia. J Am Optom Assoc. 1984;55:359–61.
9. Lovie-Kitchin JE. Validity and reliability of visual acuity measurements. Ophthalmic Physiol Opt. 1988;8:363–70.
10. Kniestedt C, Stamper RL. Visual acuity and its measurement. Ophthalmol Clin North Am. 2003;16:155–70.
11. Pandit JC. Testing acuity of vision in general practice: reaching recommended standard. BMJ. 1994;309:1408.
12. Currie Z, Bhan A, Pepper I. Reliability of Snellen charts for testing visual acuity for driving: prospective study and postal questionnaire. BMJ. 2000;321:990–2.
13. Raasch TW, Bailey IL, Bullimore MA. Repeatability of visual acuity measurement. Optom Vis Sci. 1998;75:342–8.
14. Bailey IL, Lovie JE. New design principles for visual acuity letter charts. Am J Optom Physiol Opt. 1976;53:740–5.
15. Ferris III FL, Kassoff A, Bresnick GH, et al. New visual acuity charts for clinical research. Am J Ophthalmol. 1982;94:91–6.
16. Sloan LL. New test charts for the measurement of visual acuity at far and near distances. Am J Ophthalmol. 1959;48:807–13.
17. Elliott DB, Sheridan M. The use of accurate visual acuity measurements in clinical anti-cataract formulation trials. Ophthalmic Physiol Opt. 1988;8:397–401.
18. Hudson C, Flanagan JG, Turner GS, et al. Correlation of a scanning laser derived oedema index and visual function following grid laser treatment for diabetic macular oedema. Br J Ophthalmol. 2003;87:455–61.

19. Ang GS, Vusirikala B, Mukherji S, et al. Visual acuity and diabetic maculopathy. Ann Ophthalmol. 2006;38:305–10.
20. Sakata K, Funatsu H, Harino S, et al. Relationship of macular microcirculation and retinal thickness with visual acuity in diabetic macular edema. Ophthalmology. 2007;114:2061–9.
21. Early Treatment Diabetic Retinopathy Study Research Group. hotocoagulation for diabetic macular edema. Early Treatment Diabetic Retinopathy Study report number 1. Arch Ophthalmol. 1985;103:1796–806.
22. Otani T, Kishi S, Maruyama Y. Patterns of diabetic macular edema with optical coherence tomography. Am J Ophthalmol. 1999;127:688–9.
23. Bandello F, Polito A, Del Borrello M, et al. "Light" versus "classic" laser treatment for clinically significant diabetic macular oedema. Br J Ophthalmol. 2005;89:864–70.
24. Martidis A, Duker JS, Greenberg PB, et al. Intravitreal triamcinolone for refractory diabetic macular edema. Ophthalmology. 2002;109:920–7.
25. Laursen ML, Moeller F, et al. Subthreshold micropulse diode laser treatment in diabetic macular oedema. Br J Ophthalmol. 2004;8:1173–9.
26. Massin P, Duguid G, Erginay A, et al. Optical coherence tomography for evaluating diabetic macular edema before and after vitrectomy. Am J Ophthalmol. 2003;135:169–77.
27. Diabetic Retinopathy Clinical Research Network; Browning DJ, Glassman AR, et al. Relationship between optical coherence tomography-measured central retinal thickness and visual acuity in diabetic macular edema. Ophthalmology. 2007;114:525–36.
28. Browning DJ, Apte RS, Bressler SB, et al. Association of the extent of diabetic macular edema as assessed by optical coherence tomography with visual acuity and retinal outcome variables. Retina. 2009;29:300–5.
29. Lakowski R, Aspinall PA, Kinnear PR. Association between colour vision losses and diabetes mellitus. Ophthalmol Res. 1972;4:145–59.
30. Hardy KJ, Scase MO, Foster DH, et al. Effect of short term changes in blood glucose on visual pathway function in insulin dependent diabetes. Br J Ophthalmol. 1995;79:38–41.
31. Boulton AJ, Levin S, Comstock J. A multicentre trial of the aldose-reductase inhibitor, tolrestat, in patients with symptomatic diabetic neuropathy. Diabetologia. 1990;33:431–7.
32. Florkowski CM, Rowe BR, Nightingale S, et al. Clinical and neurophysiological studies of aldose reductase inhibitor ponalrestat in chronic symptomatic diabetic peripheral neuropathy. Diabetes. 1991;40:129–33.
33. Julu PO. Essential fatty acids prevent slowed nerve conduction in streptozotocin diabetic rats. J Diabet Complications. 1988;2:185–8.
34. Tomlinson DR, Robinson JP, Compton AM, et al. Essential fatty acid treatment—effects on nerve conduction, polyol pathway and axonal transport in streptozotocin diabetic rats. Diabetologia. 1989;32:655–9.
35. Jamal GA, Carmichael H. The effect of gamma-linolenic acid on human diabetic peripheral neuropathy: a double-blind placebo-controlled trial. Diabet Med. 1990;7:319–23.
36. Dean FM, Arden GB, Dornhorst A. Partial reversal of protan and tritan colour defects with inhaled oxygen in insulin dependent diabetic subjects. Br J Ophthalmol. 1997;81:27–30.
37. Roy MS, Gunkel RD, Podgor MJ. Color vision defects in early diabetic retinopathy. Arch Ophthalmol. 1986;104:225–8.
38. Farnsworth D, editor. The Farnsworth-Munsel 100 Hue Test manual. Baltimore, MD: Munsel Color Co; 1957.
39. Francois J, Verriest G. Acquired dyschromatopsias. Ann Ocul. 1957;190:713–46.
40. Zanen J. Introduction a l'etude de dyschromatopsias retin-iennes contrales asquises. Bull Soc Belge Ophtalmol. 1953;103:3–148.
41. Bresnick GH, Condit RS, Palta M, et al. Association of hue discrimination loss and diabetic retinopathy. Arch Ophthalmol. 1985;103:1317–24.

42. Aspinall PA, Kinnear PR, Duncan LJ. Prediction of diabetic retinopathy from clinical variables and color vision data. Diabetes Care. 1983;6:144–8.
43. Kessel L, Alsing A, Larsen M. Diabetic versus non-diabetic colour vision after cataract surgery. Br J Ophthalmol. 1999;83:1042–5.
44. Kinnear PR, Aspinall PA, Lakowski R. The diabetic eye and colour vision. Trans Ophthalmol Soc U K. 1972;92:69–78.
45. Volbrecht VJ, Schneck ME, Adams AJ, et al. Diabetic short-wavelength sensitivity: variations with induced changes in blood glucose level. Invest Ophthalmol Vis Sci. 1994;35:1243–6.
46. Cho NC, Poulsen GL, Ver Hoeve JN. Selective loss of S-cones in diabetic retinopathy. Arch Ophthalmol. 2000;118:1393–400.
47. Friström B. Peripheral and central colour contrast sensitivity in diabetes. Acta Ophthalmol Scand. 1998;76:541–5.
48. North RV, Farrell U, Banford D, et al. Visual function in young IDDM patients over 8 years of age. A 4-year longitudinal study. Diabetes Care. 1997;20:1724–30.
49. Malagola R, Gargiulo P, Giusti C, et al. Screening of early colour vision defects in insulin dependent diabetic patients with background retinopathy. Invest Ophthalmol Vis Sci. 1994;35:1593.
50. Tregear SJ, Knowles PJ, Ripley LG, et al. Chromatic-contrast threshold impairment in diabetes. Eye. 1997;11:537–46.
51. Ewing FM, Deary IJ, Strachan MW, et al. Seeing beyond retinopathy in diabetes: electrophysiological and psychophysical abnormalities and alterations in vision. Endocr Rev. 1998;19:462–76.
52. Yamamoto S, Kamiyama M, Nitta K, et al. Selective reduction of the S cone electroretinogram in diabetes. Br J Ophthalmol. 1996;80:973–5.
53. Crognale MA, Switkes E, Rabin J, et al. Application of the spatiochromatic visual evoked potential to detection of congenital and acquired color-vision deficiencies. J Opt Soc Am A Opt Image Sci Vis. 1993;10:1818–25.
54. Roy MS, McCulloch C, Hanna AK, et al. Colour vision in long-standing diabetes mellitus. Br J Ophthalmol. 1984;68:215–7.
55. Hardy KJ, Lipton J, Scase MO, et al. Detection of colour vision abnormalities in uncomplicated type 1 diabetic patients with angiographically normal retinas. Br J Ophthalmol. 1992;76:461–4.
56. Kurtenbach A, Flögel W, Erb C. Anomaloscope matches in patients with diabetes mellitus. Graefes Arch Clin Exp Ophthalmol. 2002;240:79–84.
57. Verrotti A, Lobefalo L, Chiarelli F, et al. Colour vision and persistent microalbuminuria in children with type-1 (insulin-dependent) diabetes mellitus: a longitudinal study. Diabetes Res Clin Pract. 1995;30:125–30.
58. Kurtenbach A, Schiefer U, Neu A, et al. Preretinopic changes in the colour vision of juvenile diabetics. Br J Ophthalmol. 1999;83:43–6.
59. Kurtenbach A, Schiefer U, Neu A, et al. Development of brightness matching and colour vision deficits in juvenile diabetics. Vision Res. 1999;39:1221–9.
60. Giusti C. Lanthony 15-Hue desaturated test for screening of early color vision defects in uncomplicated juvenile diabetes. Jpn J Ophthalmol. 2001;45:607–11.
61. Hardy KJ, Fisher C, Heath P, et al. Comparison of colour discrimination and electroretinography in evaluation of visual pathway dysfunction in aretinopathic IDDM patients. Br J Ophthalmol. 1995;79:35–7.
62. Fong DS, Barton FB, Bresnick GH. Impaired color vision associated with diabetic retinopathy: Early Treatment Diabetic Retinopathy Study Report No. 15. Am J Ophthalmol. 1999;128:612–7.

63. Green FD, Ghafour IM, Allan D, et al. Colour vision of diabetics. Br J Ophthalmol. 1985;69:533–6.
64. Maár N, Tittl M, Stur M, et al. A new colour vision arrangement test to detect functional changes in diabetic macular oedema. Br J Ophthalmol. 2001;85:47–51.
65. Ong GL, Ripley LG, Newsom RSB, et al. Assessment of colour vision as a screening test for sight threatening diabetic retinopathy before loss of vision. Br J Ophthalmol. 2003;87: 747–52.
66. Ong GL, Ripley LG, Newsom RS, Cooper M, et al. Screening for sight-threatening diabetic retinopathy: comparison of fundus photography with automated color contrast threshold test. Am J Ophthalmol. 2004;137:445–52.
67. Fong DS, Girach A, Boney A. Visual side effects of successful scatter laser photocoagulation surgery for proliferative diabetic retinopathy: a literature review. Retina. 2007;27:816–24.
68. Di Leo MA, Caputo S, Falsini B, et al. Nonselective loss of contrast sensitivity in visual system testing in early type I diabetes. Diabetes Care. 1992;15:620–5.
69. Trick GL, Burde RM, Gordon MO, et al. The relationship between hue discrimination and contrast sensitivity deficits in patients with diabetes mellitus. Ophthalmology. 1988;95:693–8.
70. Brinchmann-Hansen O, Bangstad HJ, Hultgren S, et al. Psychophysical visual function, retinopathy, and glycemic control in insulin-dependent diabetics with normal visual acuity. Acta Ophthalmol. 1993;71:230–7.
71. Ghafour IM, Foulds WS, Allan D, et al. Contrast sensitivity in diabetic subjects with and without retinopathy. Br J Ophthalmol. 1982;66:492–5.
72. Della Sala S, Bertoni G, Somazzi L, et al. Impaired contrast sensitivity in diabetic patients with and without retinopathy: a new technique for rapid assessment. Br J Ophthalmol. 1985;69:136–42.
73. Sokol S, Moskowitz A, Skarf B, et al. Contrast sensitivity in diabetics with and without background retinopathy. Arch Ophthalmol. 1985;103:51–4.
74. Maione M, Cordella M, Franchi A, et al. Contrast sensitivity in diabetics: comparison between psychophysical and evoked potential methods. Ann Oftalmol Clin Ocul. 1986;6:445–52.
75. Arend O, Remky A, Evans D, et al. Contrast sensitivity loss is coupled with capillary dropout in patients with diabetes. Invest Ophthalmol Vis Sci. 1997;38:1819–24.
76. Midena E, Segato T, Bottin G, et al. The effect on the macular function of laser photocoagulation for diabetic macular edema. Graefes Arch Clin Exp Ophthalmol. 1992;230:162–5.
77. Talwar D, Sharma N, Pai A, et al. Contrast sensitivity following focal laser photocoagulation in clinically significant macular oedema due to diabetic retinopathy. Clin Experiment Ophthalmol. 2001;29:17–21.
78. Farahvash MS, Mahmoudi AH, Farahvash MM, et al. The impact of macular laser photocoagulation on contrast sensitivity function in patients with clinically significant macular edema. Arch Iran Med. 2008;11:143–7.
79. Midena E, Segato T, Giuliano M, et al. Macular recovery function (nyctometry) in diabetics without and with early retinopathy. Br J Ophthalmol. 1990;74:106–8.
80. Gliem H, Schulze DP. Initial phase of dark adaptation, sensibility to dazzling and diabetic retinopathy. Klin Monbl Augenheilkd. 1975;166:766–9.
81. Simonsen SE. The value of the oscillatory potential in selecting juvenile diabetics at risk of developing proliferative retinopathy. Acta Ophthalmol. 1980;58:865–78.
82. Frost-Larsen K, Larsen HW. Macular recovery time recorded by nyctometry—a screening method for selection of patients who are at risk of developing proliferative diabetic retinopathy. Results of a 5-year follow-up. Acta Ophthalmol Suppl. 1985;173:39–47.
83. Frost-Larsen K, Larsen HW, Simonsen SE. Oscillatory potential and nyctometry in insulin-dependent diabetics. Acta Ophthalmol. 1980;58:879–88.

84. Verrotti A, Lobefalo L, Chiarelli F, et al. Macular recovery time in diabetic children without retinopathy. Diabetes Res Clin Pract. 1996;32:149–55.
85. Lauritzen T, Frost-Larsen K, Larsen HW, et al. Effect of 1 year of near-normal blood glucose levels on retinopathy in insulin-dependent diabetics. Lancet. 1983;1:200–4.
86. Frost-Larsen K, Lund-Anderson C, Starup K. Macular recovery during onset and development of diabetic retinopathy in childhood and adolescence. Acta Ophthalmol (Copenh). 1989;67:401–4.
87. Frost-Larsen K, Larsen HW, Simonsen SE. Value of electroretinography and dark adaptation as prognostic tools in diabetic retinopathy. Dev Ophthalmol. 1981;2:222–34.
88. Virsu V. Retinal mechanisms of visual adaptation and afterimages. Med Biol. 1978;56: 84–96.
89. Dowling JE. The site of visual adaptation. Science. 1967;155:273–9.
90. Trick GL, Trick LR, Kilo C. Visual field defects in patients with insulin-dependent and noninsulin-dependent diabetes. Ophthalmology. 1990;97:475–82.
91. Chee CK, Flanagan DW. Visual field loss with capillary non-perfusion in preproliferative and early proliferative diabetic retinopathy. Br J Ophthalmol. 1993;77:726–30.
92. Henricsson M, Heijl A. Visual fields at different stages of diabetic retinopathy. Acta Ophthalmol. 1994;72:560–9.
93. Bek T, Lund-Andersen H. Cotton-wool spots and retinal light sensitivity in diabetic retinopathy. Br J Ophthalmol. 1991;75:13–7.
94. Zwas F, Weiss H, McKinnon P. Spectral sensitivity measurements in early diabetic retinopathy. Ophthalmic Res. 1980;12:87–96.
95. Adams AJ, Zisman F, Ai E, Bresnick G. Macular edema reduced B cone sensitivity in diabetics. Appl Opt. 1987;26:1455–7.
96. Greenstein V, Sarter B, Hood D. Hue discrimination and S cone pathway sensitivity in early diabetic retinopathy. Invest Ophthalmol Vis Sci. 1990;31:1008–14.
97. Greenstein VC, Shapiro A, Zaidi Q. Psychophysical evidence for post-receptoral sensitivity loss in diabetics. Invest Ophthalmol Vis Sci. 1992;33:2781–90.
98. Remky A, Arend O, Hendricks S. Short-wavelength automated perimetry and capillary density in early diabetic maculopathy. Invest Ophthalmol Vis Sci. 2000;41:274–81.
99. Hudson C, Flanagan JG, Turner GS, et al. Short-wavelength sensitive visual field loss in patients with clinically significant diabetic macular edema. Diabetologia. 1998;41:918–28.
100. Lobefalo L, Verrotti A, Mastropasqua L, et al. Blue-on-yellow and achromatic perimetry in diabetic children without retinopathy. Diabetes Care. 1998;21:2003–6.
101. Agardh E, Stjernquist H, Heijl A, et al. Visual acuity and perimetry as measures of visual function in diabetic macular oedema. Diabetologia. 2006;49(1):200–6.
102. Bengtsson B, Heijl A, Agardh E. Visual fields correlate better than visual acuity to severity of diabetic retinopathy. Diabetologia. 2005;48:2494–500.
103. Sunness JS, Schuchard RA, Shen N, et al. Landmark-driven fundus perimetry using the scanning laser ophthalmoscope. Invest Ophthalmol Vis Sci. 1995;36:1863–74.
104. Vujosevic S, Midena E, Pilotto E, et al. Diabetic macular edema: correlation between microperimetry and optical coherence tomography findings. Invest Ophthalmol Vis Sci. 2006;47:3044–51.
105. Midena E, Radin PP, Pilotto E, et al. Fixation pattern and macular sensitivity in eyes with subfoveal choroidal neovascularization secondary to age-related macular degeneration. A microperimetry study. Semin Ophthalmol. 2004;19:55–61.
106. Midena E, Vujosevic S, Convento E, et al. Microperimetry and fundus autofluorescence in patients with early age-related macular degeneration. Br J Ophthalmol. 2007;91: 1499–503.

107. Rohrschneider K, Springer C, Bültmann S, et al. Microperimetry—comparison between the micro perimeter 1 and scanning laser ophthalmoscope—fundus perimetry. Am J Ophthalmol. 2005;139:125–34.
108. Midena E, Vujosevic S, Cavarzeran F; for the Microperimetry Study Group. Normative age-related database for the MP1 microperimeter. Ophthalmology. 2010;117(8):1571–6.
109. Mori F, Ishiko S, Kitaya N, et al. Scotoma and fixation patterns using scanning laser ophthalmoscope microperimetry in patients with macular dystrophy. Am J Ophthalmol. 2001;132:897–902.
110. Rohrschneider K, Bültmann S, Glück R, et al. Scanning laser ophthalmoscope fundus perimetry before and after laser photocoagulation for clinically significant diabetic macular edema. Am J Ophthalmol. 2000;129:27–32.
111. Mori F, Ishiko S, Kitaya N, et al. Use of scanning laser ophthalmoscope microperimetry in clinically significant macular edema in type 2 diabetes mellitus. Jpn J Ophthalmol. 2002;46:650–5.
112. Rohrschneider K, Glück R, Becker M, et al. Scanning laser fundus perimetry before laser photocoagulation of well defined choroidal neovascularisation. Br J Ophthalmol. 1997;81:568–73.
113. Sunness JS, Applegate CA, Haselwood D, et al. Fixation patterns and reading rates in eyes with central scotomas from advanced atrophic age-related macular degeneration and Stargardt disease. Ophthalmology. 1996;103:1458–66.
114. Kube T, Schmidt S, Toonen F, et al. Fixation stability and macular light sensitivity in patients with diabetic maculopathy: a microperimetric study with a scanning laser ophthalmoscope. Ophthalmologica. 2005;219:16–20.
115. Loewenstein A, Sunness JS, Bressler NM, et al. Scanning laser ophthalmoscope fundus perimetry after surgery for choroidal neovascularization. Am J Ophthalmol. 1998;125:657–65.
116. Midena E. Psychophysical investigations in myopia. In: Midena E, editor. Myopia and related diseases. New York, NY: Ophthalmic Communications Society; 2005. p. 120–4.
117. Varano M, Parisi V, Tedeschi M, et al. Macular function after PDT in myopic maculopathy: psychophysical and electrophysiological evaluation. Invest Ophthalmol Vis Sci. 2005;46:1453–62.
118. Okada K, Yamamoto S, Mizunoya S, et al. Correlation of retinal sensitivity measured with fundus-related microperimetry to visual acuity and retinal thickness in eyes with diabetic macular edema. Eye. 2006;20:805–9.
119. Vujosevic S, Pilotto E, Bottega E, et al. Retinal fixation impairment in diabetic macular edema. Retina. 2008;28:1443–50.
120. Møller F, Bek T. The relation between visual acuity, fixation stability, and the size and location of foveal hard exudates after photocoagulation for diabetic maculopathy: a 1-year follow-up study. Graefes Arch Clin Exp Ophthalmol. 2003;241:458–62.
121. Møller F, Laursen ML, Sjølie AK. Binocular fixation topography in patients with diabetic macular oedema: possible implications for photocoagulation therapy. Graefes Arch Clin Exp Ophthalmol. 2005;243:903–10.
122. Carpineto P, Ciancaglini M, Di Antonio L, et al. Fundus microperimetry patterns of fixation in type 2 diabetic patients with diffuse macular edema. Retina. 2007;27:21–9.
123. Vujosevic S, Midena E. Diabetic retinopathy. In: Midena E, editor. Perimetry and the fundus: an introduction to microperimetry. Thorofare, NJ: Slack Incorporated; 2006. p. 177–9.
124. Vujosevic S, Bottega E, Casciano M, et al. Microperimetry and fundus autofluorescence in diabetic macular edema: subthreshold micropulse diode laser versus modified early treatment diabetic retinopathy study laser photocoagulation. Retina. 2010;30:908–16.

125. Verriest G, Van Laethem J, Uvijls A. A new assessment of the normal ranges of the Farnsworth-Munsell 100-hue test scores. Am J Ophthalmol. 1982;93:635–42.
126. Wong R, Khan J, Adewoyin T, et al. The ChromaTest, a digital color contrast sensitivity analyzer, for diabetic maculopathy: a pilot study. BMC Ophthalmol. 2008;17:8–15.
127. Virsu V, LehtiG P, Rovamo J. Contrast sensitivity in normal and pathological vision. Doc Ophthalmol Proc Ser.1981;30:263–272.
128. Regan D, Neima D. Low-contrast letter charts in early diabetic retinopathy, ocular hypertension, glaucoma, and Parkinson's disease. Br J Ophthalmol. 1984;68:885–9.
129. Khosla PK, Talwar D, Tewari HK. Contrast sensitivity changes in background diabetic retinopathy. Can J Ophthalmol. 1991;26:7–11.
130. Bangstad HJ, Brinchmann-Hansen O, Hultgren S,, et al. Impaired contrast sensitivity in adolescents and young type 1 (insulin-dependent) diabetic patients with microalbuminuria. Acta Ophthalmol (Copenh). 994;72:668–73.
131. De Marco R, Capasso L, Magli A, et al. Measuring contrast sensitivity in aretinopathic patients with Insulin Dependent Diabetes Mellitus. Doc Ophthalmol. 1996–1997;93:199–209.
132. Verrotti A, Lobefalo L, Petitti MT, et al. Relationship between contrast sensitivity and metabolic control in diabetics with and without retinopathy. Ann Med. 1998;30:369–74.
133. Mackie SW, Walsh G. Contrast and glare sensitivity in diabetic patients with and without pan-retinal photocoagulation. Ophthalmic Physiol Opt. 1998;18:173–81.
134. Lövestam-Adrian M, Svendenius N, Agardh E. Contrast sensitivity and visual recovery time in diabetic patients treated with panretinal photocoagulation. Acta Ophthalmol Scand. 2000;78:672–6.
135. Stavrou EP, Wood JM. Letter contrast sensitivity changes in early diabetic retinopathy. Clin Exp Optom. 2003;86:152–6.
136. Bek T, Lund-Andersen H. Localised blood-retinal barrier leakage and retinal light sensitivity in diabetic retinopathy. Br J Ophthalmol. 1990;74:388–92.
137. Lutze M, Bresnick GH. Lens-corrected visual field sensitivity and diabetes. Invest Ophthalmol Vis Sci. 1994;35:649–55.
138. Nomura R, Terasaki H, Hirose H, et al. Blue-on-yellow perimetry to evaluate S cone sensitivity in diabetics. Ophthalmic Res. 2000;32:69–72.
139. Verrotti A, Lobefalo L, Altobelli E, et al. Static perimetry and diabetic retinopathy: a long-term follow-up. Acta Diabetol. 2001;38:99–105.
140. Afrashi F, Erakgün T, Köse S, et al. Blue-on-yellow perimetry versus achromatic perimetry in type 1 diabetes patients without retinopathy. Diabetes Res Clin Pract. 2003;61:7–11.
141. Remky A, Weber A, Hendricks S, et al. Short-wavelength automated perimetry in patients with diabetes mellitus without macular edema. Graefes Arch Clin Exp Ophthalmol. 2003;241:468–71.
142. Han Y, Adams AJ, Bearse MA Jr, et al. Multifocal electroretinogram and short-wavelength automated perimetry measures in diabetic eyes with little or no retinopathy. Arch Ophthalmol. 2004;122:1809–15.
143. Nitta K, Saito Y, Kobayashi A, et al. Influence of clinical factors on blue-on-yellow perimetry for diabetic patients without retinopathy: comparison with white-on-white perimetry. Retina. 2006;26:797–802.
144. Pahor D. Automated static perimetry as a screening method for evaluation of retinal perfusion in diabetic retinopathy. Int Ophthalmol. 1997–1998;21:305–9.
145. Moller F, Bek T. The relation between visual acuity, fixation stability, and the size and location of foveal hard exudates after photocoagulation for diabetic maculopathy: a 1-year follow-up study. Graefes Arch Clin Exp Ophthalmol. 2003;241:458–62.

146. Unoki N, Nishijima K, Sakamoto A, et al. Retinal sensitivity loss and structural disturbance in areas of capillary nonperfusion of eyes with diabetic retinopathy. Am J Ophthalmol. 2007;144:755–760.
147. Grenga P, Lupo S, Domanico D, et al. Efficacy of intravitreal triamcinolone acetonide in long standing diabetic macular edema: a microperimetry and optical coherence tomography study. Retina. 2008;28:1270–5.

7

Mechanisms of Blood–Retinal Barrier Breakdown in Diabetic Retinopathy

Ali Hafezi-Moghadam

Contents

Keywords Vascular leakage • Leukocyte adhesion • ICAM-1 • β_2-integrin • VAP-1 • Azurocidin (AZ) • Atrial natriuretic peptide (ANP)

The consequences of the currently growing epidemic of type 2 diabetes would soon debilitate the public health [1], unless new ways are rapidly found for prevention or therapy of the various complications of the disease.

Vascular leakage is a prominent feature of diabetic retinopathy (DR), an ocular manifestation of diabetes. Vascular leakage is routinely quantified in patients as an important end point of ocular examinations and also studied at the bench in a variety of in vitro and in vivo assays. However, despite the pertinence of vascular leakage for both research and clinic, the cellular and molecular mechanisms underlying vascular leakage are not well understood.

THE PROTECTIVE BARRIERS OF THE RETINA

A barrier function in normal blood vessels of the central nervous system (CNS) was first proposed by Paul Ehrlich (1854–1915). Reese and Karnovsky [2] later showed at an ultrastructural level that tight endothelial barriers are responsible for the unique barrier properties of CNS vessels (Fig. 1). In the eye, the blood retinal barrier (BRB) describes the selective physiological barrier that protects the neural retina from molecules and

From: *Ophthalmology Research: Visual Dysfunction in Diabetes*
Edited by: J. Tombran-Tink et al. (eds.), DOI 10.1007/978-1-60761-150-9_7

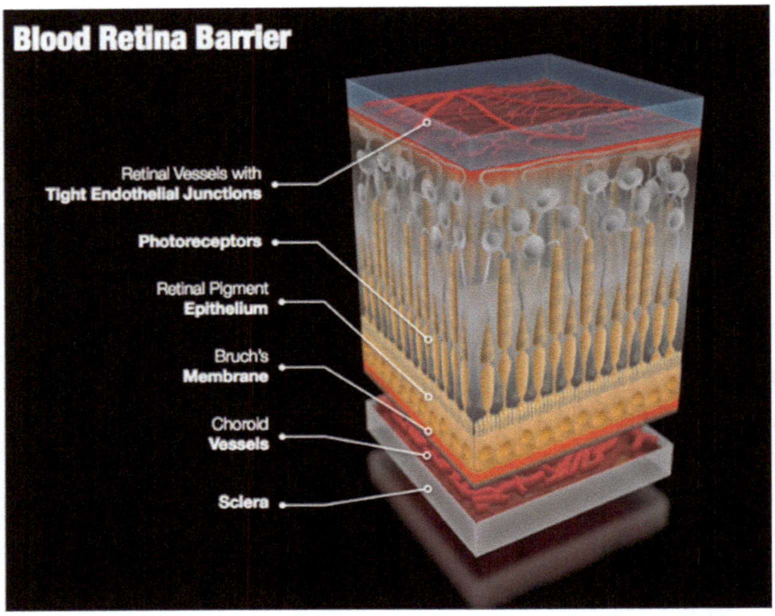

Fig. 1. Neural retina and the surrounding vasculature. The retina has two separate vascular systems: retinal and the choroidal vessels. The retinal vessels have tight endothelial barriers as also seen in the vessels of the brain, constituting the inner blood retinal barrier (BRB). In comparison, the outer BRB is comprised of the retinal pigment epithelium (RPE) together with the Bruch's membrane, separating the leaky choroidal vessels from the neural retina.

cells in the blood. The BRB acts as an active regulatory interface, where transport of fluids, proteins, and cells in both directions takes place [3]. The integrity of BRB is essential for retinal neuronal health, and a compromised BRB is seen in various ocular diseases. The inner BRB is formed by normal retinal vessels, while the outer BRB is made by the retinal pigment epithelium (RPE) (Fig. 1). Cumulatively, these barriers regulate the flow of fluid, proteins, and cells into the extracellular space of the neural retina. Active transport mechanisms in the RPE result in a net fluid flow out of the neural retina [4]. Even under pathological conditions, RPE function can compensate for part of the leakage of vessels into the extracellular environment and reduce fluid accumulation in the outer retina.

THE INNER AND THE OUTER BRB

The inner BRB of the retinal vessels is similar to that in brain microvessels (Fig. 2). Various cellular components are needed to form such a barrier [5]. A milestone was the discovery that astrocyte end feet surround microvessels and that their connection to the endothelium induces various unique barrier properties in the endothelial cells [5]. These properties include high-resistance tight junctions between the capillary endothelial cells that impede the passive diffusion of solutes from the blood into the extracellular space [5]. Since then, much of the insight gained about vascular barriers comes from cell

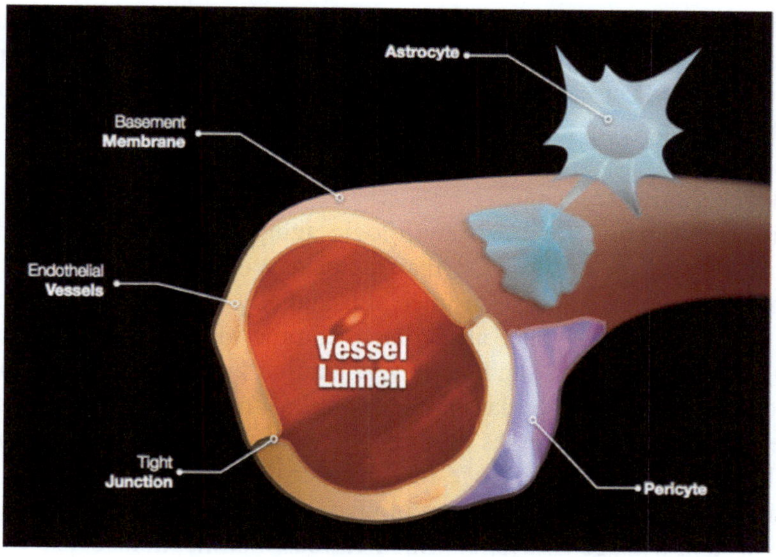

Fig. 2. Schematic of the neurovascular barrier. This is a schematic showing the tight apposition of endothelial cells lining blood vessels in the brain. This is characteristic of the selective blood–brain barrier, which separates the circulation from brain parenchyma. Pericytes sheath the basement membrane covering the vascular endothelium.

culture models, in which endothelial cells are co-cultured with astrocytes or sometimes also with pericytes.

Changes of BRB in diabetes has long been of central interest. In DR, BRB breakdown causes protein and fluid extravasation, possibly leading to acute macular edema or longer-term neuronal damage. Therefore, elucidating the factors that compromise the BRB might lead to new therapeutic approaches for DR or diabetic macular edema, which is the main cause of visual loss in diabetic patients. BRB investigations in vivo are commonly studied in the streptozotocin (STZ)-induced diabetes in rats [6]. STZ, an antibiotic produced from *Streptomyces achromogenes*, enters the cytoplasm via glucose transporter (GLUT) 2, which is the β-cell's GLUT in the pancreas [7], and reduces insulin secretion through β-cell toxicity [8]. STZ-injected animals rapidly develop hyperglycemia, resembling the conditions found in type 1 diabetes, and develop diabetic retinal vasculopathy, making them a convenient tool in the study of early diabetic changes. These animals develop some earlier vascular changes, such as increased retinal leukostasis, vascular leakage, or elevated cytokine expression. However, STZ-injected animals do not exhibit the entire pathology of the human DR. For instance, they do not show retinal neovascularization. Furthermore, the following metabolic disarray, including insulin resistance, dyslipidemia, and adipokine changes, is not truly reflected in STZ-induced diabetes. The recently introduced model of spontaneously occurring type 2 diabetes in the Nile grass rat (NGR) shows many pertinent characteristics of the human condition [9]. The hyperglycemia in NGR is accompanied by dyslipidemia and insulin resistance. Hope is great that with the help of such realistic models of human diabetes, effective mechanistic explorations as well as therapeutic advances will take place.

Due to the growing importance of age-related diseases, a large amount of interest lies in understanding the physiological changes of vascular barrier function during aging [10]. Recent work indicates a gradual and continuous decline in vascular barrier function with physiologic aging and that immune cells contribute to this process [11]. This indicates that the barrier-privileged vessels of the body, similar to other organs, are subject to changes resulting from age.

A plausible explanation for how physiologic aging might impact vascular barrier function comes from the observation that deficiency of a cholesterol transport protein, the apolipoprotein E (apoE), in mice substantially accelerates the barrier decay with age [11]. Since apoE–/– mice are prone to chronic vascular inflammation, such as accelerated atherosclerosis [12] and neurodegeneration [13], this indicates that chronic inflammation compromises vascular barrier privilege. Analogously, in normal animals, constitutive inflammatory processes during aging cause cumulative damage to the vasculature, which can be a prelude to age-related vascular diseases [11].

To investigate retinal vascular leakage in vivo, for instance in diabetes, many investigators use protein leakage assays, of which various modifications exist. These assays commonly quantify the passage of plasma albumin into the parenchyma. To do so, dyes such as Evans blue (EB) are injected into the circulation [14, 15]. Under controlled conditions, these techniques allow quantitative assessment of inner BRB leakage. However, due to the low amount of retinal tissue and the large variability between animals, albumin/protein-based leakage assays have limitations both in terms of sensitivity and in the large variability of the outcome. Therefore, there is currently a great need for more sensitive in vivo assays that can reliably quantify subtle leakage.

The outer BRB is primarily comprised of the RPE, a cellular layer that causes a tight epithelial barrier. The healthy RPE forms not only the outer BRB but also actively removes subretinal fluid, thus regulating fluid accumulation in the subretinal space. RPE function is essential to maintaining a balanced outer retinal environment. Moreover, the RPE is a principal source of angiogenic and antiangiogenic factors and also expresses the receptors for these agents.

Both acute and chronic inflammation disrupt the (BRB), as in uveitis or diabetic retinopathy, respectively [16]. These facts have led to the hypothesis that barrier changes in physiologic aging or in acute or chronic inflammation are related. Indeed, certain immune cells in the peripheral blood, neutrophils and macrophages, contain a highly potent permeability factor, azurocidin (AZ), that these cells release when interacting with the activated endothelium.

Inflammation and BRB Permeability

Leukocyte accumulation in retinal vessels is a critical early event in the pathogenesis of DR. Firm adhesion of neutrophils to the inflamed endothelium causes vascular leakage [17–19]. However, the molecular details are only beginning to be understood. Leukocyte accumulation on the inflamed endothelium of retinal vessels follows the general principles of cascade-like recruitment [20]. Leukocyte rolling, the initial step in the recruitment cascade, is followed by leukocyte activation, firm adhesion, and transmigration into the interstitial tissue [20]. The endothelium sequentially expresses adhesion molecules, such as selectins, integrins, and immunoglobulins, and presents

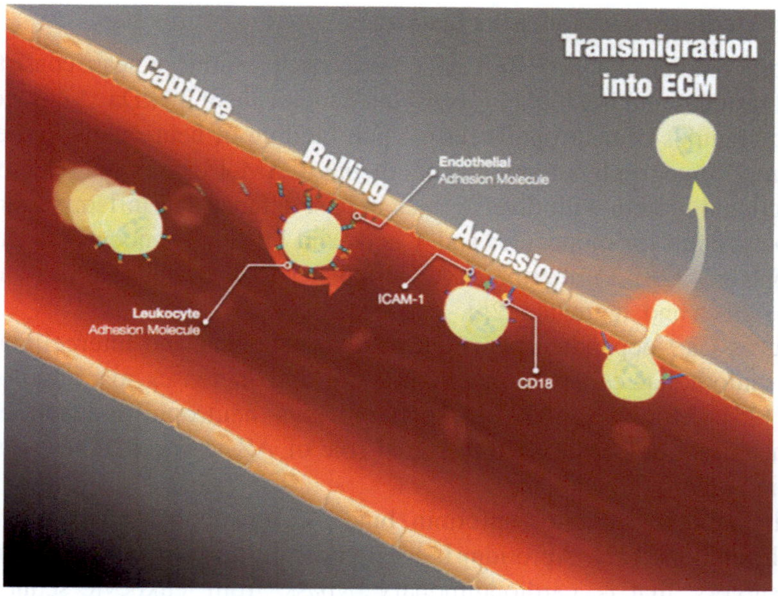

Fig. 3. Steps of inflammatory leukocyte recruitment. The transition from rolling to firm adhesion is achieved by endothelial intracellular adhesion molecule (ICAM)-1 that interacts with its leukocyte ligand, CD18 [23]. The retinal endothelium of diabetic animals expresses ICAM-1, which binds to leukocyte β_2-integrins, LFA-1 (CD18CD11a) and Mac-1 (CD18CD11b), mediating firm leukocyte adhesion. Leukocytes use their integrins to extravasate through the extracellular matrix (ECM) [103].

chemoattractants to the free-flowing leukocytes to orchestrate each stage of the recruitment process [20, 21] (Fig. 3).

Selectins mainly mediate the first steps of the leukocyte-endothelial interaction [20]. Through their lectin domain, the selectins bind to other carbohydrates presented by mucins [22]. P-selectin is the first adhesion receptor transiently upregulated on the endothelium during inflammation, which initiates leukocyte rolling [21].

Leukocyte adhesion to the retinal vessels is critical for DR pathology, as inhibition of leukocyte adhesion through intracellular adhesion molecule (ICAM)-1 or β_2-integrin blockade effectively suppresses vascular endothelial growth factor (VEGF)-induced and diabetic BRB breakdown, establishing the link between leukocyte adhesion and increased retinal vascular leakage [23, 24]. However, the molecular pathways involved in BRB breakdown downstream of leukocyte adhesion are only beginning to be understood.

When neutrophils and monocytes, two leukocyte subtypes, interact via their β_2-integrins with ICAM-1 on activated endothelium, they release the content of their azurophilic granulae. One of the protein contents of these granulae, AZ, is a potent permeability factor [25]. Interestingly, β_2-integrin expression on peripheral blood neutrophils is higher in diabetic animals [24]. Under these conditions, leukocytes are more prone to release AZ.

Leukocyte Mediators of Vascular Leakage

AZ, heparin-binding protein (HBP/CAP37) is an inactive serine protease consisting of 225 amino acid residues and is a highly glycosylated molecule of 37 kDa. AZ is stored in the azurophilic granules of neutrophils [26]. Upon binding of neutrophils to the activated endothelium, β_2-integrin ligation with endothelial ICAM-1 causes AZ release [25]. It is a multifunctional protein with diverse roles in host defense and inflammation [27]. AZ is a chemoattractant for monocytes and T cells and induces monocytes to differentiate into macrophages [28]. Furthermore, AZ stimulates endothelial cells via an unknown receptor to detach and aggregate [25].

AZ induces Ca^{2+}-dependent cytoskeletal rearrangement and intercellular gap formation in endothelial cell monolayers in vitro and increases macromolecular permeability [25]. Moreover, AZ blockade prevents neutrophil-induced endothelial hyperpermeability, emphasizing the crucial role of AZ in vascular responses during inflammation [25].

The serine protease inhibitor, aprotinin, binds AZ and abolishes its ability to disrupt endothelial junctions [29]. Aprotinin is used clinically to protect patients undergoing extensive surgery, that is, cardiopulmonary bypass, from leukocyte sequestration in organs and fluid loss from the vasculature [30]. Recent in vivo results show that aprotinin is an effective inhibitor of the AZ-induced retinal vascular leakage. Aprotinin treatment also significantly decreases VEGF-induced leakage and BRB breakdown in experimentally induced diabetes, suggesting a possible role for AZ in these events. Aprotinin, as a broad inhibitor, also blocks other serine proteases, such as neutrophil-derived elastase, cathepsin G, proteinase 3, and some proteases in coagulation and fibrinolysis pathways, including plasmin and kallikrein [29, 31, 32]. Some of these proteases may be involved in retinal vascular leakage [33], and since aprotinin does not exclusively block AZ, there is potential involvement of other proteases in the VEGF-induced retinal vascular leakage or the BRB breakdown seen in early diabetes. Furthermore, since aprotinin is known to be anti-inflammatory and an inhibitor of leukocyte recruitment, aprotinin's protective function against BRB breakdown could in part be due to its anti-inflammatory properties. Taken together, aprotinin or similar inhibitors of AZ might be useful in the treatment of retinal vascular leakage in DR.

Aprotinin is in clinical use for patients undergoing extensive cardiothoracic and orthopedic surgery. These patients often develop neutrophil sequestration in organs and massive leakage of fluid from the vasculature, and aprotinin can help to reduce blood loss and blood transfusion requirements postoperatively [30, 32]. Inhibition of AZ was proposed by Gautam et al. [25] as a possible mechanism of action for aprotinin in these clinical settings, considering the crucial role of AZ in neutrophil-evoked permeability. Two recent reports show an increased risk of renal [34] and cardiovascular toxicity, including myocardial infarction and stroke [35], following aprotinin administration in major surgeries. These systemic side effects of aprotinin might in part be due to its limited specificity in vivo. To date, there is no selective inhibitor of AZ. However, upon availability of such inhibitors, their intravitreal delivery in DR might lower the risk for systemic side effects.

OTHER MEDIATORS OF LEUKOCYTE RECRUITMENT IN DR

Due to the immune-privileged status of the eye, few immune cells transmigrate into the retina under physiological conditions. However, in DR, large numbers of immune cells cross the BRB and migrate into the neuronal retina. The infiltrating leukocytes are believed to be the cause of considerable harm to the neurons. Recent results indicate a critical role for a new molecule, the vascular adhesion protein 1 (VAP-1), in the retinas of diabetic animals [36].

VAP-1 is an endothelial adhesion molecule involved in leukocyte recruitment [37, 38]. It is a homodimeric sialylated glycoprotein expressed on the endothelium of human tissues such as skin, brain, lung, liver, and heart under both normal and inflamed conditions [39–42]. Increased levels of both soluble and membrane-associated VAP-1 are reported in diabetes [43].

In addition to being an adhesion molecule, VAP-1 is also an enzyme. Indeed, VAP-1 is the only known adhesion molecule that also has catalytic activity. It has characteristics of semicarbazide-sensitive amine oxidases (SSAO), enzymes that catalyze the deamination of primary amines such as methylamine and aminoacetone [44, 45]. SSAO's active site generates toxic formaldehyde and methylglyoxal, hydrogen peroxide and ammonia [45], reactive chemicals, and major reactive oxygen species [43].

VAP-1 is expressed on the retinal endothelium, and it plays a critical role in the recruitment of leukocytes to the eye during DR [36], acute inflammation [46], and laser-induced neovascularization [47]. The fact that VAP-1 is expressed in the human eye [48] suggests that it could become an attractive molecular target in the prevention and treatment of ocular inflammatory diseases, such as DR.

To detect molecular changes or early endothelial injury at the BRB, we recently introduced a new noninvasive molecular imaging approach [49, 50]. In this technique, fluorescent microspheres (MSs), of slightly less than cellular dimensions, are conjugated with ligands or antibodies to one or more endothelial surface molecules of interest [50, 51]. After systemic injection, the interactions of these MSs with the endothelium of the retinal and choroidal vessels of live animals is studied by scanning laser ophthalmoscopy (SLO) in normal or diabetic animals. These new approaches will likely advance our understanding of the cellular and molecular events that lead to BRB breakdown, for instance in early DR.

STRUCTURAL COMPROMISE OF THE BRB

The inner and outer BRB can also be compromised due to structural changes. A common cause of the structural damage underlying BRB breakdown is neovascularization, or the growth of new vessels. Neovascularization in the eye is a leading cause of vision loss. It occurs in the proliferative stage of diabetic retinopathy, where retinal vessels grow, likely secondary to ischemia. While normal retinal vessels have the BRB function, the neovascular vessels are leaky for proteins and also prone to bleeding, allowing accumulation of fluid within the extracellular spaces of the neurosensory retina. The ensuing damage to the cells of the neuronal retina can result in permanent vision loss. Neovascularization occurs in consequence of the intraocular release of the pro-angiogenic cytokine and vasopermeability agent, VEGF [52].

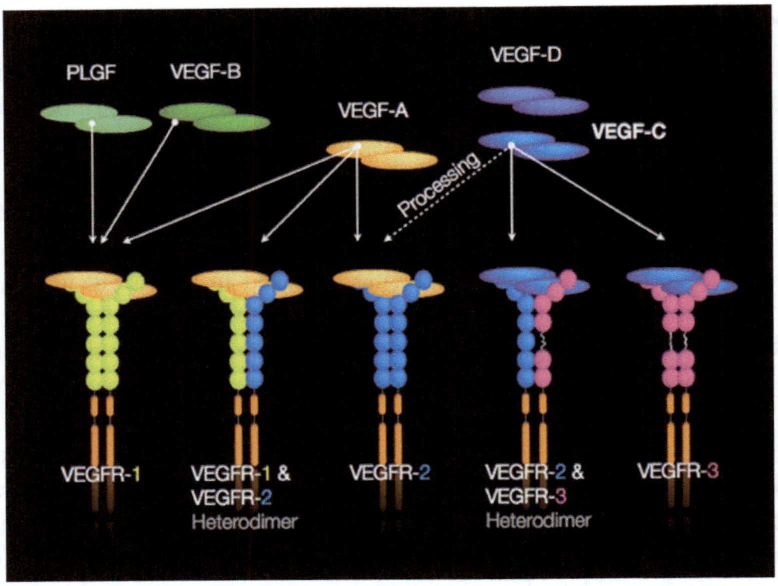

Fig. 4. Vascular endothelial growth factor (VEGF) isoforms and their endothelial receptors.

Vascular Endothelial Growth Factor

A key mediator of permeability as well as neovascularization is VEGF. VEGF expression is primarily triggered through hypoxia [53], but growth factors [54], inflammatory molecules [55], oxidative stress [56], and advanced glycation end products [57] can also induce VEGF production. A major source of VEGF in the eye is the RPE [58]. Under normal conditions, VEGF secretion from the RPE stimulates choriocapillaris endothelium development [59]. The role of VEGF in endothelial cell biology has been extensively studied. VEGF receptors activate multiple signaling pathways including survival [60], migration [61], mitogenesis [62], and permeability [63]. Recent studies show expression of VEGF receptors on RPE [64] and that VEGF modulates RPE barrier properties through the VEGFR-2 receptor [65].

The VEGF family consists of five members that bind to and activate three distinct receptors [66, 67] (Fig. 4). VEGF-A binds to both VEGFR-1 and VEGFR-2, while placental growth factor (PlGF) and VEGF-B bind only to VEGFR-1. VEGF-C and VEGF-D are the only known ligands for VEGFR-3 and do not bind to VEGFR-1 [68, 69].

VEGF-A is upregulated in various physiological and pathological conditions, causing endothelial permeability [70], lymph- [71], and angiogenesis [72]. VEGF-A induces proliferation and migration of the lymphatic endothelium through the VEGFR-2 [73]. Proteolytically processed VEGF-C binds to and activates VEGFR-2, while the unprocessed precursor form of VEGF-C signals through VEGFR-3 [74]. Both VEGF-C and VEGF-D primarily affect development of lymphatic vasculature through VEGFR-3 activation, but they also participate in angiogenesis through VEGFR-2 [75]. For instance, a soluble VEGFR-2 form that is secreted by corneal epithelial cells selectively suppresses the physiologic growth of lymphatics; however, it does not address the interdependency of lymph- and angiogenesis [76]. VEGF-C expression is higher in blood vessel endothelium than in lymphatic endothelium; conversely, VEGFR-3 expression is higher in lymphatic endothelium [73]. In comparison, VEGFR-2 expression is similar in both endothelial cell types [77]. However, the differential contribution of VEGF-C/VEGFR-2 interaction to lymph- and angiogenesis is not well understood.

VEGF is closely tied to the pathogenesis of DR. It plays a key role in the leukocyte-mediated breakdown of the BRB as well as retinal neovascularization [78]. Recent evidence ties VEGF with inflammation [79]. VEGF increases endothelial ICAM-1 expression, facilitating leukocyte adhesion [80] and BRB breakdown in diabetic retinal vessels [23].

Within the first 2 weeks of experimental diabetes in rats, retinal VEGF levels increased with associated upregulation of ICAM-1 in retinal endothelial cells and its ligands, the β_2-integrins, on the surface of peripheral blood neutrophils [81, 82]. These molecular events result in increased adhesion of leukocytes, predominantly neutrophils, with a concomitant increase in retinal vascular permeability. Analogously, intravitreal VEGF injection induces retinal vascular changes that are quite similar to those seen in experimental diabetes, namely retinal leukostasis and the concomitant BRB breakdown [78], while blockade of VEGF abolishes retinal leukostasis and vascular leakage in experimentally induced diabetes [81, 83, 84].

Recent evidence shows that in addition to being the principle cytokine in growth and leakiness of neovascular membranes, VEGF also regulates RPE function [64]. The leading treatment of neovascular diseases is based on VEGF inhibition, using monoclonal antibody fragments. These anti-VEGF therapies are efficacious not only for reducing neovascularization but also for resolving retinal edema. However, recent evidence suggests that VEGF is required for normal retinal physiology, raising concerns about the long-term use of the VEGF inhibition strategy.

This motivated a search for endogenous antagonists of VEGF. A recent study revealed natriuretic peptides (NP), cyclic peptide hormones with diuretic, natriuretic, and vasodilatory properties, which antagonize not only choroidal neovascularization but also the breakdown of the outer BRB [85]. Understanding the role of endogenous antagonists of VEGF in the retinal barrier function will help to develop new strategies in the management of DR.

ANTI-VEGF PROPERTIES OF NATRIURETIC PEPTIDES

Inhibition of VEGF is currently under investigation in clinical trials, where retinal leakage and edema is a complication [86], such as DR [87], macular edema, [88], and retinal vein occlusion [89]. The rationale in these therapies is that removal of VEGF and the edematous fluid from the intraocular environment might be beneficial. However, VEGF has also protective properties for the retina [90], suggesting that VEGF is required for normal retinal physiology. This raises concerns about the long-term use of VEGF inhibition strategy. Furthermore, the simple removal of VEGF also eliminates the potential antiproliferative effects associated with VEGFR-1 activation [91], which might explain the lack of success in some cases.

Two endogenous anti-VEGF agents have been identified in the eye. Tombran-Tink et al. [92] reported the expression of pigment epithelium-derived factor (PEDF), produced and secreted by the RPE. PEDF was initially identified as a neurotrophic factor secreted by fetal human RPE cells, but later, vascular quiescence and permeability were also found to depend on the balance between VEGF and PEDF [93]. Molecules that interfere with the VEGF signaling pathways are attractive candidates for prevention of BRB breakdown. PEDF blocks the VEGF-induced TEER breakdown via the activation of juxtamembrane proteases to digest the VEGFR-2 receptor [64]. Thus, VEGF signaling is inhibited by limiting the available VEGFR-2 receptors. PEDF's anti-VEGF and antipermeability effects in the RPE could potentially be utilized to treat retinal vascular leakage or edema.

Another endogenous anti-VEGF factor in the eye is the atrial natriuretic peptide (ANP) [85]. Natriuretic peptides are cyclic peptide hormones with diuretic, natriuretic, and vasodilatory properties. The NP family consists of three members: atrial NP (ANP), brain NP (BNP), and C-type NP (CNP). The action of NPs is mediated through two types of receptors: guanylate cyclase type A, which reacts with ANP and BNP, and guanylate cyclase type B, which is CNP specific [94, 95]. Binding of NPs to these receptors results in cGMP production, which activates protein kinase G and subsequent target genes [96]. Although primarily produced by the cardiac atria, ANPs are used in the treatment of various disorders, including hypertension, renal insufficiency, and congestive heart failure. Interestingly, ANP is also expressed in the inner plexiform layer and RPE of the human retina [97].

Recent results indicate that ANP plays an important role in neovascular diseases of the eye, as it antagonizes not only neovascularization but also the breakdown of the outer BRB [85]. VEGF-A produces a significant TEER drop in the outer BRB within 2 h posttreatment. This response reaches its peak by 5 h and lasts approximately 48 h [98]. In the presence of ANP, however, TEER levels remain at baseline values by 2 h despite VEGF administration, showing the protective function of ANP in the outer BRB. Furthermore, the ANP response is polar, as only apical but not basolateral administration of ANP reverses apical VEGF response [85]. Isatin, a universal NP receptor antagonist, completely reverses the inhibitory effects of ANP with respect to the VEGF-induced TEER reduction, indicating that ANP receptor-mediated signaling is critical in this event. These data indicate that ANP acts by inhibiting VEGF signaling pathways in RPE cells. The recent linking of the expression of natriuretic peptides and the barrier function of the RPE and the retinal vessels might lead to new therapeutic strategies in reducing retinal edema. This is because natriuretic peptides are already in use in vascular disorders, and thus, detailed knowledge of their dosage and toxicity exists. However, future work will need to address the impact of these peptides on immune regulation and other aspects of DR development.

Proposed Model of BRB Breakdown in DR

A working model for how BRB breakdown might occur in early DR involves interaction of leukocytes via their β_2-integrins to the endothelial ICAM-1. The resulting release of the serine protease, AZ, from leukocytes causes an increase in BRB permeability (Fig. 5). This is backed by the fact that recombinant AZ injected intravitreally significantly increases

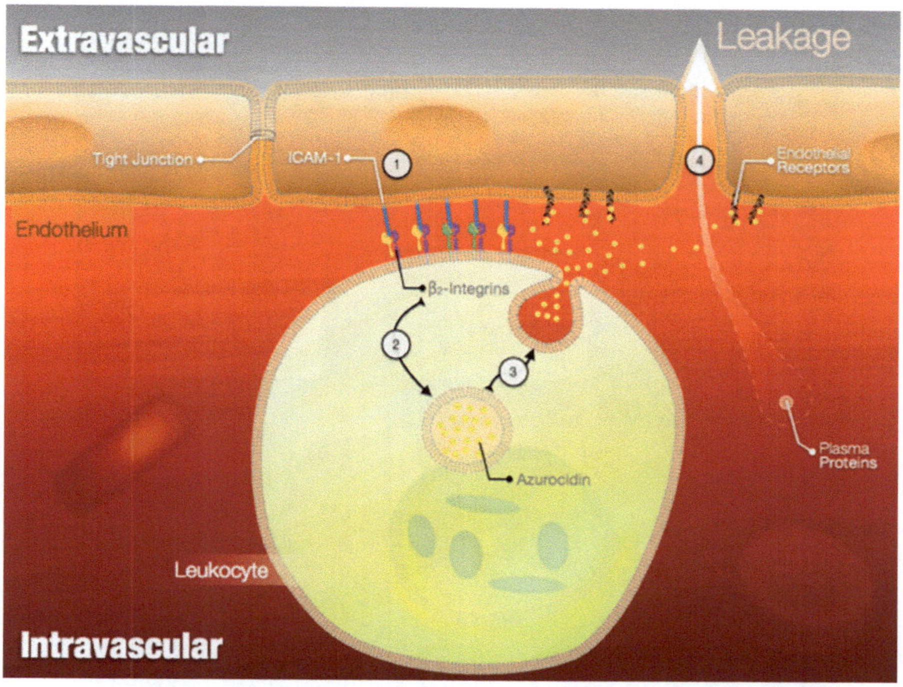

Fig. 5. Leukocyte-induced BRB permeability in early DR. β_2-integrin ligation with endothelial ICAM-1 (1) initiates signaling (2) that leads to release of AZ containing granulae (3). AZ binds to unidentified endothelial ligands and causes rapid opening of the BRB leading to leakage of plasma proteins (4).

BRB permeability as quantified by EB technique. AZ appears to be also an attractive target for controlling BRB permeability, as for instance systemic injection of aprotinin, a broad protease inhibitor, 1 h before the AZ injection completely blocks the increase in leakage. More striking is that the AZ-induced leakage is rather rapid, with a peak BRB leakage approximately 1 h after intravitreal AZ injection, suggesting a key role for AZ in diabetic BRB breakdown.

Key Role of AZ in VEGF-Induced Leakage

VEGF causes leukocyte accumulation in retinal vessels as well as protein leakage into the retinal parenchyma. Since VEGF is a key permeability factor in DR, the question arises, what portion of the VEGF-induced leakage is a direct effect of VEGF on the endothelium rather than through downstream mediators. Of course, the editors have the discretion to correct potential grammatical errors of the newly suggested sentence, however, the suggested sentence by the editors did not meet the intended scientific meaning. Whether AZ is a downstream mediator of VEGF's action is addressed by an experiment showing suppression of VEGF-induced retinal vascular leakage by AZ blockade. Intravitreal injection of VEGF together with systemic application of aprotinin completely prevents VEGF's permeability increase.

Interestingly, VEGF-induced leakage peaks around 6 h after its intravitreal injection [99]. In comparison, AZ-induced effect is more immediate, and its highest level is

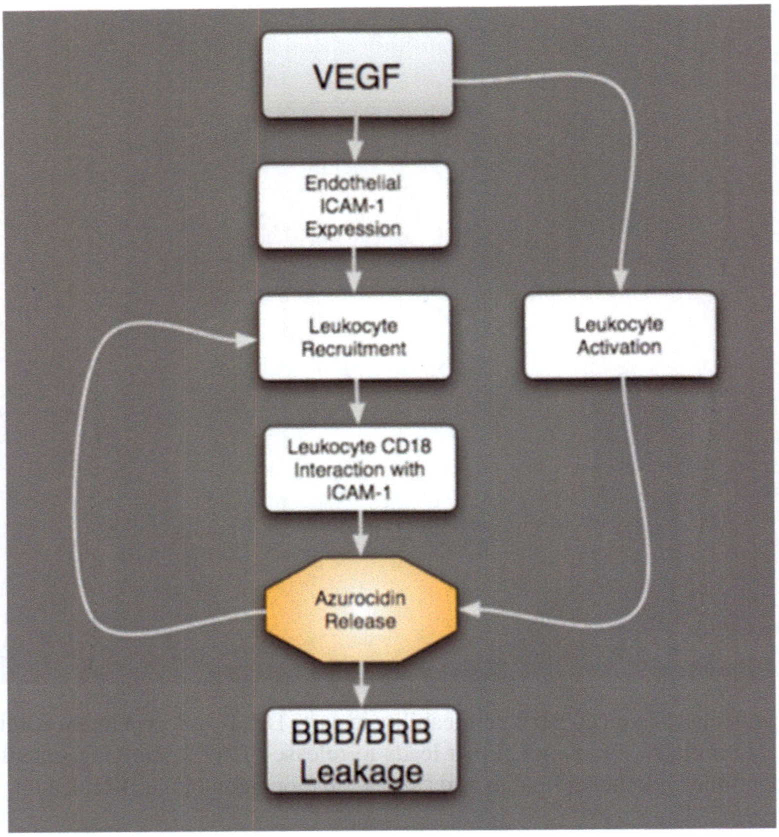

Fig. 6. Working model of VEGF-induced BRB leakage.

reached within the first hour after injection [100]. VEGF causes endothelial ICAM-1 upregulation as well as leukocyte activation [101]. The fact that AZ's effect on permeability is more rapid than that of VEGF and that leukocytes also respond to VEGF [102] makes it likely that that part of VEGF's impact on permeability in vivo is AZ mediated (Fig. 6).

How VEGF induces BRB leakage is not well understood. A novel link between VEGF and AZ suggests AZ to be a downstream effector of VEGF in causing vascular leakage:

- VEGF induces ICAM-1 expression on the endothelium of the BRB, resulting in the recruitment of leukocytes.
- Leukocyte CD18 interaction with ICAM-1 induces release of AZ.
- AZ interacts with unidentified endothelial receptors, causing the tight endothelial junctions of the BRB to open.
- AZ also acts as a chemotactic factor, recruiting additional leukocytes to the BRB, which potentiates the process.

Additionally, VEGF activates leukocytes directly, which could cause the release of AZ and thus result in amplified BRB leakage.

Azurocidin Inhibition Prevents Diabetic Retinal Vascular Leakage

Whether the newly discovered link between AZ and VEGF plays a role in DR was addressed in experiments with diabetic animals. By 2 weeks after diabetes induction with STZ, animals showed significant signs of DR, such as leukostasis and vascular leakage. However, when AZ is blocked in diabetic animals, vascular leakage is remarkably reduced and comparable to that of normal animals [100].

The results suggest that AZ plays a role in BRB breakdown induced by VEGF or in experimental diabetes. AZ release from neutrophils may be the final common pathway for a variety of upstream factors, which during DR promote neutrophil adhesion and cause BRB breakdown. These findings indicate that targeting AZ may prove beneficial in the treatment of retinal vascular leakage in experimental DR. However, the role of AZ in human DR remains to be investigated. Development of specific inhibitors of AZ might lead to a treatment option for DR.

Other ocular diseases, such as age-related macular degeneration (AMD), also have a vascular and inflammatory component, with vascular leakage being a common denominator. However, the mechanisms underlying the leakage in most cases are not understood. Better understanding of the pathogenesis of DR might not only help to find better therapies for this common and devastating disease but also for other important age-related and neurodegenerative diseases.

Vascular leakage is a critical component of DR and its management of paramount importance. VEGF is a key mediator of vascular leakage in DR, and its inhibition might become an effective strategy in reducing leakage. However, there are also concerns about long-term use of VEGF inhibitors, due to VEGF's importance in retinal health. Therefore, it is key to identify downstream mediators of VEGF that might be more specific in mediating the vascular leakage component of VEGF, while their inhibition would not affect the beneficial effects of VEGF. AZ is such a downstream mediator of VEGF-induced vascular leakage in DR. This raises the hope that retinal vascular leakage in DR could be opposed more specifically than before. However, as promising as the experimental data are, the contribution of AZ in human patients first would need to be confirmed. More importantly, more selective inhibitors of AZ need to be developed and their toxicity in humans tested.

ACKNOWLEDGMENTS

Rebecca C. Garland and Alexander Schering helped with the preparation of the manuscript and figures, respectively.

REFERENCES

1. Wang Y, Beydoun MA. The obesity epidemic in the United States—gender, age, socioeconomic, racial/ethnic, and geographic characteristics. Epidemiol Rev. 2007;29:6–28.
2. Reese TS, Karnovsky MJ. Fine structural localization of a blood-brain barrier to exogenous peroxidase. J Cell Biol. 1967;34:207–17.
3. Zlokovic BV. The blood-brain barrier in health and chronic neurodegenerative disorders. Neuron. 2008;57:178–201.
4. Marmor MF. Mechanisms of fluid accumulation in retinal edema. Doc Ophthalmol. 1999;97:239–49.

5. Janzer RC, Raff MC. Astrocytes induce blood-brain barrier properties in endothelial cells. Nature. 1987;325:253–7.
6. Fukushi S, Merola LO, Kinoshita JH. Altering the course of cataracts in diabetic rats. Invest Ophthalmol Vis Sci. 1980;19:313–5.
7. Schnedl WJ, Ferber S, Johnson JH, Newgard CB. STZ transport and cytotoxicity. Specific enhancement in GLUT2-expressing cells. Diabetes. 1994;43:1326–33.
8. Murata M, Takahashi A, Saito I, Kawanishi S. Site-specific DNA methylation and apoptosis: induction by diabetogenic streptozotocin. Biochem Pharmacol. 1999;57:881–7.
9. Noda K, Melhorn MI, Zandi S, Frimmel S, Tayyari F, Hisatomi T, et al. An animal model of spontaneous metabolic syndrome: Nile grass rat. FASEB J. 2010;24(7):2443–53.
10. Mooradian AD, Haas MJ, Chehade JM. Age-related changes in rat cerebral occludin and zonula occludens-1 (ZO-1). Mech Ageing Dev. 2003;124:143–6.
11. Hafezi-Moghadam A, Thomas KL, Wagner DD. ApoE deficiency leads to a progressive age-dependent blood-brain barrier leakage. Am J Physiol Cell Physiol. 2007;292:C1256–62.
12. Zhang SH, Reddick RL, Piedrahita JA, Maeda N. Spontaneous hypercholesterolemia and arterial lesions in mice lacking apolipoprotein E. Science. 1992;258:468–71.
13. Masliah E, Mallory M, Ge N, Alford M, Veinbergs I, Roses AD. Neurodegeneration in the central nervous system of apoE-deficient mice. Exp Neurol. 1995;136:107–22.
14. Kakinuma Y, Hama H, Sugiyama F, Yagami K, Goto K, Murakami K, et al. Impaired blood-brain barrier function in angiotensinogen-deficient mice. Nat Med. 1998;4:1078–80.
15. Yanai K, Saito T, Kakinuma Y, Kon Y, Hirota K, Taniguchi-Yanai K, et al. Renin-dependent cardiovascular functions and renin-independent blood-brain barrier functions revealed by renin-deficient mice. J Biol Chem. 2000;275:5–8.
16. Joussen AM, Poulaki V, Le ML, Koizumi K, Esser C, Janicki H, et al. A central role for inflammation in the pathogenesis of diabetic retinopathy. FASEB J. 2004;18:1450–2.
17. Kurose I, Anderson DC, Miyasaka M, Tamatani T, Paulson JC, Todd RF, et al. Molecular determinants of reperfusion-induced leukocyte adhesion and vascular protein leakage. Circ Res. 1994;74:336–43.
18. Del Maschio A, Zanetti A, Corada M, Rival Y, Ruco L, Lampugnani MG, et al. Polymorphonuclear leukocyte adhesion triggers the disorganization of endothelial cell-to-cell adherens junctions. J Cell Biol. 1996;135:497–510.
19. Bolton SJ, Anthony DC, Perry VH. Loss of the tight junction proteins occludin and zonula occludens-1 from cerebral vascular endothelium during neutrophil-induced blood-brain barrier breakdown in vivo. Neuroscience. 1998;86:1245–57.
20. Butcher EC. Leukocyte-endothelial cell recognition: three (or more) steps to specificity and diversity. Cell. 1991;67:1033–6.
21. Springer TA. Traffic signals for lymphocyte recirculation and leukocyte emigration: the multistep paradigm. Cell. 1994;76:301–14.
22. Lowe JB. Selectin ligands, leukocyte trafficking, and fucosyltransferase genes. Kidney Int. 1997;51:1418–26.
23. Miyamoto K, Khosrof S, Bursell SE, Moromizato Y, Aiello LP, Ogura Y, et al. Vascular endothelial growth factor (VEGF)-induced retinal vascular permeability is mediated by intercellular adhesion molecule-1 (ICAM-1). Am J Pathol. 2000;156:1733–9.
24. Barouch FC, Miyamoto K, Allport JR, Fujita K, Bursell SE, Aiello LP, et al. Integrin-mediated neutrophil adhesion and retinal leukostasis in diabetes. Invest Ophthalmol Vis Sci. 2000;41:1153–8.
25. Gautam N, Olofsson AM, Herwald H, Iversen LF, Lundgren-Akerlund E, Hedqvist P, et al. Heparin-binding protein (HBP/CAP37): a missing link in neutrophil-evoked alteration of vascular permeability. Nat Med. 2001;7:1123–7.

26. Pereira HA, Shafer WM, Pohl J, Martin LE, Spitznagel JK. CAP37, a human neutrophil-derived chemotactic factor with monocyte specific activity. J Clin Invest. 1990;85:1468–76.
27. Pereira HA. CAP37, a neutrophil-derived multifunctional inflammatory mediator. J Leukoc Biol. 1995;57:805–12.
28. Soehnlein O, Xie X, Ulbrich H, Kenne E, Rotzius P, Flodgaard H, et al. Neutrophil-derived heparin-binding protein (HBP/CAP37) deposited on endothelium enhances monocyte arrest under flow conditions. J Immunol. 2005;174:6399–405.
29. Petersen LC, Birktoft JJ, Flodgaard H. Binding of bovine pancreatic trypsin inhibitor to heparin binding protein/CAP37/azurocidin. Interaction between a Kunitz-type inhibitor and a proteolytically inactive serine proteinase homologue. Eur J Biochem. 1993;214:271–9.
30. Peters DC, Noble S. Aprotinin: an update of its pharmacology and therapeutic use in open heart surgery and coronary artery bypass surgery. Drugs. 1999;57:233–60.
31. Emanueli C, Salis MB, Van Linthout S, Meloni M, Desortes E, Silvestre JS, et al. Akt/protein kinase B and endothelial nitric oxide synthase mediate muscular neovascularization induced by tissue kallikrein gene transfer. Circulation. 2004;110:1638–44.
32. Engles L. Review and application of serine protease inhibition in coronary artery bypass graft surgery. Am J Health Syst Pharm. 2005;62:S9–14.
33. Guo XH, Zhao MH, Gao Y, Wang SF. Antineutrophil cytoplasmic antibody associated vasculitis induced by antithyroid agents. Zhonghua Yi Xue Za Zhi. 2003;83:932–5.
34. Karkouti K, Beattie WS, Dattilo KM, McCluskey SA, Ghannam M, Hamdy A, et al. A propensity score case-control comparison of aprotinin and tranexamic acid in high-transfusion-risk cardiac surgery. Transfusion. 2006;46:327–38.
35. Mangano DT, Tudor IC, Dietzel C. The risk associated with aprotinin in cardiac surgery. N Engl J Med. 2006;354:353–65.
36. Noda K, Nakao S, Zandi S, Engelstadter V, Mashima Y, Hafezi-Moghadam A. Vascular adhesion protein-1 regulates leukocyte transmigration rate in the retina during diabetes. Exp Eye Res. 2009;89:774–81.
37. Koskinen K, Vainio PJ, Smith DJ, Pihlavisto M, Yla-Herttuala S, Jalkanen S, et al. Granulocyte transmigration through the endothelium is regulated by the oxidase activity of vascular adhesion protein-1 (VAP-1). Blood. 2004;103:3388–95.
38. Salmi M, Jalkanen S. A 90-kilodalton endothelial cell molecule mediating lymphocyte binding in humans. Science. 1992;257:1407–9.
39. Akin E, Aversa J, Steere AC. Expression of adhesion molecules in synovia of patients with treatment-resistant lyme arthritis. Infect Immun. 2001;69:1774–80.
40. Jaakkola K, Jalkanen S, Kaunismaki K, Vanttinen E, Saukko P, Alanen K, et al. Vascular adhesion protein-1, intercellular adhesion molecule-1 and P-selectin mediate leukocyte binding to ischemic heart in humans. J Am Coll Cardiol. 2000;36:122–9.
41. Salmi M, Kalimo K, Jalkanen S. Induction and function of vascular adhesion protein-1 at sites of inflammation. J Exp Med. 1993;178:2255–60.
42. Singh B, Tschernig T, van Griensven M, Fieguth A, Pabst R. Expression of vascular adhesion protein-1 in normal and inflamed mice lungs and normal human lungs. Virchows Arch. 2003;442:491–5.
43. O'Sullivan J, Unzeta M, Healy J, O'Sullivan MI, Davey G, Tipton KF. Semicarbazide-sensitive amine oxidases: enzymes with quite a lot to do. Neurotoxicology. 2004;25:303–15.
44. Smith DJ, Salmi M, Bono P, Hellman J, Leu T, Jalkanen S. Cloning of vascular adhesion protein 1 reveals a novel multifunctional adhesion molecule. J Exp Med. 1998;188:17–27.
45. Yu PH, Wright S, Fan EH, Lun ZR, Gubisne-Harberle D. Physiological and pathological implications of semicarbazide-sensitive amine oxidase. Biochim Biophys Acta. 2003;1647:193–9.

46. Noda K, Miyahara S, Nakazawa T, Almulki L, Nakao S, Hisatomi T, et al. Inhibition of vascular adhesion protein-1 suppresses endotoxin-induced uveitis. FASEB J. 2008;22:1094–103.
47. Noda K, She H, Nakazawa T, Hisatomi T, Nakao S, Almulki L, et al. Vascular adhesion protein-1 blockade suppresses choroidal neovascularization. FASEB J. 2008;22:2928–35.
48. Almulki L, Noda K, Nakao S, Hisatomi T, Thomas KL, Hafezi-Moghadam A. Localization of vascular adhesion protein-1 (VAP-1) in the human eye. Exp Eye Res. 2010;90:26–32.
49. Miyahara S, Almulki L, Noda K, Nakazawa T, Hisatomi T, Nakao S, et al. In vivo imaging of endothelial injury in choriocapillaris during endotoxin-induced uveitis. FASEB J. 2008;22:1973–80.
50. Sun D, Nakao S, Xie F, Zandi S, Schering A, Hafezi-Moghadam A. Superior sensitivity of novel molecular imaging probe: simultaneously targeting two types of endothelial injury markers. FASEB J. 2010;24(5):1532–40.
51. Hafezi-Moghadam A, Thomas K, Prorock A, Huo Y, Ley K. L-selectin shedding regulates leukocyte recruitment. J Exp Med. 2001;193:863–72.
52. Kenyon BM, Voest EE, Chen CC, Flynn E, Folkman J, D'Amato RJ. A model of angiogenesis in the mouse cornea. Invest Ophthalmol Vis Sci. 1996;37:1625–32.
53. Shweiki D, Itin A, Soffer D, Keshet E. Vascular endothelial growth factor induced by hypoxia may mediate hypoxia-initiated angiogenesis. Nature. 1992;359:843–5.
54. Franke TF, Yang S-I, Chan TO, Datta K, Kazlauskas A, Morrison DK, et al. The protein kinase encoded by the Akt proto-oncogene is a target of the PDGF-activated phosphatidylinositol 3-kinase. Cell. 1995;81:727–36.
55. Borg SA, Kerry KE, Royds JA, Battersby RD, Jones TH. Correlation of VEGF production with IL1 alpha and IL6 secretion by human pituitary adenoma cells. Eur J Endocrinol. 2005;152:293–300.
56. Thurman JM, Renner B, Kunchithapautham K, Ferreira VP, Pangburn MK, Ablonczy Z, et al. Oxidative stress renders retinal pigment epithelial cells susceptible to complement-mediated injury. J Biol Chem. 2009;284:16939–47.
57. Novak CM, Parfitt DB, Sisk CL, Smale L. Associations between behavior, hormones, and Fos responses to novelty differ in pre- and post-pubertal grass rats. Physiol Behav. 2007;90:125–32.
58. Kvanta A. Expression and regulation of vascular endothelial growth factor in choroidal fibroblasts. Curr Eye Res. 1995;14:1015–20.
59. Marneros AG, Fan J, Yokoyama Y, Gerber HP, Ferrara N, Crouch RK, et al. Vascular endothelial growth factor expression in the retinal pigment epithelium is essential for choriocapillaris development and visual function. Am J Pathol. 2005;167:1451–9.
60. Thakker GD, Hajjar DP, Muller WA, Rosengart TK. The role of phosphatidylinositol 3-kinase in vascular endothelial growth factor signaling. J Biol Chem. 1999;274:10002–7.
61. Rousseau S, Houle F, Kotanides H, Witte L, Waltenberger J, Landry J, et al. Vascular endothelial growth factor (VEGF)-driven actin-based motility is mediated by VEGFR2 and requires concerted activation of stress-activated protein kinase 2 (SAPK2/p38) and geldanamycin-sensitive phosphorylation of focal adhesion kinase. J Biol Chem. 2000;275:10661–72.
62. Ilan N, Mahooti S, Madri JA. Distinct signal transduction pathways are utilized during the tube formation and survival phases of in vitro angiogenesis. J Cell Sci. 1998;111(Pt 24):3621–31.
63. Zachary I, Gliki G. Signaling transduction mechanisms mediating biological actions of the vascular endothelial growth factor family. Cardiovasc Res. 2001;49:568–81.
64. Ablonczy Z, Prakasam A, Fant J, Fauq A, Crosson C, Sambamurti K. Pigment epithelium-derived factor maintains retinal pigment epithelium function by inhibiting vascular endothelial growth factor-R2 signaling through gamma-secretase. J Biol Chem. 2009;284:30177–86.

65. Ablonczy Z, Crosson CE. VEGF modulation of retinal pigment epithelium resistance. Exp Eye Res. 2007;85:762–71.
66. Lohela M, Bry M, Tammela T, Alitalo K. VEGFs and receptors involved in angiogenesis versus lymphangiogenesis. Curr Opin Cell Biol. 2009;21:154–65.
67. Shibuya M, Claesson-Welsh L. Signal transduction by VEGF receptors in regulation of angiogenesis and lymphangiogenesis. Exp Cell Res. 2006;312:549–60.
68. Joukov V, Pajusola K, Kaipainen A, Chilov D, Lahtinen I, Kukk E, et al. A novel vascular endothelial growth factor, VEGF-C, is a ligand for the Flt4 (VEGFR-3) and KDR (VEGFR-2) receptor tyrosine kinases. EMBO J. 1996;15:290–8.
69. Achen MG, Jeltsch M, Kukk E, Makinen T, Vitali A, Wilks AF, et al. Vascular endothelial growth factor D (VEGF-D) is a ligand for the tyrosine kinases VEGF receptor 2 (Flk1) and VEGF receptor 3 (Flt4). Proc Natl Acad Sci U S A. 1998;95:548–53.
70. Dvorak HF, Brown LF, Detmar M, Dvorak AM. Vascular permeability factor/vascular endothelial growth factor, microvascular hyperpermeability, and angiogenesis. Am J Pathol. 1995;146:1029–39.
71. Nagy JA, Vasile E, Feng D, Sundberg C, Brown LF, Detmar MJ, et al. Vascular permeability factor/vascular endothelial growth factor induces lymphangiogenesis as well as angiogenesis. J Exp Med. 2002;196:1497–506.
72. Ferrara N. Vascular endothelial growth factor. Arterioscler Thromb Vasc Biol. 2009;29: 789–91.
73. Hirakawa S, Hong YK, Harvey N, Schacht V, Matsuda K, Libermann T, et al. Identification of vascular lineage-specific genes by transcriptional profiling of isolated blood vascular and lymphatic endothelial cells. Am J Pathol. 2003;162:575–86.
74. Joukov V, Sorsa T, Kumar V, Jeltsch M, Claesson-Welsh L, Cao Y, et al. Proteolytic processing regulates receptor specificity and activity of VEGF-C. EMBO J. 1997;16:3898–911.
75. Cao W, Henry MD, Borrow P, Yamada H, Elder JH, Ravkov EV, et al. Identification of a-dystroglycan as a receptor for lymphocytic choriomeningitis virus and lassa fever virus. Science. 1998;282:2079–81.
76. Albuquerque RJ, Hayashi T, Cho WG, Kleinman ME, Dridi S, Takeda A, et al. Alternatively spliced vascular endothelial growth factor receptor-2 is an essential endogenous inhibitor of lymphatic vessel growth. Nat Med. 2009;15:1023–30.
77. Kriehuber E, Breiteneder-Geleff S, Groeger M, Soleiman A, Schoppmann SF, Stingl G, et al. Isolation and characterization of dermal lymphatic and blood endothelial cells reveal stable and functionally specialized cell lineages. J Exp Med. 2001;194:797–808.
78. Ishida S, Usui T, Yamashiro K, Kaji Y, Amano S, Ogura Y, et al. VEGF164-mediated inflammation is required for pathological, but not physiological, ischemia-induced retinal neovascularization. J Exp Med. 2003;198:483–9.
79. Carmeliet P. Angiogenesis in health and disease. Nat Med. 2003;9:653–60.
80. Lu M, Perez VL, Ma N, Miyamoto K, Peng HB, Liao JK, et al. VEGF increases retinal vascular ICAM-1 expression in vivo. Invest Ophthalmol Vis Sci. 1999;40:1808–12.
81. Qaum T, Xu Q, Joussen AM, Clemens MW, Qin W, Miyamoto K, et al. VEGF-initiated blood-retinal barrier breakdown in early diabetes. Invest Ophthalmol Vis Sci. 2001;42:2408–13.
82. Miyamoto K, Khosrof S, Bursell SE, Rohan R, Murata T, Clermont AC, et al. Prevention of leukostasis and vascular leakage in streptozotocin-induced diabetic retinopathy via intercellular adhesion molecule-1 inhibition. Proc Natl Acad Sci U S A. 1999;96:10836–41.
83. Ishida S, Usui T, Yamashiro K, Kaji Y, Ahmed E, Carrasquillo KG, et al. VEGF164 is proinflammatory in the diabetic retina. Invest Ophthalmol Vis Sci. 2003;44:2155–62.
84. Ambati BK, Joussen AM, Ambati J, Moromizato Y, Guha C, Javaherian K, et al. Angiostatin inhibits and regresses corneal neovascularization. Arch Ophthalmol. 2002;120:1063–8.

85. Lara-Castillo N, Zandi S, Nakao S, Ito Y, Noda K, She H, et al. Atrial natriuretic peptide reduces vascular leakage and choroidal neovascularization. Am J Pathol. 2009;175:2343–50.
86. Ciulla TA, Rosenfeld PJ. Anti-vascular endothelial growth factor therapy for neovascular ocular diseases other than age-related macular degeneration. Curr Opin Ophthalmol. 2009;20:166–74.
87. Adamis AP, Altaweel M, Bressler NM, Cunningham Jr ET, Davis MD, Goldbaum M, et al. Changes in retinal neovascularization after pegaptanib (Macugen) therapy in diabetic individuals. Ophthalmology. 2006;113:23–8.
88. Cunningham Jr ET, Adamis AP, Altaweel M, Aiello LP, Bressler NM, D'Amico DJ, et al. A phase II randomized double-masked trial of pegaptanib, an anti-vascular endothelial growth factor aptamer, for diabetic macular edema. Ophthalmology. 2005;112:1747–57.
89. Stahl A, Agostini H, Hansen LL, Feltgen N. Bevacizumab in retinal vein occlusion-results of a prospective case series. Graefes Arch Clin Exp Ophthalmol. 2007;245:1429–36.
90. Nakazawa T, Takahashi H, Nishijima K, Shimura M, Fuse N, Tamai M, et al. Pitavastatin prevents NMDA-induced retinal ganglion cell death by suppressing leukocyte recruitment. J Neurochem. 2007;100:1018–31.
91. Nozaki M, Sakurai E, Raisler BJ, Baffi JZ, Witta J, Ogura Y, et al. Loss of SPARC-mediated VEGFR-1 suppression after injury reveals a novel antiangiogenic activity of VEGF-A. J Clin Invest. 2006;116:422–9.
92. Tombran-Tink J, Chader GG, Johnson LV. PEDF: a pigment epithelium-derived factor with potent neuronal differentiative activity. Exp Eye Res. 1991;53:411–4.
93. Tombran-Tink J. The neuroprotective and angiogenesis inhibitory serpin, PEDF: new insights into phylogeny, function, and signaling. Front Biosci. 2005;10:2131–49.
94. Chinkers M, Garbers DL, Chang MS, Lowe DG, Chin HM, Goeddel DV, et al. A membrane form of guanylate cyclase is an atrial natriuretic peptide receptor. Nature. 1989;338:78–83.
95. Chang MS, Lowe DG, Lewis M, Hellmiss R, Chen E, Goeddel DV. Differential activation by atrial and brain natriuretic peptides of two different receptor guanylate cyclases. Nature. 1989;341:68–72.
96. Levin ER, Gardner DG, Samson WK. Natriuretic peptides. N Engl J Med. 1998;339:321–8.
97. Rollin R, Mediero A, Roldan-Pallares M, Fernandez-Cruz A, Fernandez-Durango R. Natriuretic peptide system in the human retina. Mol Vis. 2004;10:15–22.
98. Ablonczy Z, Liu Y, Crosson C. VEGF-induced barrier breakdown in fetal human RPE cells and ARPE-19 cells. Invest Ophthalmol Vis Sci. 2010, Submitted.
99. Derevjanik NL, Vinores SA, Xiao WH, Mori K, Turon T, Hudish T, et al. Quantitative assessment of the integrity of the blood-retinal barrier in mice. Invest Ophthalmol Vis Sci. 2002;43:2462–7.
100. Skondra D, Noda K, Almulki L, Tayyari F, Frimmel S, Nakazawa T, et al. Characterization of azurocidin as a permeability factor in the retina: involvement in VEGF-induced and early diabetic blood-retinal barrier breakdown. Invest Ophthalmol Vis Sci. 2008;49:726–31.
101. Proescholdt MA, Heiss JD, Walbridge S, Muhlhauser J, Capogrossi MC, Oldfield EH, et al. Vascular endothelial growth factor (VEGF) modulates vascular permeability and inflammation in rat brain. J Neuropathol Exp Neurol. 1999;58:613–27.
102. Ferrara N, Gerber HP, LeCouter J. The biology of VEGF and its receptors. Nat Med. 2003;9:669–76.
103. Ruoslahti E, Pierschbacher MD. New perspectives in cell adhesion: RGD and integrins. Science. 1987;238:491–7.

8
Molecular Regulation of Endothelial Cell Tight Junctions and the Blood-Retinal Barrier

E. Aaron Runkle, Paul M. Titchenell, and David A. Antonetti

CONTENTS

Keywords Pericytes • Protein kinase C • Retinal pigment epithelium • Tight junction proteins • VEGF

THE BLOOD-RETINAL BARRIER

The neural retina requires metabolic support supplied by the vasculature; however, retinal function demands that these vessels yield minimal impact on light transmission. This metabolic support is provided by two independent vascular systems: the retinal and the choroidal [1–3]. The choroidal vessels include a dense, highly permeable capillary network that supports the outer retina, including the rods and cones. The BRB is maintained by a well-developed junctional complex in the retinal pigment epithelium (RPE) that controls the flux of fluid and solutes to the retina from the choroidal capillary plexus. Diffusion of metabolites and gasses across the RPE from the choroid supports the highly active outer retina. Meanwhile, the RPE controls retinal fluid by active transport of chloride followed by osmotic flow of water through aquaporins, a system regulated by lactate production in the outer retina [4]. This transcellular transport system requires the formation of the tight junction complex between RPE cells to maintain defined environments in the apical and basolateral compartments.

From: *Ophthalmology Research: Visual Dysfunction in Diabetes*
Edited by: J. Tombran-Tink et al. (eds.), DOI 10.1007/978-1-60761-150-9_8

The Retinal Vascular Barrier

The inner retina, including the ganglion cell layer, is supported by the retinal vascular system that emanates from the central retinal artery in the optic nerve and radiates to the four retinal quadrants [1, 3]. These four branches form three capillary plexuses, one in the nerve fiber layer and ganglion cell layer and two outer capillary plexuses that border the inner nuclear layer termed the shallow inner nuclear layer and deep inner nuclear layer capillary beds [5]. The inner capillary bed that resides within the nerve fiber layer and ganglion cell layer can separate to two additional beds or appear as one capillary bed. The arterioles and venules are restricted to the ganglion cell layer, and nerve fiber layer with only capillaries developed from angiogenesis, extending deeper into the retina. Primates have an avascular region known as the macula that includes the fovea, which is highly enriched with cones necessary for the high-contrast central vision [6].

The BRB controls the flux of blood-borne solutes and fluid into the retina and maintains the proper retinal environment for normal neural conduction. Low number of vesicles and fenestrae, expression of multidrug-resistance genes, and well-developed junctional complex in both the retinal vasculature and RPE combine to provide the necessary defined neural environment for proper retinal function.

Multiple cell types in the retina contribute to endothelial junctional complex formation and regulation. Investigators have demonstrated the ability of glial cells to induce vascular barrier properties in a variety of systems for both brain and retinal glia. Both astrocytes [7] and Müller cells [8] are capable of inducing barrier properties in endothelial cells, and injection of astrocytes or Müller cells into the anterior chamber of the rat eye leads to vascularization and formation of vessels with elevated barrier properties. Conversely, transplanting the avascular neural tissue of stage 13 quail brains into the coelomic cavity of 3-day chick embryos caused the invading capillaries to take on blood-brain barrier characteristics including reduced permeability to circulating dye [9].

Recent studies have identified a signal transduction adaptor molecule that promotes production of probarrier factors from astrocytes. Src-suppressed C kinase substrate or SSECKS in rodents, also termed gravin in humans, or AKAP12, coordinates signal transduction pathways by binding and organizing signaling molecules such as protein kinase C, protein kinase A, calmodulin, cyclins, and β (beta)-adrenergic receptors. In brain, SSECKS colocalizes with GFAP, indicating a glial expression pattern, and a recent report demonstrated that expression of SSECKS contributes to astrocytic induction of the blood-brain barrier [10]. Overexpression of SSECKS reduces expression of vascular endothelial growth factor (VEGF) apparently through a reduction of c-Jun and AP1 signaling and promotes angiopoietin 1 production.

In addition to astrocytes, pericytes also contribute to barrier formation by secreting an angiopoietin 1 complex, which induces occludin expression [11]. Angiopoietin 1 is a ligand for the Tie2 receptor and both stabilizes blood vessels and protects them from VEGF-induced permeability [12, 13]. Together, these studies demonstrate that glia and pericytes contribute an important role in the induction of the blood-brain and BRB. Indeed, coculture of astrocytes and pericytes with endothelial cells induces barrier properties to a greater extent than either cell type alone [14]. An understanding of the molecular mechanisms by which this differentiation proceeds is only beginning to be elucidated.

THE JUNCTIONAL COMPLEX

Formation of a well-developed tight junction creates the barrier across both the retinal vasculature and RPE as shown by electron microscopy studies. Horseradish peroxidase used as a stain in electron microscopy diffuses only to the tight junction in brain cortical capillaries, while in other tissues without tight junctions, this marker diffuses out of the vascular lumen [15]. Similar studies in the retina with dyes reveal that tight junctions mediate the BRB, preventing solute flux into the retinal parenchyma [16, 17]; however, the adherens junctions are essential to development of the barrier, and likely influence the formation of the tight junction [18–21]. Further, in blood-brain and BRBs, the tight junctions and adherens junctions are indistinguishable at the ultrastructural level [22, 23].

Tight junctions are composed of over 40 proteins encompassing transmembrane proteins, intracellular scaffolding proteins, and signaling proteins, acting in concert to influence barrier properties [24]. The transmembrane proteins include occludin, claudin family members, tricellulin, and the junctional adhesion molecules (JAMs). The transmembrane proteins are linked to the cytoskeleton via an interaction with the scaffolding protein family zonula occludens (ZO). Together, these proteins create a barrier to paracellular flux and contribute to the BRB.

ZO Proteins

The zonula occludens, or ZO, family members bind to both transmembrane structural proteins and regulatory proteins and organize the junctional complex. ZO-1 (210–225 kDa) was the first tight junction protein identified, and subsequent studies using coimmunoprecipitation identified the other ZO family members, ZO-2 (180 kDa) and ZO-3 (130 kDa) [25–29]. ZO-1 and ZO-2 also associate with the adherens junctions [30] potentially as a first step in formation of tight junctions. ZO proteins are members of the membrane-associated guanylate kinase (MAGUK) family and are characterized by the presence of three PDZ domains, one SH3 domain, and a GUK domain [31]. ZO family members are also characterized by the presence of an acidic domain, a basic domain, a leucine zipper, and a proline-rich C-terminus [25, 27, 32, 33].

The contribution of ZO-1 in junctional protein organization has been demonstrated in cell culture and gene deletion studies. The calcium switch assay allows rapid disassembly of tight junctions followed by reassembly upon return of calcium to the medium. The use of siRNA to reduce ZO-1 expression results in reduced tight junction assembly in the calcium switch assay [34, 35]. ZO-2 also contributes to junction assembly and permeability as demonstrated by ZO-2 silencing which leads to a reduction in TER values in the calcium switch assay, without affecting mature TJs and increased permeability to 70 kDa dextran [36]. Deletion of ZO-1 and ZO-2 in a cell line lacking ZO-3 led to a complete loss of tight junction formation [37]. In vivo, ZO-1 [38] and ZO-2 [39] gene deletions have been described and both are lethal very early in mouse embryogenesis. However, distinct phenotypes suggest nonredundant function for these isoforms. ZO-1 gene deletion caused developmental defects in mouse embryo, yolk sac, and allantoic membrane vasculature, suggesting a role for ZO-1 in angiogenesis [38]. Interestingly, ZO-3 deletion does not impart lethality [39].

Claudins

The claudin family consists of 24 distinct proteins that form the tight junction seal between neighboring cells particularly regulating ion flux. The claudins are 20–27 kDa and possess four membrane-spanning domains with two extracellular loops and the N- and C-terminus in the cytoplasm [40–42]. The C-terminus of claudins is essential to both their stability and their membrane targeting [43, 44]. Of important note, all claudins possess a YV sequence as the final two amino acids that is necessary for their interaction with ZO [45, 46].

Claudin family expression patterns vary from tissue to tissue, and expression of different claudins confers specificity of barrier properties. In Madin Darby canine kidney (MDCK) cells, claudin-1 overexpression increases TER values by fourfold concurrent with a decrease in permeability to small and large molecules (4 and 40 kDa FITC-dextrans) [47]. Claudins not only increase barrier properties but can also form charge-selective paracellular ion channels. For example, claudin-16 controls magnesium flux in the loop of Henle in the kidney and genetic defects of this claudin are associated with loss of magnesium [48]. Mutations of claudin-16 alter the sodium flux reducing magnesium transport potential [49]. A role for claudins in creating a charge specific barrier was demonstrated by mutational analysis. Exchanging two acidic residues in the first extracellular loop, Asp55 and Glu64, to create basic residues (D55R and E64K) in the extracellular loop of claudin-15 changes charge selectivity for paracellular permeability from sodium to chloride [50].Finally, siRNA studies altering expression of claudins-2, 4, and 7 can differentially alter cation or anion permeability [51]. Together, these studies demonstrate claudin expression, provide specific ionic barriers, and provide charge-selective paracellular channels.

A model for claudin barrier formation has recently been proposed based on transfection studies and is distinct from many of the schematics of tight junctions previously presented. Overexpression of claudin-5 in human embryonic kidney (HEK) cells, a cell type that typically does not express tight junctions, leads to formation of strands of tight junctions in the plasma membrane [52]. The investigators used mutational analysis to distinguish *trans*-interactions or interactions between claudins on adjacent cells as opposed to *cis*-interactions or interactions between claudins within a cell. By expressing fluorescent-tagged claudin-5 and performing a combination of live cell imaging, fluorescence resonance energy transfer, and scanning electron microscopy (SEM), the investigators were able to identify specific amino acids in the second extracellular loop responsible for *trans*-interactions. The position of these amino acids combined with SEM images led the investigators to propose a model in which 2 claudins first form a dimer within a membrane (*cis*-interaction). This dimer then interacts by loop 2 interactions to another dimer pair across the membrane (*trans*-interaction). This model of claudin interaction literally forms a zipper with the dimer pair of claudins interdigitating to create the tight junction seal (Fig. 1).

In the RPE, the expression of claudins-1, 2, and 5 have been detected in the developing chick embryo by embryonic day 14 [53]. Further, claudins-1, 5, and 15 are expressed in endothelial cells [45], and claudin-5 is expressed in the retinal vasculature [54]. Several studies have examined the effect of loss of claudins on barrier properties. Claudin-1 deletion in mice is lethal within 1 day postbirth as a result of excessive water loss through

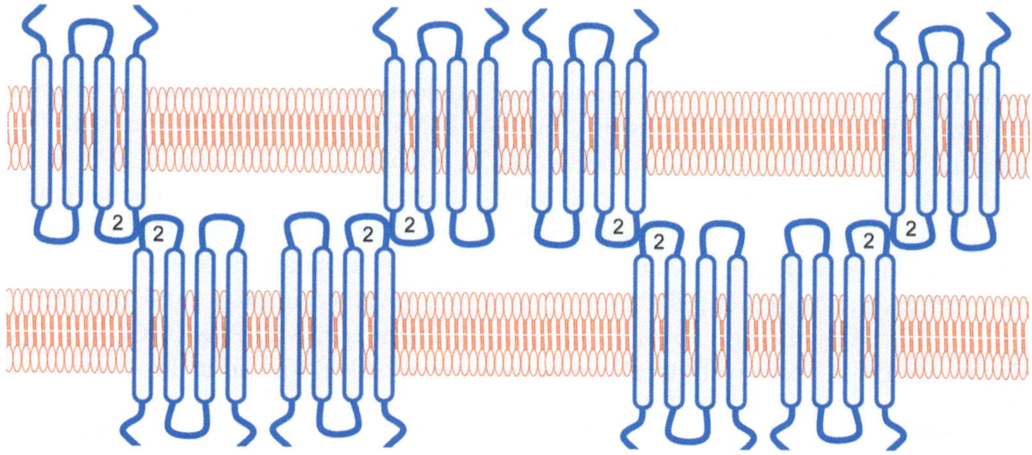

Fig. 1. Proposed interactions between claudins at the tight junction. The claudin proteins form a zipper-like structure at the tight junction by alternating *cis*- and *trans*-interactions. Claudin proteins within the same cell form a *cis*-interaction forming a dimer pair. This dimer of claudins interacts with another dimer pair in adjacent cells through loop 2 interactions forming a *trans*-interaction between the two pair. Adopted from Piontek et al. [52].

the skin [55]. A claudin-5 knockout mouse was created, which developed normally and upon birth appeared grossly normal. However, the mice died within 10 h after birth due to increased permeability of small molecules (<800 Da) across the blood-brain barrier [56]. Finally, evidence for claudin-11 in the neural tissue is found from gene deletion studies. Mice deficient for claudin-11 showed severe neurological disorders and male sterility as a result of loss of tight junctions within the CNS and Sertoli cells [57]. Unfortunately, there is little information regarding specific changes to claudins in diabetic retinopathy. A study of mRNA content of claudins-1 and 5 reveal claudin-1 mRNA first increases at 6 weeks then decreases by 12 weeks postinduction of diabetes, while claudin 5 mRNA is decreased modestly at both time points [58]. Collectively, these studies indicate that claudins are essential for tight junction function, creating charge specific barriers while providing ion selective paracellular channels across the barrier. The regulation of claudin function in diabetic retinopathy remains an area for further research.

Junctional Adhesion Molecules

The junctional adhesion molecules, or JAMs, are glycosylated single-pass transmembrane proteins, with the C-terminus located intracellularly, and an extracellular N-terminus with two immunoglobulin (Ig)-like domains [42]. The JAMs are subdivided into two groups based on sequence homology [59, 60]. The first subgroup, composed of JAM-A, JAM-B, and JAM-C, directly interacts with ZO-1 and PAR3, a protein required for cell polarity, through a C-terminal class II PDZ-binding domain motif [61–63]. The second subgroup, which is composed of coxsackie and adenovirus receptor (CAR), JAM-4, and endothelial-cell-selective adhesion molecule (ESAM) contains a class I PDZ-binding domain motif [42]. CAR and JAM-4 bind with the Ligand-of-Numb protein X1 by this PDZ-binding domain [64, 65], while JAM-4 and ESAM interact with the MAGUK protein [66, 67]. JAMs are also able to form homodimers and heterodimers through the extracellular domains. Specifically, JAM-A, JAM-B, and JAM-C interact with the integrins aLb2, a4b1, and aMb2, respectively [59, 60, 68].

JAM-A is necessary for junction resealing in both epithelial and endothelial cells. Specifically, studies demonstrate that monoclonal antibodies against JAM-A significantly inhibit junction recovery in a calcium switch assay as measured by transepithelial electrical resistance (TER) [69–71]. JAM-A also is involved in proper polarity maintenance [72] likely through its direct and specific interaction with PAR-3 [61, 62, 73]. Finally, ESAM is exclusively localized to endothelial cells [74], and its loss augments VEGF-induced permeability [75].

Occludin and Tricellulin

Occludin was the first transmembrane TJ protein discovered and is a 522-amino-acid protein of 55.9 kDa, and like the claudins, has four transmembrane domains [76]. However, the sequence and structure of occludin is sufficiently distinct from claudins, suggesting a unique role of this protein in tight junctions. A second gene product, tricellulin or MARVEL D2, has recently been identified as a 555-amino acid protein localized specifically at regions where three cells make contact [76]. Tricellulin and occludin share homology in the MARVEL domain across the tetra-transmembrane regions. *MA*L and *r*elated proteins for *ve*sicle trafficking and membrane *l*ink or MARVEL domains are present in vesicle transport proteins such as MAL, which are essential for apical trafficking of membrane and secretory proteins in epithelia and also in the neural vesicle proteins synaptophysin and synaptogyrin [77]. In epithelial cells, occludin is enriched at bicellular junctions, while tricellulin is enriched at tricellular junctions; however, upon knockdown of occludin, tricellulin can be observed at bicellular junctions, which suggests occludin normally restricts tricellulin localization [78]. Little is known about tricellulin in endothelial cells, so this chapter's primary focus will be on occludin.

Occludin content at the TJ correlates with barrier properties such that occludin is higher in cells with a tighter barrier, such as arterial endothelial cells and brain and retinal endothelium, and lower in cells known to have a more permeable barrier, such as venous endothelial cells and endothelial cells of nonneuronal tissues [79, 80] (Fig. 2). However, occludin does not provide a structural barrier for the tight junctions as do the claudins. Occludin knockout mice are viable and appear to form TJs but exhibit a number of abnormalities, including postnatal growth retardation, abnormalities in the testis leading to male infertility, and inability of females to suckle their young. Additionally, salivary gland abnormalities, thinning of compact bone, brain calcium deposits, chronic gastritis, and hyperplasia of the gastric epithelium are all a consequence of occludin gene deletion in mice [81, 82]. While a large number of additional studies including siRNA and overexpression studies suggest occludin contributes to barrier properties in a host of cell types (reviewed in ref. 2), the role of occludin in the barrier has remained difficult to understand. Recent studies suggest that occludin might not provide a direct structural component to the tight junction complex but rather act as a regulator of barrier properties.

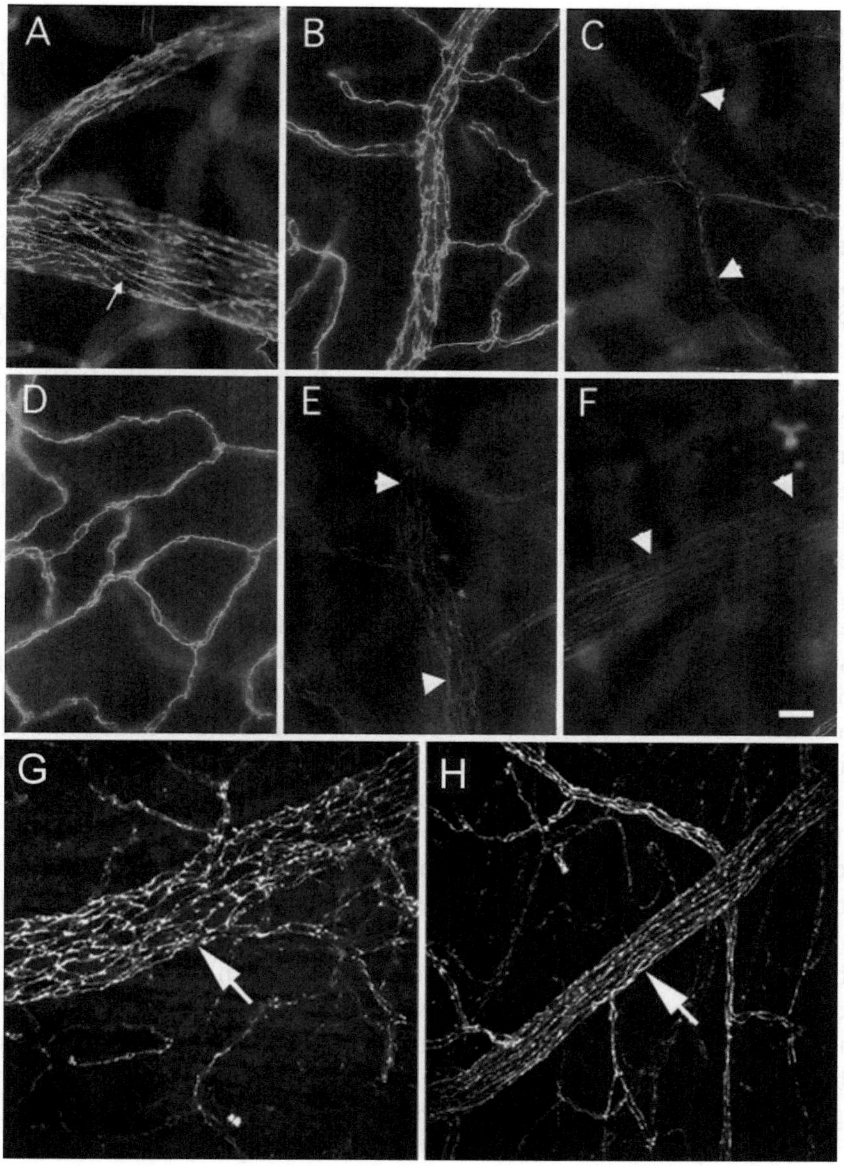

Fig. 2. Localization of occludin and claudin-5 in the vasculature of rat retina. Whole rat retinas were dissected and labeled with antibodies directed against occludin and claudin-5 for observation by confocal microscopy. These images show that occludin is differentially distributed in the blood vessels of the normal rat retina. (**A**) Occludin immunoreactivity is intense in the cell borders of main arterioles, and also can be detected as punctate immunoreactivity within cells (*arrow*). (**B**) The cell borders of smaller arterioles are also immunoreactive for occludin. (**C**) Occludin immunoreactivity in the capillaries of the inner retina (*arrowheads*) is less than that of the arterioles. (**D**) Occludin immunoreactivity of the capillaries of the outer plexiform layer is as intense as that of the arterioles. (**E**) Occludin immunoreactivity of the postcapillary venules (*arrowheads*) of the inner retina is diminished. (**F**) Immunoreactivity of the main venules (*arrowheads*) is further reduced as they approach the optic disk (*right*). In contrast, claudin-5 immunoreactivity is evenly distributed in the blood vessels of the rat retina as shown by its expression in the arteriole (**G**, *arrow*) and venule (**H**, *arrow*). Images taken from Barber and Antonetti [54].

VASCULAR PERMEABILITY IN DIABETIC RETINOPATHY

The cause of visual loss in diabetic retinopathy remains unclear but likely involves loss of proper cellular interaction between the neural retina and retinal vasculature [83]. Changes in blood vessel permeability and macular edema consistently rank as the top clinical correlates associated with loss of vision [84, 85]. Indeed, central macular thickness, as measured by optical coherence tomography, and fluorescein leakage combined with age account for 33% of the variation in visual acuity [85]. Further, the location, severity, and duration of macular edema are all linked to visual loss [86]. Alterations to the BRB are believed to contribute to retinal macular edema with increased fluorescein permeability related to the progression of macular edema [87, 88]. Collectively, these clinical studies demonstrate a strong correlation with alterations to the BRB, increased macular edema, and loss of vision in patients with diabetes. It should also be noted, however, that other factors clearly contribute to vision loss in diabetes.

Vascular changes in diabetic retinopathy are due, at least in part, to elevated VEGF expression [89–94]. Indeed, recent clinical studies using anti-VEGF antibody therapy improved visual acuity in combination with laser compared to laser treatment alone [95]. In addition to VEGF, other cytokines likely also contribute to vascular changes in diabetic retinopathy. Increased levels of interleukin-1 beta (IL-1β (beta)) and tumor necrosis factor-alpha (TNF-α (alpha)) are increased in the vitreous of diabetic patients with proliferative diabetic retinopathy [96, 97] and in diabetic rat retinas [98–100], while leukostasis has been observed in response to elevated intracellular adhesion molecule-1 expression in diabetic rodents [101]. Furthermore, proteomic analysis of vitreous from patients with diabetic retinopathy reveals increased carbonic anhydrase I likely as a result of retinal hemorrhage and erythrocyte lysis [102]. The pH increase driven by carbonic anhydrase drives kallikrein activation leading to bradykinin production and permeability of the retinal vasculature as demonstrated by carbonic anhydrase I intravitreal injection. Therefore, multiple factors contribute to the increased retinal vascular permeability in diabetic retinopathy. Changes in both growth factors and inflammatory cytokines may induce alterations in the vascular barrier properties by distinct mechanisms over the course of diabetes. Thus, understanding the mechanisms of vascular permeability in diabetic retinopathy will allow the development of rationale therapies targeting specific disease characteristics or potentially identifying common mechanisms shared by the variety of cytokines altered in diabetic retinopathy.

VEGF-Induced Regulation of Endothelial Permeability

Both VEGF treatment of endothelial cells and induction of diabetes alter occludin content and localization associated with alterations in barrier properties. Studies on rats with streptozotocin-induced diabetes with 3-month duration reveal decreased occludin content and immunostaining at cell borders concomitant with increased BRB permeability. This change in occludin content can be recapitulated in bovine retinal endothelial cells (BREC) treated with VEGF [103]. Immunohistochemical analysis of occludin in diabetes or after addition of VEGF demonstrates that occludin localization at the cell border changes specifically at regions of paracellular permeability [54]. In this study, fluorescently labeled concanavalin A was perfused through control and diabetic or control and VEGF-treated retinas that were fixed to prevent active transport and preserve protein localization. Concanavalin A does not bind endothelial cells directly but decorates regions where pores have formed that allow transport of the lectin to the endothelial basement membrane. Immunohistochemical analysis revealed that concanavalin A stained the basement membrane specifically at regions of low or absent occludin border staining, suggesting that redistribution of occludin away from the cell border created regions of paracellular permeability. Likewise, treatment of RPE cells with hepatocyte growth factor (HGF) reduced tight junctions, decreased TER, and increased diffusion of fluorescently labeled marker from the apical to basolateral membrane. After 6 h of HGF treatment, occludin, claudin-1, and α-catenin were redistributed from the membrane to the cytoplasm, and ZO-1 immunostaining was reduced [104]. Together, these studies demonstrate that changes in occludin are associated with altered permeability in the retina and suggest that occludin contributes to regulation of paracellular permeability in retinal endothelial cells.

Occludin Phosphorylation and Permeability

While gene deletion and knockdown of occudin expression reveal occludin is not necessary for formation of tight junctions, the observed changes in occludin content and localization associated with changes in barrier properties suggest occludin contributes to regulation of barrier properties. Recent studies suggest phosphorylation of occludin acts as a molecular switch to regulate endothelial barrier properties. Treatment of endothelial cells with VEGF [105, 106], cytokines [107], oxidized phospholipids [108], monocyte chemoattractant protein-1 (MCP-1 or CCL2) [109, 110], or shear stress [111] increased both serine/threonine phosphorylation of occludin and permeability. Furthermore, diabetes increases occludin phosphorylation in the rat retina similar to the VEGF-induced increase in BREC [106].

Phosphorylation of occludin leads to ubiquitination and subsequent endocytosis regulating endothelial barrier properties. The use of two-dimensional gel electrophoresis in BREC demonstrates that occludin is basally phosphorylated on two residues, and growth factor stimulation leads to phosphorylation at three additional sites [106]. Using mass spectrometry of occludin immunoprecipitated from vascular endothelial cells, Sundstrom et al. identified five putative occludin phosphosites and demonstrated at least one of these sites: Ser490 was VEGF responsive as shown by the use of a phospho-specific antibody [112]. This Ser490 phosphorylation allows subsequent ubiquitination of occludin by the E3 ligase Itch and endocytosis of the transmembrane protein by binding epsin, eps15, and Hrs, which possess ubiquitin interacting motifs and chaperon occludin through endocytosis [113]. Importantly, mutating Ser490 to alanine (S490A) prevented both occludin ubiquitination and VEGF-induced permeability, while expressing an occludin-ubiquitin chimeric protein creates leaky endothelial junctions. Thus, the carboxy-terminal tail of occludin can be phosphorylated and subsequently ubiquitinated, directing occludin into the endocytosis pathway and regulating endothelial barrier properties, potentially by controlling the localization of other junctional proteins such as the claudins.

While occludin phosphorylation and ubiquitination are necessary steps for VEGF-induced permeability, additional junction alterations are likely involved in the process. Recently, ubiquitination of claudins has also been observed in epithelial cells with the

E3 ubiquitin ligase LNX1p80 regulating claudin internalization and lysosomal degradation [114]. Further, in endothelial cells without tight junctions, the phosphorylation and endocytosis of VE-cadherin is an essential step to regulate barrier properties [115]. Additionally, the ubiquitin ligase Hakai ubiquitinates E-cadherin and induces endocytosis [116]. While the mechanisms controlling barrier properties are complex, posttranslational modifications regulating endocytosis of junctional components provide important mechanisms of permeability regulation.

Protein Kinase C in Regulation of Barrier Properties

Key mediators of BRB homeostasis and diabetes-induced vascular abnormalities include the Protein Kinase C (PKC) family [117]. Alterations of PKC isoforms during diabetes may result from hyperglycemia, de novo synthesis of diacylglycerol (DAG), advanced glycation end products (AGEs), increased expression of growth factor/inflammatory cytokines, and to a generally altered redox state [118]. As a member of the larger protein kinase AGC super family, PKC isozymes regulate essential signaling pathways in various tissues controlling proliferation, differentiation, survival, and cell growth (reviewed in [119–122]). There are three main classes of PKC isoforms based on their cofactor requirements. The classical PKC isoforms, α (alpha), βI, βII (betaI, betaII), and γ (gamma), require Ca^{2+} and diacylglycerol (DAG) for activation. Novel PKC isoforms, δ (delta), ε (epsilon), η (eta), and θ (theta), require DAG; while the atypical PKC isoforms, ζ (zeta), ι (iota) and λ (lamda), require neither DAG or Ca^{2+} to become activated [122].

Evidence for a role of PKC isoforms in vascular permeability and increased flux of macromolecules began in the late 1980s and early 1990s [123, 124]. Treatment of bovine pulmonary artery endothelial cells with phorbol 12-myristate 13-acetate (PMA), an activator of classical and novel PKC isoforms, leads to an approximately twofold increase in ^{125}I-albumin permeability [123]. Additionally, PMA and diacylglycerol treatment of bovine aortic endothelial cells alters ^{14}C-sucrose and ^{3}H-inulin flux but not ^{125}I-polyvinyl pyrrolidone (360 kDa) permeability, indicating PKC isoforms control paracellular permeability [124].

Diabetes-induced vascular permeability can be partly attributed to increased classical PKC activity. PKC activity is altered in the diabetic rat retina, BREC, and in bovine retinal pericytes (BRPs) [117]. Oral administration of LY333531, a specific PKCβ (beta) inhibitor with low nanomolar potency similar to ruboxistaurin, ameliorates the diabetes-induced effect on retinal blood flow [125]. Membrane translocation and activation of PKCα (alpha), β (beta)II, and δ (delta) isoforms in response to VEGF have been observed in vivo [126], and this translocation was blocked by oral administration of the PKCβ (beta) inhibitor [127]. Mechanistically, increased activity of classical PKC isozymes leads to tight junction deregulation, cytoskeleton rearrangements, and endothelial permeability [106, 128]. Data from our laboratory demonstrates VEGF-induced occludin phosphorylation, and ubiquitination requires PKCβ (beta) (manuscript in preparation). Furthermore, PKCα (alpha) mediates hyperglycemia-induced porcine aortic endothelial cell permeability demonstrated by RNAi knockdown [129]. Collectively, these data implicate classical PKC isoforms mediate vascular endothelial permeability induced by diabetes.

Although classical PKC isoforms contribute to VEGF-induced endothelial permeability, other signaling pathways also contribute to control of the BRB. Studies of primary retinal endothelial cell culture assays show an incomplete attenuation of VEGF-induced endothelial permeability via classical PKC inhibition [106]. In addition, tumor necrosis factor α (alpha) induces endothelial permeability over 6 h but is unaffected by classical PKC inhibitors (manuscript under review). Together, these data suggest concurrent or alternative signaling pathways may also contribute to the vascular permeability observed in diabetic retinopathy.

In addition to classical PKC isozymes, novel PKCs are implicated in mediating diabetes-induced alterations of BRB homeostasis. PKCδ (delta) translocates to the membrane fraction of retinal lysates of diabetic mice indicative of PKCδ (delta) activation [130]. Geraldes et al. identified Src homology 2 domain-containing phosphatase-1 (SHP-1), a protein tyrosine phosphatase, as a downstream target of PKCδ that leads to platelet-derived growth factor beta-receptor (PDGFβ (beta) receptor) dephosphorylation. PDGFβ (beta) is a survival signal for retinal pericytes allowing for activation of Akt, which is essential to pericyte survival [131]. Reduced PDGFβ receptor signaling results in diabetes-induced pericyte apoptosis, which increases vascular permeability in the diabetic mouse retina [130]. In addition, PKCδ mediates AGE-induced permeability in human retinal endothelial cells (HREC) as shown through the use of PKCδ small molecule inhibitors and siRNA studies which prevent the AGE-induced alterations to ZO-1 and ZO-2 protein expression [132].

In addition to the well-established contributions of classical and novel PKC isoforms to diabetes-induced junctional deregulation and vascular permeability, a role for the atypical PKC (aPKC) isoforms is emerging. The aPKC isoforms act downstream of both the phosphatidylinositol 3-kinase (PI3-K) and the small Rho GTPases family members in response to growth factors, leading to proliferation, differentiation, and cell polarity/apical-basolateral orientation [73, 133]. Additionally, aPKC isoforms are critical for the establishment of primordial junction development and the regulation of junction complexes in both endothelial and epithelial cells [134, 135]. VEGF administration leads to a twofold increase in PI3-K activity as well as transiently activating small Rho GTPases such as Cdc42, Rac1, and Rho, contributing to endothelial permeability in endothelial cells [126, 136]. Therefore, aPKC isoforms may play a critical role in the regulation of growth-factor-induced vascular permeability. Data from our laboratory demonstrates overexpression of PKCζ (zeta), an atypical PKC isoform, potentiates the effect of VEGF on permeability, whereas kinase dead-mediated competitive inhibition of PKCζ (zeta) blocks VEGF-induced permeability in BREC. Importantly, aPKC inhibition prevents TNFα-induced endothelial permeability and prevents loss of tight junction proteins claudin-5 and ZO-1 and cell border disorganization (manuscript under review). Together, these studies demonstrate aPKC isoforms contribute to VEGF and TNFα-induced permeability, elucidating a common signaling mechanism in diabetic retinopathy. Collectively, these data show a contribution of classical, novel, and atypical PKC isoforms in the control of retinal vascular permeability (Fig. 3).

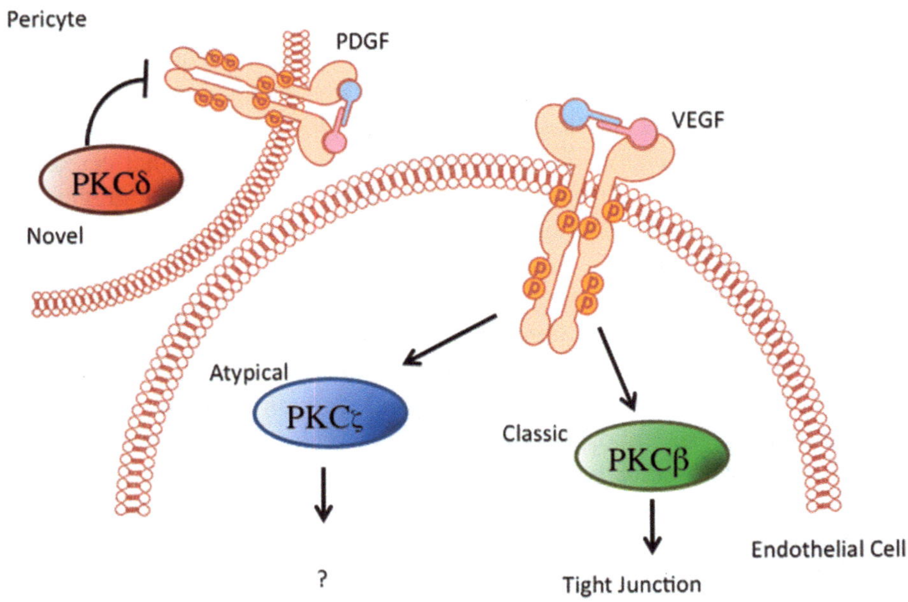

Fig. 3. PKC isozymes in the blood-retinal barrier. In endothelial cells, both classical and atypical PKC isozymes contribute to VEGF signaling. VEGF activates classical PKCs, such as PKCβ (beta) leading to phosphorylation of the tight junction protein occludin and promoting internalization and subsequent endothelial permeability. Ruboxistaurin inhibition of PKCβ (beta) prevents VEGF-induced permeability by blocking this pathway. Concurrently, atypical PKC isoforms, such as PKCζ (zeta), lead to increased endothelial permeability via unknown mechanisms. However, inhibition of atypical PKC activity effectively blocks both growth factor and inflammatory-cytokine- induced endothelial permeability. In pericytes, hyperglycemia-induced increase of novel PKCs, specifically PKCδ (delta), inhibits PDGFβ (beta) survival signaling to Akt, leading to pericyte apoptosis. Loss of pericytes, coupled with VEGF-induced endothelial permeability likely contributes to the macular edema observed in diabetic retinopathy.

CONCLUSIONS

Vascular permeability in diabetic retinopathy may be attributed to a host of changes in the retina, including increases in growth factors such as VEGF, cytokines like TNFα (alpha), or protease activation such as kallikrein/bradykinin system. Posttranslational modification of the junction proteins and regulated endocytosis is an important mechanism controlling retinal vascular permeability. Indeed, VEGF activation of PKCβ (beta) controls occludin phosphorylation and subsequent ubiquitination necessary for VEGF-induced permeability. As information regarding changes to the junction complex becomes better understood, more targeted therapies may become available, increasing our ability to maintain retinal vascular integrity and visual function in the face of diabetes.

REFERENCES

1. Bill A. Blood circulation and fluid dynamics in the eye. Physiol Rev. 1975;55:383–417.
2. Erickson KK, Sundstrom JM, Antonetti DA. Vascular permeability in ocular disease and the role of tight junctions. Angiogenesis. 2007;10:103–17.
3. Pournaras CJ, Rungger-Brandle E, Riva CE, et al. Regulation of retinal blood flow in health and disease. Prog Retin Eye Res. 2008;27:284–330.
4. Strauss O. The retinal pigment epithelium in visual function. Physiol Rev. 2005;85:845–81.
5. Gariano RF, Iruela-Arispe ML, Hendrickson AE. Vascular development in primate retina: comparison of laminar plexus formation in monkey and human. Invest Ophthalmol Vis Sci. 1994;35:3442–55.
6. Swaroop A, Branham KE, Chen W, et al. Genetic susceptibility to age-related macular degeneration: a paradigm for dissecting complex disease traits. Hum Mol Genet. 2007;16(Spec No. 2):R174–82.
7. Janzer RC, Raff MC. Astrocytes induce blood-brain barrier properties in endothelial cells. Nature. 1987;325:253–7.
8. Tout S, Chan-Ling T, Hollander H, et al. The role of Muller cells in the formation of the blood-retinal barrier. Neuroscience. 1993;55:291–301.
9. Stewart PA, Wiley MJ. Developing nervous tissue induces formation of blood-brain barrier characteristics in invading endothelial cells: a study using quail–chick transplantation chimeras. Dev Biol. 1981;84:183–92.
10. Lee SW, Kim WJ, Choi YK, et al. SSeCKS regulates angiogenesis and tight junction formation in blood-brain barrier. Nat Med. 2003;9:900–6.
11. Hori S, Ohtsuki S, Hosoya K, et al. A pericyte-derived angiopoietin-1 multimeric complex induces occludin gene expression in brain capillary endothelial cells through Tie-2 activation in vitro. J Neurochem. 2004;89:503–13.
12. Asahara T, Chen DH, Takahashi T, et al. Tie2 receptor ligands, angiopoietin-1 and angiopoietin-2, modulate VEGF-induced postnatal neovascularization. Circ Res. 1998;83:233–40.
13. Thurston G, Rudge JS, Ioffe E, et al. Angiopoietin-1 protects the adult vasculature against plasma leakage. Nat Med. 2000;6:460–3.
14. Nakagawa S, Deli MA, Kawaguchi H, et al. A new blood-brain barrier model using primary rat brain endothelial cells, pericytes and astrocytes. Neurochem Int. 2009;54:253–63.
15. Reese TS, Karnovsky MJ. Fine structural localization of a blood-brain barrier to exogenous peroxidase. J Cell Biol. 1967;34:207–17.
16. Cunha-Vaz JG, Shakib M, Ashton N. Studies on the permeability of the blood-retinal barrier. I. On the existence, development, and site of a blood-retinal barrier. Br J Ophthalmol. 1966;50:441–53.
17. Shakib M, Cunha-Vaz JG. Studies on the permeability of the blood-retinal barrier. IV. Junctional complexes of the retinal vessels and their role in the permeability of the blood-retinal barrier. Exp Eye Res. 1966;5:229–34.
18. Miyoshi J, Takai Y. Molecular perspective on tight-junction assembly and epithelial polarity. Adv Drug Deliv Rev. 2005;57:815–55.
19. Suzuki A, Ishiyama C, Hashiba K, et al. aPKC kinase activity is required for the asymmetric differentiation of the premature junctional complex during epithelial cell polarization. J Cell Sci. 2002;115:3565–73.
20. Fukuhara A, Irie K, Nakanishi H, et al. Involvement of nectin in the localization of junctional adhesion molecule at tight junctions. Oncogene. 2002;21:7642–55.
21. Fukuhara A, Irie K, Yamada A, et al. Role of nectin in organization of tight junctions in epithelial cells. Genes Cells. 2002;7:1059–72.

22. Shakib M, Cunha-Vaz JG. Studies on the permeability of the blood-retinal barrier. IV. Junctional complexes of the retinal vessels and their role in the permeability of the blood-retinal barrier. Exp Eye Res. 1966;5:229–34.
23. Cunha-Vaz JG, Shakib M, Ashton N. Studies on the permeability of the blood-retinal barrier. I. On the existence, development, and site of a blood-retinal barrier. Br J Ophthalmol. 1966;50:441–53.
24. Gonzalez-Mariscal L, Betanzos A, Nava P, et al. Tight junction proteins. Prog Biophys Mol Biol. 2003;81:1–44.
25. Haskins J, Gu L, Wittchen ES, et al. ZO-3, a novel member of the MAGUK protein family found at the tight junction, interacts with ZO-1 and occludin. J Cell Biol. 1998;141:199–208.
26. Stevenson BR, Siliciano JD, Mooseker MS, et al. Identification of ZO-1: a high molecular weight polypeptide associated with the tight junction (zonula occludens) in a variety of epithelia. J Cell Biol. 1986;103:755–66.
27. Jesaitis LA, Goodenough DA. Molecular characterization and tissue distribution of ZO-2, a tight junction protein homologous to ZO-1 and the Drosophila discs-large tumor suppressor protein. J Cell Biol. 1994;124:949–61.
28. Gumbiner B, Lowenkopf T, Apatira D. Identification of a 160-kDa polypeptide that binds to the tight junction protein ZO-1. Proc Natl Acad Sci USA. 1991;88:3460–4.
29. Balda MS, Gonzalez-Mariscal L, Matter K, et al. Assembly of the tight junction: the role of diacylglycerol. J Cell Biol. 1993;123:293–302.
30. Itoh M, Morita K, Tsukita S. Characterization of ZO-2 as a MAGUK family member associated with tight as well as adherens junctions with a binding affinity to occludin and alpha catenin. J Biol Chem. 1999;274:5981–6.
31. Woods DF, Bryant PJ. ZO-1, DlgA and PSD-95/SAP90: homologous proteins in tight, septate and synaptic cell junctions. Mech Dev. 1993;44:85–9.
32. Willott E, Balda MS, Heintzelman M, et al. Localization and differential expression of two isoforms of the tight junction protein ZO-1. Am J Physiol. 1992;262:C1119–24.
33. Beatch M, Jesaitis LA, Gallin WJ, et al. The tight junction protein ZO-2 contains three PDZ (PSD-95/Discs-Large/ZO-1) domains and an alternatively spliced region. J Biol Chem. 1996;271:25723–6.
34. Umeda K, Matsui T, Nakayama M, et al. Establishment and characterization of cultured epithelial cells lacking expression of ZO-1. J Biol Chem. 2004;279:44785–94.
35. McNeil E, Capaldo CT, Macara IG. Zonula occludens-1 function in the assembly of tight junctions in Madin-Darby canine kidney epithelial cells. Mol Biol Cell. 2006;17:1922–32.
36. Hernandez S, Chavez Munguia B, Gonzalez-Mariscal L. ZO-2 silencing in epithelial cells perturbs the gate and fence function of tight junctions and leads to an atypical monolayer architecture. Exp Cell Res. 2007;313:1533–47.
37. Umeda K, Ikenouchi J, Katahira-Tayama S, et al. ZO-1 and ZO-2 independently determine where claudins are polymerized in tight-junction strand formation. Cell. 2006;126:741–54.
38. Katsuno T, Umeda K, Matsui T, et al. Deficiency of zonula occludens-1 causes embryonic lethal phenotype associated with defected yolk sac angiogenesis and apoptosis of embryonic cells. Mol Biol Cell. 2008;19:2465–75.
39. Xu J, Kausalya PJ, Phua DC, et al. Early embryonic lethality of mice lacking ZO-2, but Not ZO-3, reveals critical and nonredundant roles for individual zonula occludens proteins in mammalian development. Mol Cell Biol. 2008;28:1669–78.
40. Van Itallie CM, Anderson JM. Claudins and epithelial paracellular transport. Annu Rev Physiol. 2006;68:403–29.
41. Tsukita S, Furuse M, Itoh M. Multifunctional strands in tight junctions. Nat Rev Mol Cell Biol. 2001;2:285–93.

42. Chiba H, Osanai M, Murata M, et al. Transmembrane proteins of tight junctions. Biochim Biophys Acta. 2008;1778:588–600.
43. Ruffer C, Gerke V. The C-terminal cytoplasmic tail of claudins 1 and 5 but not its PDZ-binding motif is required for apical localization at epithelial and endothelial tight junctions. Eur J Cell Biol. 2004;83:135–44.
44. Arabzadeh A, Troy TC, Turksen K. Role of the Cldn6 cytoplasmic tail domain in membrane targeting and epidermal differentiation in vivo. Mol Cell Biol. 2006;26:5876–87.
45. Morita K, Furuse M, Fujimoto K, et al. Claudin multigene family encoding four-transmembrane domain protein components of tight junction strands. Proc Natl Acad Sci USA. 1999;96:511–6.
46. Itoh M, Furuse M, Morita K, et al. Direct binding of three tight junction-associated MAGUKs, ZO-1, ZO-2, and ZO-3, with the COOH termini of claudins. J Cell Biol. 1999;147:1351–63.
47. Inai T, Kobayashi J, Shibata Y. Claudin-1 contributes to the epithelial barrier function in MDCK cells. Eur J Cell Biol. 1999;78:849–55.
48. Simon DB, Lu Y, Choate KA, et al. Paracellin-1, a renal tight junction protein required for paracellular Mg2+ resorption. Science. 1999;285:103–6.
49. Hou J, Paul DL, Goodenough DA. Paracellin-1 and the modulation of ion selectivity of tight junctions. J Cell Sci. 2005;118:5109–18.
50. Colegio OR, Van Itallie CM, McCrea HJ, et al. Claudins create charge-selective channels in the paracellular pathway between epithelial cells. Am J Physiol Cell Physiol. 2002;283:C142–7.
51. Hou J, Gomes AS, Paul DL, et al. Study of claudin function by RNA interference. J Biol Chem. 2006;281:36117–23.
52. Piontek J, Winkler L, Wolburg H, et al. Formation of tight junction: determinants of homophilic interaction between classic claudins. FASEB J. 2008;22:146–58.
53. Rahner C, Fukuhara M, Peng S, et al. The apical and basal environments of the retinal pigment epithelium regulate the maturation of tight junctions during development. J Cell Sci. 2004;117:3307–18.
54. Barber AJ, Antonetti DA. Mapping the blood vessels with paracellular permeability in the retinas of diabetic rats. Invest Ophthalmol Vis Sci. 2003;44:5410–6.
55. Furuse M, Hata M, Furuse K, et al. Claudin-based tight junctions are crucial for the mammalian epidermal barrier: a lesson from claudin-1-deficient mice. J Cell Biol. 2002;156:1099–111.
56. Nitta T, Hata M, Gotoh S, et al. Size-selective loosening of the blood-brain barrier in claudin-5-deficient mice. J Cell Biol. 2003;161:653–60.
57. Gow A, Southwood CM, Li JS, et al. CNS myelin and sertoli cell tight junction strands are absent in Osp/claudin-11 null mice. Cell. 1999;99:649–59.
58. Klaassen I, Hughes JM, Vogels IM, et al. Altered expression of genes related to blood-retina barrier disruption in streptozotocin-induced diabetes. Exp Eye Res. 2009;89:4–15.
59. Bazzoni G. The JAM family of junctional adhesion molecules. Curr Opin Cell Biol. 2003;15:525–30.
60. Ebnet K, Suzuki A, Ohno S, et al. Junctional adhesion molecules (JAMs): more molecules with dual functions? J Cell Sci. 2004;117:19–29.
61. Ebnet K, Suzuki A, Horikoshi Y, et al. The cell polarity protein ASIP/PAR-3 directly associates with junctional adhesion molecule (JAM). EMBO J. 2001;20:3738–48.
62. Itoh M, Sasaki H, Furuse M, et al. Junctional adhesion molecule (JAM) binds to PAR-3: a possible mechanism for the recruitment of PAR-3 to tight junctions. J Cell Biol. 2001;154:491–7.

63. Ebnet K, Aurrand-Lions M, Kuhn A, et al. The junctional adhesion molecule (JAM) family members JAM-2 and JAM-3 associate with the cell polarity protein PAR-3: a possible role for JAMs in endothelial cell polarity. J Cell Sci. 2003;116:3879–91.
64. Sollerbrant K, Raschperger E, Mirza M, et al. The Coxsackievirus and adenovirus receptor (CAR) forms a complex with the PDZ domain-containing protein ligand-of-numb protein-X (LNX). J Biol Chem. 2003;278:7439–44.
65. Kansaku A, Hirabayashi S, Mori H, et al. Ligand-of-Numb protein X is an endocytic scaffold for junctional adhesion molecule 4. Oncogene. 2006;25:5071–84.
66. Hirabayashi S, Tajima M, Yao I, et al. JAM4, a junctional cell adhesion molecule interacting with a tight junction protein, MAGI-1. Mol Cell Biol. 2003;23:4267–82.
67. Wegmann F, Ebnet K, Du Pasquier L, et al. Endothelial adhesion molecule ESAM binds directly to the multidomain adaptor MAGI-1 and recruits it to cell contacts. Exp Cell Res. 2004;300:121–33.
68. Weber C, Fraemohs L, Dejana E. The role of junctional adhesion molecules in vascular inflammation. Nat Rev Immunol. 2007;7:467–77.
69. Martin-Padura I, Lostaglio S, Schneemann M, et al. Junctional adhesion molecule, a novel member of the immunoglobulin superfamily that distributes at intercellular junctions and modulates monocyte transmigration. J Cell Biol. 1998;142:117–27.
70. Liu Y, Nusrat A, Schnell FJ, et al. Human junction adhesion molecule regulates tight junction resealing in epithelia. J Cell Sci. 2000;113(Pt 13):2363–74.
71. Mandell KJ, McCall IC, Parkos CA. Involvement of the junctional adhesion molecule-1 (JAM1) homodimer interface in regulation of epithelial barrier function. J Biol Chem. 2004;279:16254–62.
72. Rehder D, Iden S, Nasdala I, et al. Junctional adhesion molecule-a participates in the formation of apico-basal polarity through different domains. Exp Cell Res. 2006;312:3389–403.
73. Macara IG. Parsing the polarity code. Nat Rev Mol Cell Biol. 2004;5:220–31.
74. Hirata K, Ishida T, Penta K, et al. Cloning of an immunoglobulin family adhesion molecule selectively expressed by endothelial cells. J Biol Chem. 2001;276:16223–31.
75. Wegmann F, Petri B, Khandoga AG, et al. ESAM supports neutrophil extravasation, activation of Rho, and VEGF-induced vascular permeability. J Exp Med. 2006;203:1671–7.
76. Furuse M, Hirase T, Itoh M, et al. Occludin: a novel integral membrane protein localizing at tight junctions. J Cell Biol. 1993;123:1777–88.
77. Sanchez-Pulido L, Martin-Belmonte F, Valencia A, et al. MARVEL: a conserved domain involved in membrane apposition events. Trends Biochem Sci. 2002;27:599–601.
78. Ikenouchi J, Sasaki H, Tsukita S, et al. Loss of occludin affects tricellular localization of tricellulin. Mol Biol Cell. 2008;19:4687–93.
79. Hirase T, Staddon JM, Saitou M, et al. Occludin as a possible determinant of tight junction permeability in endothelial cells. J Cell Sci. 1997;110(Pt 14):1603–13.
80. Kevil CG, Payne DK, Mire E, et al. Vascular permeability factor/vascular endothelial cell growth factor-mediated permeability occurs through disorganization of endothelial junctional proteins. J Biol Chem. 1998;273:15099–103.
81. Saitou M, Furuse M, Sasaki H, et al. Complex phenotype of mice lacking occludin, a component of tight junction strands. Mol Biol Cell. 2000;11:4131–42.
82. Schulzke JD, Gitter AH, Mankertz J, et al. Epithelial transport and barrier function in occludin-deficient mice. Biochim Biophys Acta. 2005;1669:34–42.
83. Antonetti DA, Barber AJ, Bronson SK, et al. Diabetic retinopathy: seeing beyond glucose-induced microvascular disease. Diabetes. 2006;55:2401–11.
84. Moss SE, Klein R, Klein BE. The 14-year incidence of visual loss in a diabetic population. Ophthalmology. 1998;105:998–1003.

85. Browning DJ, Glassman AR, Aiello LP, et al. Relationship between optical coherence tomography-measured central retinal thickness and visual acuity in diabetic macular edema. Ophthalmology. 2007;114:525–36.
86. Gardner TW, Larsen M, Girach A, et al. Diabetic macular oedema and visual loss: relationship to location, severity and duration. Acta Ophthalmol. 2009;87(7):709–13.
87. Cunha-Vaz JG, Faria de Abreu JR, Campos AJ. Early breakdown of the blood-retinal barrier in diabetes. Br J Ophthalmol. 1975;59:649–56.
88. Sander B, Thornit DN, Colmorn L, et al. Progression of diabetic macular edema: correlation with blood retinal barrier permeability, retinal thickness, and retinal vessel diameter. Invest Ophthalmol Vis Sci. 2007;48:3983–7.
89. Adamis AP, Miller JW, Bernal MT, et al. Increased vascular endothelial growth factor levels in the vitreous of eyes with proliferative diabetic retinopathy. Am J Ophthalmol. 1994;118:445–50.
90. Aiello LP. Vascular endothelial growth factor and the eye - Past, present, and future. Arch Ophthalmol. 1996;114:1252–4.
91. Aiello LP, Avery RL, Arrigg PG, et al. Vascular endothelial growth factor in ocular fluid of patients with diabetic retinopathy and other retinal disorders. N Eng J Med. 1994;331:1480–7.
92. Caldwell RB, Bartoli M, Behzadian MA, et al. Vascular endothelial growth factor and diabetic retinopathy: pathophysiological mechanisms and treatment perspectives. Diabetes Metab Res Rev. 2003;19:442–55.
93. Hammes HP, Lin J, Bretzel RG, et al. Upregulation of the vascular endothelial growth factor/vascular endothelial growth factor receptor system in experimental background diabetic retinopathy of the rat. Diabetes. 1998;47:401–6.
94. Neufeld G, Cohen T, Gengrinovitch S, et al. Vascular endothelial growth factor (VEGF) and its receptors. FASEB J. 1999;13:9–22.
95. Elman MJ, Aiello LP, Beck RW, et al. Randomized trial evaluating ranibizumab plus prompt or deferred laser or triamcinolone plus prompt laser for diabetic macular edema. Ophthalmology. 2010;117:1064–77. e35.
96. Demircan N, Safran BG, Soylu M, et al. Determination of vitreous interleukin-1 (IL-1) and tumour necrosis factor (TNF) levels in proliferative diabetic retinopathy. Eye (Lond). 2006;20:1366–9.
97. Abu el Asrar AM, Maimone D, Morse PH, et al. Cytokines in the vitreous of patients with proliferative diabetic retinopathy. Am J Ophthalmol. 1992;114:731–6.
98. Carmo A, Cunha-Vaz JG, Carvalho AP, et al. L-arginine transport in retinas from streptozotocin diabetic rats: correlation with the level of IL-1 beta and NO synthase activity. Vision Res. 1999;39:3817–23.
99. Krady JK, Basu A, Allen CM, et al. Minocycline reduces proinflammatory cytokine expression, microglial activation, and caspase-3 activation in a rodent model of diabetic retinopathy. Diabetes. 2005;54:1559–65.
100. Joussen AM, Poulaki V, Mitsiades N, et al. Nonsteroidal anti-inflammatory drugs prevent early diabetic retinopathy via TNF-alpha suppression. FASEB J. 2002;16:438–40.
101. Joussen AM, Poulaki V, Qin W, et al. Retinal vascular endothelial growth factor induces intercellular adhesion molecule-1 and endothelial nitric oxide synthase expression and initiates early diabetic retinal leukocyte adhesion in vivo. Am J Pathol. 2002;160:501–9.
102. Gao BB, Clermont A, Rook S, et al. Extracellular carbonic anhydrase mediates hemorrhagic retinal and cerebral vascular permeability through prekallikrein activation. Nat Med. 2007;13:181–8.
103. Antonetti DA, Barber AJ, Khin S, et al. Vascular permeability in experimental diabetes is associated with reduced endothelial occludin content: vascular endothelial growth factor

decreases occludin in retinal endothelial cells. Penn State Retina Research Group. Diabetes. 1998;47:1953–9.
104. Jin M, Barron E, He S, et al. Regulation of RPE intercellular junction integrity and function by hepatocyte growth factor. Invest Ophthalmol Vis Sci. 2002;43:2782–90.
105. Antonetti DA, Barber AJ, Hollinger LA, et al. Vascular endothelial growth factor induces rapid phosphorylation of tight junction proteins occludin and zonula occluden 1. A potential mechanism for vascular permeability in diabetic retinopathy and tumors. J Biol Chem. 1999;274:23463–7.
106. Harhaj NS, Felinski EA, Wolpert EB, et al. VEGF activation of protein kinase C stimulates occludin phosphorylation and contributes to endothelial permeability. Invest Ophthalmol Vis Sci. 2006;47:5106–15.
107. Hirase T, Kawashima S, Wong EY, et al. Regulation of tight junction permeability and occludin phosphorylation by Rhoa-p160ROCK-dependent and -independent mechanisms. J Biol Chem. 2001;276:10423–31.
108. DeMaio L, Rouhanizadeh M, Reddy S, et al. Oxidized phospholipids mediate occludin expression and phosphorylation in vascular endothelial cells. Am J Physiol Heart Circ Physiol. 2006;290:H674–83.
109. Stamatovic SM, Dimitrijevic OB, Keep RF, et al. Protein kinase Calpha-RhoA cross-talk in CCL2-induced alterations in brain endothelial permeability. J Biol Chem. 2006;281:8379–88.
110. Stamatovic SM, Shakui P, Keep RF, et al. Monocyte chemoattractant protein-1 regulation of blood-brain barrier permeability. J Cereb Blood Flow Metab. 2005;25:593–606.
111. DeMaio L, Chang YS, Gardner TW, et al. Shear stress regulates occludin content and phosphorylation. Am J Physiol Heart Circ Physiol. 2001;281:H105–13.
112. Sundstrom JM, Tash BR, Murakami T, et al. Identification and Analysis of Occludin Phosphosites: A Combined Mass Spectrometry and Bioinformatics Approach. J Proteome Res. 2009;8(2):808–17.
113. Murakami T, Felinski EA, Antonetti DA. Occludin phosphorylation and ubiquitination regulate tight junction trafficking and vascular endothelial growth factor (VEGF)-induced permeability. J Biol Chem. 2009;284(31):21036–46.
114. Takahashi S, Iwamoto N, Sasaki H, et al. The E3 ubiquitin ligase LNX1p80 promotes the removal of claudins from tight junctions in MDCK cells. J Cell Sci. 2009;122:985–94.
115. Gavard J, Gutkind JS. VEGF controls endothelial-cell permeability by promoting the beta-arrestin-dependent endocytosis of VE-cadherin. Nat Cell Biol. 2006;8:1223–34.
116. Fujita Y, Krause G, Scheffner M, et al. Hakai, a c-Cbl-like protein, ubiquitinates and induces endocytosis of the E-cadherin complex. Nat Cell Biol. 2002;4:222–31.
117. Das Evcimen N, King GL. The role of protein kinase C activation and the vascular complications of diabetes. Pharmacol Res. 2007;55:498–510.
118. Geraldes P, King GL. Activation of protein kinase C isoforms and its impact on diabetic complications. Circ Res. 2010;106:1319–31.
119. Hofmann J. The potential for isoenzyme-selective modulation of protein kinase C. FASEB J. 1997;11:649–69.
120. Kanashiro CA, Khalil RA. Signal transduction by protein kinase C in mammalian cells. Clin Exp Pharmacol Physiol. 1998;25:974–85.
121. Liu WS, Heckman CA. The sevenfold way of PKC regulation. Cell Signal. 1998;10:529–42.
122. Steinberg SF. Structural basis of protein kinase C isoform function. Physiol Rev. 2008;88:1341–78.
123. Lynch JJ, Ferro TJ, Blumenstock FA, et al. Increased endothelial albumin permeability mediated by protein kinase C activation. J Clin Invest. 1990;85:1991–8.

124. Oliver JA. Adenylate cyclase and protein kinase C mediate opposite actions on endothelial junctions. J Cell Physiol. 1990;145:536–42.
125. Ishii H, Jirousek MR, Koya D, et al. Amelioration of vascular dysfunctions in diabetic rats by an oral PKC beta inhibitor. Science. 1996;272:728–31.
126. Xia P, Aiello LP, Ishii H, et al. Characterization of vascular endothelial growth factor's effect on the activation of protein kinase C, its isoforms, and endothelial cell growth. J Clin Invest. 1996;98:2018–26.
127. Aiello LP, Bursell SE, Clermont A, et al. Vascular endothelial growth factor-induced retinal permeability is mediated by protein kinase C in vivo and suppressed by an orally effective beta-isoform-selective inhibitor. Diabetes. 1997;46:1473–80.
128. Stasek Jr JE, Patterson CE, Garcia JG. Protein kinase C phosphorylates caldesmon77 and vimentin and enhances albumin permeability across cultured bovine pulmonary artery endothelial cell monolayers. J Cell Physiol. 1992;153:62–75.
129. Hempel A, Maasch C, Heintze U, et al. High glucose concentrations increase endothelial cell permeability via activation of protein kinase C alpha. Circ Res. 1997;81:363–71.
130. Geraldes P, Hiraoka-Yamamoto J, Matsumoto M, et al. Activation of PKC-delta and SHP-1 by hyperglycemia causes vascular cell apoptosis and diabetic retinopathy. Nat Med. 2009;15:1298–306.
131. Enge M, Bjarnegard M, Gerhardt H, et al. Endothelium-specific platelet-derived growth factor-B ablation mimics diabetic retinopathy. EMBO J. 2002;21:4307–16.
132. Kim JH, Jun HO, Yu YS, et al. Inhibition of protein kinase C delta attenuates blood-retinal barrier breakdown in diabetic retinopathy. Am J Pathol. 2010;176:1517–24.
133. Chou MM, Hou W, Johnson J, et al. Regulation of protein kinase C zeta by PI 3-kinase and PDK-1. Curr Biol. 1998;8:1069–77.
134. Davis GE, Koh W, Stratman AN. Mechanisms controlling human endothelial lumen formation and tube assembly in three-dimensional extracellular matrices. Birth Defects Res C Embryo Today. 2007;81:270–85.
135. Suzuki A, Yamanaka T, Hirose T, et al. Atypical protein kinase C is involved in the evolutionarily conserved par protein complex and plays a critical role in establishing epithelia-specific junctional structures. J Cell Biol. 2001;152:1183–96.
136. Beckers CM, van Hinsbergh VW. van Nieuw Amerongen GP. Driving Rho GTPase activity in endothelial cells regulates barrier integrity. Thromb Haemost. 2010;103:40–55.

9
Capillary Degeneration in Diabetic Retinopathy

Timothy S. Kern

Contents

Keywords Diabetic retinopathy • Vasoocclusion • Nonperfusion • Pathogenesis

Capillary degeneration is a required step during normal development [1–5]. Capillary degeneration also has serious and undesirable consequences in several ischemic diseases, including retinopathy of prematurity, sickle-cell retinopathy [6–9], and diabetic retinopathy. This review will focus on causes of vascular nonperfusion and capillary degeneration in the retina, and their relation to diabetic retinopathy.

Vascular pathology in the early stages of diabetic retinopathy is characterized histologically by the presence of saccular capillary microaneurysms, pericyte-deficient capillaries, and nonperfused and degenerate capillaries in patients (Fig. 1). Capillary nonperfusion and/or degeneration are particularly important lesions of the early retinopathy [10, 11]. The area of nonperfusion in the retina is significantly correlated with the mean severity grade of the retinopathy [12], and it is generally accepted that capillary nonperfusion and degeneration play major and causal roles in the progression to preretinal neovascularization that develops in some diabetic patients [13]. The extent of capillary nonperfusion in diabetic retinopathy has been found to correlate with the amount and localization of neovascularization [13]. As more and more capillaries become nonperfused or occluded, local areas of the retina likely become deprived of oxygen and nutrients, thus stimulating production of one or more ischemia-driven growth factors, such as vascular endothelial

From: *Ophthalmology Research: Visual Dysfunction in Diabetes*
Edited by: J. Tombran-Tink et al. (eds.), DOI 10.1007/978-1-60761-150-9_9

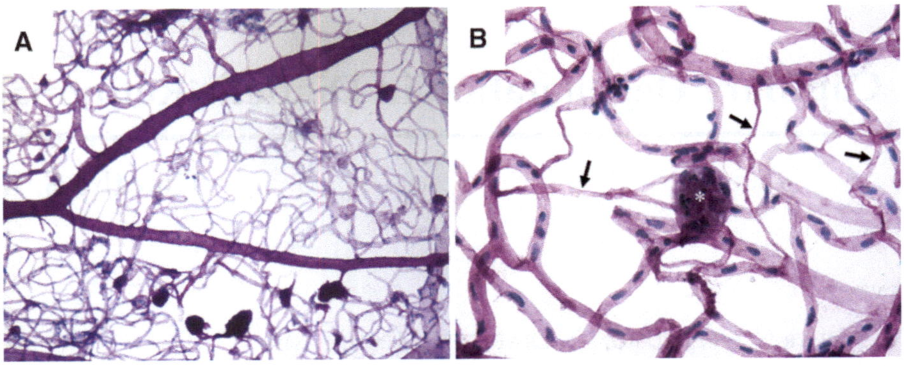

Fig. 1. (**A**) Low-power view of retinal histopathology in a patient having nonproliferative diabetic retinopathy. There is a large area of capillary degeneration in the photo, indicated by the absence of dark nuclear stain in most vessels. Numerous microaneurysms are along the top and bottom of the micrograph. (**B**) Close-up of vascular histopathology in a diabetic patient. Degenerate capillaries are indicated by *arrows*, and saccular capillary microaneurysm is indicated by *asterisk* (*).

growth factor (VEGF). VEGF is known to be a key molecule leading to retinal permeability and neovascularization in diabetes and other retinal diseases [14–16].

VASCULAR NONPERFUSION IN DIABETES: MECHANISMS

Capillary nonperfusion can be due either to temporary or permanent occlusion/degeneration. Degenerate capillaries that are detected via histologic preparations of the isolated vasculature (trypsin digest or elastase methods) apparently once were functional capillaries that degenerated until only a basement membrane tube remains. These degenerate capillaries are no longer perfused, and have been used as histologic markers of nonperfused capillaries [10]. Although devoid of nuclei, these degenerate vessels sometimes are not truly acellular, and may be filled with cytoplasmic processes of glial cells [17].

Nonperfusion of capillaries also might be temporary. Temporary occlusions do not always cause damage to the capillary or nearby tissue, but repeated ischemic insults in a chronic disease like diabetes likely could cause progressive injury. Moreover, the neural retina of diabetic animals has been shown to be more sensitive to ischemia [18]. Small nonperfused areas observed in some retinas of diabetic patients later were found to be reperfused, and even the entire fundus became reperfused in a small number of other diabetic patients [19]. It is not clear if the reperfusion occurred in vessels that originally were occluded, or if other patent vessels took their place to supply blood to the ischemic region.

Mechanisms believed to contribute to the nonperfusion and degeneration of retinal capillaries in diabetes include occlusion of the vascular lumen by white blood cells, platelets, or other cells (notably glial cell processes), or altered hemodynamics. These mechanisms are not mutually exclusive.

1. Vasoocclusion by white blood cells. Using either ex vivo or in vivo techniques, diabetes increases adhesion of leukocytes to the vascular wall in diabetic animals [20–34]. Moreover, instances have been reported where the circulation of fluorescent dye injected into

the blood or using in situ (whole mount) perfusion methods is blocked by an immobile leukocyte, suggesting that the leukostasis is contributing to the capillary nonperfusion in diabetic retinopathy [27, 35]. Although individual instances of temporary capillary occlusion by a blood cell might be short-lived, cumulative effects of such repeated ischemia/reperfusion injuries over a prolonged interval are not known. Leukocyte stiffness has been reported to be increased in diabetes, thus making the cells less filterable and more likely to occlude retinal vessels [21, 36]. Abnormal leukocyte adherence to retinal vessels in diabetes occurs via expression of ICAM-1 and other adhesion molecules on the endothelial surface. Diabetes increases expression of ICAM-1 and other adhesion molecules in retinas of animals and humans [24, 28, 37–39], and interaction of this adhesion molecule with the CD18 adhesion molecule on leukocytes contributes to the diabetes-induced increase in adherence of white blood cells to the vascular wall in retinal vessels [24]. Diabetic mice lacking ICAM-1 and CD18 do not develop either the diabetes-induced increase in leukostasis, vascular permeability, or degeneration of retinal capillaries [33], providing strong evidence that white blood cells likely contribute to the eventual capillary damage and degeneration that is characteristic of diabetic retinopathy. Leukocytes have been found to be associated with capillary closure in retinas of spontaneously diabetic monkeys [40].

Although evidence suggesting a role for white blood cells in the development of the retinopathy is accumulating [33, 41, 42], whether or not leukostasis [23, 24, 26, 27, 33, 39, 43, 44] per se is a good parameter of the process of leading to capillary degeneration or diabetic retinopathy is less clear. A disconnect between leukostasis per se and the degeneration of retinal capillaries in diabetes was suggested by evidence that 12-lipoxygenase$^{-/-}$ diabetic mice did not develop the diabetes-induced increase in leukostasis, but nevertheless developed the capillary degeneration of diabetic retinopathy [45].

2. Vasoocclusion by platelets. Platelet microthrombi have been detected in the retinas of diabetic rats and humans, and have been spatially associated with apoptotic endothelial cells [46, 47]. Nevertheless, the selective antiplatelet drug (clopidogrel) did not prevent neuronal apoptosis, glial reactivity, capillary cell apoptosis, or degeneration of retinal capillaries in diabetic rats [48], thus providing no support for a postulated role of platelet aggregation in the development of capillary occlusion in diabetes. Moreover, aspirin (delivered at low doses that should have inhibited platelet aggregation) did not [49] or only modestly [50] inhibited the progression of diabetic retinopathy in clinical trials.
3. Hemodynamics. Many studies of diabetes indicate that there are alterations in blood flow to the retina [51–54]. Reduction in flow might be due to diabetes-induced increase in vascular resistance or viscosity, or to a reduction in metabolic activity in the retina which thus reduces the metabolic demand for flow. Whatever the cause, subsequent impairments to flow, even if slight, have been speculated to allow temporary stasis until backpressure increases.
4. Invasion of the vascular lumen by other cell types. Cellular processes from retinal glial cells have been found inside of occasional degenerate capillaries (identified from the basement membrane tube that surrounds vessels) [17, 55, 56]. It is not clear whether this glial invasion precedes and *causes* the capillary to degenerate or is a *result* of the capillary cells dying (thus opening spaces for the glial cell to expand into).

5. Growth factor withdrawal. Intravitreal administration of VEGF antagonists has been reported to cause apparent nonperfusion or regression of neovascular tufts in diabetic retinopathy [57, 58]. The later reappearance of the neovascular tufts in the same area of retina in some patients [57], however, suggests that the treatment had reduced perfusion of the vessels, but apparently had not caused regression.

MOLECULAR CAUSES OF CAPILLARY DEGENERATION

The molecular mechanisms by which capillary degeneration occurs in diabetes have not been studied in humans, human studies instead focusing on the retinopathy as a whole. Thus, the primary focus of the present discussion on molecular causes of diabetes-induced degeneration of retinal capillaries will focus largely on animal studies. Factors or pathways involved in the capillary degeneration in early stages of diabetic retinopathy have been identified primarily using pharmacologic inhibitors or genetically modified animals.

Metabolic control. Intensive insulin therapy, blood pressure medications, and lipid-lowering therapy all have been shown to inhibit the development of diabetic retinopathy in patients [59–63]. Consistent with this, animal studies have demonstrated that these therapies likewise inhibited degeneration of the retinal vasculature in diabetes [64–67], and they demonstrate that the therapies did inhibit degeneration of the retinal vasculature. Likewise, lipid levels have been shown to influence the development or progression of the retinopathy in diabetic animals [68, 69].

Pathways secondary to poor metabolic control of diabetes. Metabolic sequelae of hyperglycemia have been extensively studied to identify potential causes responsible for the development of diabetic retinopathy and its associated vascular abnormalities. A variety of therapies have reduced the number of TUNEL-positive capillary cells or degenerate capillaries compared to control [27, 33, 39, 44, 48, 67, 70–77], suggesting that related metabolic abnormalities also contribute to the capillary cell death. Tables 1 and 2 summarize a number of therapies or genetic modifications that have been reported to inhibit degeneration of retinal capillaries in diabetic animals. TUNEL-positive retinal capillary cells are a much less reproducible finding in diabetic mice than in diabetic rats (Kern, unpublished).

UNEXPLAINED ASPECTS OF DIABETES-INDUCED DEGENERATION OF RETINAL CAPILLARIES

Nonuniform degeneration of capillaries within the same retina. Despite the evidence indicating that hyperglycemia is a (or the) major determinant of capillary degeneration in diabetic retinopathy, capillary degeneration (like other lesions of the retinopathy) does not develop uniformly across even the same retina of diabetic dogs or patients [78, 79]. The superior temporal portion of retina develops significantly more pathology than, for example, inferior nasal retina. Likewise, midperipheral retina is more prone to undergo capillary nonperfusion in diabetic retinopathy than is the posterior or anterior retina [13].

Why does it take so long for capillary degeneration to become apparent in diabetic retinopathy? As mentioned earlier, vascular remodeling is a normal process, and so all

Table 1. Pharmacologic inhibition of capillary degeneration in retinas from diabetic animals

Presumed target	Drug	Presumed pathway	References	Other possible mechanisms	References
Angiotensin converting enzyme	Captopril	Blood pressure	[67]	Inhibition of glucose uptake into retina	[93]
Caspase-1	Minocycline	Inflammation	[94]	Inhibition of microglia	[95]
Cyclooxygenase	Nepafenac	Inflammation	[76]		
Poly(ADP-ribose) polymerase	PARP inhibitor	Inflammation	[39]		
p38	p38 inhibitor	Inflammation	[96]		
Inflammation	Salicylates	Inflammation	[48, 77, 97]		
TNFα (alpha)[a]	Pegsunercept	Inflammation	[98]		
FOXO1	siRNA against FOXO1	Cell signaling	[99]		
RAGE	sRAGE	Inflammation	[69]		
Aldose reductase[b]	Aldose reductase inhibitor	Metabolic abnormality	[74, 100]	Inflammation (independent of hyperglycemia)	[101–104]
Transketolase	Benfotiamine	Metabolic abnormality	[105]		
Glycation, lipoxidation	Pyridoxamine	Metabolic abnormality	[73]	CD36	[106]
iNOS	Aminoguanidine	Metabolic abnormality	[71]	Inhibit formation advanced glycation endproducts	[107, 108]
AGE formation	Tenilsetam	AGE formation	[109]		
Oxidative stress	Antioxidants	Oxidative stress	[72, 75, 110]		
Oxidative stress	AREDS diet	Oxidative stress	[111]		
TrkA	Nerve growth factor	Indirect action via nonvascular cell	[112], Kern, unpublished	Neuroprotection	[113]

Although not studied in diabetic animals, inhibition of TNFα[a] or aldose reductase[b] inhibited capillary degeneration in galactose-fed animals [114–116]

Table 2. Genetic modifications found to inhibit diabetes-induced capillary degeneration in animals

Gene modification	Target	Presumed action	References
Cu/Zn superoxide dismutase$^{-/-}$	Superoxide dismutase	Oxidative stress	[117]
Mn superoxide dismutase$^{-/-}$	Superoxide dismutase	Oxidative stress	[118]
CD-18$^{-/-}$	Leukocyte adherence	Inflammation	[33]
ICAM-1$^{-/-}$	Leukocyte adherence	Inflammation	[33]
IL-1β receptor$^{-/-}$	IL-1β signaling	Inflammation	[94]
iNOS$^{-/-}$	Nitric oxide production	Inflammation	[44]
5-Lipoxygenase$^{-/-}$	Eicosenoid production	Inflammation	[45]

retinas have at least some degenerate, nonperfused remnants of vessels, and occasional vessels likely become occluded or nonfunctional also throughout life. Diabetes greatly accelerates this process, but prolonged exposure to the abnormal milieu of diabetes (approximately 6 months in rodents, 3 or more years in dogs and other larger mammals, and many years in patients) still is required before the diabetes-induced vasoobliteration becomes clearly greater than normal. Why this prolonged interval is required before the degenerative process become apparent remains a mystery, but understanding it likely will provide valuable information into understanding the pathogenesis of the retinopathy.

Metabolic memory. Capillary degeneration has been observed to continue on for at least some interval after restoration of euglycemia in diabetic dogs [65] and rats [80]. Although not specifically focusing on capillary degeneration, retinopathy likewise was found to progress for about a year in diabetic patients after reinstitution of glycemic control [59]. Various molecular changes also have been found to show this "memory" after prior exposure to elevated glucose concentration [81–84], but their relevance to the vascular degeneration of diabetic retinopathy has not been clearly established.

WHAT IS THE RELATION BETWEEN THE RETINAL VASCULATURE AND NEURONAL RETINA STRUCTURE AND FUNCTION IN DIABETES?

Like other neural tissues, metabolic demand by the neural retina influences the retinal vasculature to regulate blood flow to that tissue [85–87]. Conversely, delivery of oxygen and nutrients and removal of waste by the vasculature influence function of the neural retina. Thus, there is neurovascular coupling which can become altered in diabetes [88]. In a very interesting study, the disruption of that coupling led to protection of the retinal vasculature in diabetes. As expected, the density of the retinal vasculature in wild-type control animals diabetic for many months became subnormal, but this capillary degeneration did not develop in diabetic mice having photoreceptor degeneration (rhodopsin knockout mice) [89]. Thus, the outer retina seemed to somehow influence the retinal vasculature (possibly via effects on metabolism by the inner retina), and loss of the outer retina protected the vasculature despite continued exposure to hyperglycemia. In contrast to this report, however, degeneration of retinal neurons in a different model of retinal degeneration (transgenic rat overexpressing a mutant cilia gene polycystin-2) resulted in *increased* degeneration of retinal capillaries [90]. The neuronal and vascular components of the retina clearly are interactive, but the extent of that interaction under pathologic circumstances requires additional investigation.

Although diabetes-induced defects in metabolism of retinal neurons might impair vision in diabetes independent of vascular disease, occlusion of retinal vessels is closely associated with detrimental effects of diabetes on visual function. The extent of retinal capillary nonperfusion detected by fluorescein angiography has been associated with a reduction in retinal sensitivity as assessed by microperimetry [91]. In addition, diabetic patients with extensive retinal arteriolar and capillary obstruction developed an ischemic maculopathy that resulted in severe loss of visual acuity in some eyes [92]. Whether this is due solely to the vascular nonperfusion or associated neuronal dysfunction is not clear.

CONCLUSION

Vascular abnormalities are major contributors to the morbidity resulting from long-standing diabetes in patients, and capillary nonperfusion and degeneration are especially important in the progression the advanced, proliferative stages of the retinopathy. Clinical attention has focused especially on inhibiting the abnormalities (such as neovascularization and retinal edema) that can have immediate effects on visual impairment, but appreciable retinal vascular damage already will have occurred by focusing on these late stages of the disease. Success is now being made also in the earlier stages of the retinopathy to inhibit capillary degeneration, with hopes that inhibiting the early damage will inhibit development of the more advanced stages of the retinopathy.

ACKNOWLEDGMENT

This work was funded by PHS grant EY00300 and a grant from the Medical Research Service of the Department of Veteran Affairs.

REFERENCES

1. Engerman RL, Meyer RK. Development of retinal vasculature in rats. Am J Ophthalmol. 1965;60:628–41.
2. Hughes S, Chang-Ling T. Roles of endothelial cell migration and apoptosis in vascular remodeling during development of the central nervous system. Microcirculation. 2000;7:317–33.
3. Ishida S et al. Leukocytes mediate retinal vascular remodeling during development and vaso-obliteration in disease. Nat Med. 2003;9:781–8.
4. Hughes S et al. Altered pericyte-endothelial relations in the rat retina during aging: implications for vessel stability. Neurobiol Aging. 2006;27:1838–47.
5. Dorrell MI, Friedlander M. Mechanisms of endothelial cell guidance and vascular patterning in the developing mouse retina. Prog Retin Eye Res. 2006;25:277–95.

6. Smith LE et al. Oxygen-induced retinopathy in the mouse. Invest Ophthalmol Vis Sci. 1994;35:101–11.
7. Madan A, Penn JS. Animal models of oxygen-induced retinopathy. Front Biosci. 2003;8:d1030–43.
8. Smith LE. Pathogenesis of retinopathy of prematurity. Semin Neonatol. 2003;8:469–73.
9. Heidary G, Vanderveen D, Smith LE. Retinopathy of prematurity: current concepts in molecular pathogenesis. Semin Ophthalmol. 2009;24:77–81.
10. Kohner EM, Henkind P. Correlation of fluorescein angiogram and retinal digest in diabetic retinopathy. Am J Ophthalmol. 1970;69:403–14.
11. de Venecia G, Davis MD, Engerman RL. Clinicopathologic correlations in diabetic retinopathy. 1. Histology and fluorescein angiography of microaneurysms. Arch Ophthalmol. 1976;94:1766–73.
12. Sleightholm MA, Aldington SJ, Arnold J, Kohner EM. Diabetic retinopathy: II. Assessment of severity and progression from fluorescein angiograms. J Diabet Complications. 1988;2:117–20.
13. Shimizu K, Kobayashi Y, Muraoka K. Midperipheral fundus involvement in diabetic retinopathy. Ophthalmology. 1981;88:601–12.
14. Witmer AN, Vrensen GF, Van Noorden CJ, Schlingemann RO. Vascular endothelial growth factors and angiogenesis in eye disease. Prog Retin Eye Res. 2003;22:1–29.
15. Jardeleza MS, Miller JW. Review of anti-VEGF therapy in proliferative diabetic retinopathy. Semin Ophthalmol. 2009;24:87–92.
16. Schlingemann RO, Witmer AN. Treatment of retinal diseases with VEGF antagonists. Prog Brain Res. 2009;175:253–67.
17. Engerman RL. Pathogenesis of diabetic retinopathy. Diabetes. 1989;38:1203–6.
18. Kawai SI et al. Modeling of risk factors for the degeneration of retinal ganglion cells after ischemia/reperfusion in rats: effects of age, caloric restriction, diabetes, pigmentation, and glaucoma. FASEB J. 2001;15:1285–7.
19. Takahashi K, Kishi S, Muraoka K, Shimizu K. Reperfusion of occluded capillary beds in diabetic retinopathy. Am J Ophthalmol. 1998;126:791–7.
20. Schroder S, Palinski W, Schmid-Schonbein GW. Activated monocytes and granulocytes, capillary nonperfusion, and neovascularization in diabetic retinopathy. Am J Pathol. 1991;139:81–100.
21. Harris AG, Skalak TC, Hatchell DL. Leukocyte-capillary plugging and network resistance are increased in skeletal muscle of rats with streptozotocin-induced hyperglycemia. Int J Microcirc Clin Exp. 1994;14:159–66.
22. Hatchell DL, Wilson CA, Saloupis P. Neutrophils plug capillaries in acute experimental retinal ischemia. Microvasc Res. 1994;47:344–54.
23. Miyamoto K, Hiroshiba N, Tsujikawa A, Ogura Y. In vivo demonstration of increased leukocyte entrapment in retinal microcirculation of diabetic rats. Invest Ophthalmol Vis Sci. 1998;39:2190–4.
24. Miyamoto K et al. Prevention of leukostasis and vascular leakage in streptozotocin-induced diabetic retinopathy via intercellular adhesion molecule-1 inhibition. Proc Natl Acad Sci USA. 1999;96:10836–41.
25. Nonaka A et al. PKC-beta inhibitor (LY333531) attenuates leukocyte entrapment in retinal microcirculation of diabetic rats. Invest Ophthalmol Vis Sci. 2000;41:2702–6.
26. Ogura Y. In vivo evaluation of leukocyte dynamics in the retinal and choroidal circulation. Jpn J Ophthalmol. 2000;44:322–3.
27. Joussen AM et al. Leukocyte-mediated endothelial cell injury and death in the diabetic retina. Am J Pathol. 2001;158:147–52.

28. Joussen AM et al. Nonsteroidal anti-inflammatory drugs prevent early diabetic retinopathy via TNF-alpha suppression. FASEB J. 2002;16:438–40.
29. Kinoshita N et al. Effective and selective prevention of retinal leukostasis in streptozotocin-induced diabetic rats using gliclazide. Diabetologia. 2002;45:735–9.
30. Mori F et al. Inhibitory effect of losartan, an AT1 angiotensin II receptor antagonist, on increased leucocyte entrapment in retinal microcirculation of diabetic rats. Br J Ophthalmol. 2002;86:1172–4.
31. Moore TC et al. The role of advanced glycation end products in retinal microvascular leukostasis. Invest Ophthalmol Vis Sci. 2003;44:4457–64.
32. Tadayoni R, Paques M, Gaudric A, Vicaut E. Erythrocyte and leukocyte dynamics in the retinal capillaries of diabetic mice. Exp Eye Res. 2003;77:497–504.
33. Joussen AM et al. A central role for inflammation in the pathogenesis of diabetic retinopathy. FASEB J. 2004;18:1450–2.
34. Tamura H et al. Intravitreal injection of corticosteroid attenuates leukostasis and vascular leakage in experimental diabetic retina. Invest Ophthalmol Vis Sci. 2005;46:1440–4.
35. Kinukawa Y, Shimura M, Tamai M. Quantifying leukocyte dynamics and plugging in retinal microcirculation of streptozotosin-induced diabetic rats. Curr Eye Res. 1999;18:49–55.
36. Kelly LW, Barden CA, Tiedeman JS, Hatchell DL. Alterations in viscosity and filterability of whole blood and blood cell subpopulations in diabetic cats. Exp Eye Res. 1993;56:341–7.
37. Lefer DJ, McLeod DS, Merges C, Lutty GA. Immunolocalization of ICAM-1 (CD54) in the posterior eye of sickle cell and diabetic patients. Invest Ophthalmol Vis Sci. 1993;34:1206.
38. McLeod DS, Lefer DJ, Merges C, Lutty GA. Enhanced expression of intercellular adhesion molecule-1 and P-selectin in the diabetic human retina and choroid. Am J Pathol. 1995;147:642–53.
39. Zheng L, Szabo C, Kern TS. Poly(ADP-ribose) polymerase is involved in the development of diabetic retinopathy via regulation of nuclear factor-kappaB. Diabetes. 2004;53:2960–7.
40. Kim SY et al. Neutrophils are associated with capillary closure in spontaneously diabetic monkey retinas. Diabetes. 2005;54:1534–42.
41. Kern TS. Contributions of inflammatory processes to the development of the early stages of diabetic retinopathy. Exp Diabetes Res. 2007;2007:95103.
42. Adamis AP, Berman AJ. Immunological mechanisms in the pathogenesis of diabetic retinopathy. Semin Immunopathol. 2008;30:65–84.
43. Hirata F, Yoshida M, Ogura Y. High glucose exacerbates neutrophil adhesion to human retinal endothelial cells. Exp Eye Res. 2006;82:179–82.
44. Zheng L et al. Critical role of inducible nitric oxide synthase in degeneration of retinal capillaries in mice with streptozotocin-induced diabetes. Diabetologia. 2007;50:1987–96.
45. Gubitosi-Klug RA, Talahalli R, Du Y, Nadler JL, Kern TS. 5-Lipoxygenase, but not 12/15-lipoxygenase, contributes to degeneration of retinal capillaries in a mouse model of diabetic retinopathy. Diabetes. 2008;57:1387–93.
46. Boeri D, Maiello M, Lorenzi M. Increased prevalence of microthromboses in retinal capillaries of diabetic individuals. Diabetes. 2001;50:1432–9.
47. Yamashiro K et al. Platelets accumulate in the diabetic retinal vasculature following endothelial death and suppress blood-retinal barrier breakdown. Am J Pathol. 2003;163:253–9.
48. Sun W, Gerhardinger C, Dagher Z, Hoehn T, Lorenzi M. Aspirin at low-intermediate concentrations protects retinal vessels in experimental diabetic retinopathy through non-platelet-mediated effects. Diabetes. 2005;54:3418–26.
49. Early Treatment Diabetic Retinopathy Research Group. Effects of aspirin treatment on diabetic retinopathy. Ophthalmology. 1991;98:757–65.

50. DAMAD Study Group. Effect of aspirin alone and aspirin plus dipyridamole in early diabetic retinopathy: a multicenter randomized controlled clinical trial. Diabetes. 1989;38:491–8.
51. Grunwald JE, DuPont J, Riva CE. Retinal haemodynamics in patients with early diabetes mellitus. Br J Ophthalmol. 1996;80:327–31.
52. Konno S et al. Retinal blood flow changes in type I diabetes. A long-term follow-up study. Invest Ophthalmol Vis Sci. 1996;37:1140–8.
53. Clermont AC, Bursell SE. Retinal blood flow in diabetes. Microcirculation. 2007;14:49–61.
54. Pemp B, Schmetterer L. Ocular blood flow in diabetes and age-related macular degeneration. Can J Ophthalmol. 2008;43:295–301.
55. Bek T. Immunohistochemical characterization of retinal glial cell changes in areas of vascular occlusion secondary to diabetic retinopathy. Acta Ophthalmol Scand. 1997;75:388–92.
56. Bek T. Glial cell involvement in vascular occlusion of diabetic retinopathy. Acta Ophthalmol Scand. 1997;75:239–43.
57. Schmidinger G, Maar N, Bolz M, Scholda C, Schmidt-Erfurth U. Repeated intravitreal bevacizumab (Avastin(R)) treatment of persistent new vessels in proliferative diabetic retinopathy after complete panretinal photocoagulation. Acta Ophthalmol. 2011;89:76–81.
58. Mendrinos E, Donati G, Pournaras CJ. Rapid and persistent regression of severe new vessels on the disc in proliferative diabetic retinopathy after a single intravitreal injection of pegaptanib. Acta Ophthalmol. 2009;87:683–4.
59. Diabetes Control and Complications Trial Research Group. The effect of intensive treatment of diabetes on the development of long-term complications in insulin-dependent diabetes mellitus. N Engl J Med. 1993;329:977–86.
60. United Kingdom Prospective Diabetes Study. Intensive blood-glucose control with sulphonylureas or insulin compared with conventional treatment and risk of complications in patients with type 2 diabetes. Lancet. 1998;352:837–53.
61. Chaturvedi N et al. Effect of lisinopril on progression of retinopathy in normotensive people with type 1 diabetes. The EUCLID Study Group. EURODIAB Controlled Trial of Lisinopril in Insulin-Dependent Diabetes Mellitus. Lancet. 1998;351:28–31.
62. UK Prospective Diabetes Study Group. Tight blood pressure control and risk of macrovascular and microvascular complications in type 2 diabetes: UKPDS 38. UK Prospective Diabetes Study Group. BMJ. 1998;317:703–13.
63. Keech AC et al. Effect of fenofibrate on the need for laser treatment for diabetic retinopathy (FIELD study): a randomised controlled trial. Lancet. 2007;370:1687–97.
64. Engerman RL, Bloodworth Jr JMB, Nelson S. Relationship of microvascular disease in diabetes to metabolic control. Diabetes. 1977;26:760–9.
65. Engerman RL, Kern TS. Progression of incipient diabetic retinopathy during good glycemic control. Diabetes. 1987;36:808–12.
66. Hammes H-P et al. Islet transplantation inhibits diabetic retinopathy in the sucrose-fed diabetic Cohen diabetic rat. Invest Ophthalmol Vis Sci. 1993;34:2092–6.
67. Zhang JZ, Xi X, Gao L, Kern TS. Captopril inhibits capillary degeneration in the early stages of diabetic retinopathy. Curr Eye Res. 2007;32:883–9.
68. Hammes HP et al. Acceleration of experimental diabetic retinopathy in the rat by omega-3 fatty acids. Diabetologia. 1996;39:251–5.
69. Barile GR et al. The RAGE axis in early diabetic retinopathy. Invest Ophthalmol Vis Sci. 2005;46:2916–24.
70. Mizutani M, Kern TS, Lorenzi M. Accelerated death of retinal microvascular cells in human and experimental diabetic retinopathy. J Clin Invest. 1996;97:2883–90.
71. Kern TS et al. Response of capillary cell death to aminoguanidine predicts the development of retinopathy: comparison of diabetes and galactosemia. Invest Ophthalmol Vis Sci. 2000;41:3972–8.

72. Kowluru RA, Tang J, Kern TS. Abnormalities of retinal metabolism in diabetes and experimental galactosemia. VII. Effect of long-term administration of antioxidants on the development of retinopathy. Diabetes. 2001;50:1938–42.
73. Stitt A et al. The AGE inhibitor pyridoxamine inhibits development of retinopathy in experimental diabetes. Diabetes. 2002;51:2826–32.
74. Asnaghi V, Gerhardinger C, Hoehn T, Adeboje A, Lorenzi M. A role for the polyol pathway in the early neuroretinal apoptosis and glial changes induced by diabetes in the rat. Diabetes. 2003;52:506–11.
75. Kowluru RA, Odenbach S. Effect of long-term administration of alpha-lipoic acid on retinal capillary cell death and the development of retinopathy in diabetic rats. Diabetes. 2004;53:3233–8.
76. Kern TS et al. Topical administration of nepafenac inhibits diabetes-induced retinal microvascular disease and underlying abnormalities of retinal metabolism and physiology. Diabetes. 2007;56:373–9.
77. Zheng L, Howell SJ, Hatala DA, Huang K, Kern TS. Salicylate-based anti-inflammatory drugs inhibit the early lesion of diabetic retinopathy. Diabetes. 2007;56:337–45.
78. Kern TS, Engerman RL. Vascular lesions in diabetes are distributed non-uniformly within the retina. Exp Eye Res. 1995;60:545–9.
79. Tang J, Mohr S, Du Y, Kern TS. Non-uniform distribution of lesions and biochemical abnormalities within the retina of diabetic humans. Curr Eye Res. 2003;27:7–13.
80. Su EN et al. Continued progression of retinopathy despite spontaneous recovery to normoglycemia in a long-term study of streptozotocin-induced diabetes in rats. Graefes Arch Clin Exp Ophthalmol. 2000;238:163–73.
81. Kowluru RA. Effect of reinstitution of good glycemic control on retinal oxidative stress and nitrative stress in diabetic rats. Diabetes. 2003;52:818–23.
82. Kowluru RA, Chakrabarti S, Chen S. Re-institution of good metabolic control in diabetic rats and activation of caspase-3 and nuclear transcriptional factor (NF-kappaB) in the retina. Acta Diabetol. 2004;41:194–9.
83. Kowluru RA, Kanwar M, Kennedy A. Metabolic memory phenomenon and accumulation of peroxynitrite in retinal capillaries. Exp Diabetes Res. 2007;2007:21976.
84. El-Osta A et al. Transient high glucose causes persistent epigenetic changes and altered gene expression during subsequent normoglycemia. J Exp Med. 2008;205:2409–17.
85. Feke GT, Zuckerman R, Green GJ, Weiter JJ. Response of human retinal blood flow to light and dark. Invest Ophthalmol Vis Sci. 1983;24:136–41.
86. Ferrez PW, Chamot SR, Petrig BL, Pournaras CJ, Riva CR. Effect of visual stimulation on blood oxygenation in the optic nerve head of miniature pigs: a pilot study. Klin Monbl Augenheilkd. 2004;221:364–6.
87. Hardarson SH et al. Oxygen saturation in human retinal vessels is higher in dark than in light. Invest Ophthalmol Vis Sci. 2009;50:2308–11.
88. Riva CE, Logean E, Falsini B. Visually evoked hemodynamical response and assessment of neurovascular coupling in the optic nerve and retina. Prog Retin Eye Res. 2005;24:183–215.
89. de Gooyer TE et al. Retinopathy is reduced during experimental diabetes in a mouse model of outer retinal degeneration. Invest Ophthalmol Vis Sci. 2006;47:5561–8.
90. Feng Y et al. Vasoregression linked to neuronal damage in the rat with defect of polycystin-2. PLoS One. 2009;4:e7328.
91. Unoki N et al. Retinal sensitivity loss and structural disturbance in areas of capillary nonperfusion of eyes with diabetic retinopathy. Am J Ophthalmol. 2007;144:755–60.
92. Bresnick GH, De Venecia G, Myers FL, Harris JA, Davis MD. Retinal ischemia in diabetic retinopathy. Arch Ophthalmol. 1975;93:1300–10.

93. Zhang J-Z, Kern TS. Captopril inhibits intracellular glucose accumulation in retinal cells in diabetes. Invest Ophthalmol Vis Sci. 2003;44:4001–5.
94. Vincent JA, Mohr S. Inhibition of caspase-1/Interleukin-1β signaling prevents degeneration of retinal capillaries in diabetes and galactosemia. Diabetes. 2007;56:224–30.
95. Krady JK et al. Minocycline reduces proinflammatory cytokine expression, microglial activation, and caspase-3 activation in a rodent model of diabetic retinopathy. Diabetes. 2005;54:1559–65.
96. Du, Y. et al. Inhibition of p38 MAPK inhibits early stages of diabetic retinopathy. 2010;51:2158–64.
97. Sun W, Hoenh T, Gerhardinger C, Lorenzi M. Antiplatelet/anti-inflammatory drugs do not prevent early neuroretinal apoptosis and glial changes in diabetic rats (American Diabetes Association abstract). Diabetes 2004;899-P.
98. Behl Y et al. Diabetes-enhanced tumor necrosis factor-alpha production promotes apoptosis and the loss of retinal microvascular cells in type 1 and type 2 models of diabetic retinopathy. Am J Pathol. 2008;172:1411–8.
99. Behl Y, Krothapalli P, Desta T, Roy S, Graves DT. FOXO1 plays an important role in enhanced microvascular cell apoptosis and microvascular cell loss in type 1 and type 2 diabetic rats. Diabetes. 2009;58:917–25.
100. Dagher Z et al. Studies of rat and human retinas predict a role for the polyol pathway in human diabetic retinopathy. Diabetes. 2004;53:2404–11.
101. Ramana KV, Bhatnagar A, Srivastava SK. Inhibition of aldose reductase attenuates TNF-alpha-induced expression of adhesion molecules in endothelial cells. FASEB J. 2004;18:1209–18.
102. Ramana KV, Friedrich B, Srivastava S, Bhatnagar A, Srivastava SK. Activation of nuclear factor-kappaB by hyperglycemia in vascular smooth muscle cells is regulated by aldose reductase. Diabetes. 2004;53:2910–20.
103. Ramana KV et al. Endotoxin-induced cardiomyopathy and systemic inflammation in mice is prevented by aldose reductase inhibition. Circulation. 2006;114:1838–46.
104. Tammali R, Ramana KV, Singhal SS, Awasthi S, Srivastava SK. Aldose reductase regulates growth factor-induced cyclooxygenase-2 expression and prostaglandin e2 production in human colon cancer cells. Cancer Res. 2006;66:9705–13.
105. Hammes HP et al. Benfotiamine blocks three major pathways of hyperglycemic damage and prevents experimental diabetic retinopathy. Nat Med. 2003;9:294–9.
106. Murakoshi M et al. Pleiotropic effect of pyridoxamine on diabetic complications via CD36 expression in KK-Ay/Ta mice. Diabetes Res Clin Pract. 2009;83:183–9.
107. Hammes H-P, Martin S, Federlin K, Geisen K, Brownlee M. Aminoguanidine treatment inhibits the development of experimental diabetic retinopathy. Proc Natl Acad Sci USA. 1991;88:11555–8.
108. Hammes H-P et al. Aminoguanidine inhibits the development of accelerated diabetic retinopathy in the spontaneous hypertensive rat. Diabetologia. 1994;37:32–5.
109. Hoffmann J et al. Tenilsetam prevents early diabetic retinopathy without correcting pericyte loss. Thromb Haemost. 2006;95:689–95.
110. Hammes HP, Bartmann A, Engel L, Wulfroth P. Antioxidant treatment of experimental diabetic retinopathy in rats with nicanartine. Diabetologia. 1997;40:629–34.
111. Kowluru RA, Kanwar M, Chan PS, Zhang JP. Inhibition of retinopathy and retinal metabolic abnormalities in diabetic rats with AREDS-based micronutrients. Arch Ophthalmol. 2008;126:1266–72.

112. Hammes H-P, Federoff HJ, Brownlee M. Nerve growth factor prevents both neuroretinal programmed cell death and capillary pathology in experimental diabetes. Mol Med. 1995;1:527–34.
113. Sofroniew MV, Howe CL, Mobley WC. Nerve growth factor signaling, neuroprotection, and neural repair. Annu Rev Neurosci. 2001;24:1217–81.
114. Robison Jr WG, Tillis TN, Laver N, Kinoshita JH. Diabetes-related histopathologies of the rat retina prevented with an aldose reductase inhibitor. Exp Eye Res. 1990;50:355–66.
115. Robison Jr WG, Laver NM, Jacot JL, Glover JP. Sorbinil prevention of diabetic-like retinopathy in the galactose-fed rat model. Invest Ophthalmol Vis Sci. 1995;36:2368–80.
116. Joussen AM et al. TNF-alpha mediated apoptosis plays an important role in the development of early diabetic retinopathy and long-term histopathological alterations. Mol Vis. 2009;15:1418–28.
117. Berkowitz BA, Gradianu M, Bissig D, Kern TS, Roberts R. Retinal ion regulation in a mouse model of diabetic retinopathy: natural history and the effect of Cu/Zn superoxide dismutase overexpression. Invest Ophthalmol Vis Sci. 2009;50:2351–8.
118. Kanwar M, Chan PS, Kern TS, Kowluru RA. Oxidative damage in the retinal mitochondria of diabetic mice: possible protection by superoxide dismutase. Invest Ophthalmol Vis Sci. 2007;48:3805–11.

10
Proteases in Diabetic Retinopathy

Sampathkumar Rangasamy, Paul McGuire, and Arup Das

CONTENTS

Keywords Urokinase Plasminogen activator (uPA) • Matrix Metalloproteinases (MMPs) • Tissue inhibitors of metalloproteinases (TIMPs) • Plasminogen activator inhibitors (PAI)

PROTEASES IN RETINAL VASCULATURE

The human retinal structure along with the neuronal component develops from a single layer of undifferentiated neuroepithelial cells during embryonic ontogenesis. During this process, retinal vasculature develops to form an elaborate vascular tree that matches the metabolic need of tissues. Retinal vascular development involves a complex process of vasculogenesis and angiogenesis. Vasculogenesis describes the de novo formation of vessels from vascular endothelial precursor cells (angioblasts), which migrate to or differentiate at the location of future vessels, coalesce into cords, and differentiate into endothelial cells leading to the formation of ultimate vessels [1]. Angiogenesis is a multistep process that requires degradation of the basement membrane, endothelial cell migration and proliferation, and the capillary tube formation, which results in sprouting of new capillaries from the existing blood vessels. Also, new evidence indicates that the bone marrow–derived endothelial progenitor cells contribute to the postnatal neovascularization.

From: *Ophthalmology Research: Visual Dysfunction in Diabetes*
Edited by: J. Tombran-Tink et al. (eds.), DOI 10.1007/978-1-60761-150-9_10

Angiogenesis plays a central part not only in the development of retina but also in the visual impairment attributable to retinopathy in diabetes, retinal vascular occlusion, retinopathy of prematurity, sickle cell disease, and in age-related macular degeneration. The process of angiogenesis in the retina and other tissues is characterized by distinct phases or activities including an initial response to locally produced angiogenic factors and signals. This event is followed by a rapid upregulation of matrix-degrading enzymes or extracellular proteases (extracellular proteolytic mediators) that facilitate the breakdown of the capillary basal lamina and migration and subsequent invasion of activated endothelial cells into the surrounding extracellular tissues [2, 3]. Extracellular proteases help not only in the degradation of interstitial extracellular matrices (ECMs) and basement membranes but also in the recruitment of progenitor cells into the ECM during tissue remodeling. Proteases are expressed by normal cells in tissue remodeling events and also during pathological events such as tumor angiogenesis and metastasis. This chapter will review these extracellular proteases and discuss their potential roles in diabetic retinopathy and the development of therapeutic strategies targeting these molecules in preventing retinal neovascularization and diabetic macular edema.

Extracellular Proteases

The ECM is a complex assembly of proteins and polysaccharides which provides the physical support and organization to tissues. Cell-surface receptors on the plasma membrane bind to ECM and regulate intracellular signaling pathways that control cell migration and proliferation. Cell migration often involves the coordination of ECM proteolysis, adhesion, and signaling. The important enzymes that are primarily involved in the process of ECM proteolysis are the serine proteases that include (1) urokinase plasminogen activator (uPA) and (2) members of the family of zinc-dependent endopeptidases called matrix metalloproteinases (MMPs).

Urokinase Plasminogen Activator System (uPA/uPAR System)

The proteolytically active urokinase (uPA) on the endothelial cell surface is critical for cell migration. The uPA is produced as an inactive single-chain protein known as pro-uPA, which binds to uPAR (uPA receptor) and is activated by plasmin [4]. Receptor-bound pro-uPA is more rapidly cleaved by plasmin than the unbound form. The uPA is present in cells in two molecular forms, a 54 kDa high-molecular-weight form and a 32 kDa low-molecular-weight form which lacks the amino-terminal fragment (ATF) of the protein [5–7]. The ATF contains the growth factor and kringle domains of the protein that mediate binding to uPAR and play an important role in cell proliferation [8]. The main function of the uPA is to convert the inactive zymogen form of the enzyme plasminogen to plasmin, a broad spectrum of proteinase, which can cleave a variety of ECM components including collagen IV, fibronectin, and elastin including uPA (Fig. 1). The invasive and migratory potential of endothelial cells is largely determined upon the pool of active urokinase available on the cell surface. The uPA has also shown to directly activate the prohepatocyte growth factor/scatter factor (HGF/SF), and it also cleaves fibronectin and its own inhibitor, plasminogen activator inhibitor-1 (PAI-1), in

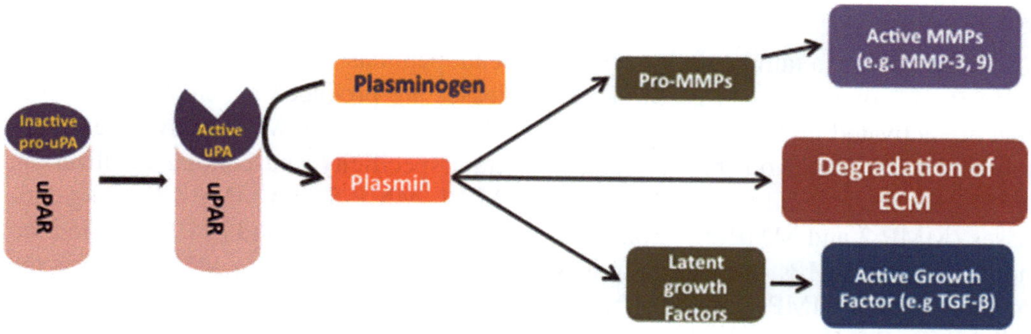

Fig. 1. uPA/uPAR in the degradation of ECM. Binding of inactive urokinase (pro-uPA) to urokinase receptor (uPAR) activates uPA. Active uPA proteolytically converts the inactive zymogen plasminogen to active plasmin, which then breaks down ECM components or activates latent growth factors such as transforming growth factor 1 (TGF-1). Plasmin can also degrade the ECM indirectly through activation of promatrix metalloproteinases (pro-MMPs).

a plasminogen-independent manner. The uPA/uPAR interaction represents a sensitive and flexible system to regulate proteolytic potential in endothelial cells. The uPAR is a cell-surface molecule that interacts with many potential ligands including uPA and vitronectin. The uPAR has also shown to be associated with several members of the integrin family which plays an important role in cell adhesion and migration [9]. This process is mediated through the low-density-lipoprotein-receptor-related protein (LRP), a multiligand receptor that can interact with both PAI-1 and uPAR. The uPA system also plays an important role in the activation of several MMPs and in the release and activation of growth factors stored in the ECM [10]. The contribution of the uPA/uPAR system to angiogenesis has been studied in several animal models of tumor angiogenesis, choroidal angiogenesis, and retinal angiogenesis. Many studies show that in addition to regulating proteolysis, uPAR is a signaling receptor that promotes cell motility, invasion, proliferation, and survival. Signaling through uPAR has been shown to activate many pathways involving kinases such as Ras–mitogen-activated protein kinase (MAPK) pathway [11]. These signaling events have been shown to involve the binding of its ligand such as uPA (independent of uPA proteolytic activity) and vitronectin.

The uPAR is a member of the lymphocyte antigen 6 (Ly-6) superfamily of proteins that are characterized by the Ly-6 and uPAR (LU) domain, also called the three-finger fold [12]. The LU domain folds into a globular structure with 5–6 antiparallel β-strands linked by 4–5 disulfide bonds [12, 13]. The uPAR contains three LU domains, designated D1–D3, connected by short linker regions, and these three domains pack together into a concave structure [14–16] in which the ligands such as uPA and vitronectin bind. Recent studies have indicated the importance of uPAR in human diseases, including many cancers. Hence, therapeutic targeting of uPAR is considered as an important concept to interrupt proteolytic cascades and block intracellular signaling in disease pathogenesis [17].

Matrix Metalloproteinases

The MMPs are a family of zinc-containing endopeptidases that are capable of degrading various components of ECM. These proteases are produced as latent proenzymes that are activated proteolytically. At least 21 different types of MMPs have been identified to date. Based on their structure/substrate specificity and cellular localization, MMPs are grouped into the collagenases (MMP-1, MMP-8, and MMP-13), the gelatinases (MMP-2 and MMP-9), stromelysins (MMP-3, MMP-10, and MMP-11), and the nontraditional MMPs (matrilysin or MMP-7 and metalloelastase or MMP-12) and the membrane-type MMPs (MT-MMPs) [3, 18]. There are at least five distinct types of MT-MMPs (MMP-12, -15, -16, -17 and -21), and these MMPs are bound to cell surface through C-terminal transmembrane domain or glycosylphosphatidylinositol anchor. The MT-MMPs can degrade gelatin, fibronectin, and other ECM substrates [19, 20].

The basic structure of the MMPs contains the following domains that include (a) pre- or signal-peptide domain that directs MMPs to the secretory or plasma membrane insertion pathway; (b) prodomain that confers latency to the enzymes by occupying the active-site zinc, making the catalytic enzyme inaccessible to substrates; (c) zinc-containing catalytic domain; and (d) hemopexin domain or the C-terminal domain which mediates interactions with substrates and confers specificity of the enzymes, and also, it is connected to the catalytic domain by a flexible hinge region or linker region [21] (Fig. 2).

Various members of the MMPs have been implicated in a wide range of physiological and pathological processes, including wound healing, angiogenesis, inflammation, and tumor metastases [22–24]. During the physiological and pathological processes, the MMP functions included the proteolytic cleavage of ECM structures and destruction of cell-surface proteins and proteinase inhibitors. In addition to their capacity to degrade a large variety of ECM molecules, MMPs are known to process a number of bioactive molecules, and in many cases, MMP action leads to the proteolytic activation or release of latent signaling molecules and proteases including cytokines [25]. MMPs regulate a variety of cell behaviors such as cell proliferation, migration, differentiation, apoptosis, and host defense (Fig. 3).

Studies have shown that MMPs are one of the important molecules in the cascade of angiogenesis process and can be considered as proangiogenic agents. Specific MMPs have been shown to induce angiogenesis by detaching the pericytes from vessel wall and thereby releasing ECM-bound angiogenic growth factors. Also, this process has been implicated in the exposure of cryptic proangiogenic integrins binding sites in the ECM through the cleavage of endothelial cell–cell adhesion [26, 27]. Degradation of ECM releases ECM/basement membrane–sequestered angiogenic factors such as VEGF, bFGF, and TGF-β [28]. MMPs have been shown to have multiple effects on endothelial cells themselves. As mentioned earlier, MMPs facilitate endothelial cell migration and tube formation [29, 30]. Exogenous MMP-9 has been shown to enhance endothelial cell growth in vitro [31]. The cleavage of the ectodomain of VE-cadherin by MMPs is considered as an important event in the breaking of cell–cell adhesions [32]. MMPs involved in angiogenesis have been shown to originate from the infiltrating inflammatory cells or from endothelial cells. MMPs are synthesized in response

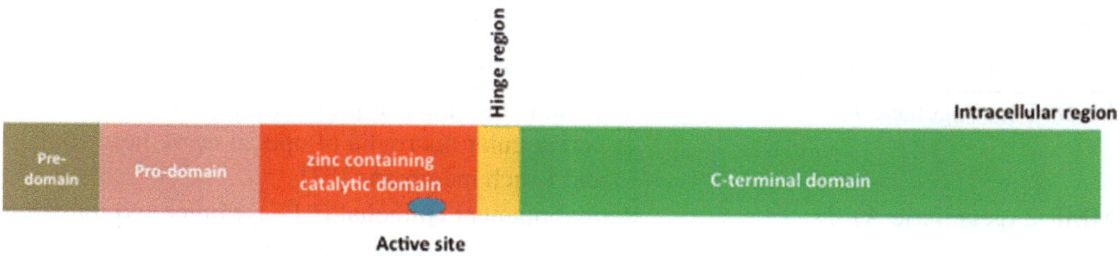

Fig. 2. Basic domain structure of MMPs. The domain structure of MMPs includes (a) pre- or signal-peptide domain that directs MMPs to the secretory or plasma membrane insertion pathway, (b) prodomain, (c) zinc-containing catalytic domain, and (d) hemopexin domain or the C-terminal domain. The catalytic domain is connected to the C-terminal domain by a flexible hinge region. The C-terminal domain has structural similarity to the serum protein hemopexin and is also called as hemopexin domain.

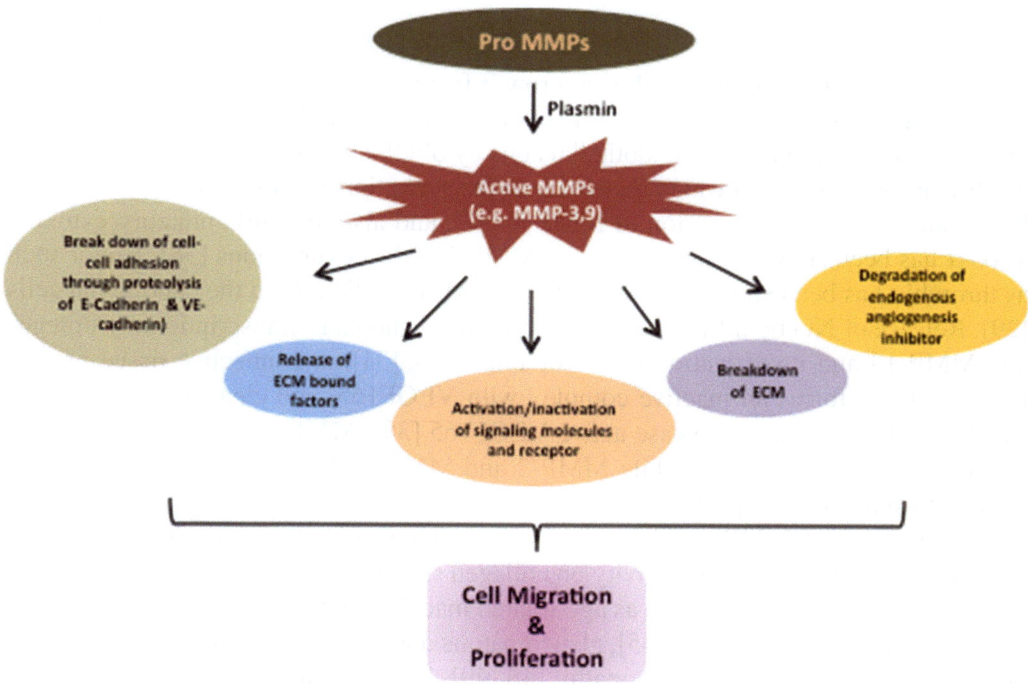

Fig. 3. Matrix metalloproteinases cellular function. Activation of MMPs leads to the proteolytic degradation of various cellular substrates. Also, MMPs induce the release of ECM-bound growth factors and the degradation of angiogenesis inhibitors. Through the coordinate action including activation of many molecules, MMPs promote cell growth, migration, and proliferation resulting in angiogenesis.

Table 1. Different types of MMPs expressed in the retina

Matrix metalloproteinases (MMPs)	Retinal expression
MMP-1 (collagenase 1)	Inner and outer nuclear plexiform layers [51], perivascular microglia of optical nerve head [52], and Bruch membrane [53]
MMP-2 (gelatinase A)	Retinal pigment epithelial (RPE), Muller and retinal capillaries, perivascular microglia of nerve head [52], and Bruch membrane [53]
MMP-3 (stromelysin-1)	Perivascular microglia of nerve head [52] and Bruch membrane [53]
MMP-9 (gelatinase A)	Retinal pigment epithelial (RPE), Muller cells, retinal capillaries [54, 55], and Bruch membrane [53]
MMP-14 (membrane-type MMP)	Perivascular microglia of nerve head [53]
ADAM15 (disintegrin and metalloproteinase domain–containing protein 15)	Retinal capillaries [56]

to diverse stimuli including cytokines, growth factors, hormones, and oxidative stress [33, 34]. Basic fibroblast growth factor (bFGF) induces endothelial MMP-9 expression via AP-1 [35]. Stimulation of endothelial cells by bFGF also upregulates the expression of uPA and integrin $\alpha v\beta 3$ which then leads to the activation MMPs [36]. VEGF has also been indicated in the expression of MMP-1 [37], and also, the inflammatory cytokine TNF-α has been shown to upregulate the MMP-2 and -9 expressions [38]. Factor such as thrombin has been shown to activate the pro-MMP-2 directly in the endothelial cells [29]. Release of NO by inflammatory cells has been shown to transcriptionally upregulate MMP-13 and its activation by endothelial cells [34]. A connective tissue growth factor (CTGF) forms an inactive complex with VEGF165, and cleavage of CTGF by MMPs has been shown to release active VEGF165 [39]. MMP-2 has been indicated in the release of latent TGF-1, while MMP-2 and MMP-9 cleave the latency-associated peptide to activate TGF-β1 [40, 41].

The presence of MMPs in the eye has been demonstrated as early as 1968 in the cornea through its proteolytic activity on collagen substrate [42]. MMPs have been indicated in many eye disorders such as age-related macular degeneration [43], proliferative diabetic retinopathy (PDR) [44, 45], glaucomatous optic nerve head damage [46], vitreal liquification [47], and vitreoretinopathy [48, 49]. The cellular origin of the MMPs in these studies is still not clear, but it is likely that the expression would come from the resident cells, invading vasculature, and the inflammatory cells [50]. The importance of MMPs in the retinal pathology is currently well known, and many recent studies have demonstrated the presence of various MMPs such as MMP-1, MMP-2, MMP-3, MMP-9, MMP-13, and MMP-14 that are expressed at different retinal tissues (Table 1). Regardless of the sources in the retina, MMPs are considered as an attractive therapeutic target to treat proliferative diabetic retinopathy (PDR) and diabetic macular edema (DME).

Endogenous Inhibitors of Proteases

Tissue Inhibitors of Metalloproteinases (TIMPs)

Tissue inhibitors of metalloproteinases (TIMPs) are specific endogenous inhibitors that bind MMPs in 1:1 stoichiometry. Four types of TIMPs have been identified, which have overlapping activities with respect to their inhibition of most soluble MMPs [3, 57]. TIMP molecules block the activity of MMPs by binding to the active-site cleft, in a manner similar to their substrates. TIMP-2 and TIMP-3 are good inhibitors of MT1-MMP, MMP-19, and ADAM-17, while TIMP-1 is poor in this respect. TIMPs also have biological effects unrelated to metalloproteinase inhibition. TIMP-4 has been shown to be localized mostly in the vascular tissues. TIMP-3, but neither TIMP-1 nor TIMP-2, is involved in binding to the VEGF receptor-2 (KDR) and competes for debinding of VEGF to this receptor. Overexpression of TIMP-3 can induce apoptosis. TIMP-1 and TIMP-2 display antiapoptotic properties and indirectly induce cell signaling. A point mutation of the TIMP-3 gene has been implicated in Sorsby's fundus dystrophy, an autosomal dominant macular disease similar to wet macular degeneration, but with earlier onset of symptoms [58, 59]. Studies have also characterized protein molecules such as RECK, α2-macroglobulin (α2M), and tissue factor pathway inhibitor-2 to inhibit the MMP activity [60–62].

Plasminogen Activator Inhibitors (PAI)

Plasminogen activator inhibitors (PAIs), which are members of the serine proteinase inhibitor (SERPIN) family, regulate the proteolytic activity of uPA through the inhibition of uPA and plasmin formation. PAI-1 and PAI-2 have been found to interact with urokinase in 1:1 ratio to inhibit enzyme activity and cause enzyme/inhibitor internalization and turnover [63]. These serine protease inhibitors (SERPINS) bind covalently to their targets, inhibiting proteolytic activity [64]. PAI-1 has been shown to be a strong prognostic marker for several cancer types [65].

PROTEASES IN RETINAL NEOVASCULARIZATION

Significant upregulation of uPA (both the 54- and 32-kDa isoforms) along with increases in secretion and activation of MMP-2 and -9 was observed in the retinas of animals with neovascularization [45]. These results suggest that proteolytic activity and its regulatory mechanisms might play an important role in the angiogenic process. Various studies have shown that the plasma levels and the fibrovascular epiretinal membranes MMP-2 and -9 levels were significantly elevated in patients with PDR [66–71]. Examination of proteases in epiretinal neovascular membranes removed surgically from humans with PDR showed a similar increase in the levels of uPA and proform and active form of MMP-2 and -9 as compared to normal retinas [72]. It has been also shown that the pro-MMP-2 is efficiently activated in the fibrovascular tissues of PDR through interaction with MT1-MMP and TIMP-2 [73], indicating its increased role in PDR. Further, MMP-2 deficiency in mice has been shown to reduce the retinal angiogenesis [74]. Type 1 diabetes subjects with retinopathy have displayed elevated systemic levels of

MMP-9 and MMP-9/TIMP-1 ratio. This increased level of MMP-9 has been suggested to be surrogate biomarkers of retinopathy in type 1 diabetic patients free of other vascular complications [75].

Further characterization of roles of MMPs in diabetic retinopathy revealed various molecular aspects of its activity and regulation. Hyperglycemic condition induces the increased activation of retinal capillary MMP-2 and MT1-MMP and decreases in TIMP-2. This activation was inhibited by superoxide scavengers, and their accelerated apoptosis was prevented by the inhibitors of MMP-2 [76]. The hyperglycemia-induced activation of MMP-9 accelerates apoptosis of retinal capillary cells, a phenomenon that predicts the development of diabetic retinopathy, and the activation of MMP-9 is downstream of H-Ras. Also, inhibition of high glucose–activated MMP-9 by pharmacologic inhibitor or siRNA ameliorated accelerated apoptosis in the retinal endothelial cells [77]. Interestingly, the human retinal pericytes treated with high glucose levels have been shown to have increased MMP-2 activity leading to increased ECM turnover while there was no MMP-9 activity observed in those cells. Thiamine and benfotiamine correct the increase in MMP-2 activity due to high glucose in HRP, while increasing TIMP-1 levels in the pericytes [78]. MMP inhibitor such as α2M has been shown to play a key role in the control and regulation of the retinal neovascularization involved in the pathogenesis of PDR [79]. The transcription factor such as AP-1 and JUN has been shown to regulate retinal MMP synthesis during neovascularization. The importance of transcriptional factor as a therapeutic target that regulates the expression of MMPs such as MMP-2 in microvascular endothelial growth and retinal neovascularization is also considered [80, 81].

Angiogenesis and matrix degradation are an important step in endothelial cell migration and proliferation. Evidence has indicated the role of serine proteases, such as tissue plasminogen activator (TPA), urokinase-type plasminogen activator (UPA), and PAI, in the retinal neovascularization. In PDR, the vitreous levels of these proteases are increased [82]. At cellular level, hyperglycemic condition has been shown to alter the levels of t-PA and PAI in the retinal microvascular endothelial cells [83]. In an animal model of hypoxia-induced retinal neovascularization, it was found that the expression of the urokinase receptor (uPAR) was required to mediate an angiogenic response. uPAR$^{-/-}$ mice demonstrated normal retinal vascularity but showed a significant reduction (by 73%) in the extent of pathological neovascularization as compared to wild-type controls (Fig. 4). The expression of uPAR mRNA was upregulated in experimental animals during the active phase of angiogenesis, and uPAR protein was localized to endothelial cells in the superficial layers of the retina [84–86].

TISSUE INHIBITOR OF MATRIX METALLOPROTEINASES IN RETINAL NEOVASCULARIZATION

In the retinas of normal mice, TIMP-2 mRNA and protein levels have been found to increase steadily between postnatal days 13 and 17. This was in contrast to retinas of mice with hypoxia-induced retinal angiogenesis, in which TIMP-2 mRNA and protein remained low and significantly less than in retinas of "room air" controls [84]. Thus, a temporal correlation between proteases (MMP-2 and -9 and MT1-MMP) and TIMP-2 was seen in retinas with neovascularization as compared to controls [85].

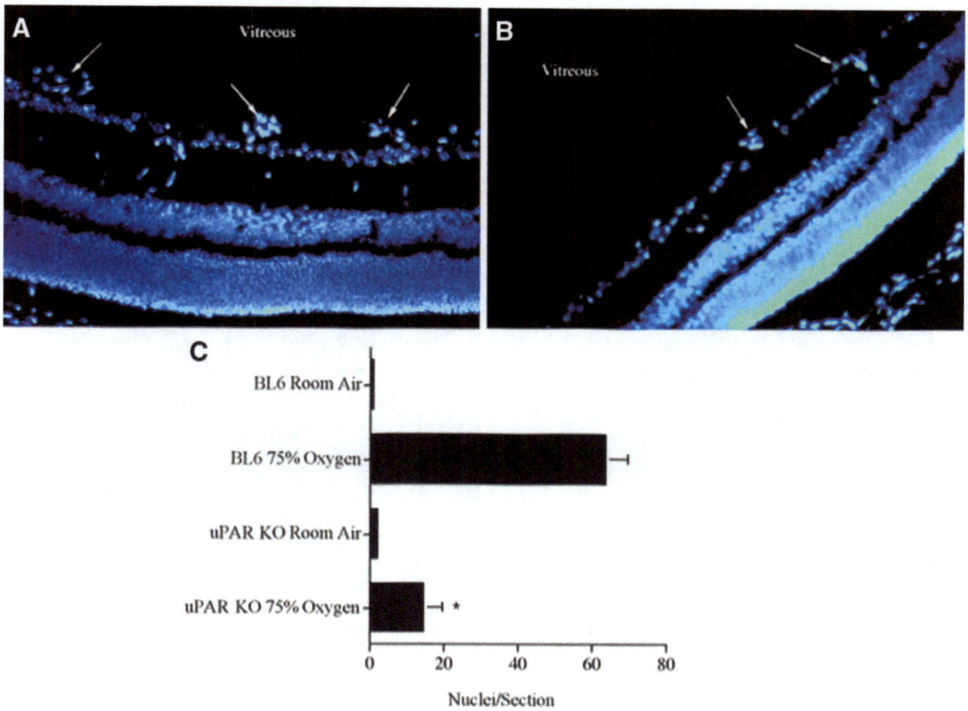

Fig. 4. Absence of the urokinase receptor uPAR reduced the extent of retinal neovascularization in the mouse. (**A**) Representative section of the retina from an experimental oxygen-treated P17 C57BL6 mouse demonstrating numerous neovascular tufts on the surface of the retina (*arrows*). (**B**) A similar section from an experimental oxygen-treated P17 uPAR$^{-/-}$ mouse with many fewer vascular tufts (*arrows*). (**C**) Quantitation of neovascularization in C57BL6 and uPAR$^{-/-}$ mice. The uPAR$^{-/-}$ mice demonstrated 73% less neovascularization compared with the normal C57BL6 mice. Values are the mean ± SEM for $n=4$ mice in each group (eight eyes, 15–20 sections/eye). *Significantly less than in C57BL6 mice, $P<0.01$ (reproduced with permission from McGuire et al. [84]).

Inhibition of Retinal Angiogenesis by MMP Inhibitors

Inhibition of MMP activity is considered an important therapeutic option in the prevention of diabetic retinopathy. Preclinical studies have shown the importance of these inhibitors in the prevention of retinal neovascularization. Systemic injection of a broad-spectrum MMP inhibitor, BB-94 (1 mg/kg), in the murine model has been shown to suppress retinal neovascularization by 72% [45] (Fig. 5). The retinas of BB-94-treated animals demonstrated a significant decrease in the levels of active forms of MMP-2 and MMP-9 compared to controls. In a mouse model of OIR, the extent of preretinal neovascularization was drastically reduced in MMP-2$^{-/-}$ (75%) and MMP-9$^{-/-}$ mice (44%) at postnatal day 19, compared to wild-type control mice [45]. The functional association of MMP-2 and $\alpha v\beta 3$ on the cell surface of angiogenic blood vessels points to the ability of MMPs to regulate cell adhesion and integrin-mediated behavior.

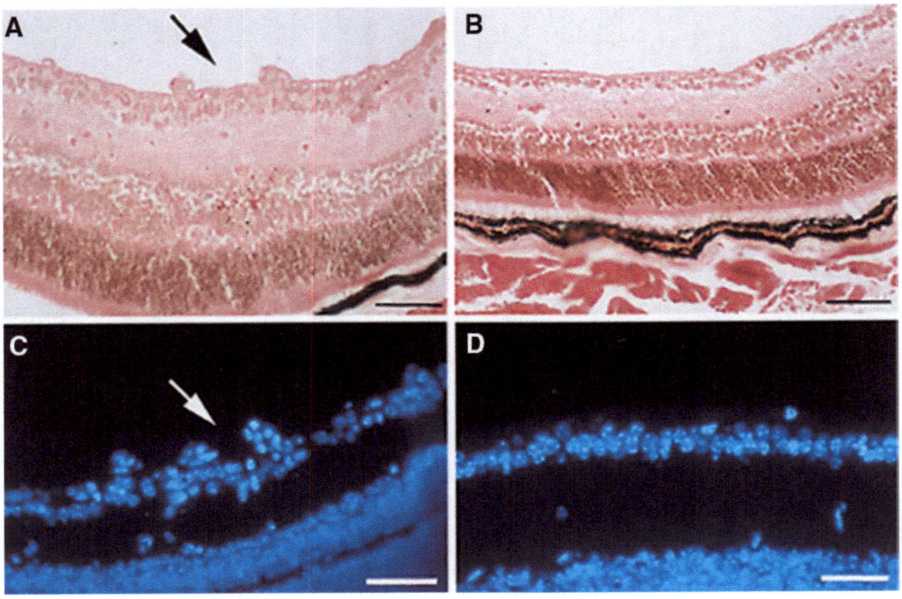

Fig. 5. Use of a matrix metalloproteinase inhibitor suppresses the development of retinal neovascularization. (**A**) Hematoxylin–eosin-stained cross section from the retina of a mouse exposed to 75% oxygen for 5 days followed by room air for an additional 5 days. Capillary tufts are present on the vitreal side of the inner limiting membrane, characteristic of the angiogenic response in this tissue *(arrow)*. (**B**) Representative hematoxylin-and-eosin-stained section from the retina of an experimental mouse treated with BB-94 1 mg/kg, on postnatal days 12, 14, and 16. (**C**) Similar section from an experimental animal stained with diamidinophenylindole showing individual endothelial cell nuclei that belong to new vessels *(arrow)*. (**D**) Similar section from the retina of a BB-94-treated mouse stained with diamidinophenylindole showing a significant reduction in the number of neovascular nuclei. Only a single endothelial cell nucleus is present on the vitreal side of the inner limiting membrane. Scale bars: (**A, B**) 166 μm; (**C, D**) 113 μm (reproduced with permission from Das et al. [45]).

Inhibition of Retinal Angiogenesis by Inhibitors of the uPA/uPAR System

A peptide inhibitor of the urokinase system, A6 (an octapeptide that inhibits the interaction of uPA with uPAR), was able to reduce the extent of retinal neovascularization and uPAR expression in the experimental animals. Intravitreal injection of an adenoviral vector carrying the murine ATF has been shown to inhibit retinal neovascularization by 78% in the oxygen-induced retinopathy level [84]. These results suggest that inhibition of the urokinase receptor might be a promising target for antiangiogenic therapy in the retina.

PROTEASES IN DIABETIC MACULAR EDEMA

The vitreous level of MMP-9 has been shown to be higher in diabetic subjects with DR than with the diabetic subjects without DR. This study indicates a potential role of MMPs in the pathogenesis of DR [87]. Furthermore, in an animal model of diabetes, both MMP-2 and MMP-9 were elevated in the retinas [88]. The MT1-MMP was also increased along with MMP-2 in the diabetic animals, and concomitant to this, there was an increased apoptosis of pericytes in the diabetic retina when compared to the normal retina. This may further accelerate the BRB alteration in the diabetic state. We have shown that retinal vascular permeability was significantly increased in rats following 2 weeks of diabetes coincident with a decrease of VE-cadherin expression. This increased vascular permeability could be inhibited with an MMP inhibitor [89]. Treatment of endothelial cells with AGE-BSA led to a reduction of VE-cadherin staining on the cell surface and increased permeability, which was MMP-mediated. This suggests that MMPs have a direct role in the alteration of endothelial permeability [89]. Treatment of cells with specific MMPs or AGEs resulted in cleavage of VE-cadherin from the cell surface. These observations suggest a possible mechanism by which diabetes contributes to BRB breakdown through proteolytic degradation of VE-cadherin. The ability of a broad-spectrum MMP inhibitor in the breakdown of BRB suggests a potential alternative therapeutic strategy to the treatment of diabetic macular edema. High glucose can activate many soluble mediators such as AGE, ROS, and inflammatory cytokines, which can increase MMP expression and activity in the diabetic state.

The role of MMP-9 is implicated in the alteration of barrier function which is shown to be mediated by TGF-β [90]. Studies have hinted that diabetes causes retinal inflammation which unleashes a sequelae of events resulting in the vascular leakage. Retinal inflammation attracts increased leukocytes to the retina which then bind to the vascular endothelium. The binding of leukocyte to the endothelial cells can also activate cellular proteases that may clip off VE-cadherin and its associated protein from the cell surface resulting in endothelial monolayer alteration.

Inhibition of Proteases in the Prevention of Blood–Retinal Barrier in Diabetes

MMPs have emerged as regulators of endothelial barrier function in several tissues. Studies have demonstrated an increased expression of MMPs in the retinas of diabetic animals. The proteolytic degradation of vascular endothelial (VE)-cadherin from the surface of cultured endothelial cells by MMP-9 has been shown to increase the vascular monolayer permeability. An inhibitor of MMP-9 (Batimastat (BB-94)) was able to block diabetes-induced vascular permeability and prevented the loss of VE-cadherin in the retinal vasculature. These study result indicates a role for extracellular proteases in the alteration of the BRB seen in diabetic retinopathy and can be a potential therapeutic target for treating DME [89].

We have shown that the increased retinal vascular permeability in diabetic rats was associated with a decrease in vascular endothelial (VE)-cadherin expression in retinal vessels. Treatment with the uPA/uPAR-inhibiting peptide (A6) was shown to reduce diabetes-induced permeability and the loss of VE-cadherin [91]. The increased permeability of cultured cells in response to advanced glycation end products (AGEs) was also significantly inhibited with A6. Treatment of endothelial cells with specific MMPs or AGEs resulted in loss of VE-cadherin from the cell surface, which could be inhibited by A6. uPA/uPAR physically interacts with AGEs/receptor for advanced glycation end products on the cell surface and regulates its activity. uPA and its receptor uPAR play important roles in the alteration of the blood–retinal barrier through proteolytic degradation of VE-cadherin. The ability of A6 to block retinal vascular permeability in diabetes suggests a potential therapeutic approach for the treatment of diabetic macular edema.

CONCLUSION

Emerging evidence indicates that extracellular proteases play a role in both retinal angiogenesis and diabetic macular edema. Also, studies have shown the beneficial effect of protease inhibitors in the prevention of retinal-barrier breakdown and in retinal angiogenesis. Thus, proteases may serve as an attractive therapeutic target in diabetic retinopathy. Currently, the majority of the clinical trials in retinal diseases have targeted the VEGF molecule. Clinical trials have shown that at least for diabetic macular edema, anti-VEGF agents may not be sufficient to prevent the leakage, and repeated injections are needed. Probably, factors other than VEGF, like proteases may play a role in diabetic macular edema. The urokinase inhibitor, A6 (Angstrom Pharmaceuticals, San Diego, CA), which is currently in a Phase II clinical trial in ovarian carcinoma patients, has been shown to be a promising agent in both retinal angiogenesis and macular edema in the preclinical studies. Several MMP inhibitors are currently in clinical trials for different types of cancer, and many of these agents have been shown to be effective in retinal angiogenesis as well. Factors other than VEGF are critical in the development of diabetic retinopathy, and targeting these "other" molecules will probably result in better clinical outcome. A combination therapy with proteinase inhibitors with the currently used anti-VEGF agents may be an effective alternative strategy which needs to be further explored. Such an option may reduce the number of intravitreal injections that is often needed to control the extent of neovascularization and edema.

ACKNOWLEDGMENT

Supported by NIH Grant RO1 EY 12604 and Juvenile Diabetes Research Foundation (JDRF).

REFERENCES

1. Fruttiger M. Development of the mouse retinal vasculature: angiogenesis versus vasculogenesis. Invest Ophthalmol Vis Sci. 2002;43:522–7.
2. Das A, McGuire PG. Retinal and choroidal angiogenesis: pathophysiology and strategies for inhibition. Prog Retin Eye Res. 2003;22:721–48.
3. Pepper MS. Role of the matrix metalloproteinase and plasminogen activator-plasmin systems in angiogenesis. Arterioscler Thromb Vasc Biol. 2001;21:1104–17.
4. Nielsen LS et al. Purification of zymogen to plasminogen activator from human glioblastoma cells by affinity chromatography with monoclonal antibody. Biochemistry. 1982;21:6410–5.
5. Quax P, van Muijen G, Weening-Verhoeff E, et al. Metastatic behavior of human melanoma cell lines in nude mice correlates with urokinase type plasminogen activator, its type inhibitor, urokinase-mediated matrix degradation. J Cell Biol. 1991;115:191–9.
6. Manchanda N, Schwartz BS. Single chain urokinase: augmentation of enzymatic activity upon binding to monocytes. J Biol Chem. 1991;266:14580–4.

7. Rabbani SA, Desjardins J, Bell AW, et al. An amino terminal fragment of urokinase isolated from a prostate cancer cell line is mitogenic for osteoblast-like cells. Biochem Biophys Res Commun. 1990;173:1058–64.
8. Blasi R. Urokinase and urokinase receptor: a paracrine/autocrine system regulating cell migration and invasiveness. Bioassays. 1993;15:105–11.
9. Blasi F, Carmeliet P. uPA: a versatile signaling orchestrator. Nat Rev Mol Cell Biol. 2002;3:931–43.
10. Houck KA, Leung DW, Rowland AM, Winer J, Ferrara N. Dual regulation of vascular endothelial growth factor bioavailability by genetic and proteolytic mechanisms. J Biol Chem. 1992;267:26031–7.
11. Smith HW, Marshall CJ. Regulation of cell signalling by uPAR. Nat Rev Mol Cell Biol. 2010;11:23–36.
12. Ploug M, Ellis V. Structure-function relationships in the receptor for urokinase-type plasminogen activator. Comparison to other members of the Ly-6 family and snake venom α-neurotoxins. FEBS Lett. 1994;349:163–8.
13. Kjaergaard M, Hansen LV, Jacobsen B, Gardsvoll H, Ploug M. Structure and ligand interactions of the urokinase receptor (uPAR). Front Biosci. 2008;13:5441–61.
14. Huai Q et al. Crystal structures of two human vitronectin, urokinase and urokinase receptor complexes. Nat Struct Mol Biol. 2008;15:422–3.
15. Huai Q et al. Structure of human urokinase plasminogen activator in complex with its receptor. Science. 2006;311:656–9.
16. Llinas P et al. Crystal structure of the human urokinase plasminogen activator receptor bound to an antagonist peptide. EMBO J. 2005;24:1655–63.
17. Mazar AP. Urokinase plasminogen activator receptor choreographs multiple ligand interactions: implications for tumor progression and therapy. Clin Cancer Res. 2008;14:5649–55.
18. Stetler-Stevenson WG. The role of matrix metalloproteinases in tumor invasion, metastasis and angiogenesis. Surg Oncol Clin N Am. 2001;10:383–92.
19. Hotary K, Allen E, Punturieri A, Yana I, Weiss SJ. Regulation of cell invasion and morphogenesis in a three-dimensional type I collagen matrix by membrane-type matrix metalloproteinases 1, 2, and 3. J Cell Biol. 2000;149:1309–23.
20. Sato H, Okada Y, Seiki M. Membrane-type matrix metalloproteinases (MT-MMPs) in cell invasion. Thromb Haemost. 1997;78:497–500.
21. Massova I, Kotra LP, Fridman R, Mobashery S. Matrix metalloproteinases: structures, evolution, and diversification. FASEB J. 1998;12:1075–95.
22. Werb Z. ECM and cell surface proteolysis: regulating cellular ecology. Cell. 1997;91:439–42.
23. Werb Z, Vu TH, Rinkenberger JL, Coussens LM. Matrix-degrading proteases and angiogenesis during development and tumor formation. APMIS. 1999;107(1):11–8.
24. Nagase H, Woessner Jr JF. Matrix metalloproteinases. J Biol Chem. 1999;274(31):21491–4.
25. Page-McCaw A, Ewald AJ, Werb Z. Matrix metalloproteinases and the regulation of tissue remodelling. Nat Rev Mol Cell Biol. 2007;8:221–33.
26. Deryugina EI, Ratnikov BI, Postnova TI, Rozanov DV, Strongin AY. Processing of integrin alpha(v) subunit by membrane type 1 matrix metalloproteinase stimulates migration of breast carcinoma cells on vitronectin and enhances tyrosine phosphorylation of focal adhesion kinase. J Biol Chem. 2002;277:9749–56.
27. Xu J et al. Proteolytic exposure of a cryptic site within collagen type IV is required for angiogenesis and tumor growth in vivo. J Cell Biol. 2001;154:1069–79.
28. Kalluri R. Basement membranes: structure, assembly and role in tumour angiogenesis. Nat Rev Cancer. 2003;3:422–33.

29. Nguyen M, Arkell J, Jackson CJ. Human endothelial gelatinases and angiogenesis. Int J Biochem Cell Biol. 2001;33:960–70.
30. Hiraoka N, Allen E, Apel IJ, Gyetko MR, Weiss SJ. Matrix metalloproteinases regulate neovascularization by acting as pericellular fibrinolysins. Cell. 1998;95:365–77.
31. Pozzi A, LeVine WF, Gardner HA. Low plasma levels of matrix metalloproteinase 9 permit increased tumor angiogenesis. Oncogene. 2002;21:272–81.
32. Herren B, Levkau B, Raines EW, Ross R. Cleavage of beta-catenin and plakoglobin and shedding of VE-cadherin during endothelial apoptosis: evidence for a role for caspases and metalloproteinases. Mol Biol Cell. 1998;9:1589–601.
33. Zaragoza C et al. Activation of the mitogen-activated protein kinase extracellular signal-regulated kinase 1 and 2 by the nitric oxide-cGMP-cGMP-dependent protein kinase axis regulates the expression of matrix metalloproteinase 13 in vascular endothelial cells. Mol Pharmacol. 2002;62:927–35.
34. Zaragoza C, Balbín M, López-Otín C, Lamas S. Nitric oxide regulates matrix metalloprotease-13 expression and activity in endothelium. Kidney Int. 2002;61:804–8.
35. Mohan R et al. Curcuminoids inhibit the angiogenic response stimulated by fibroblast growth factor-2, including expression of matrix metalloproteinase gelatinase B. J Biol Chem. 2000;275:10405–12.
36. Powers CJ, McLeskey SW, Wellstein A. Fibroblast growth factors, their receptors and signaling. Endocr Relat Cancer. 2000;7:165–97.
37. Ferrara N. Molecular and biological properties of vascular endothelial growth factor. J Mol Med. 1999;77:527–43.
38. Ries C, Egea V, Karow M, Kolb H, Jochum M, Neth P. MMP-2, MT1-MMP, and TIMP-2 are essential for the invasive capacity of human mesenchymal stem cells:differential regulation by inflammatory cytokines. Blood. 2007;109:4055–63.
39. Hashimoto G, Inoki I, Fujii Y, Aoki T, Ikeda E, Okada Y. Matrix metalloproteinases cleave connective tissue growth factor and reactivate angiogenic activity of vascular endothelial growth factor 165. J Biol Chem. 2002;277:36288–95.
40. Imai K, Hiramatsu A, Fukushima D, Pierschbacher MD, Okada Y. Degradation of decorin by matrix metalloproteinases: identification of the cleavage sites, kinetic analyses and transforming growth factor-beta1 release. Biochem J. 1997;322:809–14.
41. Yu Q, Stamenkovic I. Cell surface-localized matrix metalloproteinase-9 proteolytically activates TGF-beta and promotes tumor invasion and angiogenesis. Genes Dev. 2000;14:163–76.
42. Slansky HH, Freeman MI, Itoi M. Collagenolytic activity in bovine corneal epithelium. Arch Ophthalmol. 1968;80:496–8.
43. Plantner JJ, Jiang C, Smine A. Increase in interphotoreceptor matrix gelatinase a (MMP-2) associated with age-related macular degeneration. Exp Eye Res. 1998;67:637–45.
44. Brown D, Hamdi H, Bahri S, Kenney MC. Characterization of an endogenous metalloproteinase in human vitreous. Curr Eye Res. 1994;13:639–47.
45. Das A, McLamore A, Song W, McGuire PG. Retinal neovascularization is suppressed with a matrix metalloproteinase inhibitor. Arch Ophthalmol. 1999;117:498–503.
46. Yan X, Tezel G, Wax MB, Edward DP. Matrix metalloproteinases and tumor necrosis factor alpha in glaucomatous optic nerve head. Arch Ophthalmol. 2000;118:666–73.
47. Vaughan-Thomas A, Gilbert SJ, Duance VC. Elevated levels of proteolytic enzymes in the aging human vitreous. Invest Ophthalmol Vis Sci. 2000;41:3299–304.
48. Webster L, Chignell AH, Limb GA. Predominance of MMP-1 and MMP-2 in epiretinal and subretinal membranes of proliferative vitreoretinopathy. Exp Eye Res. 1999;68:91–8.
49. Kon CH, Occleston NL, Charteris D, Daniels J, Aylward GW, Khaw PT. A prospective study of matrix metalloproteinases in proliferative vitreoretinopathy. Invest Ophthalmol Vis Sci. 1998;39:1524–9.

50. Sivak JM, Fini ME. MMPs in the eye: emerging roles for matrix metalloproteinases in ocular physiology. Prog Retin Eye Res. 2002;21:1–14.
51. Salzmann J et al. Matrix metalloproteinases and their natural inhibitors in fibrovascular membranes of proliferative diabetic retinopathy. Br J Ophthalmol. 2000;84:1091–6.
52. Yuan L, Neufeld AH. Activated microglia in the human glaucomatous optic nerve head. J Neurosci Res. 2001;64:523–32.
53. Ahir A, Guo L, Hussain AA, Marshall J. Expression of metalloproteinases from human retinal pigment epithelial cells and their effects on the hydraulic conductivity of Bruch's membrane. Invest Ophthalmol Vis Sci. 2002;43:458–65.
54. Limb GA et al. Differential expression of matrix metalloproteinases 2 and 9 by glial Müller cells: response to soluble and extracellular matrix-bound tumor necrosis factor-alpha. Am J Pathol. 2002;160:1847–55.
55. De La Paz MA, Itoh Y, Toth CA, Nagase H. Matrix metalloproteinases and their inhibitors in human vitreous. Invest Ophthalmol Vis Sci. 1998;39:1256–60.
56. Xie B et al. An Adam15 amplification loop promotes vascular endothelial growth factor-induced ocular neovascularization. FASEB J. 2008;22:2775–83.
57. Baramova E, Foidart JM. Matrix metalloproteinase family. Cell Biol Int. 1995;19:239–42.
58. Weber BHF, Vogt G, Pruett RC, et al. Mutations in the tissue inhibitor of metalloproteinases-3 (TIMP-3) in patients with Sorsby's fundus dystrophy. Nat Genet. 1994;8:352–6.
59. Farris RN, Apte SS, Luhert PJ, et al. Accumulations of tissue inhibitor metalloproteinases-3 in human eyes with Sorsby's fundus dystrophy or retinitis pigmentosa. Br J Ophthalmol. 1998;82:1329–34.
60. Oh J et al. The membrane-anchored MMP inhibitor RECK is a key regulator of extracellular matrix integrity and angiogenesis. Cell. 2001;107:789–800.
61. Cawston TE, Mercer E. Preferential binding of collagenase to alpha 2-macroglobulin in the presence of the tissue inhibitor of metalloproteinases. FEBS Lett. 1986;209:9–12.
62. Herman MP et al. Tissue factor pathway inhibitor-2 is a novel inhibitor of matrix metalloproteinases with implications for atherosclerosis. J Clin Invest. 2001;107:1117–26.
63. Andreasen PA, Georg B, Lund LR, Riccio A, Stacey SN. Plasminogen activator inhibitors: hormonally regulated serpins. Mol Cell Endocrinol. 1990;68:1–19.
64. Ye S, Goldsmith EJ. Serpins and other covalent protease inhibitors. Curr Opin Struct Biol. 2001;11:740–5.
65. Thorgeirson UP, Linsay CK, Cottam DW, Gomez DE. Tumor invasion, proteolysis, angiogenesis. J Neurooncol. 1994;18:89–103.
66. Beránek M et al. Genetic variations and plasma levels of gelatinase A (matrix metalloproteinase-2) and gelatinase B (matrix metalloproteinase-9) in proliferative diabetic retinopathy. Mol Vis. 2008;14:1114–21.
67. Ishizaki E et al. Correlation between angiotensin-converting enzyme, vascular endothelial growth factor, and matrix metalloproteinase-9 in the vitreous of eyes with diabetic retinopathy. Am J Ophthalmol. 2006;141:129–34.
68. Patel JI, Tombran-Tink J, Hykin PG, Gregor ZJ, Cree IA, et al. Vitreous and aqueous concentrations of proangiogenic, antiangiogenic factors and other cytokines in diabetic retinopathy patients with macular edema: implications for structural differences in macular profiles. Exp Eye Res. 2006;82:798–806.
69. Jin M, Kashiwagi K, Iizuka Y, Tanaka Y, Imai M, Tsukahara S. Matrix metalloproteinases in human diabetic and nondiabetic vitreous. Retina. 2001;21:28–33.
70. Salzmann J et al. Matrix metalloproteinases and their natural inhibitors in fibrovascular membranes of proliferative diabetic retinopathy. Br J Ophthalmol. 2000;84:1091–6.
71. Kosano H et al. ProMMP-9 (92 kDa gelatinase) in vitreous fluid of patients with proliferative diabetic retinopathy. Life Sci. 1999;64:2307–15.

72. Das A, McGuire PG, Eriqat C, et al. Human diabetic neovascular membranes contain high levels of urokinase and metalloproteinase enzymes. Invest Ophthalmol Vis Sci. 1999;40:809–13.
73. Noda K et al. Production and activation of matrix metalloproteinase-2 in proliferative diabetic retinopathy. Invest Ophthalmol Vis Sci. 2003;44:2163–70.
74. Ohno-Matsui K et al. Reduced retinal angiogenesis in MMP-2-deficient mice. Invest Ophthalmol Vis Sci. 2003;44:5370–5.
75. Jacqueminet S et al. Elevated circulating levels of matrix metalloproteinase-9 in type 1 diabetic patients with and without retinopathy. Clin Chim Acta. 2006;367:103–7.
76. Kowluru RA, Kanwar M. Oxidative stress and the development of diabetic retinopathy: contributory role of matrix metalloproteinase-2. Free Radic Biol Med. 2009;46:1677–85.
77. Kowluru RA, Kanwar M. Translocation of H-Ras and its implications in the development of diabetic retinopathy. Biochem Biophys Res Commun. 2009;387:461–6.
78. Tarallo S, Beltramo E, Berrone E, Dentelli P, Porta M. Effects of high glucose and thiamine on the balance between matrix metalloproteinases and their tissue inhibitors in vascular cells. Acta Diabetol. 2010;47(2):105–11.
79. Sánchez MC et al. Effect of retinal laser photocoagulation on the activity of metalloproteinases and the alpha(2)-macroglobulin proteolytic state in the vitreous of eyes with proliferative diabetic retinopathy. Exp Eye Res. 2007;85:644–50.
80. Zhang G, Fahmy RG, diGirolamo N, Khachigian LM. JUN siRNA regulates matrix metalloproteinase-2 expression, microvascular endothelial growth and retinal neovascularisation. J Cell Sci. 2006;119:3219–26.
81. Iwai S et al. Activation of AP-1 and increased synthesis of MMP-9 in the rabbit retina induced by lipid hydroperoxide. Curr Eye Res. 2006;31:337–46.
82. Hattenbach LO, Allers A, Gümbel HO, Scharrer I, Koch FH. Vitreous concentrations of TPA and plasminogen activator inhibitor are associated with VEGF in proliferative diabetic vitreoretinopathy. Retina. 1999;19:383–9.
83. Grant MB, Guay C. Plasminogen activator production by human retinal endothelial cells of nondiabetic and diabetic origin. Invest Ophthalmol Vis Sci. 1991;32:53–64.
84. McGuire PG, Jones TR, Talarico N, et al. The urokinase/urokinase receptor system in retinal neovascularization: inhibition by A6 suggests a new therapeutic target. Invest Ophthalmol Vis Sci. 2003;44:2736–42.
85. Majka S, McGuire PG, Colombo S, Das A. The balance between proteinases and inhibitors in a murine model of proliferative retinopathy. Invest Ophthalmol Vis Sci. 2001;42:210–5.
86. Le Gat L, Gogat K, Bouquet C, et al. In vivo adenovirus-mediated delivery of a uPA/uPAR antagonist reduces retinal neovascularization in a mouse model of retinopathy. Gene Ther. 2003;10:2098–103.
87. Jin M, Kashiwagi K, Iizuka Y, Tanaka Y, Imai M, Tsukahara S. Matrix metalloproteinases in human diabetic and nondiabetic vitreous. Retina. 2001;21(1):28–33.
88. Giebel SJ, Menicucci G, McGuire PG, Das A. Matrix metalloproteinases in early diabetic retinopathy and their role in alteration of the blood-retinal barrier. Lab Invest. 2005;85(5):597–607.
89. Navaratna D, McGuire PG, Menicucci G, Das A. Proteolytic degradation of VE-cadherin alters the blood-retinal barrier in diabetes. Diabetes. 2007;56:2380–7.
90. Behzadian MA, Wang XL, Windsor LJ, et al. TGF-beta increases retinal endothelial cell permeability by increasing MMP-9: possible role of glial cells in endothelial barrier function. Invest Ophthalmol Vis Sci. 2001;42:853–9.
91. Navaratna D, Menicucci G, Maestas J, Srinivasan R, McGuire P, Das A. A peptide inhibitor of the urokinase/urokinase receptor system inhibits alteration of the blood-retinal barrier in diabetes. FASEB J. 2008;22:3310–7.

11

Proteomics in the Vitreous of Diabetic Retinopathy Patients

Edward P. Feener

CONTENTS

Keywords Diabetic retinopathy • Mass spectrometry • Proteomics • Retina • Vitreous

INTRODUCTION

Vision loss cause by diabetic retinopathy is primarily associated with advanced stages of this disease, including proliferative diabetic retinopathy (PDR) and diabetic macular edema (DME). While abnormalities in microvascular functions and structure appear central to the progression of diabetic retinopathy [1], the specific factors that modulate the transition to the advanced sight-threatening stages of this disease are not fully understood. Moreover, since animal models do not reproduce many of the specific pathologies associated with PDR and DME, further characterization of ocular biochemical changes from patients with diabetic retinopathy is needed to identify factors that could be associated with the advance stages of this disease and vision loss. Analyses of vitreous fluid obtained during pars plana vitrectomy have provided opportunities to identify factors that may contribute to, or protect against, advanced stages of diabetic retinopathy. This chapter examines the methodologies for vitreous proteomics and the findings that are beginning to emerge from studies using this approach.

Characterization of vitreous from patients with diabetic retinopathy compared with vitreous from nondiabetic subjects has revealed a variety of differences in intraocular

From: *Ophthalmology Research: Visual Dysfunction in Diabetes*
Edited by: J. Tombran-Tink et al. (eds.), DOI 10.1007/978-1-60761-150-9_11

protein abundance, modification, and activities. Over the past several decades, a variety of biochemical and immunological techniques have been used to characterize specific candidate proteins and protein functions from vitreous samples. While this approach continues to provide important new information, especially for low-abundance proteins, emerging opportunities utilizing omic technologies are rapidly expanding our understanding of the complexity of vitreous fluid. Proteomic approaches have identified specific proteins in vitreous that are associated with diabetic retinopathy, and a limited number of these proteins have been shown to induce functional and structural changes in the retina in animal models that are consistent with diabetic retinopathy. Moreover, recent advances in proteomics and bioinformatics are creating opportunities to characterize biological processes that may contribute to diabetic retinopathy and identify biomarkers that further characterize differences in disease progression and responses to therapeutic interventions among patients with seemingly similar disease characteristics. While vitreous proteomics holds exciting potential for expanding understanding of the molecular mechanisms and complexities of diabetic retinopathies, these studies will require methods to integrate the rapidly expanding volume of proteomic data with basic science and clinical aspects of vitreous biology and diabetic retinopathy.

VITREOUS ANATOMY

The vitreous is an optically transparent gel-like fluid that provides both structural and biochemical functions in ocular physiology. The gel-like composition of the vitreous is derived mainly from a hydrated network of fibular macromolecules, including glycosaminoglycans (GAG), proteoglycans, and collagen fibrils. Within this fluid and lattice of macromolecules there is a metabolically active and dynamic biochemical milieu. Soluble proteins can diffuse between the vitreous and retinal interstitial fluid across of the inner limiting membrane (ILM), suggesting that the vitreous may contain information derived from retinal disorders, and proteins in the vitreous can feedback to influence retinal functions and pathologies.

The normal adult vitreous is largely acellular and organized with collagen fibrils oriented along an anterior to posterior axis [2]. The interface between the vitreous and retina involves the posterior vitreous cortex and ILM, which mediate regions of vitreoretinal adhesion. The concentrations of collagen isoforms, including types II, V, IX, and XI, are higher in the vitreous cortex compared to central vitreous [3]. Intravitreal localization of other major component molecules, such as hyaluronan, also varies according to their anatomical distribution within the vitreous. These extracellular matrix (ECM) molecules provide a scaffold that binds ions, water, and soluble proteins, which can influence diffusion within the vitreous compartment. The organization of ECM molecules within the vitreous suggests the possibility that soluble proteins that bind to ECM may also be spatially organized or heterogeneously distributed within this compartment.

The vitreous often undergoes a liquefaction process during aging, which alters the biochemical and anatomical heterogeneity of this structure and can alter oxygen consumption and gradients [4]. Liquefaction of vitreous together with the age-related weakening of adhesion between the vitreous cortex and ILM contributes to vitreoretinal disorders, including rhegmatogenous retinal detachment (RRD) [2, 5]. Changes in vitreous ECM, liquefaction of vitreous during aging, and effects of vitreoretinal traction could influence the diffusion, retention, and localization of proteins in the vitreous. Thus, the composition of vitreous samples collected from control subjects and subject patients with diabetic retinopathies is likely influenced by coexisting vitreous disorders.

A CANDIDATE APPROACH

Studies of vitreous during the 1970s and 1980s revealed a number of proteins and biochemical activities within this fluid. These early studies of vitreous identified iron-binding proteins, including transferrin [6], which were suggested to provide a protective role for the retina against the detrimental effects of iron, resultant of vitreous hemorrhage [7, 8]. The vitreous was also shown to contain fibrinolytic activity and complement components, which were implicated as clearance mechanisms for hemorrhage and infection [9, 10]. Further early investigations of vitreous activities identified growth factors, potential regulators of growth factor action, and proteins involved in remodeling [11–15]. These findings, and others, revealed that proteins within the vitreous provide a plethora of biochemical functions in ocular physiology. Moreover, this early work suggested that vitreous may not only contain protein involved in the maintenance of normal ocular physiology but may also contain factors that contribute to retinal diseases, including diabetic retinopathy [16, 17].

A series of reports in 1994 revealed increased abundance of vascular endothelial growth factor (VEGF) in vitreous during ocular neovascularization, experimentally induced retinal ischemia, and PDR [18–20]. Subsequent reports demonstrated that intravitreal injection of VEGF induces retinal vascular permeability (RVP) [21], intravitreal VEGF levels are elevated in DME [22], and inhibition of the VEGF pathway ameliorates DME [23, 24]. These findings have revealed that the vitreous, at least in a subgroup of patients with diabetic retinopathy, contains a key mediator of PDR and DME, namely VEGF.

Over the past 2–3 decades, multiple studies have utilized similar candidate molecule approaches to further characterize changes in proteins, including a variety of chemokines, hormones, growth factors, inflammatory molecules, as well as angiogenic and anti-angiogenic factors, in vitreous from patients with diabetic retinopathy. Funatsu et al. reported that in DME VEGF levels in vitreous correlate with elevated levels of intercellular adhesion molecule-1 (ICAM-1), interleukin (IL)-6, and monocyte chemotactic protein-1 (MCP-1) [25], suggesting a link between VEGF and inflammation. Moreover, elevated levels of these factors correlated with increased RVP and retinal thickness [22, 25]. Yoshimura et al. has shown that IL-6, IL-8, and MCP-1 are elevated in vitreous from PDR and DME compared with nondiabetic (NDM) controls [26], and increases in levels of these inflammatory factors correlated with elevated VEGF levels in vitreous. Platelet-derived growth factor (PDGF)-AA, PDGF-AB, PDGF-BB isoform levels were shown to be elevated in vitreous from subjects with PDR, and increasing concentration of these PDGF isoforms was also shown to correlate with VEGF levels [27]. Moreover, changes in intravitreal levels of insulin-like growth factor-I (IGF-I) and IGF-binding proteins in people with diabetic retinopathy have also been reported [28]. This growing body of work has provided insights into the complexity and heterogeneity of potential hormonal, growth factor, and cytokine influences of the vitreous on diabetic retinopathy. While these finding suggest a variety of protein and pathways that may contribute to diabetic retinopathy, limitations of this candidate protein approach are that it is often directed by preexisting theories and the relatively small number of molecules that have been examined.

PROTEOMIC APPROACHES

Proteomics is the large-scale analysis of proteins, which often includes a combination of characterization of amino acid sequence, quantification, modifications, and interactions. Advances in proteomics over that past decade have created opportunities to use rapid de novo protein discovery methods to further characterize the composition of vitreous and identify protein changes associated with diabetes and diabetic retinopathy. Most of this work has utilized mass spectrometry as the centerpiece for protein identification; however, markedly different proteomic methods have been utilized, which limit the comparison of studies and findings across studies. As such, the present status of vitreous proteomic data in diabetes is a collection of somewhat unique studies. The following describes the proteomic approaches that have been used for the analysis of vitreous and discusses findings that are beginning to emerge from this work.

Proteomics is a multistep process with variety of options available at each step. Although the utilization of diverse methods yields important information and differences in perspectives, the lack of uniformity limits the assimilation of data among different studies. Further understanding of the differences among experimental design and data output is critical for interpreting the current body of vitreous proteomic information. The workflow of vitreous proteomics can be separated into a series of steps, including (1) vitreous acquisition, (2) sample fractionation, (3) mass spectrometry, (4) spectral analysis, and (5) data analysis (Fig. 1). The following section describes options and parameters that have been applied to vitreous proteomics within each of these steps.

Vitreous Acquisition

Study design has an overarching influence on the information generated from vitreous proteomics. Vitreous fluid is usually obtained during pars plana vitrectomy for treatment of specific retinal and vitreoretinal disorders, including, but not limited to, epiretinal membrane (ERM), macular hole (MH), vitreoretinal traction, and non-clearing vitreous hemorrhage. The potential influences of these surgical indications, apart from the influences of diabetes and diabetic retinopathy, on the vitreous proteome are unknown.

Additional factors that could influence the vitreous proteome at a given stage of diabetic retinopathy include patient demographics, rate of disease progression, disease duration, and treatment history, including, for example, laser photocoagulation and pharmacotherapy. A growing number of studies have shown that levels of specific proteins can differ markedly within a selected group of patients, for example, VEGF levels can differ markedly among individual PDR vitreous samples [20, 26]. Since obtaining multiple vitreous samples for longitudinal studies is generally not feasible, large numbers of samples from well-characterized patients will be needed to examine protein correlations with retinopathy stage.

Increases in total protein concentration in the vitreous in diabetic retinopathy are well documented (Table 1). Most studies have reported that vitreous protein levels are about

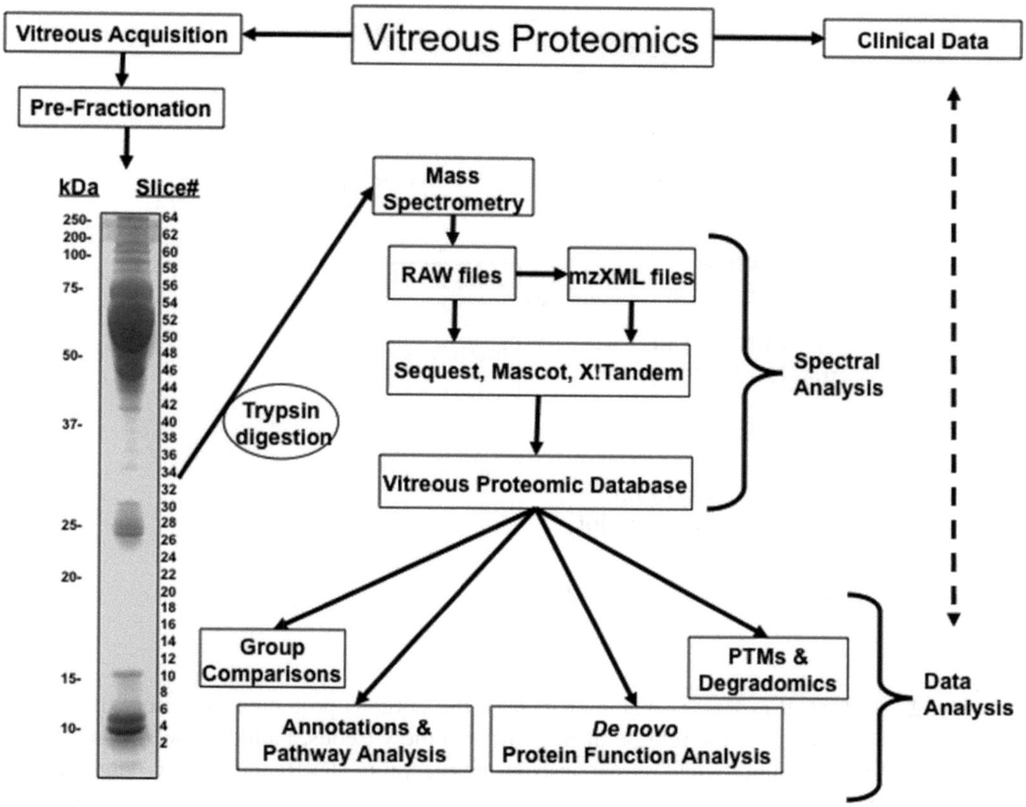

Fig. 1. Vitreous proteomics. An example of a work flow for 1-DE-based proteomics is shown. The major steps include vitreous acquisition, which involves both study design and clinical characteristics. Sample pre-fractionation in performed by separation by 1-D SDS-PAGE followed by fractionation into gel slices. Mass spectrometry involves analysis of *m*/*z* of peptides and fragmentation ions. Spectral analysis involves matching spectra with amino acid sequences using search algorithms such as Sequest, Mascot, and X!Tandem, and quantitative analysis based on spectral parameters. Data analysis enables proteome comparisons, analysis of pathways and functions, and posttranslational modifications (PTMs) and proteolysis.

Table 1. Total protein concentration in control and PDR vitreous

NDM vitreous (mg/mL)	PDR vitreous (mg/mL)	References
0.4668, MH ($n=26$)	4.129 ($n=33$)	[29]
1.96 ± 0.5, MH ($n=10$)	4.45 ± 1.4 ($n=8$)	[30]
0.77 ± 0.47, MH, ERM ($n=13$)	4.21 ± 2.2 ($n=16$)	[31]
0.67, MH, ERM ($n=30$)	3.97 ($n=28$)	[32]

fourfold higher in PDR compared with vitreous obtained from NDM subjects with MH. A primary cause for increased total protein in diabetic retinopathy is due to elevated RVP, which occurs early in diabetic retinopathy and increases further during disease progression [33, 34]. These additional increases in protein content in advanced diabetic

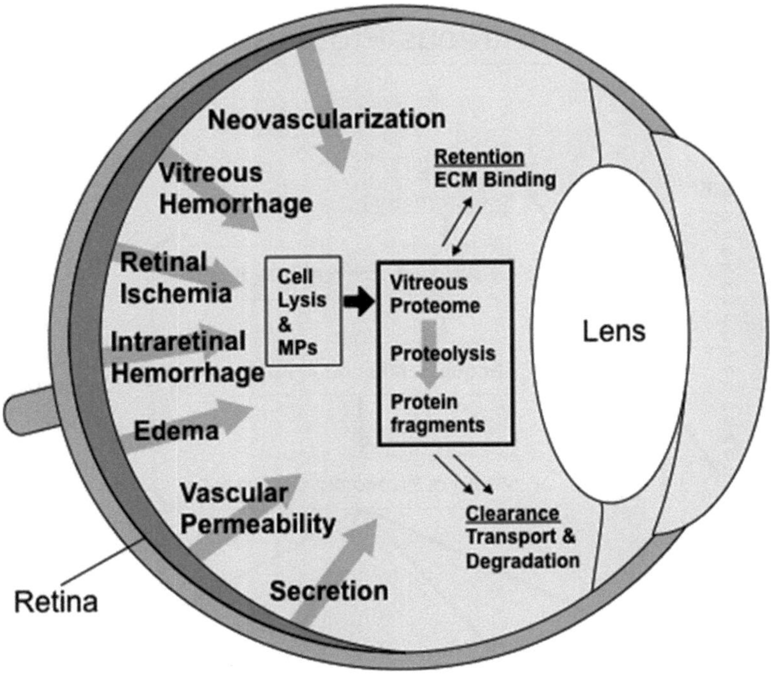

Fig. 2. Biological processes that contribute to the vitreous proteome. A variety of biological processes contribute to the release of proteins into the vitreous, including secretion, plasma extravagation due to pathological retinal vascular permeability (RVP) and edema, hemorrhages, release of microparticles (MP) and cell lysis, and release due to retinal ischemia and clearance of blood cells. Vitreous proteins can be retained in the vitreous by binding to extracellular matrix (ECM) or removed by active or passive transport mechanisms. Vitreous proteins also undergo proteolysis, which may modify their activities and facilitate clearance.

retinopathy are likely the results of a combination of factors including increased RVP, vitreous and intraretinal hemorrhage, tissue damage associated with retinal ischemia, and neovascularization (Fig. 2).

Sample Pre-Fractionation

Sample fractionation provides opportunities to further characterize the vitreous proteins based on physiochemical properties and improves detection sensitivity. One of the goals of most pre-fractionation methods is to separate high-abundance proteins, such as serum albumin, from lower-abundance proteins to improve their detection. Most proteomic analyses of vitreous have utilized protein fractionation based on either 1-dimensional (1-D) or 2-D gel electrophoresis. 1-D SDS-PAGE provides a preparative method of fractionation that enables downstream mass spectrometry of the entire sample separated according to molecular weight (mw, mobility in SDS-PAGE). In 1-DE gel protein staining is typically performed using Coomassie Brilliant Blue stain, and quantitative comparison of proteins among samples utilizes mass spectrometry data. In 2-DE, samples are fractionated by isoelectric focusing (IEF) followed by SDS-PAGE and protein staining. This results in a 2-D display of vitreous proteins, and relative protein staining can be used as a semiquantitative measure of abundance. 2-DE provides both isoelectric point and mw information, and selected proteins and protein isoforms, resolved by IEF, can be isolated for analysis. Quantifying proteins from 2-DE gel staining is complicated by protein isoform separation into multiple spots along an IEF gradient and the possibility that a single spots can contain multiple proteins. Albumin and IgG affinity chromatography has been used in a limited number of studies, prior to separation by gel electrophoresis, to increase detection sensitivity for low-abundance proteins [30, 35]. In solution digestion of protein methods for vitreous proteomics could provide opportunities for increasing throughput; however, this approach does not provide protein mw data, and high levels of glycated macromolecules could potentially interfere with digestion and downstream separation methods.

Mass Spectrometry

A variety of mass spectrometry platforms have been used for vitreous proteomics, which can be separated into two groups based on ionization source, including electrospray ionization (ESI) liquid chromatography–tandem mass spectrometry (LC MS/MS) [35–40] and matrix-assisted laser desorption ionization mass spectrometry (MALDI MS) [29–31, 35, 38, 41]. In addition, the parameters for a given mass spectrometry platform can have a major impact on instrument sensitivity and performance with complex samples. Kim et al. performed side-by-side analyses of vitreous proteins using LC-MALDI-MS/MS and LC-ESI-MS/MS systems [35]. This study reports that MALDI and ESI systems identified 83 and 518 proteins, respectively, which resulted in 531 proteins in the merged datasets. While these findings demonstrate that different mass spectrometry platforms can provide complementary protein datasets, these results also show the limitations in comparing results from different systems. Since there are a number of inherent differences among mass spectrometry platform [42], in addition to user defined parameters, the assimilation of data across studies requires downstream solutions directed at spectral and data analysis.

Spectral Analysis

Spectra generated by mass spectrometry is matched to amino acid sequences using a variety of algorithms, including Sequest, Mascot, and X!Tandem. Gao et al. compared Sequest and X!Tandem analyses of LC-MS/MS data from a set of human vitreous samples [37]. This study generated 231 and 213 proteins using X!Tandem and Sequest, respectively, with 192 proteins identified by both algorithms and a total of 252 proteins in the merged dataset. As described above, these data show that different platforms provide complimentary data that increase the number of proteins identified. However, some low-abundance protein matches were limited to single search algorithms. The criteria used to identify a match are user-defined and instrument-dependent. Parameters and thresholds used to identify proteins are a balance between optimizing detection sensitivity and minimizing the false positive rate (FDR), which is determined by searches against a reversed or randomized protein database [43]. The criteria used to identify a protein vary among vitreous proteomic studies. For example, in two studies using a similar LC-ESI-MS/MS platform, Gao et al. used two unique peptides identified from the same or adjacent gel slice in at least two independent vitreous samples to generate

252 proteins [37], and Kim et al. used a single peptide match as a minimum criteria to identify 518 protein matches [35]. In the latter study, a single unique peptide spectral was detected for about 100 proteins, which has a higher FDR compared with proteins detected based on at least two unique spectral-peptide matches. Moreover, the Gao et al. study use individual samples whereas the Kim et al. study used pooled samples and both nondepleted and immunoaffinity-depleted preparations. Thus, comparisons of protein lists from different studies should take into account both protein identification criteria and sample preparation.

Spectral data provides multiple options for both relative and absolute quantification of protein levels. The most widely used method for vitreous proteomics has been based on label-free measurements of spectral-peptide matches, using either the number of unique [36] or total spectral matches [37] for a given protein. Addition label-free options the use of multiple reaction monitoring [44] and analyses of ion intensity and spectral peak area [45]. The use of high mass accuracy and resolution mass spectrometers not only improves the sensitivity of these label-free methods but also creates more robust quantitative options that involve isotope-labeling techniques [46]. Quantitative proteomic methods are of central importance to characterizing the changing in proteins in diabetes and diabetic retinopathy, and the topic of quantitative proteomics has been extensively reviewed elsewhere [47, 48].

Data Analysis

Vitreous proteomics from multiple laboratories has generated lists of proteins detected in vitreous fluid along with quantitative data used for comparisons of protein levels among patients with or without diabetic retinopathy. As describe above, the parameters used to collect these data differ at multiple levels. Thus, while these studies provide different perspectives of the vitreous proteome, the assimilation of data from different reports is complex and often relies on manual techniques. The in-depth comprehension and comparison of proteomic dataset from different groups will likely require integration of these data with emerging bioinformatics tools and strategies [49].

In contrast, there are multiple options available for data analysis within a given proteomic database. Vitreous proteomic databases have been used for quantitative comparisons of protein abundance among groups of subjects, analysis of amino acid modifications and protein fragments, and grouping of proteins according to gene ontology and functional networks [37]. One important limitation of this bioinformatics approach in further understanding the vitreous proteome is that many of the proteins that have been identified in this fluid are not well characterized. Moreover, the functions of these proteins, as well as other more full-characterized proteins, in the vitreous compartment are largely unknown. Thus, in addition to the organization of vitreous proteome using computer algorithms and databases, it is likely that functional studies will be needed to assess the actions of individual proteins within the vitreous milieu.

THE VITREOUS PROTEOME

Two main proteomic approaches, based on 2-D and 1-D gel pre-fractionation, have been used to characterize protein composition of the human vitreous and identify changes associated with diabetic retinopathy. Although differences in experimental methods (as described above) complicate the comparison of these studies and data, a number of findings from vitreous proteomics have emerged.

2-DE-Based Proteomics

The earliest comparative proteomic studies were performed using vitreous samples separated by two-dimensional electrophoresis (2-DE). Nakanishi et al. [38] compared silver-stained proteins separated by 2-D electrophoresis of vitreous obtained from subjects with MH and diabetic retinopathy. This study analyzed proteins from 412 spots separated by 2-DE of diabetic retinopathy vitreous and identified proteins in 113 of these spots, which represented 50 different proteins. Comparison of vitreous was normalized to 100 μg of dialyzed protein, and the authors reported that Ig, α1-antitrypsin, α2-HS glycoprotein, and complement factor 4, and pigmented epithelial-derived factor (PEDF) were elevated in vitreous from diabetic retinopathy. While this study, and others that visualized vitreous proteins by 2-DE, detected several hundred spots of protein staining, these include a large fraction of spots corresponding to protein isoforms separated along the IEF gradient.

A report by Yamane et al. [29] using 2-DE detected more than 400 silver-stains spots and identified 78 proteins in vitreous from patients with MH and 600 spots and identified 141 in vitreous from patients with PDR. This study showed that vitreous (both MH and PDR) and plasma displayed similar patterns of proteins, and most proteins that were identified to be increased in PDR compared with MH were also found in serum. Comparisons of vitreous were normalized to 40 μL of undiluted vitreous volume. The authors concluded that the increases in proteins in the PDR vitreous were the result of increased RVP and hemorrhage. Four proteins, including PEDF, prostaglandin-D2-synthase, plasma glutathione peroxidase, and IRBP were identified in MH vitreous but not in serum, suggesting that these proteins are locally produced in the eye [29]. An analysis of relative protein-staining intensity among gel spots indicates that the most highly abundant proteins in the vitreous include serum albumin, PEDF, α1-antitrypsin, prostaglandin-D2-synthase, apolipoprotein A1, and transthyretin. Ouchi et al. detected over 200 spots using SYPRO Ruby staining of vitreous and identified proteins in 72 spots from vitreous from non-proliferative diabetic retinopathy (NPDR) with DME and 64 spots from vitreous from subjects with NPDR without DME [40]. Comparisons were normalized to 15 μg of total protein. ApoH was detected in non-DME vitreous but not in DME vitreous. PEDF, plasma retinol-binding protein (PRBP), apo A4, apo A1, Trip-11, and vitamin D–binding protein were reported to be elevated in DME vitreous [40].

Garcia-Ramirez et al. [30] compared vitreous proteomes from PDR and MH subjects using fluorescence-based labeling differences in 2-DE. Vitreous samples were subjected to affinity depletion to removed albumin and IgG, and comparisons were normalized to 2-mg/mL protein eluate. This study reported that levels of eight proteins were increased in PDR vitreous, including zinc α2-glycoprotein, apo A1 and apoH, fibrinogen A, complement proteins C3, C4b, C9, and factor B. In addition, three proteins were identified to be decreased in PDR vitreous, including PEDF, IRBP, and inter-α-trypsin inhibitor heavy chain. Subsequent studies from this group further characterized the decrease in IRBP [50] and increased in apo A1 and apoH [51] in diabetic retinopathy.

Kim et al. [42] compare vitreous from subjects with MH and PDR. In this study, compared with MH, prostaglandin-H2 d-isomerase and PEDF were elevated, and α1-antitrypsin and beta V spectrin were reduced in PDR. Shitama et al. [31] compared the relative abundance of 105 proteins among approximately 400 spots visualized by 2-DE of vitreous samples collected from control subjects or patients with NPDR, PDR, RRD, or proliferative vitreoretinopathy. This study identified about ten proteins that were elevated in NPDR and PDR compared with control vitreous, including apo A4, complement C3, α1-B-glycoprotein, α1-antitrypsin, zinc α2-glycoprotein 1, vitamin D–binding protein, and fibrinogen γ.

1-DE-Based Proteomics

Preparative 1-DE was also used in early studies to characterize the vitreous composition however comparative analyses of groups of samples required the development of databases and spectral-based quantitative methods. Koyama et al. [39] characterized the vitreous protein, separated by 1-DE, from a single subject with diabetic retinopathy. This report cataloged 84 different proteins in this vitreous sample.

Gao et al. [36] compared vitreous from three groups of subjects, including NDM, diabetes with no diabetic retinopathy (DM noDR), and PDR. This study identified 117 proteins, including 27 proteins that were elevated in vitreous from PDR compared with vitreous from NDM. This report revealed that PDR vitreous contains increased levels of a number of intracellular and plasma proteins, suggesting that retinal hemorrhage and increased RVP have a major impact on the composition of vitreous in diabetic retinopathy. A key observation generated from this work was that the effects of these newly discovered vitreous proteins on ocular functions were not readily apparent from previous descriptions of protein activities and subcellular locations. This report demonstrated that intravitreal injection of carbonic anhydrase I (CA-I) into rat vitreous increased RVP and retinal thickness via activation of the plasma kallikrein system [36]. The findings suggested a new pathway contributing to diabetic retinopathy which involved intraocular hemorrhage, lysis of erythrocytes to release intracellular CA-I, followed by activation of the kallikrein kinin system (Fig. 3). Moreover, beyond this specific pathway, this report demonstrated that the functions of proteins in the vitreous may not be readily inferred by previous descriptions of protein annotations, and that direct functional analyses of protein actions within the vitreous milieu may be needed to elucidate protein actions from the information generated by vitreous proteomics. Kim et al. [35] used both 2-DE and 1-DE fractionation methods to characterize both non-depleted and albumin/IgG-depleted vitreous from PDR and MH. Pooled samples were used, and comparisons of PDR and MH were normalized to 500 μg per lane for 1-DE. This study generated used multiple pre-fractionation methods and mass spectrometry platforms to generate the largest number of proteins identified from vitreous from diabetic retinopathy; however, the study was not designed to enable statistical comparisons among conditions.

Gao et al. [37] expanded the analyses of NDM, DM noDR, and PDR vitreous that was initiated previously [36]. This report identified 252 proteins in vitreous and used spectral-peptide counts to characterize the vitreous proteome. This analysis showed that albumin represents about 40% of the total soluble protein content (Fig. 4), and that the total spectral peptide content for albumin in PDR vitreous is increased by about

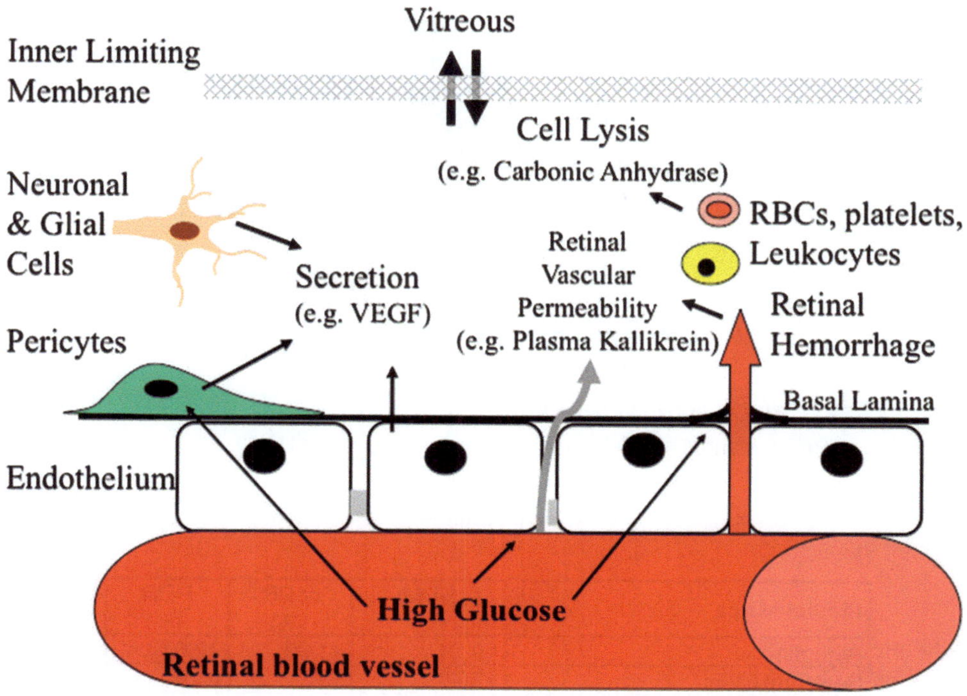

Fig. 3. Origins of vitreous proteins that been implicated in diabetic retinopathy progression. Diabetic retinopathy induces the release of active proteins into the vitreous by secretion (for example, VEGF), RVP (for example, plasma kallikrein), and retinal hemorrhages and cell lysis (for example, carbonic anhydrase I).

two- and fourfold compared with noDR and NDM vitreous, respectively. In addition to transport proteins, this analysis revealed that the protease inhibitor α1-antitrypsin, the anti-angiogenic factor PEDF, and complement C3 are highly abundant in PDR vitreous. This report also identified 56 proteins which differed in abundance in noDR and PDR compared with NDM. The majority of these changes were increases by two- to fourfold, which were comparable with increases in serum albumin (Fig. 5). For example, angiotensinogen (AGT) was show to be increased by two- to threefold in DM noDR and PDR vitreous. In addition, small subsets of proteins were increased by over tenfold or were decreased in noDR and PDR compared with NDM vitreous. As previously reported with CA-I, the functions of most of the vitreous proteins may require further study to evaluate their effects in the vitreous. This proteomic study also revealed that groups of proteins from the complement cascade, coagulation system, and kallikrein kinin system are present in the vitreous, suggesting that the vitreous proteome contains biochemical systems [37]. Further analyses revealed that a number of individual proteins existed as protein fragments, suggesting that the vitreous is proteolytically active, and certain protein functions may be associated with these fragments, as previously described for the anti-angiogenic factor endostatin, which is generated from the limited proteolysis of collagen XVIII [52].

A

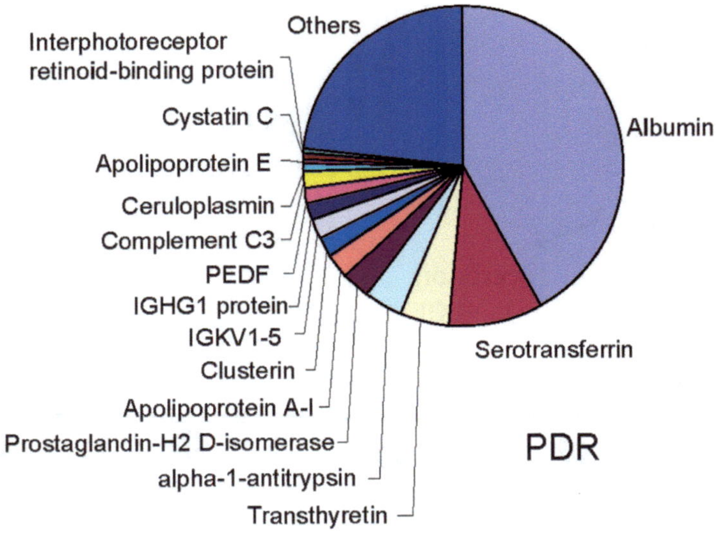

B

Protein Name	NDM (%)	noDR (%)	PDR (%)
Albumin	40.24	41.90	42.07
Serotransferrin	12.78	9.82	9.29
Transthyretin	4.77	2.88	5.12
Clusterin	4.42	2.13	1.99
IGKV1-5	3.70	2.96	1.93
Prostaglandin-H2 D-isomerase	2.76	1.77	2.59
alpha-1-antitrypsin	2.66	2.73	3.88
PEDF	2.11	1.94	1.57
Interphotoreceptor retinoid-binding protein	1.85	3.25	0.24
Ceruloplasmin	1.57	1.41	0.81
Complement C3	1.55	2.24	1.50
Cystatin C	1.40	0.51	0.63
IGHG1 protein	1.26	2.02	1.91
Apolipoprotein E	1.06	0.61	0.70
Apolipoprotein A-I	1.04	1.41	2.51
Others	16.81	22.45	23.27

Fig. 4. Fractional distribution of the most abundant proteins in human vitreous. (**A**) Chart showing a summary of the relative amounts of highly abundant proteins in PDR vitreous. (**B**) Table showing the mean percent of number of total peptides for the 15 most abundant proteins identified in NDM, noDR, and PDR samples relative to the number of total peptides detected from respective samples. Reprinted with permission from Gao et al. [37]. Copyright 2008 American Chemical Society.

SUMMARY AND CONCLUSIONS

Mass spectrometry–based proteomics has identified at least several hundred proteins from human vitreous. Diabetic retinopathy is associated with about a fourfold increase in total vitreous protein content and increased protein diversity compared

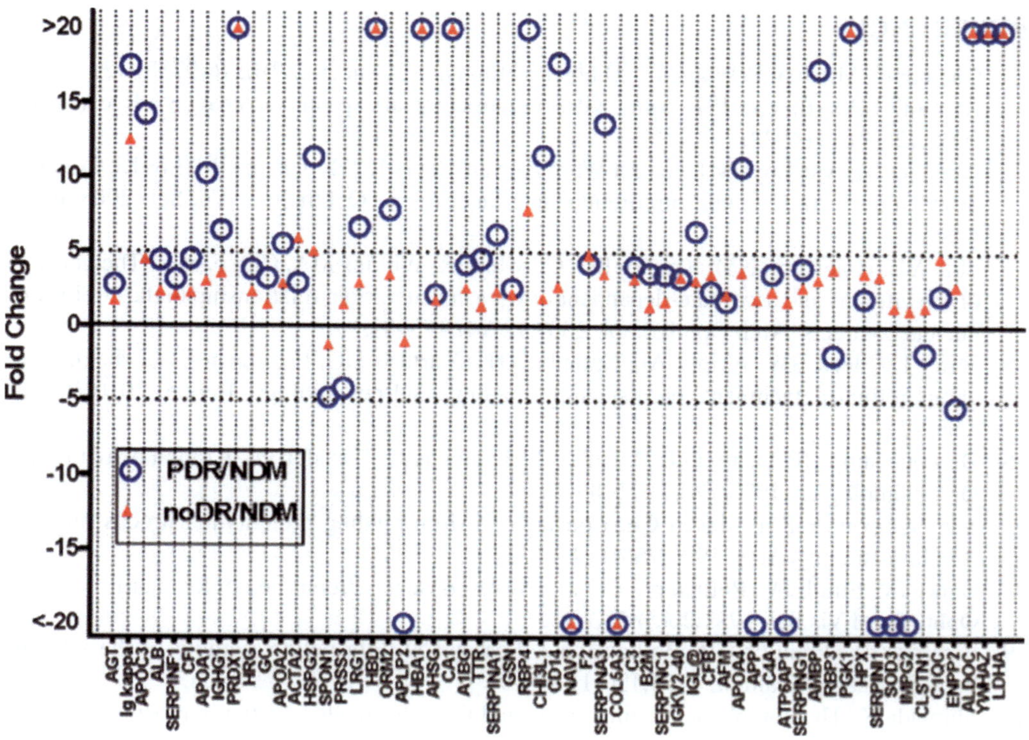

Fig. 5. Comparison of proteins abundance in noDR and PDR vitreous relative to NDM vitreous. Ratio of the mean total peptides detected in noDR or PDR groups relative to the NDM group. The absence of protein detection in a group is indicated by >20-fold. Reprinted with permission from Gao et al. [37]. Copyright 2008 American Chemical Society.

with NDM control vitreous. Most of these increases in protein in diabetic retinopathy appear to be due to the infiltration of plasma proteins and contributions from intraocular hemorrhage and cell lysis. Once in the vitreous, a limited number of these plasma and intracellular proteins have been shown to exert potent effects on retinal functions. These findings suggest that the loss of blood retinal barrier function in diabetes may promote further increases in RVP as diabetic retinopathy progresses. While the number of proteins identified by vitreous proteomics is increasing rapidly, the relative significance and biological functions of most of these proteins within the vitreous milieu are unknown. Direct functional analyses of protein action in the vitreous are needed to elucidate their potential effects in diabetic retinopathy. In addition, further characterization of the vitreous proteome may reveal biomarkers that correlate with clinical characteristics and could provide new insights into disease progression and responses to therapies.

ACKNOWLEDGMENTS

This work was supported in part by the US National Institutes of Health (grants EY019029, DK 36836) and the Juvenile Diabetes Research Foundation.

REFERENCES

1. Cheung N, Mitchell P, Wong TY. Diabetic retinopathy. Lancet. 2010;376(9735):124–36.
2. Le Goff MM, Bishop PN. Adult vitreous structure and postnatal changes. Eye (Lond). 2008;22(10):1214–22.
3. Ponsioen TL, van Luyn MJ, van der Worp RJ, van Meurs JC, Hooymans JM, Los LI. Collagen distribution in the human vitreoretinal interface. Invest Ophthalmol Vis Sci. 2008;49(9): 4089–95.
4. Shui YB, Holekamp NM, Kramer BC, Crowley JR, Wilkins MA, Chu F, et al. The gel state of the vitreous and ascorbate-dependent oxygen consumption: relationship to the etiology of nuclear cataracts. Arch Ophthalmol. 2009;127(4):475–82.
5. Mitry D, Fleck BW, Wright AF, Campbell H, Charteris DG. Pathogenesis of rhegmatogenous retinal detachment: predisposing anatomy and cell biology. Retina. 2010;30(10): 1561–72.
6. Dernouchamps JP, Vaerman JP, Michiels J, Heremans JF. Transferrins in rabbit ocular fluids. Ophthalmologica. 1975;170(1):72–83.
7. Van Bockxmeer FM, Martin CE, Constable IJ. Iron-binding proteins in vitreous humour. Biochim Biophys Acta. 1983;758(1):17–23.
8. Burke JM, Smith JM. Retinal proliferation in response to vitreous hemoglobin or iron. Invest Ophthalmol Vis Sci. 1981;20(5):582–92.
9. Forrester JV, Prentice CR, Williamson J, Forbes CD. Fibrinolytic activity of the vitreous body. Invest Ophthalmol. 1974;13(11):875–9.
10. Shimada K. The complement components and their inactivators in the intraocular fluids of the guinea pig. Invest Ophthalmol. 1970;9(4):307–15.
11. Raymond L, Jacobson B. Isolation and identification of stimulatory and inhibitory cell growth factors in bovine vitreous. Exp Eye Res. 1982;34(2):267–86.
12. Jacobson B, Dorfman T, Basu PK, Hasany SM. Inhibition of vascular endothelial cell growth and trypsin activity by vitreous. Exp Eye Res. 1985;41(5):581–95.
13. Taylor CM, Weiss JB. Partial purification of a 5.7K glycoprotein from bovine vitreous which inhibits both angiogenesis and collagenase activity. Biochem Biophys Res Commun. 1985;133(3):911–6.
14. Preis I, Langer R, Brem H, Folkman J. Inhibition of neovascularization by an extract derived from vitreous. Am J Ophthalmol. 1977;84(3):323–8.
15. Glaser BM, D'Amore PA, Michels RG. The effect of human intraocular fluid on vascular endothelial cell migration. Ophthalmology. 1981;88(9):986–91.
16. Glaser BM, D'Amore PA, Michels RG, Brunson SK, Fenselau AH, Rice T, et al. The demonstration of angiogenic activity from ocular tissues. Preliminary report. Ophthalmology. 1980;87(5):440–6.
17. Glaser BM, D'Amore PA, Lutty GA, Fenselau AH, Michels RG, Patz A. Chemical mediators of intraocular neovascularization. Trans Ophthalmol Soc U K. 1980;100(3):369–73.
18. Miller JW, Adamis AP, Shima DT, D'Amore PA, Moulton RS, O'Reilly MS, et al. Vascular endothelial growth factor/vascular permeability factor is temporally and spatially correlated with ocular angiogenesis in a primate model. Am J Pathol. 1994;145(3):574–84.
19. Adamis AP, Miller JW, Bernal MT, D'Amico DJ, Folkman J, Yeo TK, et al. Increased vascular endothelial growth factor levels in the vitreous of eyes with proliferative diabetic retinopathy. Am J Ophthalmol. 1994;118(4):445–50.
20. Aiello LP, Avery RL, Arrigg PG, Keyt BA, Jampel HD, Shah ST, et al. Vascular endothelial growth factor in ocular fluid of patients with diabetic retinopathy and other retinal disorders. N Engl J Med. 1994;331(22):1480–7.

21. Aiello LP, Bursell SE, Clermont A, Duh E, Ishii H, Takagi C, et al. Vascular endothelial growth factor-induced retinal permeability is mediated by protein kinase C in vivo and suppressed by an orally effective beta-isoform-selective inhibitor. Diabetes. 1997;46(9):1473–80.
22. Funatsu H, Yamashita H, Sakata K, Noma H, Mimura T, Suzuki M, et al. Vitreous levels of vascular endothelial growth factor and intercellular adhesion molecule 1 are related to diabetic macular edema. Ophthalmology. 2005;112(5):806–16.
23. Diabetic Retinopathy Clinical Research Network, Elman MJ, Aiello LP, Beck RW, Bressler NM, Bressler SB, et al. Randomized trial evaluating ranibizumab plus prompt or deferred laser or triamcinolone plus prompt laser for diabetic macular edema. Ophthalmology. 2010;117(6):1064–77.
24. Nguyen QD, Shah SM, Khwaja AA, Channa R, Hatef E, Do DV, et al. Two-year outcomes of the ranibizumab for edema of the mAcula in diabetes (READ-2) study. Ophthalmology. 2010;117(11):2146–51.
25. Funatsu H, Noma H, Mimura T, Eguchi S, Hori S. Association of vitreous inflammatory factors with diabetic macular edema. Ophthalmology. 2009;116(1):73–9.
26. Yoshimura T, Sonoda KH, Sugahara M, Mochizuki Y, Enaida H, Oshima Y, et al. Comprehensive analysis of inflammatory immune mediators in vitreoretinal diseases. PLoS One. 2009;4(12):e8158.
27. Praidou A, Klangas I, Papakonstantinou E, Androudi S, Georgiadis N, Karakiulakis G, et al. Vitreous and serum levels of platelet-derived growth factor and their correlation in patients with proliferative diabetic retinopathy. Curr Eye Res. 2009;34(2):152–61.
28. Simo R, Hernandez C, Segura RM, Garcia-Arumi J, Sararols L, Burgos R, et al. Free insulin-like growth factor 1 in the vitreous fluid of diabetic patients with proliferative diabetic retinopathy: a case-control study. Clin Sci (Lond). 2003;104(3):223–30.
29. Yamane K, Minamoto A, Yamashita H, Takamura H, Miyamoto-Myoken Y, Yoshizato K, et al. Proteome analysis of human vitreous proteins. Mol Cell Proteomics. 2003;2(11):1177–87.
30. Garcia-Ramirez M, Canals F, Hernandez C, Colome N, Ferrer C, Carrasco E, et al. Proteomic analysis of human vitreous fluid by fluorescence-based difference gel electrophoresis (DIGE): a new strategy for identifying potential candidates in the pathogenesis of proliferative diabetic retinopathy. Diabetologia. 2007;50(6):1294–303.
31. Shitama T, Hayashi H, Noge S, Uchio E, Oshima K, Haniu H, et al. Proteome profiling of vitreoretinal diseases by cluster analysis. Proteomics Clin Appl. 2008;2(9):1265–80.
32. Simo R, Vidal MT, Garcia-Arumi J, Carrasco E, Garcia-Ramirez M, Segura RM, et al. Intravitreous hepatocyte growth factor in patients with proliferative diabetic retinopathy: a case-control study. Diabetes Res Clin Pract. 2006;71(1):36–44.
33. Krogsaa B, Lund-Andersen H, Mehlsen J, Sestoft L, Larsen J. The blood-retinal barrier permeability in diabetic patients. Acta Ophthalmol (Copenh). 1981;59(5):689–94.
34. Plehwe WE, Sleightholm MA, Kohner EM. Does vitreous fluorophotometry reflect severity of early diabetic retinopathy? Br J Ophthalmol. 1989;73(4):255–60.
35. Kim T, Kim SJ, Kim K, Kang UB, Lee C, Park KS, et al. Profiling of vitreous proteomes from proliferative diabetic retinopathy and nondiabetic patients. Proteomics. 2007;7(22):4203–15.
36. Gao BB, Clermont A, Rook S, Fonda SJ, Srinivasan VJ, Wojtkowski M, et al. Extracellular carbonic anhydrase mediates hemorrhagic retinal and cerebral vascular permeability through prekallikrein activation. Nat Med. 2007;13(2):181–8.
37. Gao BB, Chen X, Timothy N, Aiello LP, Feener EP. Characterization of the vitreous proteome in diabetes without diabetic retinopathy and diabetes with proliferative diabetic retinopathy. J Proteome Res. 2008;7(6):2516–25.

38. Nakanishi T, Koyama R, Ikeda T, Shimizu A. Catalogue of soluble proteins in the human vitreous humor: comparison between diabetic retinopathy and macular hole. J Chromatogr B Analyt Technol Biomed Life Sci. 2002;776(1):89–100.
39. Koyama R, Nakanishi T, Ikeda T, Shimizu A. Catalogue of soluble proteins in human vitreous humor by one-dimensional sodium dodecyl sulfate-polyacrylamide gel electrophoresis and electrospray ionization mass spectrometry including seven angiogenesis-regulating factors. J Chromatogr B Analyt Technol Biomed Life Sci. 2003;792(1):5–21.
40. Ouchi M, West K, Crabb JW, Kinoshita S, Kamei M. Proteomic analysis of vitreous from diabetic macular edema. Exp Eye Res. 2005;81(2):176–82.
41. Kim SJ, Kim S, Park J, Lee HK, Park KS, Yu HG, et al. Differential expression of vitreous proteins in proliferative diabetic retinopathy. Curr Eye Res. 2006;31(3):231–40.
42. Aebersold R, Mann M. Mass spectrometry-based proteomics. Nature. 2003;422(6928): 198–207.
43. Gao BB, Phipps JA, Bursell D, Clermont AC, Feener EP. Angiotensin AT1 receptor antagonism ameliorates murine retinal proteome changes induced by diabetes. J Proteome Res. 2009;8(12):5541–9.
44. Kim K, Kim SJ, Yu HG, Yu J, Park KS, Jang IJ, et al. Verification of biomarkers for diabetic retinopathy by multiple reaction monitoring. J Proteome Res. 2010;9(2):689–99.
45. Gao BB, Stuart L, Feener EP. Label-free quantitative analysis of one-dimensional PAGE LC/MS/MS proteome: application on angiotensin II-stimulated smooth muscle cells secretome. Mol Cell Proteomics. 2008;7(12):2399–409.
46. Mann M, Kelleher NL. Precision proteomics: the case for high resolution and high mass accuracy. Proc Natl Acad Sci USA. 2008;105(47):18132–8.
47. Mueller LN, Brusniak MY, Mani DR, Aebersold R. An assessment of software solutions for the analysis of mass spectrometry based quantitative proteomics data. J Proteome Res. 2008;7(1):51–61.
48. Bantscheff M, Schirle M, Sweetman G, Rick J, Kuster B. Quantitative mass spectrometry in proteomics: a critical review. Anal Bioanal Chem. 2007;389(4):1017–31.
49. Kumar C, Mann M. Bioinformatics analysis of mass spectrometry-based proteomics data sets. FEBS Lett. 2009;583(11):1703–12.
50. Garcia-Ramirez M, Hernandez C, Villarroel M, Canals F, Alonso MA, Fortuny R, et al. Interphotoreceptor retinoid-binding protein (IRBP) is downregulated at early stages of diabetic retinopathy. Diabetologia. 2009;52(12):2633–41.
51. Simo R, Higuera M, Garcia-Ramirez M, Canals F, Garcia-Arumi J, Hernandez C. Elevation of apolipoprotein A-I and apolipoprotein H levels in the vitreous fluid and overexpression in the retina of diabetic patients. Arch Ophthalmol. 2008;126(8):1076–81.
52. Bhutto IA, Kim SY, McLeod DS, Merges C, Fukai N, Olsen BR, et al. Localization of collagen XVIII and the endostatin portion of collagen XVIII in aged human control eyes and eyes with age-related macular degeneration. Invest Ophthalmol Vis Sci. 2004;45(5):1544–52.

12

Neurodegeneration in Diabetic Retinopathy

Alistair J. Barber, William F. Robinson, and Gregory R. Jackson

CONTENTS

Keywords Neurodegeneration • Retinal ganglion cell • Nerve fiber layer • Caspases • Scotopic threshold response

INTRODUCTION

Neurodegeneration can be defined as a chronic, progressive loss of neuronal function and structural integrity, which usually includes the death and removal of neurons at an accelerated rate. In neurodegenerative diseases, the loss of neurons occurs gradually over a protracted period of time, such as the kind of neural loss that occurs in Parkinson's or Alzheimer's disease. The term neurodegeneration is used frequently in discussions of many disease pathologies that primarily affect neurons; however, neurodegenerative diseases have been more accurately defined as, "…neurological disorders with heterogeneous clinical and pathological expressions affecting specific subsets of neurons in specific functional anatomic systems; they arise for unknown reasons and progress in a relentless fashion" [1]. By this strict definition, neuronal loss in Alzheimer's disease is classed as neurodegeneration; while acute loss of neurons in a stroke is not, although neurodegeneration is often modeled using experimentally induce ischemia to accelerate neuronal cell death.

Neurodegenerative diseases are commonly thought of as affecting the brain or peripheral nervous system, but this chapter will consider diabetic retinopathy as a candidate neurodegenerative disease of the retina. There are a series of features that are gener-

From: *Ophthalmology Research: Visual Dysfunction in Diabetes*
Edited by: J. Tombran-Tink et al. (eds.), DOI 10.1007/978-1-60761-150-9_12

ally associated with neurodegenerative diseases, which can be broadly categorized into histological, biochemical, and functional pathologies. This chapter will present evidence for retinal neurodegeneration in diabetes, segregated according to these three categories, and will finish by including a brief summary of the theorized mechanisms.

HISTOLOGICAL EVIDENCE

Early Pathology Studies

Early efforts to characterize the histology of diabetic retinopathy were the first to identify potential neuropathy accompanying the vascular changes. An early study of histological sections from postmortem specimens noted atrophy of retinal ganglion cells (RGCs) as one of the pathological changes that accompanied vascular lesions, and suggested that diabetes may induce a gradual loss of neurons as the disease progresses [2]. A similar study on a larger number of specimens also identified degeneration of the inner plexiform and ganglion cell layers as common features in humans with diabetic retinopathy [3]. Later, a paper by Bresnick suggested that neurodegeneration could possibly be viewed as a neurosensory disorder that involved degeneration of the neural retina, possibly preceding the vascular lesions [4]. One common feature of diabetic retinopathy that can be recognized by clinical observation is the appearance of "cotton wool spots" which are thought to be the axoplasmic debris from atrophied neurons in the nerve fiber layer (NFL) [5], and can appear as an early pathological feature in some patients [6].

Histological Evidence of Apoptosis

Studies of tissue from human and animals with diabetes identified apoptotic cells in the retina. In some cases, these included RGCs reviewed recently by Kern and Barber [7]. Many histological studies of apoptosis have used the classic technique of DNA terminal dUTP nick end labeling (TUNEL), which most commonly uses terminal transferase to label nuclei-containing DNA nicks in fixed tissue sections [8–10]. An early study using TUNEL identified apoptotic cells in cross sections of retinas from streptozotocin (STZ)-diabetic rats, although quantification of the numbers of neurons was not possible in this study [11]. Another study, using the trypsin-digest approach to specifically examine the vasculature of rat retinas, indicated a modest increase in TUNEL-labeled nuclei in rats after 6–8 months of STZ diabetes, suggesting that vascular cells also underwent apoptosis [12]; a finding that has been confirmed by others [13–15].

While trypsin digest makes it possible to specifically examine the vasculature of the retina, TUNEL labeling in intact flat-mounted retinas from STZ-diabetic rats made it possible to quantify the numbers of cells undergoing apoptosis in the entire retina. Using this approach, it was found that diabetic rats had significantly more TUNEL positive cells with a similar rate of cell death in groups of rats after 1, 3, 6, and 12 months of hyperglycemia (Fig. 1). The absolute number of positive cells was greater than in the trypsin-digest studies, suggesting that neurons and glial cells were also involved [16]. Others showed that TUNEL labeling was also increased in mouse models of both type I and type II diabetes [17, 18], and quantification of TUNEL labeling in whole retinas from Ins2^{Akita} mice, a spontaneously diabetic genetic model, found a frequency of apoptosis similar to the STZ-diabetic rats [19].

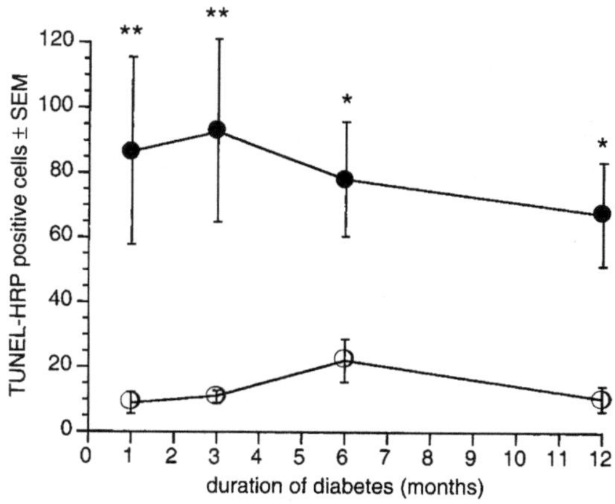

Fig. 1. Diabetes increased apoptosis in whole rat retinas. Apoptotic cells were identified by TUNEL in whole retinas of STZ (streptozotocin)-diabetic rats after 1, 3, 6, and 12 months of hyperglycemia. The total number of positive nuclei in each retina was counted by microscopy. There were significantly more apoptotic cells in the retinas from diabetic rats (*black circles*) compared to controls (*white circles*), $*p<0.01$, $**p<0.001$, 1-way ANOVA with Newman-Keuls test. Taken from Barber et al. [16].

A variety of other histological studies have confirmed the increase in TUNEL labeling in diabetic animals, although the types of cells and the degree of apoptosis vary widely. An early phase of TUNEL labeling in photoreceptors was indicated in one study, accompanied by several indications of degeneration in amacrine, horizontal, and ganglion cells [20], although one study reported no significant increase in apoptosis of nonvascular cells in STZ-mouse retinas [21]. The rate of retinal apoptosis in diabetic rats was further increased by experimentally induced intraocular hypertension, similar to that in glaucoma [22].

As an alternative or additional approach to using TUNEL to detect cells undergoing apoptosis, some investigators have used antibodies to the activated form of caspase enzymes in histological sections of retina. Caspases-3 and -7 are often referred to as "executioner enzymes" because they cleave target proteins at specific aspartate recognition sequences. Antibodies raised to identify only the active form of caspase-3 can be used for immunohistochemical detection of cells undergoing apoptosis at the time of tissue fixation [23]. Using this approach, the number of cells positive for active caspase-3 was found to be elevated in the ganglion cell layer of retinas from mice after 2, 6, and 12 weeks of STZ diabetes [17]. A similar approach labeling for active caspase-3 in whole-mount retinas from Ins2^{Akita} mice found that, after 4 weeks of hyperglycemia, there were significantly more positive cells compared to nondiabetic age–matched litter mates [24]. There is also evidence that caspase-3 is activated in ganglion cells of postmortem retinas from subjects with diabetes [25]. Similarly, caspase-3 and -9 immunohistochemistry in human postmortem retinas colocalized with Fluoro-Jade B, an indicator of degenerating neurons, and was most abundant in cell bodies in the RGC layer [26]. Caspase-3 immunoreactivity was also found to colocalize with several other neuronal markers in flat-mount retinas of Ins2^{Akita} diabetic mice, suggesting that the cells undergoing apop-

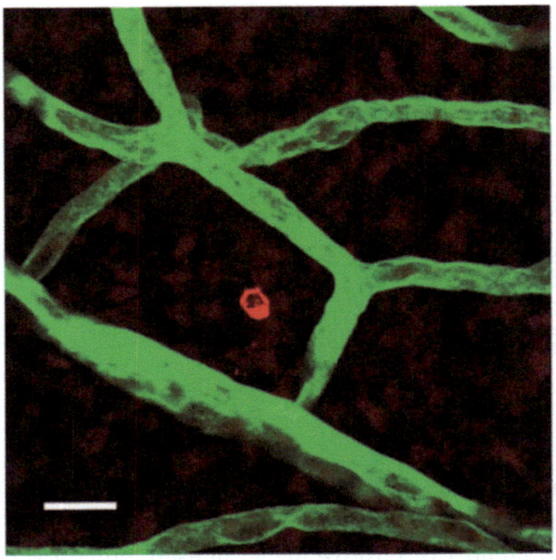

Fig. 2. Immunoreactivity for active caspase-3 did not localize to the vasculature. Whole retinas from STZ-diabetic rats were labeled by immunofluorescence for agrin, a vascular basement membrane glycoprotein (*green*) and active caspase-3 (*red*). The majority of caspase-3 positive cells were located away from blood vessels, suggesting that they were neural in origin, scale bar = 50 μm. Taken from Gastinger et al. [38].

tosis included RGCs, amacrine cells, and photoreceptors. Quantification of caspase-3 positive cells in these mice yielded a similar estimate of the total number of apoptotic cells compared to data obtained by TUNEL, and the majority of caspase-3 positive cells did not colocalize with agrin immunoreactivity in vascular basement membrane, indicating that the dying cells were mostly not vascular in origin [19] (Fig. 2).

Gross Morphological Changes in the Retina

Many studies investigating loss of neurons in the retina use measures of the thickness of retinal layers as a measure of cell loss [27]. STZ-diabetic rats had significantly reduced thickness of the inner plexiform and inner nuclear layers, 7.5 months after the onset of hyperglycemia [16] (Fig. 3), suggesting that the increased apoptosis identified in this study leads to an accumulated loss of cells making up the inner retina. A different study suggested that reduction in the thickness of the inner plexiform layer was accompanied by loss of the outer nuclear layers and increased TUNEL labeling primarily among photoreceptors in STZ rats after 6 months of diabetes [20]. Decreased thickness of the outer retina was also noted after 12 and 24 weeks of diabetes in STZ-diabetic rats [28].

Similar reductions in retinal layer thickness were measured in diabetic mice. The inner retina was reduced in $Ins2^{Akita}$ diabetic mice after 5 months of hyperglycemia, although the reduction was limited to the peripheral retina, suggesting that the cell loss may occur more slowly in the central region of the retina in this mouse model [24]. The reduction in inner retina thickness was comparable to previous observations in STZ-diabetic mice that were diabetic for 10 weeks [17].

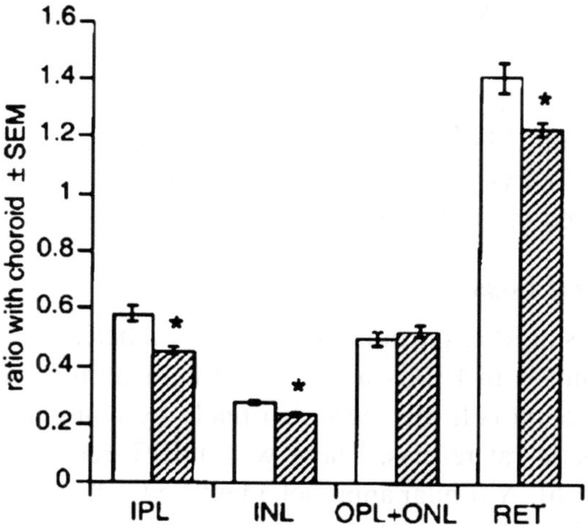

Fig. 3. Diabetes reduced the thickness of the inner retina in rat. The thickness of the inner plexiform layer (IPL), inner nuclear layer (INL), combined outer plexiform and outer nuclear layers (OPL + ONL), and entire retina (RET) were measured as a ratio with the choroid in H&E sections of eyes from rats after 7.5 months of diabetes (*shaded bars*) and compared to eyes from control rats (*white bars*). The IPL and INL were significantly thinner in diabetic rats ($*p<0.001$) compared to controls. Taken from Vanguilder et al. [56].

Several studies have reported cell loss by measuring cell layer thicknesses in rodent models of diabetes; however, there are disparities in the rate of cell loss and whether the degeneration is predominantly inner or outer retina. The differences between these studies are difficult to explain but may be due to variations in the degree of induced diabetes, genetic background of the animals, and variations in animal husbandry, including the fat content of the diet [29], differences in handling, or exposure to environmental pathogens.

Reductions in Numbers of Surviving Amacrine Cells

Results of several morphological studies indicate that diabetes may deplete the number of amacrine cells in the retina. Tyrosine hydroxylase immunoreactivity, a marker of dopaminergic neurons, was reduced in amacrine cells of the obese sand rat, which becomes moderately hyperglycemic [30]. Necrosis of amacrine cells was also reported in STZ rats, along with photoreceptors, ganglion cells, and other neurons [20]. Tyrosine hydroxylase protein levels were also depleted by approximately 50%, accompanied by significant reduction in the density of dopaminergic amacrine cells [31]. Labeling of neuronal nitric oxide synthase (nNOS)-positive amacrine cells was also reduced in diabetic rats, suggesting a down regulation in the enzyme expression, or a loss of amacrine cells [32]. While the morphology of surviving amacrine cells appeared to be normal in whole-mount retinas from diabetic $Ins2^{Akita}$ mice after 6 months of hyperglycemia, they were reduced in number by 16–20% [19].

Reductions in neurotransmitters and associated enzyme activity also imply a loss of amacrine cells. The concentration of dopamine was significantly reduced in STZ-diabetic rats after 3 weeks of hyperglycemia, while there was no change in tyrosine hydroxylase activity or the uptake in tyrosine, suggesting a potential increase in the dopaminergic efflux due to diabetes [33]. In a more recent study, butyrylcholinesterase activity was reduced 30% in retinas of STZ-diabetic rats, implying a potential loss of subtypes of amacrine cells [34]. A similar study confirmed the depletion of the butyrylcholinesterase enzyme activity in STZ rats [35]. Similarly, NADPH diaphorase immunoreactivity was reduced in the processes of amacrine cells [36].

Retinal Ganglion Cell Loss

Evidence for the loss of RGCs in animal models of diabetes was recently reviewed [7]. Apoptosis had been noted in RGCs using TUNEL in retinal cross sections [11]. An approximation of ganglion cell loss was also made by counting large nuclei in H&E sections of STZ-diabetic rat retinas. There was a 10% reduction in these cells after 7 months of diabetes [16]. A similar approach identified a 20–25% loss of ganglion cells from retinas of STZ-diabetic mice after 14 weeks of hyperglycemia [17]. The number of ganglion cells in the retinas of Brown-Norway STZ-diabetic rats was also found to be reduced by about 16% within 4–5 weeks, using a fluorescent retrograde labeling technique [37]. Another study on STZ-diabetic mice, however, reported no change in H&E sections [21]. Ganglion cell bodies can be easily confused with astrocytes and amacrine cells in this type of preparation, which may account for the variation in results using this method. To overcome this potential confounding variable, RGCs were quantified in whole retinas of mice-expressing endogenous ganglion cell markers. $Ins2^{Akita}$ mice were crossed with Thy1-CFP transgenic mice, which express an endogenous fluorescent protein in the cell body of the majority of RGCs. There was a 16% reduction in RGCs in the peripheral retina within 3 months of the onset of diabetes in these mice [38].

Abnormalities in Ganglion Cell Morphology

Accelerated loss of RGCs due to diabetes has been suggested in a number of studies, but it is likely that the apoptosis is accompanied by other pathological features in these important neurons. A study of flat-mounted retinas from $Ins2^{Akita}$ diabetic mice with endogenously fluorescent markers under the Thy1 promoter revealed several abnormal features in subsets of RGCs [38]. Cell bodies were enlarged, and there were numerous axonal swellings, often associated with a constriction between the cell body and the swelling. The morphology of dendrites was also altered, most dramatically in the large On-ganglion cells. The dendritic fields of these neurons tended to be more complex, possessing more branches and terminals. It was suggested that the reason for this apparent increase in plasticity among certain ganglion cells was in compensation for loss of input from bipolar or amacrine cells or an attempt to compensate for the loss of neighboring ganglion cells (Fig. 4).

A morphological study of RGCs in rats used DiI (1,1′-dioctadecyl-3,3,3′,3′-tetramethylindocarbocyanine perchlorate) applied to whole retinas of STZ-diabetic rats with a gene gun [39]. The largest subtype of cell defined in this study appeared to have enlarged dendritic field areas in the diabetic animals compared to controls. A similar study in a small sample of human postmortem samples with advanced stages of diabetic retinopathy also suggested abnormal morphology of some parasol and midget ganglion cells, including axon swelling and beading, while the dendritic field areas of

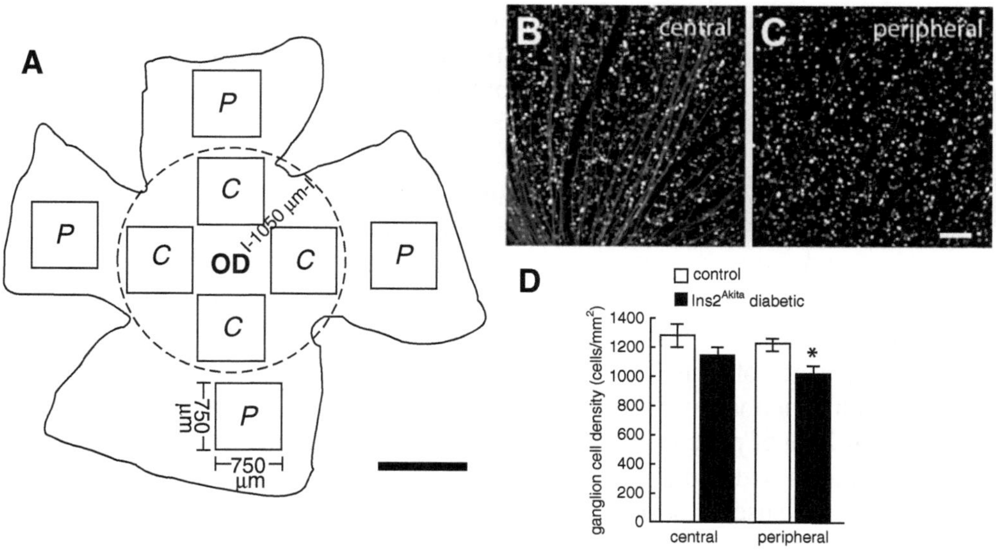

Fig. 4. Diabetes reduced the number of Thy1-CFP-positive ganglion cells in the retinas of Ins2^{Akita} mice Ins2^{Akita} mice were crossed with transgenic mice-expressing CFP under the Thy1 promoter, which is specific to RGCs. (**A**) Retinas were flat mounted, and the number of CFP-positive cells was counted in four inner and four outer regions as illustrated (scale bar = 1 mm). (**B**, **C**) The CFP-positive cell bodies were identified in confocal maximum projections from confocal *z*-scans (scale bar = 100 µm). (**D**) In mice that were hyperglycemic for 3 months, the average ganglion cell density was significantly lower in the peripheral regions of retinas compared to age-matched littermates ($n = 5$ controls, $n = 7$ diabetic, $*p < 0.05$). Taken from Santiago et al. [130].

these cells appeared reduced [40]. In all three studies, the morphology of some classes of RGCs was found to be changed by diabetes, but the size of the dendritic field and the density of dendrites were dissimilar. It is unclear why there is disagreement in the results between mice, rats and human retinas, but the degree of retinal pathology and methods of identifying and classifying the RGCs may be responsible for at least some of the discrepancies.

This small number of studies on RGC morphology suggests common features in which abnormal swellings occurs on ganglion cell axons; although it appears that the human study suggested a general decline in the size and complexity of the dendritic field, while the rodent studies suggested that the fields of surviving cells become more complex. These changes may also represent precursors to apoptosis, or alternatively could be plastic responses to the loss of neighboring ganglion cells, or to the loss of input from bipolar and amacrine cells.

Centrifugal Axon Abnormalities

An area of potential neural damage that is not often considered is the centrifugal axon. This mysterious group of projections was described in early histological studies of humans, primates, and rodents and was proposed to play a role in regulating blood flow [41]. These axons can be identified by immunohistochemistry of histamine [42].

A histological study of STZ-diabetic rat retinas indicated that histaminergic centrifugal axons had several pathological abnormalities including swellings [43], which may indicate potential retrograde transport problems leading to distal accumulation of materials.

Nerve Fiber Layer Thickness

The evidence of histological changes in RGCs in humans is difficult to interpret because of the difficulties in collecting postmortem specimens, and there is currently no way to image ganglion cell structure in vivo. Several studies have, however, used clinical imaging techniques to measure the thickness of the NFL, which is presumably relative to the abundance of ganglion cells. A defect in the thickness of the NFL was found in a population of humans with type II diabetes using red-free fundus photography. It was present in 20% of patients with only mild retinopathy (no microaneurysms) and in 57–78% of patients with more severe vascular abnormalities [44]. Scanning laser polarimetry has also been used in several studies to assess changes in the NFL due to diabetes. In age-matched patients grouped according to their glycosylated hemoglobin levels, the NFL was significantly thinner in patients with diabetes and poor blood glucose control (HbA1c > 8%) and in patients with nonproliferative retinopathy [45]. The NFL thickness in patients in this study who maintained HbA1c < 8%, however, was not different from normal. A similar study identified an asymmetric NFL loss in the superior segment of patients with type I diabetes [46]. A further study on type II patients also indicated that NFL thickness decreased with increasing severity of diabetic retinopathy [47]. Together, these clinical studies suggest a strong link between NFL thinning, glycemic control of diabetes, increasing duration, and degree of retinopathy. Recently, a further study on NFL thickness measured by optical coherence tomography showed that panretinal photocoagulation can exaggerate the thinning of the NFL in diabetes, suggesting that laser surgery may induce further atrophy of RGC axons [48].

Thickness of the NFL in rodents has not been measured; however, one study on cross sections of optic nerve from STZ-diabetic rats indicated a reduction in the density of axons, accompanied by increases in the number of glial cells, suggesting denervation or loss of ganglion cells [49].

BIOCHEMICAL EVIDENCE OF NEURODEGENERATION AND CELL DEATH

The predominant evidence for diabetes-induced neurodegeneration in the retina comes from histological studies; however, other studies present biochemical evidence of apoptosis and neuronal dysfunction. Immunohistochemical studies suggest reductions in Bcl-2, which could be linked to increases in apoptosis [25, 50]. Further evidence includes increased activity of several caspase enzymes. A comprehensive assessment of activity of caspases in rat retinas showed that caspases-1, -2, -6, -8, and -9 become active within 2 months of the onset of diabetes in STZ rats. Similar activities were found in postmortem tissue donated from humans with diabetes. In this study, the executioner caspases-3 and -6 became active later in the course of diabetes corresponding to a period when capillary cells are expected to be lost [51]. Similarly, caspase-3 enzyme activity was increased in the retinas of alloxan-diabetic rats after 14 months, but not 2 months, of hyperglycemia [52]. In rats after 3 months of STZ diabetes, the increased caspase-3 activity was reversed by the anti-inflammatory drug, minocycline, suggesting the possibility that caspase-3 dependent apoptosis is due to an inflammatory signal [53]. Minocycline also reduced caspase-1 activity in STZ-diabetic mouse retinas [54]. Further evidence of a link between inflammatory signaling and caspase enzyme activation is provided by a study with nepafenac, a COX-1/-2 inhibitor, given topically to the eye. In this study, the anti-inflammatory treatment inhibited the increase in caspases-3 and -6 after 9 months of diabetes [55].

While the evidence for increases in apoptosis-associated enzymes is compelling, the cell types in which these changes take place are not easily determined. It is arguable that these changes occur in vascular cells as well as, or to the exclusion of, neurons. Other biochemical evidence for changes in neurons comes from measurements of synapse-specific proteins such as postsynaptic density 95 (PSD95), and synaptic vesicle-associated proteins such as synaptophysin. The retinal content of several synaptic proteins was found to be decreased after the first month of hyperglycemia in STZ-diabetic rats [56] (Fig. 5). These changes were accompanied by a further depletion in the content of phosphorylated synapsin 1, suggesting a reduction in the mobilization of neurotransmitter vesicles. Interestingly, the content reduction in synaptophysin was reversed by angiotensin II receptor blockers [57].

Other biochemical changes that could be associated with neurodegeneration include increases in nNOS, which increased in both protein content and activity in retinas from STZ rats [58]. It was proposed that nNOS provided a regulatory link between neurons and vascular blood flow and that the number of nNOS-positive neurons was depleted by diabetes [59]. A similar study confirmed the increase in nNOS expression and identified multiple subtypes of nNOS-containing neurons, including amacrine, bipolar, and horizontal cells, that were damaged by diabetes [28, 32]. Elevated levels of nNOS are accompanied by increases in the production of nitric oxide, especially in the plexiform layers, measured by a novel in situ immunohistochemical imaging technique [60]. Elevated levels of nitric oxide could have a dramatic influence on neuronal function, including altered glutamate receptor signaling [61], increased peroxynitrite production associated with excitotoxicity [62], and altering RGC axon morphology [63].

FUNCTIONAL EVIDENCE OF NEURODEGENERATIVE CHANGES

There is an abundant variety of electrophysiological studies indicating that diabetes induces functional changes in the retina. Many of these studies will be reviewed elsewhere in this volume; however, some electrophysiological studies specifically indicate a neurodegenerative mechanism.

Electrophysiological Evidence for Neurodegeneration

The electroretinogram (ERG) is frequently used to measure the electrical responses originating from the retina due to light stimulus. The response, recorded by an electrode placed on the cornea, produces a waveform with several components, provided by

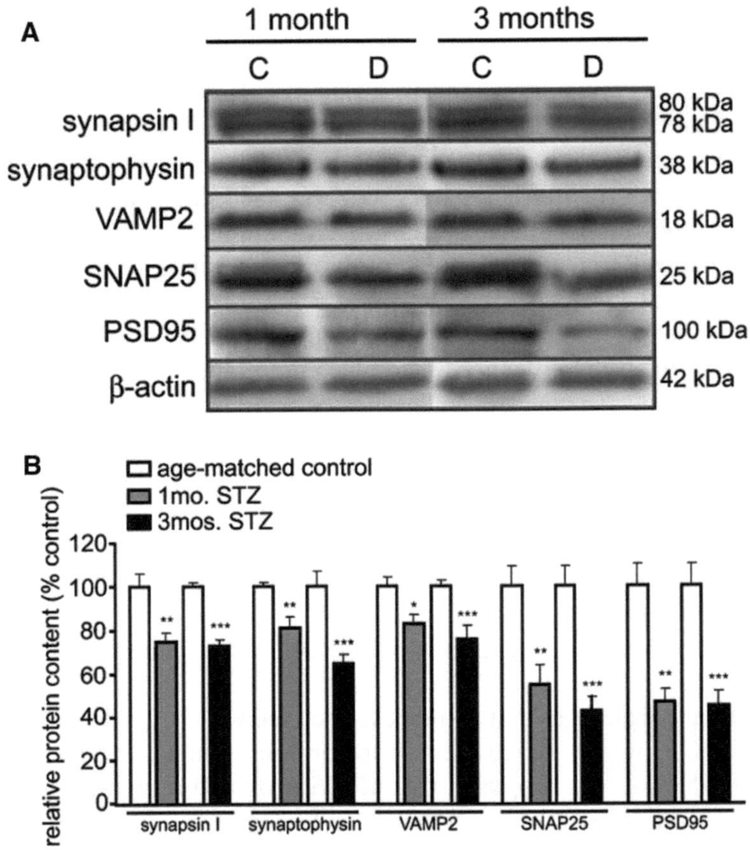

Fig. 5. Diabetes decreased the content of synaptic proteins in rat retinas. Synaptic proteins were quantified by western blot in the retinal homogenates from STZ-diabetic and control rats after 1 and 3 months of hyperglycemia. (**A**) Protein bands were apparent at the predicted molecular weight for each synaptic protein, and band densities were standardized to β-actin in the same sample. (**B**) Relative protein content was obtained as % control. There was a significant reduction in each of the proteins measured in the retinas from STZ-diabetic rats ($n = 8$ per group, $*p < 0.05$, $**p < 0.01$, $*p < 0.001$). Taken from Vanguilder et al. [56].

different cell types from the neural retina. Immediately following light stimulus, the a-wave is a negative deflection produced by the photoreceptors. The postreceptor b-wave response is a large positive deflection originating primarily from the ON-center bipolar cells [64, 65] modified by input from OFF-center bipolar and horizontal cells [66]. The oscillatory potentials (OPs) are small, higher frequency wavelets on the ascending portion of the b-wave, are thought to represent the modulation of interactions between bipolar, amacrine, and ganglion cells [67, 68], and are often analyzed in clinical and research studies of diabetic retinopathy.

Clinical studies of patients with diabetes were concerned with OP changes associated with diabetic retinopathy. In 1962, Yonemura et al. reported deterioration of oscillatory potentials not only in patients with diabetes, most of whom had been diagnosed with retinopathy, but also in a smaller number of patients without ophthalmoscopic evidence of retinopathy [69]. Additional clinical studies followed, establishing that humans with diabetic retinopathy have specific alterations in ERG response, including reduced OP amplitude [70–72] and increased OP latency [71]. Juen and Kieselbach noted that patients of 18–33 years old had significant loss of OP amplitude after being diagnosed with diabetes for an average of only 7 years, prior to the advent of any notable vascular changes associated with diabetic retinopathy [72]. Alterations in OPs have been correlated with loss of visual acuity, hue discrimination [71, 73], and contrast sensitivity [70]. Other studies showing that the ERG was altered within a few years of the onset of juvenile diabetes suggested that this could be used as an early diagnostic approach [74, 75].

Much of the research into the effects of diabetes on the a-wave has been performed in the STZ-diabetic rat model. Several studies reported a reduction of the a-wave amplitude by 12 weeks after the onset of diabetes, suggesting loss of photoreceptor function [76–78]. Hancock and Kraft also found a delay in the a-wave implicit time [79], a result which had also been reported in humans with diabetes [71, 80]. Animal studies consistently demonstrate a decrease in b-wave amplitude by 12 weeks after the onset of diabetes [79, 81, 82]. Phipps et al. recorded decreased b-wave amplitudes as early as 2 days after STZ injection [77]. The same study also determined that there was no change in b-wave latency in the diabetic rat model, a finding that was replicated in another animal study [83]. The b-wave latency is increased in patients with diabetic retinopathy [84, 85]. Taken together, the results of many ERG studies provide evidence of loss of function in photoreceptors, amacrine cells, bipolar and horizontal cells. The mechanism for these electrophysiological changes is unclear. The small amounts of cell death are unlikely to give rise to these large changes in the electrophysiological output of the retina. A more likely possibility is that changes in the amount of neurotransmitter release, or transmembrane ionic currents could account for the electrophysiological deficits.

The scotopic threshold response (STR) is another component of the ERG, considered to be an indicator of RGC function [68, 86, 87]. While this response is often not measured, because it can only be determined in response to very low intensity flashes of light, there is good evidence that it is reduced in both humans and rats with diabetes [83, 88, 89]. Furthermore, electrophysiological ganglion cell function was found to be compromised even in children with diabetes [90]. These data provide functional evidence of diabetes-induced RGC degeneration.

Optic Nerve Retrograde Transport

Several studies have determined that diabetes causes functional reductions in retrograde transport along the optic nerve. Retrograde axonal transport of fluorogold into medium and large RGCs was reduced in STZ-diabetic rats (but not a type II animal model) [91, 92]. The effect was reduced by treating rats with an aldose reductase inhibitor, suggesting that the loss of function in diabetes may be due to activation of the polyol pathway [93]. Loss of the visually evoked potential, accompanied by optic nerve pathology, in spontaneously diabetic BB/W rats is a further indication that optic nerve function is compromised by diabetes [94].

Other Changes in Visual Function

There have been a variety of studies applying psychophysical testing on humans with diabetes, recording a number of deficits that could be explained by altered neural function or neurodegeneration. A study of visual evoked potential, a measure of the visual cortex response to a flash of light, found that the evoked response was reduced and delayed in juvenile patients with diabetes [95]. Furthermore, the evoked response to stimuli with low contrast sensitivity was reduced more than stimuli with greater contrast sensitivity in patients with type 1 diabetes but no evidence of vascular retinopathy [96].

Reduced night vision is often associated with the early stages on diabetic retinopathy [97]. Loss of night vision may be associated with reduced contrast sensitivity and prolonged dark adaptation, and patients with maculopathy are often aware of peripheral field defects and color vision abnormalities [98].

Contrast sensitivity has also been studied extensively in diabetic patients. In a larger study, a group of non-insulin-dependent diabetic subjects with minimal visible fundus signs of diabetic retinopathy had abnormal contrast sensitivity at one or more spatial frequencies [99]. In another study of type 1 diabetic subjects with no retinopathy, there was a reduction in contrast sensitivity at multiple spatial frequencies between 1.0 and 9.6 cycles/degree [100]. A similar study indicated that presence of microalbuminuria predicted a reduction in contrast sensitivity in type 1 patients [101]. Subjects with insulin resistance and dyslipidemia also have significant reductions in mesopic and low photopic contrast sensitivity, suggesting that this loss of function is not limited to those with severe insulin-dependent diabetes [102].

Color vision defects can also occur in humans with diabetes [103]. A histological study on postmortem retinas found a selective reduction in the number of S-cones in samples from donors with diabetes, possibly by apoptosis [104], which could explain the tritan color confusion and loss of sensitivity to blue light that is known to occur in diabetic retinopathy [105–107].

The studies on visual function in humans with diabetes indicate that there are specific deficits that are measurable early on in the course of the disease, often in the absence of gross vascular defects evident by fundus examination. These data suggest that changes in neural function begin early in diabetes. It is important that we examine the cellular substrate for various elements of vision, such as contrast sensitivity, dark adaptation, and color contrast, in order to develop better ways to protect vision in diabetes. In order to develop better treatments for neurodegenerative changes in the retina, a number of theories for the causative mechanisms have evolved, which we will attempt to summarize next.

POTENTIAL MECHANISMS OF RETINAL NEURODEGENERATION IN DIABETES

The relationship between vascular permeability and retinal neurodegeneration in diabetes is still unclear. It is reasonable to assume, however, that a breach in the blood-retinal barrier will give rise to local changes in neural function that could result in necrosis or apoptosis of neurons. There is a clinical link between macular edema and loss of visual acuity, although correlations with more sensitive measures of visual function have not been attempted [108]. Equally, reductions in contrast sensitivity have been correlated with reductions in capillary density, as an index of ischemia in the retina [109].

Related to the concept that the neural retina is compromised by ischemia is the proposal that the retina becomes hypoxic in diabetes. One proposed mechanism for hypoxia is that poor blood flow to the inner retina, in concert with the heavy metabolic demand from photoreceptors under dark-adapted conditions, leads to tissue oxygen depletion. This is based on observations that diabetic retinopathy is limited in situations where the photoreceptors are lost, like retinitis pigmentosa or in animal models such as the rhodopsin knockout mouse ($Rho^{-/-}$) [110, 111].

Glutamate excitotoxicity is a commonly considered mechanism for many diseases involving neurodegeneration and has been suggested to occur in diabetes [112]. GABA and glutamate levels were increased in vitreous of 22 patients with proliferative diabetic retinopathy, compared to a similar set of nondiabetic patients who had pars plana vitrectomy [113]. Similar increases have been measured in rats [114, 115]. Elevated concentrations of glutamate and GABA increased immunoreactivity for glutamate receptors NMDA and GluR2/3, accompanied by increased expression of calcium-binding proteins calbindin and parvalbumin in ganglion, amacrine, and bipolar cells [116]. Furthermore, glutamate oxidation was 62% less than controls in retina explants from STZ-diabetic rats, related to the reduction in the activity and content of glutamine synthetase, suggesting a reduced ability to process glutamate in the retina [117]. Reductions in the uptake rate of glutamate into Müller cells have also been measured [118, 119], along with alterations in the expression of some glutamate receptor subunits [120, 121]. The weak NMDA receptor antagonist has been reported to correct electroretinographic changes, prevent loss of RGCs, and reduce the amount of retinal vascular permeability in diabetic Brown-Norway rats, suggesting that this class of drugs may represent a useful therapeutic to prevent loss of function in diabetic retinopathy [37]. There is an intimate relationship between oxidative stress, nitric oxide toxicity, and glutamate excitotoxicity, and diabetes may induce all these biochemical processes in the retina [115].

The role of advanced glycation end-products (AGEs) in diabetic complications and retinopathy in particular has been discussed widely, especially since the AGE receptor was discovered [122]. The specific effect of AGEs on neurons in the retina has not been as well defined. Many studies have shown that treatment with aminoguanidine, an inhibitor of AGE formation, can rescue the vascular changes in diabetes [13, 123, 124]. This drug also reduced the loss of nNOS-containing neurons in STZ rats [59]; however, it failed to improve the abnormal ERG response in diabetic rats [76]. It may be that the effect of AGEs in neurons is indirect, acting by inducing an inflammatory response in glial cells and the vasculature [29].

Another mechanism that may be responsible for pathological changes to neurons in diabetes is loss of growth factor signaling, either through reduction in abundance of the growth factors or through loss of receptor sensitivity and second messenger signaling. BDNF was depleted from both brain and retina of diabetic rats [31, 125]. Loss of tyrosine hydroxylase-positive amacrine cells was prevented by injection of exogenous BDNF in rats [31]. There are also reductions in the kinase activity of components of

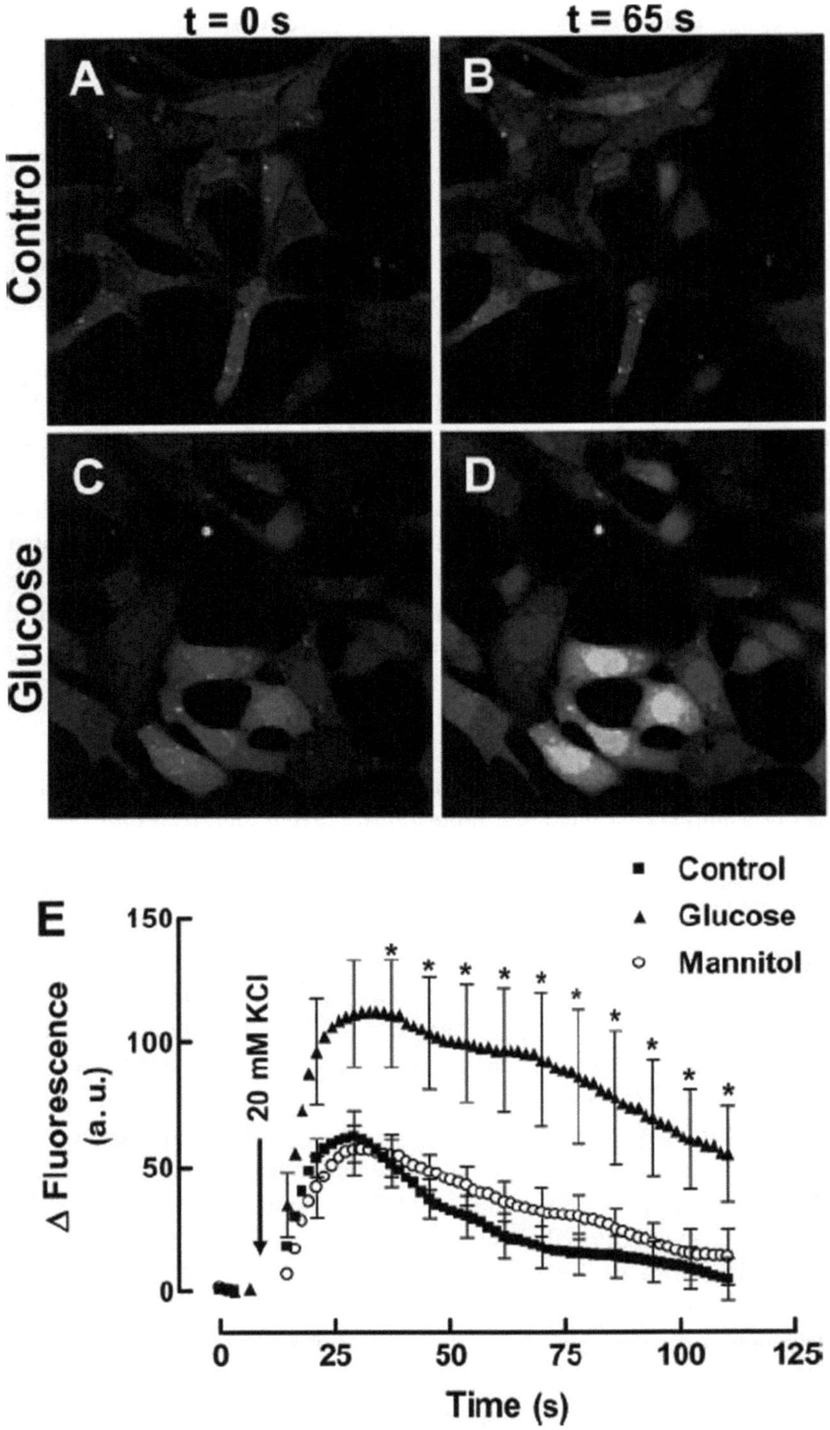

Fig. 6. Glucose elevated the intracellular calcium response to membrane depolarization in cell culture model of retinal neurons. Immortalized retinal neurons (R28 cells) were grown in 5 mM glucose, 20 mM glucose, or 15 mM mannitol with 5 mM glucose, for 2 days. Intracellular calcium was detected by fluo-4, a compound that becomes more fluorescent in the presence of

the PI3kinase-Akt pathway in STZ-diabetic rat retinas, accompanied by reduced insulin receptor kinase activity [126]. Systemic administration of IGF-1 was found to reduce the amount of apoptosis measured by TUNEL and caspase-3 activity in STZ rats, suggesting that increased growth factor signaling may protect the retina [127].

A less widely considered explanation of neuronal cell death and dysfunction is a change in the way intracellular calcium concentration is regulated. Calcium is an especially potent signal in neurons, responsible for initiating many metabolic events, including plastic changes at the synapse [128, 129]. Some in vitro studies indicate that elevated levels of glucose augmented the intracellular calcium response to membrane depolarization [130] (Fig. 6).

SUMMARY AND CONCLUSIONS

Diabetic retinopathy is considered to be a vascular disease of the retina, because clinically identifiable signs of the disease include vascular lesions such as microaneurysms and loss of the blood-retinal barrier leading to macular edema (nonproliferative stage). Later in the disease, there can be vascular proliferation and ischemia (proliferative stage) resulting in profound vision loss, although progression to this stage is less common [131, 132]. Clinical detection of diabetic retinopathy is almost exclusively through recognition of the vascular indications of the disease. These symptoms are accompanied by loss of visual acuity [133], and the patient usually recognizes the effects of the disease as a reduction in quality of life due to gradual deterioration of functional vision [134]. There is little doubt that diabetes reduces the ability of the retina to function correctly, but retinal function is difficult to measure in the clinic, so the fundus examination is regarded as the standard method to diagnose and map the progress of diabetic retinopathy. The gradual loss of retinal structure and function can, however, be interpreted as the most basic indication that neurodegeneration of the retina, leading to compromised visual function, is a prevalent component of diabetic retinopathy. Future advances in diagnosis and treatment of diabetic retinopathy will likely include consideration of this important aspect of the disease.

Fig. 6. (continued) calcium. The live cells were imaged by confocal microscopy during membrane depolarization by addition of 20 mM KCl. (**A**) Five seconds of baseline images of cells were recorded, followed by depolarization with 20 mM KCl. (**B**) In control cells with 5 mM glucose, the intracellular fluorescence increased transiently and returned almost to baseline within 65 s. (**C**) Cells grown with 20 mM glucose had baseline fluorescence similar to control cells. (**D**) Cells grown with 20 mM glucose displayed a more dramatic increase in fluorescence in response to KCl, and this did not return to baseline within 65 s. (**E**) Relative quantification of whole cell fluorescence (cytoplasmic and nuclear), by digital image analysis, indicated that there was a significant increase in calcium-induced fluorescence in the cells grown with 20 mM glucose compared to those grown with 5 mM glucose. Addition of mannitol did not alter the calcium response compared to the control cells, indicating that the effect was not due to osmotic changes in the media ($*p<0.05$). Similar results were obtained from primary cultures of retinal cells. Taken from Santiago et al. [130].

REFERENCES

1. Przedborski S, Vila M, et al. Neurodegeneration: what is it and where are we? J Clin Invest. 2003;111(1):3–10.
2. Wolter JR. Diabetic retinopathy. Am J Ophthalmol. 1961;51:1123–39.
3. Bloodworth Jr JM. Diabetic retinopathy. Diabetes. 1962;11:1–22.
4. Bresnick GH. Diabetic retinopathy viewed as a neurosensory disorder. Arch Ophthalmol. 1986;104:989–90.
5. Schmidt D. The mystery of cotton-wool spots—a review of recent and historical descriptions. Eur J Med Res. 2008;13(6):231–66.
6. Roy MS, Rick ME, et al. Retinal cotton-wool spots: an early finding in diabetic retinopathy? Br J Ophthalmol. 1986;70(10):772–8.
7. Kern TS, Barber AJ. Retinal ganglion cells in diabetes. J Physiol. 2008;586(Pt 18):4401–8.
8. Iseki S. DNA strand breaks in rat tissues as detected by in situ nick translation. Exp Cell Res. 1986;167(2):311–26.
9. Gavrieli Y, Sherman Y, et al. Identification of programmed cell death in situ via specific labeling of nuclear DNA fragmentation. J Cell Biol. 1992;119(3):493–501.
10. Wijsman JH, Jonker RR, et al. A new method to detect apoptosis in paraffin sections: in situ end-labeling of fragmented DNA. J Histochem Cytochem. 1993;41(1):7–12.
11. Hammes HP, Federoff HJ, et al. Nerve growth factor prevents both neuroretinal programmed cell death and capillary pathology in experimental diabetes. Mol Med. 1995;1(5):527–34.
12. Mizutani M, Kern TS, et al. Accelerated death of retinal microvascular cells in human and experimental diabetic retinopathy. J Clin Invest. 1996;97(12):2883–90.
13. Kern TS, Tang J, et al. Response of capillary cell death to aminoguanidine predicts the development of retinopathy: comparison of diabetes and galactosemia. Invest Ophthalmol Vis Sci. 2000;41(12):3972–8.
14. Kowluru RA, Odenbach S. Role of interleukin-1beta in the development of retinopathy in rats: effect of antioxidants. Invest Ophthalmol Vis Sci. 2004;45(11):4161–6.
15. Sugiyama T, Kobayashi M, et al. Enhancement of P2X(7)-induced pore formation and apoptosis: an early effect of diabetes on the retinal microvasculature. Invest Ophthalmol Vis Sci. 2004;45(3):1026–32.
16. Barber AJ, Lieth E, et al. Neural apoptosis in the retina during experimental and human diabetes. Early onset and effect of insulin. J Clin Invest. 1998;102(4):783–91.
17. Martin PM, Roon P, et al. Death of retinal neurons in streptozotocin-induced diabetic mice. Invest Ophthalmol Vis Sci. 2004;45(9):3330–6.
18. Ning X, Baoyu Q, et al. Neuro-optic cell apoptosis and microangiopathy in KKAY mouse retina. Int J Mol Med. 2004;13(1):87–92.
19. Gastinger MJ, Singh RS, et al. Loss of cholinergic and dopaminergic amacrine cells in streptozotocin-diabetic rat and Ins2Akita-diabetic mouse retinas. Invest Ophthalmol Vis Sci. 2006;47(7):3143–50.
20. Park SH, Park JW, et al. Apoptotic death of photoreceptors in the streptozotocin-induced diabetic rat retina. Diabetologia. 2003;46(9):1260–8.
21. Feit-Leichman RA, Kinouchi R, et al. Vascular damage in a mouse model of diabetic retinopathy: relation to neuronal and glial changes. Invest Ophthalmol Vis Sci. 2005;46(11): 4281–7.
22. Kanamori A, Nakamura M, et al. Diabetes has an additive effect on neural apoptosis in rat retina with chronically elevated intraocular pressure. Curr Eye Res. 2004;28(1):47–54.
23. Srinivasan A, Roth KA, et al. In situ immunodetection of activated caspase-3 in apoptotic neurons in the developing nervous system. Cell Death Differ. 1998;5(12):1004–16.

24. Barber AJ, Antonetti DA, et al. The Ins2Akita mouse as a model of early retinal complications in diabetes. Invest Ophthalmol Vis Sci. 2005;46(6):2210–8.
25. Abu-El-Asrar AM, Dralands L, et al. Expression of apoptosis markers in the retinas of human subjects with diabetes. Invest Ophthalmol Vis Sci. 2004;45(8):2760–6.
26. Oshitari T, Yamamoto S, et al. Mitochondria- and caspase-dependent cell death pathway involved in neuronal degeneration in diabetic retinopathy. Br J Ophthalmol. 2008;92(4):552–6.
27. Hughes WF. Quantitation of ischemic damage in the rat retina. Exp Eye Res. 1991;53(5): 573–82.
28. Park JW, Park SJ, et al. Up-regulated expression of neuronal nitric oxide synthase in experimental diabetic retina. Neurobiol Dis. 2006;21(1):43–9.
29. Barile GR, Pachydaki SI, et al. The RAGE axis in early diabetic retinopathy. Invest Ophthalmol Vis Sci. 2005;46(8):2916–24.
30. Larabi Y, Dahmani Y, et al. Tyrosine hydroxylase immunoreactivity in the retina of the diabetic sand rat *Psammomys obesus*. J Hirnforsch. 1991;32(4):525–31.
31. Seki M, Tanaka T, et al. Involvement of brain-derived neurotrophic factor in early retinal neuropathy of streptozotocin-induced diabetes in rats: therapeutic potential of brain-derived neurotrophic factor for dopaminergic amacrine cells. Diabetes. 2004;53(9):2412–9.
32. Goto R, Doi M, et al. Contribution of nitric oxide-producing cells in normal and diabetic rat retina. Jpn J Ophthalmol. 2005;49(5):363–70.
33. Nishimura C, Kuriyama K. Alterations in the retinal dopaminergic neuronal system in rats with streptozotocin-induced diabetes. J Neurochem. 1985;45(2):448–55.
34. Sanchez-Chavez G, Salceda R. Effect of streptozotocin-induced diabetes on activities of cholinesterases in the rat retina. IUBMB Life. 2000;49(4):283–7.
35. Sanchez-Chavez G, Salceda R. Acetyl- and butyrylcholinesterase in normal and diabetic rat retina. Neurochem Res. 2001;26(2):153–9.
36. Li Q, Zemel E, et al. NADPH diaphorase activity in the rat retina during the early stages of experimental diabetes. Graefes Arch Clin Exp Ophthalmol. 2003;241(9):747–56.
37. Kusari J, Zhou S, et al. Effect of memantine on neuroretinal function and retinal vascular changes of streptozotocin-induced diabetic rats. Invest Ophthalmol Vis Sci. 2007;48(11): 5152–9.
38. Gastinger MJ, Kunselman AR, et al. Dendrite remodeling and other abnormalities in the retinal ganglion cells of Ins2 Akita diabetic mice. Invest Ophthalmol Vis Sci. 2008;49(6): 2635–42.
39. Qin Y, Xu G, et al. Dendritic abnormalities in retinal ganglion cells of three-month diabetic rats. Curr Eye Res. 2006;31(11):967–74.
40. Meyer-Rusenberg B, Pavlidis M, et al. Pathological changes in human retinal ganglion cells associated with diabetic and hypertensive retinopathy. Graefes Arch Clin Exp Ophthalmol. 2007;245(7):1009–18.
41. Wolter JR. Centrifugal nerve fibers in the adult human optic nerve: 16 days after enucleation. Trans Am Ophthalmol Soc. 1978;76:140–55.
42. Gastinger MJ, O'Brien JJ, et al. Histamine immunoreactive axons in the macaque retina. Invest Ophthalmol Vis Sci. 1999;40(2):487–95.
43. Gastinger MJ, Barber AJ, et al. Abnormal centrifugal axons in streptozotocin—diabetic rat retinas. Invest Ophthalmol Vis Sci. 2001;42(11):2679–85.
44. Chihara E, Matsuoka T, et al. Retinal nerve fiber layer defect as an early manifestation of diabetic retinopathy. Ophthalmology. 1993;100(8):1147–51.
45. Ozdek S, Lonneville YH, et al. Assessment of nerve fiber layer in diabetic patients with scanning laser polarimetry. Eye. 2002;16(6):761–5.

46. Lopes de Faria JM, Russ H, et al. Retinal nerve fibre layer loss in patients with type 1 diabetes mellitus without retinopathy. Br J Ophthalmol. 2002;86(7):725–8.
47. Takahashi H, Goto T, et al. Diabetes-associated retinal nerve fiber damage evaluated with scanning laser polarimetry [see comment]. Am J Ophthalmol. 2006;142(1):88–94.
48. Lim MC, Tanimoto SA, et al. Effect of diabetic retinopathy and panretinal photocoagulation on retinal nerve fiber layer and optic nerve appearance. Arch Ophthalmol. 2009;127(7):857–62.
49. Scott TM, Foote J, et al. Vascular and neural changes in the rat optic nerve following induction of diabetes with streptozotocin. J Anat. 1986;144:145–52.
50. Mizutani M, Gerhardinger C, et al. Muller cell changes in human diabetic retinopathy. Diabetes. 1998;47(3):445–9.
51. Mohr S, Xi X, et al. Caspase activation in retinas of diabetic and galactosemic mice and diabetic patients. Diabetes. 2002;51(4):1172–9.
52. Kowluru RA, Koppolu P. Diabetes-induced activation of caspase-3 in retina: effect of antioxidant therapy. Free Radic Res. 2002;36(9):993–9.
53. Krady JK, Basu A, et al. Minocycline reduces proinflammatory cytokine expression, microglial activation, and caspase-3 activation in a rodent model of diabetic retinopathy. Diabetes. 2005;54(5):1559–65.
54. Vincent JA, Mohr S. Inhibition of caspase-1/interleukin-1beta signaling prevents degeneration of retinal capillaries in diabetes and galactosemia. Diabetes. 2007;56(1):224–30.
55. Kern TS, Miller CM, et al. Topical administration of nepafenac inhibits diabetes-induced retinal microvascular disease and underlying abnormalities of retinal metabolism and physiology. Diabetes. 2007;56(2):373–9.
56. Vanguilder HD, Brucklacher RM, et al. Diabetes downregulates presynaptic proteins and reduces basal synapsin I phosphorylation in rat retina. Eur J Neurosci. 2008;28(1):1–11.
57. Kurihara T, Ozawa Y, et al. Angiotensin II type 1 receptor signaling contributes to synaptophysin degradation and neuronal dysfunction in the diabetic retina. Diabetes. 2008;57(8):2191–8.
58. do Carmo A, Lopes C, et al. Nitric oxide synthase activity and L-arginine metabolism in the retinas from streptozotocin-induced diabetic rats. Gen Pharmacol. 1998;30(3):319–24.
59. Roufail E, Soulis T, et al. Depletion of nitric oxide synthase-containing neurons in the diabetic retina: reversal by aminoguanidine. Diabetologia. 1998;41(12):1419–25.
60. Giove TJ, Deshpande MM, et al. Increased neuronal nitric oxide synthase activity in retinal neurons in early diabetic retinopathy. Mol Vis. 2009;15:2249–58.
61. Yu HM, Xu J, et al. Coupling between neuronal nitric oxide synthase and glutamate receptor 6-mediated c-Jun N-terminal kinase signaling pathway via S-nitrosylation contributes to ischemia neuronal death. Neuroscience. 2008;155(4):1120–32.
62. Leist M, Nicotera P. Apoptosis, excitotoxicity, and neuropathology. Exp Cell Res. 1998;239(2):183–201.
63. Cogen J, Cohen-Cory S. Nitric oxide modulates retinal ganglion cell axon arbor remodeling in vivo. J Neurobiol. 2000;45(2):120–33.
64. Green DG, Kapousta-Bruneau NV. A dissection of the electroretinogram from the isolated rat retina with microelectrodes and drugs. Vis Neurosci. 1999;16(4):727–41.
65. Karwoski CJ, Xu X. Current source-density analysis of light-evoked field potentials in rabbit retina. Vis Neurosci. 1999;16(2):369–77.
66. Sieving PA, Murayama K, et al. Push-pull model of the primate photopic electroretinogram: a role for hyperpolarizing neurons in shaping the b-wave. Vis Neurosci. 1994;11(3):519–32.
67. Wachtmeister L. Oscillatory potentials in the retina: what do they reveal. Prog Retin Eye Res. 1998;17(4):485–521.

68. Bui BV, Fortune B. Ganglion cell contributions to the rat full-field electroretinogram. J Physiol. 2004;555(Pt 1):153–73.
69. Yonemura D, Aoki T, et al. Electroretinogram in diabetic retinopathy. Arch Ophthalmol. 1962;68:19–24.
70. Kawasaki K, Yonemura K, et al. Correlation between ERG oscillatory potential and psychophysical contrast sensitivity in diabetes. Doc Ophthalmol. 1986;64(2):209–15.
71. Bresnick GH, Palta M. Temporal aspects of the electroretinogram in diabetic retinopathy. Arch Ophthalmol. 1987;105(5):660–4.
72. Juen S, Kieselbach GF. Electrophysiological changes in juvenile diabetics without retinopathy. Arch Ophthalmol. 1990;108(3):372–5.
73. Bresnick GH, Palta M. Oscillatory potential amplitudes. Relation to severity of diabetic retinopathy. Arch Ophthalmol. 1987;105(7):929–33.
74. Simonsen SE. Prognostic value of ERG (oscillatory potential) in juvenile diabetics. Acta Ophthalmol Suppl. 1974;123:223–4.
75. Simonsen SE. The value of the oscillatory potential in selecting juvenile diabetics at risk of developing proliferative retinopathy. Acta Ophthalmol. 1980;58(6):865–78.
76. Bui BV, Armitage JA, et al. ACE inhibition salvages the visual loss caused by diabetes. Diabetologia. 2003;46(3):401–8.
77. Phipps JA, Fletcher EL, et al. Paired-flash identification of rod and cone dysfunction in the diabetic rat. Invest Ophthalmol Vis Sci. 2004;45(12):4592–600.
78. Phipps JA, Yee P, et al. Rod photoreceptor dysfunction in diabetes: activation, deactivation, and dark adaptation. Invest Ophthalmol Vis Sci. 2006;47(7):3187–94.
79. Hancock HA, Kraft TW. Oscillatory potential analysis and ERGs of normal and diabetic rats. Invest Ophthalmol Vis Sci. 2004;45(3):1002–8.
80. Liu W, Deng Y. The analysis of electroretinography of diabetes mellitus. Yan Ke Xue Bao 2001;17(3):173–5, 179.
81. Li Q, Zemel E, et al. Early retinal damage in experimental diabetes: electroretinographical and morphological observations. Exp Eye Res. 2002;74(5):615–25.
82. Zhang Y, Wang Q, et al. Protection of exendin-4 analogue in early experimental diabetic retinopathy. Graefes Arch Clin Exp Ophthalmol. 2008;247(5):699–706.
83. Kohzaki K, Vingrys AJ, et al. Early inner retinal dysfunction in streptozotocin-induced diabetic rats. Invest Ophthalmol Vis Sci. 2008;49(8):3595–604.
84. Chung NH, Kim SH, et al. The electroretinogram sensitivity in patients with diabetes. Korean J Ophthalmol. 1993;7(2):43–7.
85. Zakareia FA, Alderees AA, et al. Correlation of electroretinography b-wave absolute latency, plasma levels of human basic fibroblast growth factor, vascular endothelial growth factor, soluble fatty acid synthase, and adrenomedullin in diabetic retinopathy. J Diabetes Complications. 2009;24(3):179–85.
86. Sieving PA, Frishman LJ, et al. Scotopic threshold response of proximal retina in cat. J Neurophysiol. 1986;56(4):1049–61.
87. Naarendorp F, Sieving PA. The scotopic threshold response of the cat ERG is suppressed selectively by GABA and glycine. Vision Res. 1991;31(1):1–15.
88. Abraham FA, Haimovitz J, et al. The photopic and scotopic visual thresholds in diabetics without diabetic retinopathy. Metab Pediatr Syst Ophthalmol. 1988;11(1–2):76–7.
89. Aylward GW. The scotopic threshold response in diabetic retinopathy. Eye. 1989;3(Pt 5):626–37.
90. Greco AV, Di Leo MA, et al. Early selective neuroretinal disorder in prepubertal type 1 (insulin-dependent) diabetic children without microvascular abnormalities. Acta Diabetol. 1994;31(2):98–102.

91. Zhang L, Inoue M, et al. Alterations in retrograde axonal transport in optic nerve of type I and type II diabetic rats. Kobe J Med Sci. 1998;44(5–6):205–15.
92. Zhang LX, Ino-ue M, et al. Retrograde axonal transport impairment of large- and medium-sized retinal ganglion cells in diabetic rat. Curr Eye Res. 2000;20(2):131–6.
93. Ino-Ue M, Zhang L, et al. Polyol metabolism of retrograde axonal transport in diabetic rat large optic nerve fiber. Invest Ophthalmol Vis Sci. 2000;41(13):4055–8.
94. Sima AA, Zhang WX, et al. Impaired visual evoked potential and primary axonopathy of the optic nerve in the diabetic BB/W-rat. Diabetologia. 1992;35(7):602–7.
95. Papakostopoulos D, Hart JC, et al. The scotopic electroretinogram to blue flashes and pattern reversal visual evoked potentials in insulin dependent diabetes. Int J Psychophysiol. 1996;21(1):33–43.
96. Lopes de Faria JM, Katsumi O, et al. Neurovisual abnormalities preceding the retinopathy in patients with long-term type 1 diabetes mellitus. Graefes Arch Clin Exp Ophthalmol. 2001;239(9):643–8.
97. Klein R. Age-related eye disease, visual impairment, and driving in the elderly. Hum Factors. 1991;33(5):521–5.
98. Bailey CC, Sparrow JM. Visual symptomatology in patients with sight-threatening diabetic retinopathy. Diabet Med. 2001;18(11):883–8.
99. Sokol S, Moskowitz A, et al. Contrast sensitivity in diabetics with and without background retinopathy. Arch Ophthalmol. 1985;103(1):51–4.
100. Di Leo MA, Caputo S, et al. Nonselective loss of contrast sensitivity in visual system testing in early type I diabetes. Diabetes Care. 1992;15(5):620–5.
101. Bangstad HJ, Brinchmann-Hansen O, et al. Impaired contrast sensitivity in adolescents and young type 1 (insulin-dependent) diabetic patients with microalbuminuria. Acta Ophthalmol. 1994;72(6):668–73.
102. Dosso AA, Yenice-Ustun F, et al. Contrast sensitivity in obese dyslipidemic patients with insulin resistance. Arch Ophthalmol. 1998;116(10):1316–20.
103. Roy MS, Gunkel RD, et al. Color vision defects in early diabetic retinopathy. Arch Ophthalmol. 1986;104(2):225–8.
104. Cho NC, Poulsen GL, et al. Selective loss of S-cones in diabetic retinopathy. Arch Ophthalmol. 2000;118(10):1393–400.
105. Daley ML, Watzke RC, et al. Early loss of blue-sensitive color vision in patients with type I diabetes. Diabetes Care. 1987;10(6):777–81.
106. Rockett M, Anderle D, et al. Blue-yellow vision deficits in patients with diabetes. West J Med. 1987;146(4):431–3.
107. Ong GL, Ripley LG, et al. Assessment of colour vision as a screening test for sight threatening diabetic retinopathy before loss of vision. Br J Ophthalmol. 2003;87(6):747–52.
108. Moss SE, Klein R, et al. The 14-year incidence of visual loss in a diabetic population. Ophthalmology. 1998;105(6):998–1003.
109. Arend O, Remky A, et al. Contrast sensitivity loss is coupled with capillary dropout in patients with diabetes. Invest Ophthalmol Vis Sci. 1997;38(9):1819–24.
110. Arden GB. The absence of diabetic retinopathy in patients with retinitis pigmentosa: implications for pathophysiology and possible treatment. Br J Ophthalmol. 2001;85(3):366–70.
111. de Gooyer TE, Stevenson KA, et al. Retinopathy is reduced during experimental diabetes in a mouse model of outer retinal degeneration. Invest Ophthalmol Vis Sci. 2006;47(12):5561–8.
112. Barber AJ. A new view of diabetic retinopathy: a neurodegenerative disease of the eye. Prog Neuropsychopharmacol Biol Psychiatry. 2003;27(2):283–90.

113. Ambati J, Chalam KV, et al. Elevated gamma-aminobutyric acid, glutamate, and vascular endothelial growth factor levels in the vitreous of patients with proliferative diabetic retinopathy. Arch Ophthalmol. 1997;115(9):1161–6.
114. Lieth E, Barber AJ, et al. Glial reactivity and impaired glutamate metabolism in short-term experimental diabetic retinopathy. Diabetes. 1998;47(5):815–20.
115. Kowluru RA, Engerman RL, et al. Retinal glutamate in diabetes and effect of antioxidants. Neurochem Int. 2001;38(5):385–90.
116. Ng YK, Zeng XX, et al. Expression of glutamate receptors and calcium-binding proteins in the retina of streptozotocin-induced diabetic rats. Brain Res. 2004;1018(1):66–72.
117. Lieth E, LaNoue KF, et al. Diabetes reduces glutamate oxidation and glutamine synthesis in the retina. Exp Eye Res. 2000;70(6):723–30.
118. Puro DG. Diabetes-induced dysfunction of retinal Muller cells. Trans Am Ophthalmol Soc. 2002;100:339–52.
119. Ward MM, Jobling AI, et al. Glutamate uptake in retinal glial cells during diabetes. Diabetologia. 2005;48(2):351–60.
120. Santiago AR, Hughes JM, et al. Diabetes changes ionotropic glutamate receptor subunit expression level in the human retina. Brain Res. 2008;1198:153–9.
121. Santiago AR, Gaspar JM, et al. Diabetes changes the levels of ionotropic glutamate receptors in the rat retina. Mol Vis. 2009;15:1620–30.
122. Schmidt AM, Yan SD, et al. Activation of receptor for advanced glycation end products: a mechanism for chronic vascular dysfunction in diabetic vasculopathy and atherosclerosis [review] [89 refs]. Circ Res. 1999;84(5):489–97.
123. Tilton RG, Chang K, et al. Prevention of diabetic vascular dysfunction by guanidines. Inhibition of nitric oxide synthase versus advanced glycation end-product formation. Diabetes. 1993;42(2):221–32.
124. Kern TS, Engerman RL. Pharmacological inhibition of diabetic retinopathy: aminoguanidine and aspirin. Diabetes. 2001;50(7):1636–42.
125. Nitta A, Murai R, et al. Diabetic neuropathies in brain are induced by deficiency of BDNF. Neurotoxicol Teratol. 2002;24(5):695–701.
126. Reiter CE, Wu X, et al. Diabetes reduces basal retinal insulin receptor signaling: reversal with systemic and local insulin. Diabetes. 2006;55(4):1148–56.
127. Seigel GM, Lupien SB, et al. Systemic IGF-I treatment inhibits cell death in diabetic rat retina. J Diabetes Complications. 2006;20(3):196–204.
128. Verkhratsky A. Physiology and pathophysiology of the calcium store in the endoplasmic reticulum of neurons. Physiol Rev. 2005;85(1):201–79.
129. Verkhratsky A, Shmigol A. Calcium-induced calcium release in neurones. Cell Calcium. 1996;19(1):1–14.
130. Santiago AR, Rosa SC, et al. Elevated glucose changes the expression of ionotropic glutamate receptor subunits and impairs calcium homeostasis in retinal neural cells. Invest Ophthalmol Vis Sci. 2006;47(9):4130–7.
131. Bloodworth Jr JM, Molitor DL. Ultrastructural aspects of human and canine diabetic retinopathy. Invest Ophthalmol. 1965;4(6):1037–48.
132. Aiello LP, Gardner TW, et al. Diabetic retinopathy. Diabetes Care. 1998;21(1):143–56.
133. Moss SE, Klein R, et al. The incidence of vision loss in a diabetic population. Ophthalmology. 1988;95(10):1340–8.
134. Association AD. Economic costs of diabetes in the U.S. in 2007. Diabetes Care. 2008;31(3):596–615.

13 Glucose-Induced Cellular Signaling in Diabetic Retinopathy

Zia A. Khan and Subrata Chakrabarti

CONTENTS

Keywords Diabetes • Retinopathy • Complications • Endothelial cells • Pericytes • Angiogenesis • Extracellular matrix • Cellular signaling

INTRODUCTION

Diabetic retinopathy (DR) is a microvascular complication of diabetes. It is the most common cause of blindness in the working population. Nearly all people with diabetes, both type 1 and type 2, will eventually develop some form of retinopathy [1]. Clinical trials have consistently shown that good glycemic control can reduce the development of retinopathy in both type 1 and type 2 diabetic patients [2, 3]. Other factors such as hyperlipidemia and hyperinsulinemia may also be involved. However, the major contributor does seem to be excess blood glucose levels. Sustained hyperglycemia leads to a sequence of adverse events in the retina (summarized in Fig. 1). Early events include altered expression of vasoactive factors and basement membrane (BM) proteins [4–6]. This manifests as loss of vasoregulation, thickening of the BM, and increased permeability. Increased permeability may also cause macular edema and significant vision loss. With continued hyperglycemic insult, the vascular cells exhibit exhaustion and degeneration leading to the formation of acellular capillaries [7–9]. All these functional and structural changes then converge to create an ischemic retina. Elaboration of growth factors to induce new blood vessel formation then proceeds. This sequence of events, continued insult, and continued adaptation, ultimately causes unregulated angiogenesis

From: *Ophthalmology Research: Visual Dysfunction in Diabetes*
Edited by: J. Tombran-Tink et al. (eds.), DOI 10.1007/978-1-60761-150-9_13

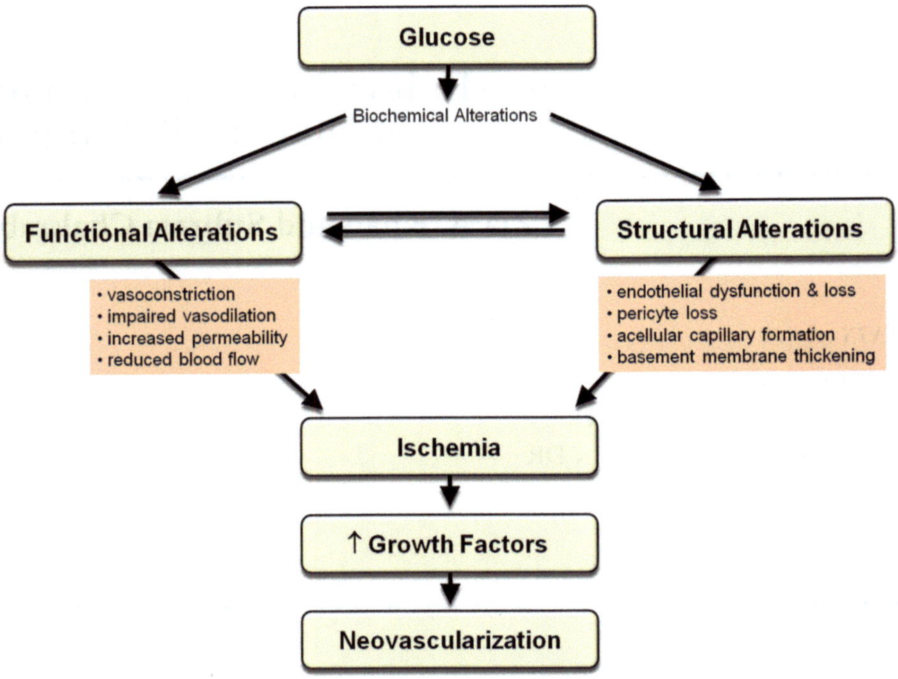

Fig. 1. Key events in the development and progression of DR. High plasma glucose levels lead to biochemical dysfunction in the retinal vascular cells. These changes result in structural and functional alterations at the vascular unit level. Reduced blood flow to the retina produces an ischemic environment which dictates elaboration of various angiogenic factors. These continued insults to the retinal tissue ultimately lead to EC hyperplasia and unregulated angiogenesis.

and blindness in diabetic patients. It is well accepted that understanding the molecular basis of endothelial cell (EC) dysfunction and loss will provide better therapeutic targets for DR. In this chapter, we review the cellular and molecular (signaling) mechanisms that ultimately lead to the development of DR.

CELLULAR TARGETS IN DR

In order to gain insight into the pathogenetic mechanisms underlying any disease, the first step is to develop in vitro and in vivo models that provide a phenocopy or at least exhibit the key structural and functional features of the disease. A prerequisite, therefore, is to identify the target cellular population. In the case of DR, retinal fluorescein angiography has provided important information about the primary cellular target [10, 11]. These studies show numerous areas of nonperfusion in the retina. The underlying cause of nonperfusion seems to be loss of vascular cells [12, 13]. These vascular cells include both ECs and pericytes that eventually succumb to glucotoxicity.

Endothelial Cell (EC) Dysfunction

Retinal angiography and digest studies show that normal retinal vascular perfusion is dependent on intact endothelium [14, 15]. The working hypothesis is that high levels

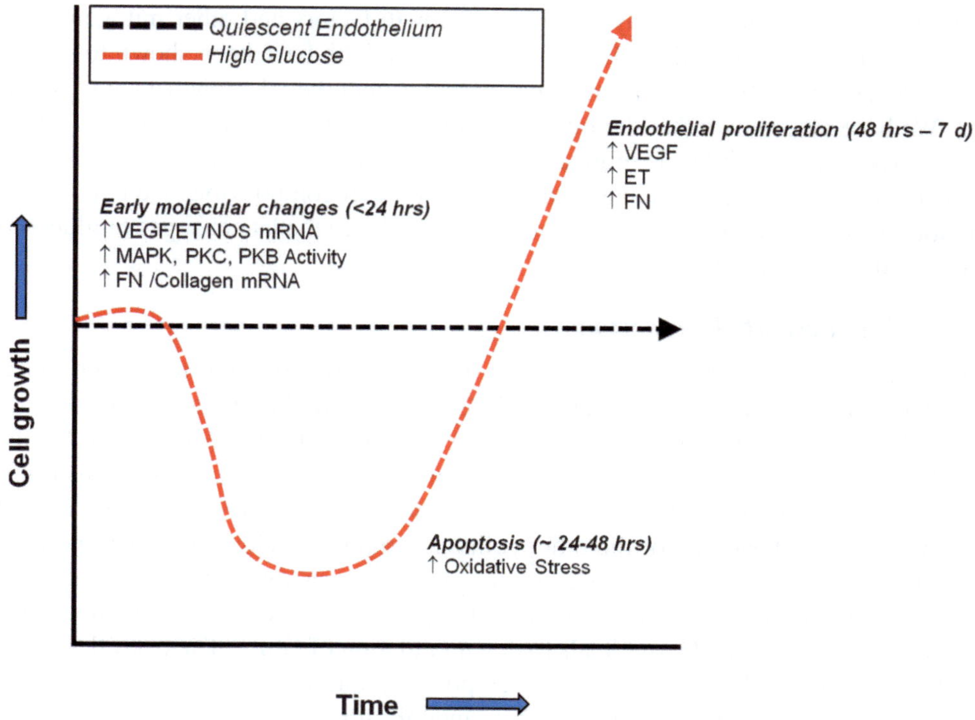

Fig. 2. Molecular and phenotypic changes in ECs exposed to high levels of glucose. Studies from our labs and others have shown that acute exposure to high glucose causes reduced viability and increased apoptosis in the ECs. However, with continued exposure, the ECs proliferate which is associated with increased matrix protein and VEGF production. *ET* endothelin; *FN* fibronectin; *MAPK* mitogen-activated protein kinase; *NOS* nitric oxide synthase; *PKB* protein kinase B; *PKC* protein kinase C; *VEGF* vascular endothelial growth factor.

of glucose lead to EC dysfunction and loss. An important assumption, therefore, is that high levels of extracellular glucose equate to high levels of intracellular glucose. In other words, there is no adaptive transport mechanism in the ECs. This certainly seems to be the case. ECs incorporate glucose via facilitative diffusion without significant alterations of glucose transporter-1 (Glut1) levels [16]. Therefore, continued exposure of ECs to high glucose leads to continued intracellular glucose accumulation. When assayed in culture, exposure of ECs to high glucose causes activation and dysfunction which is reflected by increased extracellular matrix (ECM) protein production and altered cellular activities [17–22]. Data from our laboratories and others show that this simple in vitro model illustrates most of the molecular changes that we see in clinical DR (Fig. 2). Early changes in the ECs following glucose exposure include reduced viability and increased apoptosis [23]. Interestingly, these changes are followed by increased proliferation [24]. This biphasic effect is reminiscent of EC changes in the early and advanced DR. Although the mechanism of this biphasic effect is not clear, we hypothesize that the mechanism of the late proliferative response is a change in the microenvironment—this is expected to occur in vivo as well. ECs rest on a scaffold of ECM proteins called the BM. This matrix serves as a reservoir of growth factors and other signaling proteins. With continued exposure to high glucose levels, the ECs may accumulate growth factors and other mitogens in the matrix. In fact, ECs exposed to glucose for more than 72 h have been shown to increase protein levels of an EC-specific mitogen, vascular endothelial growth factor (VEGF) [25]. We have also shown that the mRNA of VEGF is upregulated as early as 24 h following exposure to high levels of glucose [26]. In addition, the matrix itself is expected to change in terms of the protein amount and the protein composition (see below). This potentially creates a permissive environment that mediates the late changes of glucose in culture and in advanced clinical DR.

Endothelial-Pericyte Interactions

Pericytes are the contractile cells present in microvessels (similar to smooth muscle cells in larger vessels). These cells are in close contact with the ECs and form a discontinuous layer. The physiological function of the pericytes is to stabilize vessels, regulate vessel contraction, and keep the endothelium in a quiescent state. This intimate relationship between the vascular cells suggests that aberration in one cell type will lead to alterations in the phenotype of the other cellular component. However, when pericytes are cultured in high levels of glucose, we see an interesting contrast to ECs. Both pericytes and smooth muscle cells exhibit an autoregulatory glucose transport mechanism [16], that is, exposure to glucose leads to downregulation of Glut1. The overall transport of glucose seems to be higher in pericytes possibly due to greater biosynthetic ability. Therefore, these perivascular cells also undergo glucose-induced dysfunction and loss. In fact, loss of pericytes is considered one of the structural hallmarks of DR [7–9]. Pericyte loss is implicated in contributing to acellular capillary formation and may also be important in late stages of DR. Evidence for this comes from studies in platelet-derived growth factor-B knockout mice that lack pericytes in the brain capillaries [27]. These animals develop microaneurysms, acellular capillaries, and EC hyperplasia. These results are exacerbated when PDGF-deficient animals are made diabetic [12] suggesting an important role of pericyte-EC interaction in advanced DR.

The biochemical mechanisms underlying pericyte loss seem to be similar to ECs with the same players emerging (metabolic distress, vasoactive factors, protein kinase activation). In addition, it has been shown that an abrupt drop in glucose levels causes pericyte apoptosis [28]. Another mechanism may involve the angiopoietin system. Hyperglycemia has been shown to increase the expression of angiopoietin-2 in the retina that leads to pericyte dropout [29]. Furthermore, angiopoietin deficiency in the diabetic animals prevented pericyte loss and subsequent acellular capillary formation.

Endothelial-Matrix Interactions

Neovascularization, formation of a complete vascular unit either through angiogenesis or vasculogenesis, is a multistep process. Both endothelial and perivascular cells undergo a number of structural and functional changes to form a blood vessel. These cellular activities include endothelial proliferation and migration, formation of cell-cell contacts and tubules, recruitment of pericytes, and contribution to the ECM. In addition to providing a scaffold for the organization of the vascular cells, the ECM has been implicated in providing critical cues for proper blood vessel formation [30, 31]. The BM (sheet of ECM proteins) of normal microvessels predominantly contains laminin, collagen, and nidogen (entactin) [32]. A consistent feature of DR is (a) an increase in the ECM proteins; (b) a switch in the type of ECM proteins, that is, composition; and (c) posttranslational modifications of ECM proteins such as glycation.

In cultured retinal ECs, high levels of glucose can increase mRNA expression of both collagen and fibronectin (FN) [19, 33, 34]. The retinal BM of diabetic animals also shows increased expression of collagen, laminin, and FN [35]. These are early molecular changes and are evident in approximately 8 weeks following diabetes induction [35]. We have previously shown that FN is upregulated in the retinal tissues of diabetic rats in 1 month [36]. This increased expression continues for up to several months. The upregulated matrix protein expression then manifests as thickening of the BM in animal models [37]. In addition to collagen and FN, tenascin has been found in retinal vessels of diabetic patients and animals [38, 39]. It is important to note that this does not represent a general phenomenon of BM duplication/expansion but is a selective upregulation of key ECM proteins. For example, no difference in the amount of proteoglycans in diabetic patients has been reported [40]. This suggests that the composition of the BM may be important in providing critical cues to the vascular cells [30–32, 41, 42]. In support, we have recently shown that FN undergoes alterative splicing in DR to produce an embryonic isoform, ED-B + FN (also known as oncofetal FN) [26, 43]. Increased levels of this isoform are evident in vitreous of patients with advanced DR [43, 44] and retinal tissues of diabetic rats [43]. In cultured vascular ECs, we have shown that ED-B + FN is increased following exposure to high levels of glucose and that this FN isoform is involved in VEGF expression and EC proliferation.

Functionally, FN in the matrix may play a critical role in DR. FN is highly expressed in developing vessels as compared to stable quiescent vessels [45, 46]. During vascular remodeling (e.g., during wound healing or tumorigenesis), FN is upregulated [47, 48]. Further support of a functional role of FN in the retina comes from studies that show expression of FN in the active zones of vascularization [49]. FN also provides critical survival and proliferative signals to brain capillary ECs [50]. ECs express a number of ECM protein receptors, and function-blocking antibodies against FN integrins lead to reduced EC proliferation [50].

SIGNALING MECHANISMS IN DR

Altered Vasoactive Factors

DR is a culmination of numerous biochemical alterations that take place in the vascular tissue of the retina. An important physiological function of the endothelium is the regulation of regional blood flow. This is achieved by creating a balance between vasoconstricting factors and vasodilating factors. Diabetes leads to a disruption of this balance, and these altered vasoactive molecules play a role in both the early and the late stages of DR. Increased vasoconstriction and impaired endothelium-dependent vasodilation has been reported in diabetes [37, 51–55]. This vasoregulatory impairment has been shown to precede the structural changes in the vasculature [52, 54, 56–59]. The mechanistic basis of impaired endothelium-dependent vasodilatory responses has been extensively researched in diabetic patients, animal models, and cultured cells. This mechanism involves increased expression of endothelin-1 (ET-1), the most potent vasoconstrictor

[60]. ETs are short peptides that are secreted by ECs and mediate vasoconstriction by binding to ET receptors on the perivascular cells. Increased ET has been shown to cause vasoconstriction and reduced blood flow in diabetes [60]. Interestingly, improvement of the vasodilator responses have also been noted in diabetic patients that were administered an ET receptor antagonist [61]. In streptozotocin-induced diabetic rats, we have reported that diabetes-induced retinal capillary vasoconstriction is normalized with an ET receptor antagonist (Bosentan) [37]. We have also shown that high levels of glucose increase ET-1 and mediate increased EC permeability and ECM protein expression in cultured cells [26, 62, 63]. ET may also function as a mitogen for both perivascular cells [64, 65] and ECs [66, 67] which may be important in the late stages of DR.

It is expected that increased ET-1 levels may accompany decreased vasodilator levels (such as nitric oxide; NO). NO is produced by a family of enzymes called NO synthases (NOS). Studies have shown increased levels of both endothelial (e-) and inducible (i-) NOS enzymes in response to high levels of glucose [68–71]. This is also seen in animal models and human diabetes [69]. A number of signaling pathways that are activated in diabetes may lead to increased expression of NOS. These pathways may include VEGF [72] and protein kinase pathways [72–74]. The reason for this apparent discrepancy has been recently hypothesized to be an increased scavenging and reduced bioavailability of NO. In diabetes, NO levels may be reduced through sequestration by reactive oxygen species (ROS). It is also important to note that increased NOS expression may not lead to increased NO production. Acute exposure of ECs to glucose decreases NO generation by agonists including bradykinin [75]. These effects were shown to be the direct result of high glucose levels. Purified eNOS, when assayed in the presence of glucose, shows significantly lower NO production [75]. This suggests that increasing NO production/availability may undo some of the glucose-induced changes. When diabetic animals are treated with an NO donor, molsidomine, the diabetes-induced vasoconstriction in the retina is normalized [76].

Alteration of Metabolic Pathways

Polyol Pathway

Physiologic metabolism of glucose is accomplished mainly by the glycolytic pathway. However, in diabetes, increased flux and shunting of glucose through alternative pathways take place (Fig. 3). One such pathway is the polyol pathway [77, 78]. In this pathway, glucose is metabolized to sorbitol by aldose reductase (AR) [78]. Sorbitol itself may cause cellular damage [78, 79] which may be prevented by myo-inositol supplementation [80]. However, the major contribution of the polyol pathway to the adverse effects of high glucose levels seems to be the alteration in enzyme cofactor levels. The first enzymatic reaction that converts glucose to sorbitol requires NADPH. An increase in glucose flux is expected to decrease NADPH levels. NADPH is also a cofactor for antioxidant enzyme system (reduced glutathione) and, therefore, contributes to impairment of cellular antioxidant system. The second reaction that converts sorbitol to fructose requires NAD+ and generates NADH. It is believed that increased NADH production leads to augmented levels of glyceraldehyde 3-phosphate. Increased glyceraldehyde 3-phosphate may then increase advanced glycation end product formation through methylglyoxal [81].

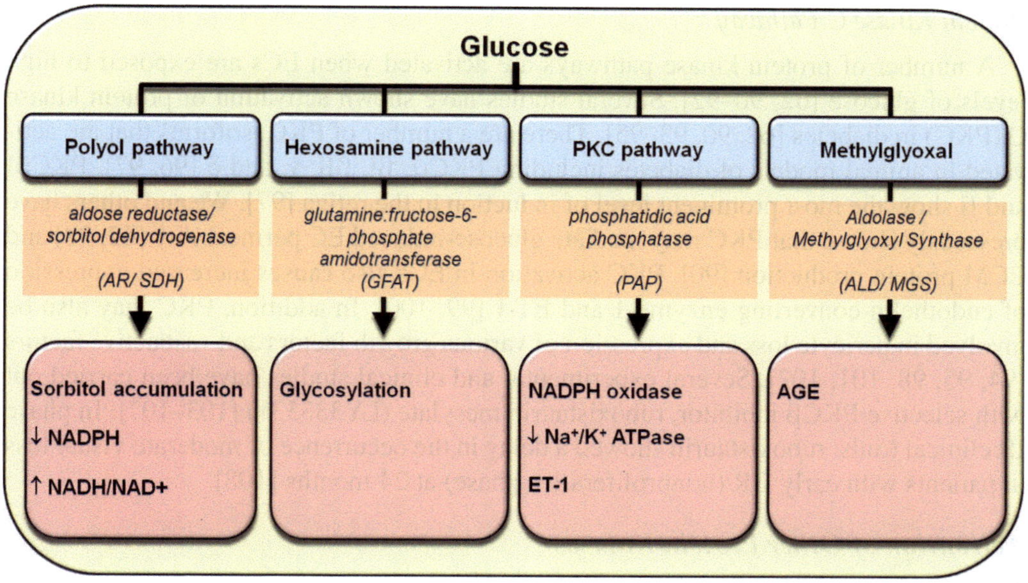

Fig. 3. Early metabolic/biochemical changes in ECs exposed to high levels of glucose. Increased flux of cytosolic glucose through the polyol, hexosamine, protein kinase C, and methylglyoxal pathways represents early alteration in the ECs. Activation of these pathways paves the path for EC dysfunction and loss through elaboration of reactive oxygen species (ROS), loss of vasoregulatory function (endothelin/nitric oxide imbalance), and modification of proteins. Key enzymes involved in these pathways are also indicated. *AGE* advanced glycation end products; *ET* endothelin; *PKC* protein kinase C; *NAD+* nicotinamide adenine dinucleotide; *NADH* nicotinamide adenine dinucleotide, reduced; *NADPH* nicotinamide adenine dinucleotide phosphate, reduced.

Clinical studies show that polymorphisms in AR gene may be linked to increased susceptibility of microvascular complications [82–84]. Although inhibition of AR has not provided any conclusive results, one recent trial with the AR inhibitor sorbinil showed slower rate of microaneurysms in the retina [85]. A new class of AR inhibitors was recently tested in streptozotocin-induced diabetic rats [86], but whether this selective AR inhibitor (ARI-809) produces favorable results in clinical trials remains to be determined.

Hexosamine Pathway

Metabolites of the glycolytic pathway may also be shunted through the hexosamine pathway in diabetes [87]. This pathway produces uridine diphosphate *N*-acetylglucosamine (UDP-GlcNAc), substrate for *O*-linked glycosylation of serine/threonine-containing proteins and proteoglycan synthesis. Studies have shown that inhibition of the key enzyme in this pathway, glutamine:fructose 6-phosphate amidotransferase (GFAT), reduces hyperglycemia-induced fibrogenic protein expression in aortic ECs [88]. In addition, a large number of proteins that are implicated in the development of diabetic complications are modified by *O*-linked glycosylation. These include protein kinases, growth factors, and transcription factors [89].

Protein Kinase C Pathway

A number of protein kinase pathways are activated when ECs are exposed to high levels of glucose [62, 90–92]. Several studies have shown activation of protein kinase C (PKC) in diabetes [62, 90, 93–95]. There are a number of PKC isoforms that are activated in animal models of diabetes including PKCα, βI, βII, γ, and δ [96, 97]. PKCβI and II show the most prominent level of induction in the retina [97]. We and others have previously shown that PKC may mediate glucose-induced EC permeability [62, 98] and ECM protein production [90]. PKC activation in ECs also causes increased expression of endothelin-converting enzyme-1 and ET-1 [99, 100]. In addition, PKC may also be involved in pericyte loss and expression of various growth factors and vasoactive factors [94, 95, 98, 101, 102]. Several experimental and clinical studies have been carried out with selective PKCβ inhibitor, ruboxistaurin mesylate (LY333531) [103–107]. In phase III clinical trials, ruboxistaurin showed a delay in the occurrence of moderate visual loss in patients with early DR (nonproliferative phase) at 24 months [108].

Activation of Other Protein Kinases

Mitogen-Activated Protein Kinase (MAPK)

Recently, studies have reported an important role of mitogen-activated protein kinase (MAPK) pathway in the diabetic complications [109, 110]. The MAPK family consists of extracellular signal-regulated kinase (ERK) and stress-activated components, namely c-jun N-terminal kinase (JNK) and p38 [110, 111]. We have shown that glucose-induced ECM protein synthesis in cultured ECs is mediated by the activation of the MAPK pathway [90]. We have further demonstrated that MAPK activity leads to activation of transcription factors, nuclear factor-κB (NF-κB), and activating protein-1 (AP-1) [90]. Inhibition of either MAPK or PKC is able to normalize the effects of high levels of glucose. Furthermore, inhibiting PKC in cells exposed to high glucose reduces MAPK activation suggesting an important cross-regulation between PKC and MAPK pathways. It is possible that MAPK activation may also occur in vascular ECs via a PKC-independent pathway [112]. Oxidative stress may cause MAPK activation by ERK5 (big MAPK1/BMK1) [113]. Knocking out BMK1 results in angiogenic defect and embryonic lethality [114]. BMK1, however, differs from other MAPK as it contains a transcriptional activation domain, mediating protein–protein interaction with several other factors [114, 115]. Whether such pathways are also activated in DR remains to be determined.

Protein Kinase B and Serum- and Glucocorticoid-Regulated Kinase (SGK-1)

Cultured ECs challenged with high levels of glucose also show an important role of protein kinase B (PKB) [92] and serum- and glucocorticoid-regulated kinase-1 (SGK-1) [91]. Several growth factors stimulate the activation of PKB. There are three major PKB isoforms α, β, γ. These isoforms belong to a subfamily of protein kinases named AGC protein kinases and include PKC and PKA. PKB can regulate the function of cytoplasmic as well as nuclear proteins [116, 117]. We have shown rapid glucose-induced activation of PKB [92] and SGK-1 [91]. Inhibiting PKB and SGK-1 either by dominant negative transfections and/or small interfering RNA causes complete normalization of

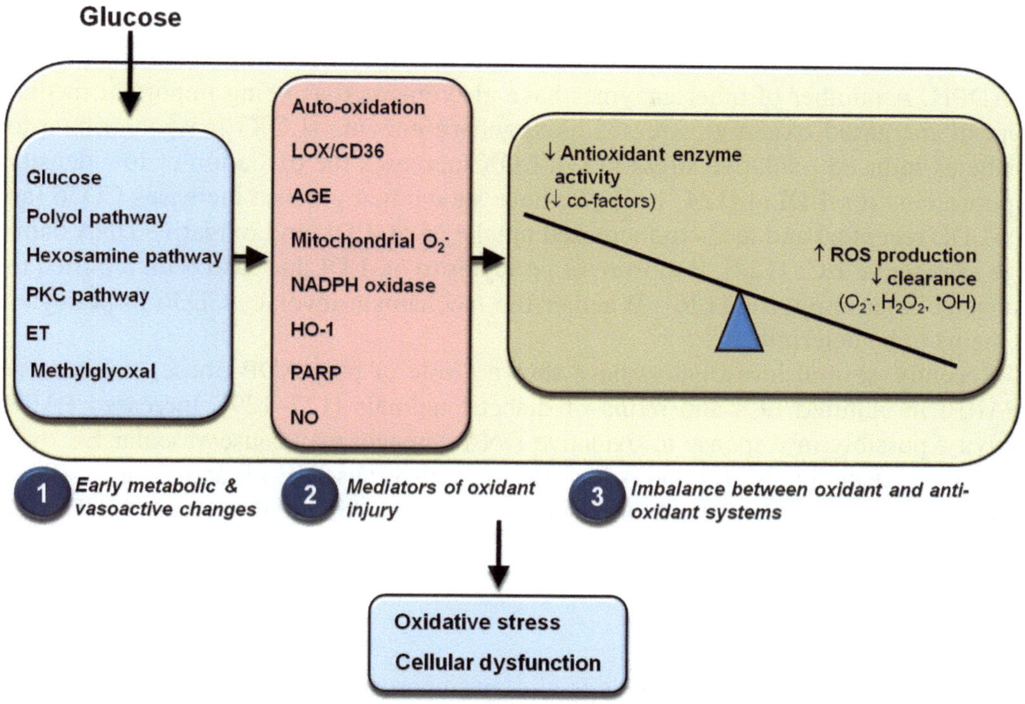

Fig. 4. Mechanisms causing hyperglycemia-induced oxidative stress. High glucose levels directly increase ROS production by autoxidation. Increased flux through the polyol, hexosamine, PKC, and methylglyoxal pathways may also lead to increased oxidative stress. In addition, hyperglycemia may increase ROS indirectly by increasing the activity of various enzymes that lead to oxidative stress. *AGE* advanced glycation end products; *ET* endothelin; *HO* heme oxygenase; *PKC* protein kinase C; *LOX* lectin-like oxidized LDL receptor; *NO* nitric oxide; *PARP* poly(ADP-ribose) polymerase; *SOD* superoxide dismutase.

high glucose-induced FN expression in the vascular ECs. Interestingly, this role of PKB in ECM protein expression is also regulated by both MAPK and PKC [92]. We have further shown that PKB phosphorylation can lead to the activation of NF-κB and AP-1 [92]. These studies suggest that multiple pathways converge on NF-κB and AP-1 to mediate increased ECM protein synthesis.

Increased Oxidative Stress

Increased glucose-induced oxidative stress is another early event in the ECs. There are multiple pathways that increase oxidative stress (Fig. 4). Acute exposure of vascular cells to high ambient glucose causes glucose autoxidation [87] and mitochondrial superoxide production [118–120]. Inhibiting mitochondrial superoxide production has been shown to be beneficial for DR by blocking major pathogenetic pathways [118]. Oxidative stress in diabetes may also be induced by indirect means, which include the NADPH oxidase enzyme [121, 122]. NADPH oxidase may increase superoxide production and through induction of xanthine oxidase. This pathway may also inhibit superoxide dismutase. Impairment of antioxidant enzymes could also be carried out by increased AR activity through the imbalance between NADP+ and NADPH. A number of other enzymes have also emerged as being important mediators of increased oxidative stress. Lipoxygenase enzyme (LOX) may contribute to diabetes-induced oxidative stress [123]. LOX increases the oxidation of low density lipoproteins (ox-LDLs) [124, 125]. We have shown that glucose increases CD36 (an ox-LDL receptor) and leads to increased uptake of ox-LDL and oxidative DNA damage in vascular ECs [124]. Exposure of pericytes to ox-LDL has also been reported to cause cellular apoptosis [126]. Whether the mechanism involves CD36 in pericytes remains to be determined.

Recently, several investigators have shown a role of poly(ADP-ribose) polymerase (PARP) in cultured ECs and retina of diabetic animals [127–129]. Increased PARP activity, possibly in response to oxidative DNA damage, may cause vascular EC dysfunction by depleting NAD+ and ATP. PARP may also cause NF-κB activation [130]. In a nondiabetic system, PARP activation has been linked to histone deacetylases (HDACs) and transcription coactivator p300 [131, 132]. Whether a similar pathway may also be involved in DR requires further investigation.

Protein Glycation

Accelerated glycation of proteins is also an important mechanism leading to cellular dysfunction in diabetes. High levels of glucose may cause nonenzymatic glycation of both intracellular and extracellular proteins [133, 134]. These modified proteins are recognized by AGE receptors (RAGEs) and possibly other scavenger receptors. Studies have shown that retinal vascular tissue and cultured ECs express both RAGEs and CD36 (a scavenger receptor) [135–139]. Although the mechanisms of AGE-mediated cellular dysfunction are currently being elucidated [140–142], aberrant modification of proteins is expected to alter the function of the proteins. In the case of extracellular proteins, glycation may also lead to aberrant outside-in signaling. Evidence for this comes from studies that show that injecting exogenous AGEs in diabetic animals causes retinal pericyte loss [143]. Interestingly, when retinal ECs are exposed to glycated BM proteins [144], the cells proliferate. A specific inhibitor of nonenzymatic glycation, aminoguanidine, has been shown to prevent retinal microaneurysms, acellular capillaries, and pericyte loss in the diabetic dogs [145]. In clinical trials, however, modest beneficial effects were noted [146].

Aberrant Expression of Growth Factors

A number of growth factors have been implicated in the pathogenesis of DR. Growth factor alterations are believed to mediate BM thickening, EC hyperplasia, and unregulated angiogenesis [147]. The list of growth factors that exhibit altered expression in vitreous of diabetic patients or retinal tissues of diabetic rats is a long one [147]. Important growth factors include insulin-like growth factor-1 [148], platelet-derived growth factor [149], basic fibroblast growth factor [150], transforming growth factor-β [151], and VEGF [152]. These growth factors have been shown to induce EC proliferation, ECM synthesis (especially in the case of TGF-β), and cause retinopathy-like lesions in animals [147].

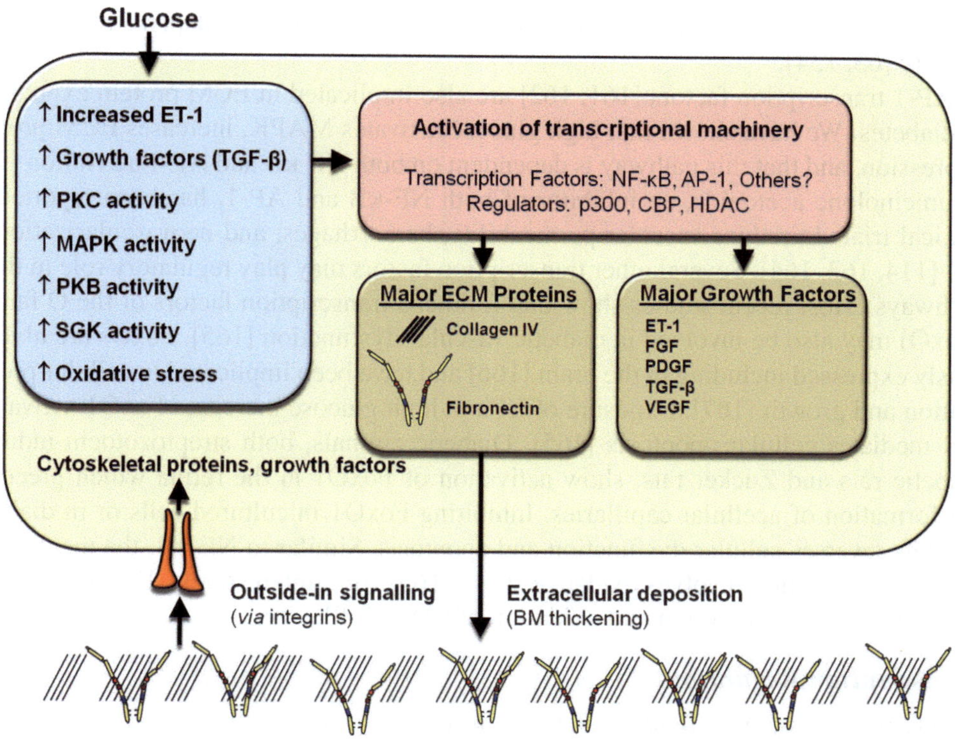

Fig. 5. Mechanisms of glucose-induced growth factor and ECM protein expression in ECs. High levels of glucose lead to activation of a number of intracellular signaling proteins. These signaling proteins mediate the effects of glucose by activating transcription factors and altering other transcriptional regulators (coactivators/corepressors). Transcription factor activity then leads to increased expression of key ECM proteins and growth factors. *AP-1* activating protein-1; *BM* basement membrane; *ET* endothelin; *FGF* fibroblast growth factor; *MAPK* mitogen-activated protein kinase; *NF-kB* nuclear factor-κB; *PKB* protein kinase B; *PKC* protein kinase C; *PDGF* platelet-derived growth factor; *SGK* serum- and glucocorticoid-regulated kinase; *VEGF* vascular endothelial growth factor.

Transcription Factors

All glucose-induced signals converge on transcription factors to regulate expression of key genes involved in vascular function (Fig. 5). Two main transcription factors with wide range of activities are NF-κB and AP-1. NF-κB is a redox-sensitive transcription factor. In quiescent cells, NF-κB exists as an inactive dimer bound to an inhibitory protein, IκB. Upon stimulation, IκB is degraded and NF-κB translocates to the nucleus [153]. In diabetes, NF-κB is believed to be activated by a number of factors including ROS and ET-1 [63, 154]. Interestingly, ET-1 expression may also be regulated by NF-κB activity [155]. Studies have reported nuclear NF-κB immunoreactivity (activated state) in the pericytes but not ECs of human diabetic eyes [156]. In experimental diabetes, however, NF-κB activity is evident in retinal vessel ECs [130, 157–159]. Furthermore, cultured ECs show increased NF-κB activity and downstream effects when exposed to high levels of glucose [63, 128, 130, 137, 139, 160]. We have also shown that

ECM protein expression in ECs and retinas of diabetic animals is dependent on NF-κB activity [63, 154].

AP-1 transcription factors [161, 162] are also implicated in ECM protein expression in diabetes. We have shown that high glucose activates MAPK, increases ECM protein expression, and that this pathway is dependent on both NF-κB and AP-1 activation [90]. Triamcinolone acetonide, an inhibitor of both NF-κB and AP-1, has been reported in clinical trials to reduce vascular permeability, hemorrhages, and neovascularization in DR [114, 163, 164]. Several other transcription factors may play regulatory role in these pathways. Most recent studies show that forkhead transcription factors of the O family (FoxO) may also be involved in diabetic vascular dysfunction [165]. FoxOs are ubiquitously expressed including in the brain [166] and have been implicated in cellular proliferation and growth [167]. Exposure of ECs to high glucose increases FoxO1 activation and mediates cellular apoptosis [165]. Diabetic animals, both streptozotocin-induced diabetic rats and Zucker rats, show activation of FoxO1 in the retina which precedes the formation of acellular capillaries. Inhibiting FoxO1 in cultured cells or in diabetic animals reverses cellular dysfunction and apoptosis. Similar to NF-κB, the mechanism of FoxO1 activation involves oxidative stress [165, 168]. Interestingly, FoxO1 may also facilitate eNOS dysfunction and oxidation of LDL [168].

Transcription Regulators

One of the emerging fields in diabetes research is the epigenetic regulation of gene expression. Chromatin structure and access to transcription factors is regulated by a number of modifications including acetylation, methylation, and phosphorylation [169]. One of the extensively studied processes is the acetylation and deacetylation of histone residues. Two main classes of proteins, acting in opposing manner, regulate acetylation and deacetylation. Histone acetyltransferases (HATs) and HDACs control several cellular processes through regulating transcription factors [170]. The best characterized HATs are p300 and CREB-binding protein (CBP) [170]. These HATs add an acetyl group on lysine residues of histones 3 and 4 (H3 and H4). It is believed that addition of acetyl groups leads to chromatin relaxation and access to transcription factors. Involvement of HATs and HDACs in diabetic complications becomes evident when we consider that transcription factors such as NF-κB remain inactive even after nuclear translocation without the association of p300 [170, 171]. We and others have also shown that NF-κB activity in diabetes is regulated by p300 [128, 172]. In addition, FN expression, in both cultured ECs and the retina of diabetic rats, is mediated by p300 induction [128]. Whether HDACs also modulate these pathways is not clear.

Another mode of chromatin remodeling is regulated by enzymes that add or remove a methyl group. Similar to acetylation/deacetylation, methylation/demethylation may also lead to increased or decreased expression of the target genes. Recently, Reddy et al. [173] showed that smooth muscle cells isolated from diabetic animals exhibit increased monocyte chemotactic protein-1 and interleukin expression via methylation of histone-3 lysine-4 (H3K4). Interestingly, this methylation was found near the NF-κB response element. The same group has also shown reduced histone-3 lysine-9 trimethylation at the promoter region of these target genes [174]. A similar phenomenon is also evident in ECs [175, 176]. A brief exposure of aortic ECs to high glucose levels was associated with increased NF-κB p65 expression and H3K4 monomethylation at the NF-κB p65 promoter region [176]. What is fascinating is that these modifications produce long-term phenotypic changes in the cultured cells even following removal of the high glucose stimulus. This has lead to the concept that histone modification may indeed dictate diabetic/metabolic/hyperglycemic memory.

CONCLUDING REMARKS

Diabetes leads to vascular disruption in selected organs that include the retina.

Experimental evidence from animal models and cultured cells suggests that various signaling pathways in concert lead to the pathogenetic changes in the retinal vascular bed. Early adverse effects of high glucose levels may be mediated by metabolic changes (polyol pathway, hexosamine pathway), vasoactive factors (ET and NO), and oxidative stress (leading to EC dysfunction and loss). Aberrations in EC function may then be perpetuated by continued activation of intracellular signaling proteins such as PKC, PKB, MAPK/ERK, and transcriptional regulators (NF-κB and AP-1, p300). Further investigation as to how these signaling pathways interact is timely. Recent evidence of epigenetic changes producing the "diabetic phenotype" supports the notion that a solid understanding of the hyperglycemia-induced transcription machinery is the only means to identifying the molecular signature and point of convergence in DR.

ACKNOWLEDGMENTS

The authors acknowledge grant supports from the Canadian Diabetes Association (SC; ZAK), Canadian Institutes of Health Research (SC), and Lawson Health Research Institute (ZAK). ZAK is a recipient of the New Investigator Award from the Heart & Stroke Foundation of Canada.

REFERENCES

1. Fong DS, Aiello L, Gardner TW, et al. Diabetic retinopathy. Diabetes Care. 2003;26:226–9.
2. The Diabetes Control and Complications Trial Research Group. The effect of intensive treatment of diabetes on the development and progression of long-term complications in insulin-dependent diabetes mellitus. N Engl J Med. 1993;329:977–86.
3. UK Prospective Diabetes Study (UKPDS) Group. Intensive blood-glucose control with sulphonylureas or insulin compared with conventional treatment and risk of complications in patients with type 2 diabetes (UKPDS 33). Lancet. 1998;352:837–53.
4. Khan ZA, Chakrabarti S. Therapeutic targeting of endothelial dysfunction in chronic diabetic complications. Recent Pat Cardiovasc Drug Discov. 2006;1:167–75.
5. Khan ZA, Chakrabarti S. Cellular signaling and potential new treatment targets in diabetic retinopathy. Exp Diabetes Res. 2007;2007:31867.
6. Khan ZA, Farhangkhoee H, Chakrabarti S. Towards newer molecular targets for chronic diabetic complications. Curr Vasc Pharmacol. 2006;4:45–57.
7. Archer DB. Bowman lecture 1998. Diabetic retinopathy: some cellular, molecular and therapeutic considerations. Eye. 1999;13(Pt 4):497–523.
8. Feman SS. The natural history of the first clinically visible features of diabetic retinopathy. Trans Am Ophthalmol Soc. 1994;92:745–73.
9. Lorenzi M, Gerhardinger C. Early cellular and molecular changes induced by diabetes in the retina. Diabetologia. 2001;44:791–804.
10. Chee CK, Flanagan DW. Visual field loss with capillary non-perfusion in preproliferative and early proliferative diabetic retinopathy. Br J Ophthalmol. 1993;77:726–30.

11. Kohner EM, Henkind P. Correlation of fluorescein angiogram and retinal digest in diabetic retinopathy. Am J Ophthalmol. 1970;69:403–14.
12. Hammes HP, Lin J, Renner O, et al. Pericytes and the pathogenesis of diabetic retinopathy. Diabetes. 2002;51:3107–12.
13. Murata M, Ohta N, Fujisawa S, et al. Selective pericyte degeneration in the retinal capillaries of galactose-fed dogs results from apoptosis linked to aldose reductase-catalyzed galactitol accumulation. J Diabetes Complications. 2002;16:363–70.
14. Cai J, Boulton M. The pathogenesis of diabetic retinopathy: old concepts and new questions. Eye. 2002;16:242–60.
15. Ciulla TA, Harris A, Latkany P, et al. Ocular perfusion abnormalities in diabetes. Acta Ophthalmol Scand. 2002;80:468–77.
16. Mandarino LJ, Finlayson J, Hassell JR. High glucose downregulates glucose transport activity in retinal capillary pericytes but not endothelial cells. Invest Ophthalmol Vis Sci. 1994;35:964–72.
17. Baumgartner-Parzer SM, Wagner L, Pettermann M, Grillari J, Gessl A, Waldhausl W. High-glucose—triggered apoptosis in cultured endothelial cells. Diabetes. 1995;44:1323–7.
18. Boeri D, Almus FE, Maiello M, Cagliero E, Rao LV, Lorenzi M. Modification of tissue-factor mRNA and protein response to thrombin and interleukin 1 by high glucose in cultured human endothelial cells. Diabetes. 1989;38:212–8.
19. Cagliero E, Maiello M, Boeri D, Roy S, Lorenzi M. Increased expression of basement membrane components in human endothelial cells cultured in high glucose. J Clin Invest. 1988;82:735–8.
20. Graier WF, Grubenthal I, Dittrich P, Wascher TC, Kostner GM. Intracellular mechanism of high D-glucose-induced modulation of vascular cell proliferation. Eur J Pharmacol. 1995;294:221–9.
21. Maiello M, Boeri D, Podesta F, et al. Increased expression of tissue plasminogen activator and its inhibitor and reduced fibrinolytic potential of human endothelial cells cultured in elevated glucose. Diabetes. 1992;41:1009–15.
22. McGinn S, Saad S, Poronnik P, Pollock CA. High glucose-mediated effects on endothelial cell proliferation occur via p38 MAP kinase. Am J Physiol Endocrinol Metab. 2003;285: E708–17.
23. Chen YH, Guh JY, Chuang TD, et al. High glucose decreases endothelial cell proliferation via the extracellular signal regulated kinase/p15(INK4b) pathway. Arch Biochem Biophys. 2007;465:164–71.
24. Roy S, Roth T. Proliferative effect of high glucose is modulated by antisense oligonucleotides against fibronectin in rat endothelial cells. Diabetologia. 1997;40:1011–7.
25. Hsu CC, Yin MC, Tian R. Ascorbic acid and uric acid suppress glucose-induced fibronectin and vascular endothelial growth factor production in human endothelial cells. J Diabetes Complications. 2005;19:96–100.
26. Khan ZA, Chan BM, Uniyal S, et al. EDB fibronectin and angiogenesis—a novel mechanistic pathway. Angiogenesis. 2005;8:183–96.
27. Hellstrom M, Gerhardt H, Kalen M, et al. Lack of pericytes leads to endothelial hyperplasia and abnormal vascular morphogenesis. J Cell Biol. 2001;153:543–53.
28. Li W, Liu X, Yanoff M, Cohen S, Ye X. Cultured retinal capillary pericytes die by apoptosis after an abrupt fluctuation from high to low glucose levels: a comparative study with retinal capillary endothelial cells. Diabetologia. 1996;39:537–47.
29. Hammes HP, Lin J, Wagner P, et al. Angiopoietin-2 causes pericyte dropout in the normal retina: evidence for involvement in diabetic retinopathy. Diabetes. 2004;53:1104–10.

30. Davis GE, Senger DR. Endothelial extracellular matrix: biosynthesis, remodeling, and functions during vascular morphogenesis and neovessel stabilization. Circ Res. 2005; 97:1093–107.
31. Davis GE, Senger DR. Extracellular matrix mediates a molecular balance between vascular morphogenesis and regression. Curr Opin Hematol. 2008;15:197–203.
32. Hynes RO. Cell-matrix adhesion in vascular development. J Thromb Haemost. 2007;5 Suppl 1:32–40.
33. Cagliero E, Roth T, Roy S, Lorenzi M. Characteristics and mechanisms of high-glucose-induced overexpression of basement membrane components in cultured human endothelial cells. Diabetes. 1991;40:102–10.
34. Hua H, Goldberg HJ, Fantus IG, Whiteside CI. High glucose-enhanced mesangial cell extracellular signal-regulated protein kinase activation and alpha1(IV) collagen expression in response to endothelin-1: role of specific protein kinase C isozymes. Diabetes. 2001;50:2376–83.
35. Nishikawa T, Giardino I, Edelstein D, Brownlee M. Changes in diabetic retinal matrix protein mRNA levels in a common transgenic mouse strain. Curr Eye Res. 2000;21:581–7.
36. Evans T, Deng DX, Chen S, Chakrabarti S. Endothelin receptor blockade prevents augmented extracellular matrix component mRNA expression and capillary basement membrane thickening in the retina of diabetic and galactose-fed rats. Diabetes. 2000;49:662–6.
37. Deng D, Evans T, Mukherjee K, Downey D, Chakrabarti S. Diabetes-induced vascular dysfunction in the retina: role of endothelins. Diabetologia. 1999;42:1228–34.
38. Ljubimov AV, Burgeson RE, Butkowski RJ, et al. Basement membrane abnormalities in human eyes with diabetic retinopathy. J Histochem Cytochem. 1996;44:1469–79.
39. Spirin KS, Saghizadeh M, Lewin SL, Zardi L, Kenney MC, Ljubimov AV. Basement membrane and growth factor gene expression in normal and diabetic human retinas. Curr Eye Res. 1999;18:490–9.
40. Witmer AN, van den Born J, Vrensen GF, Schlingemann RO. Vascular localization of heparan sulfate proteoglycans in retinas of patients with diabetes mellitus and in VEGF-induced retinopathy using domain-specific antibodies. Curr Eye Res. 2001;22:190–7.
41. Nikolova G, Strilic B, Lammert E. The vascular niche and its basement membrane. Trends Cell Biol. 2007;17:19–25.
42. Rhodes JM, Simons M. The extracellular matrix and blood vessel formation: not just a scaffold. J Cell Mol Med. 2007;11:176–205.
43. Khan ZA, Cukiernik M, Gonder JR, Chakrabarti S. Oncofetal fibronectin in diabetic retinopathy. Invest Ophthalmol Vis Sci. 2004;45:287–95.
44. George B, Chen S, Chaudhary V, Gonder J, Chakrabarti S. Extracellular matrix proteins in epiretinal membranes and in diabetic retinopathy. Curr Eye Res. 2009;34:134–44.
45. Peters JH, Chen GE, Hynes RO. Fibronectin isoform distribution in the mouse. II. Differential distribution of the alternatively spliced EIIIB, EIIIA, and V segments in the adult mouse. Cell Adhes Commun. 1996;4:127–48.
46. Peters JH, Hynes RO. Fibronectin isoform distribution in the mouse. I. The alternatively spliced EIIIB, EIIIA, and V segments show widespread codistribution in the developing mouse embryo. Cell Adhes Commun. 1996;4:103–25.
47. Astrof S, Crowley D, George EL, et al. Direct test of potential roles of EIIIA and EIIIB alternatively spliced segments of fibronectin in physiological and tumor angiogenesis. Mol Cell Biol. 2004;24:8662–70.
48. Singh P, Reimer CL, Peters JH, Stepp MA, Hynes RO, Van De Water L. The spatial and temporal expression patterns of integrin alpha9beta1 and one of its ligands, the EIIIA segment of fibronectin, in cutaneous wound healing. J Invest Dermatol. 2004;123:1176–81.

49. Jiang B, Liou GI, Behzadian MA, Caldwell RB. Astrocytes modulate retinal vasculogenesis: effects on fibronectin expression. J Cell Sci. 1994;107(Pt 9):2499–508.
50. Wang J, Milner R. Fibronectin promotes brain capillary endothelial cell survival and proliferation through alpha5beta1 and alphavbeta3 integrins via MAP kinase signalling. J Neurochem. 2006;96:148–59.
51. Dogra G, Rich L, Stanton K, Watts GF. Endothelium-dependent and independent vasodilation studies at normoglycaemia in type I diabetes mellitus with and without microalbuminuria. Diabetologia. 2001;44:593–601.
52. Johnstone MT, Creager SJ, Scales KM, Cusco JA, Lee BK, Creager MA. Impaired endothelium-dependent vasodilation in patients with insulin-dependent diabetes mellitus. Circulation. 1993;88:2510–6.
53. Lambert J, Aarsen M, Donker AJ, Stehouwer CD. Endothelium-dependent and -independent vasodilation of large arteries in normoalbuminuric insulin-dependent diabetes mellitus. Arterioscler Thromb Vasc Biol. 1996;16:705–11.
54. McVeigh GE, Brennan GM, Johnston GD, et al. Impaired endothelium-dependent and independent vasodilation in patients with type 2 (non-insulin-dependent) diabetes mellitus. Diabetologia. 1992;35:771–6.
55. van de Ree MA, Huisman MV, de Man FH, van der Vijver JC, Meinders AE, Blauw GJ. Impaired endothelium-dependent vasodilation in type 2 diabetes mellitus and the lack of effect of simvastatin. Cardiovasc Res. 2001;52:299–305.
56. Nitenberg A, Valensi P, Sachs R, Dali M, Aptecar E, Attali JR. Impairment of coronary vascular reserve and ACh-induced coronary vasodilation in diabetic patients with angiographically normal coronary arteries and normal left ventricular systolic function. Diabetes. 1993;42:1017–25.
57. Saenz de Tejada I, Goldstein I, Azadzoi K, Krane RJ, Cohen RA. Impaired neurogenic and endothelium-mediated relaxation of penile smooth muscle from diabetic men with impotence. N Engl J Med. 1989;320:1025–30.
58. Steinberg HO, Chaker H, Leaming R, Johnson A, Brechtel G, Baron AD. Obesity/insulin resistance is associated with endothelial dysfunction. Implications for the syndrome of insulin resistance. J Clin Invest. 1996;97:2601–10.
59. Ting HH, Timimi FK, Boles KS, Creager SJ, Ganz P, Creager MA. Vitamin C improves endothelium-dependent vasodilation in patients with non-insulin-dependent diabetes mellitus. J Clin Invest. 1996;97:22–8.
60. Khan ZA, Chakrabarti S. Endothelins in chronic diabetic complications. Can J Physiol Pharmacol. 2003;81:622–34.
61. Cardillo C, Campia U, Bryant MB, Panza JA. Increased activity of endogenous endothelin in patients with type II diabetes mellitus. Circulation. 2002;106:1783–7.
62. Chen S, Apostolova MD, Cherian MG, Chakrabarti S. Interaction of endothelin-1 with vasoactive factors in mediating glucose-induced increased permeability in endothelial cells. Lab Invest. 2000;80:1311–21.
63. Chen S, Khan ZA, Cukiernik M, Chakrabarti S. Differential activation of NF-kappa B and AP-1 in increased fibronectin synthesis in target organs of diabetic complications. Am J Physiol Endocrinol Metab. 2003;284:E1089–97.
64. Yamagishi S, Hsu CC, Kobayashi K, Yamamoto H. Endothelin 1 mediates endothelial cell-dependent proliferation of vascular pericytes. Biochem Biophys Res Commun. 1993;191:840–6.
65. Weissberg PL, Witchell C, Davenport AP, Hesketh TR, Metcalfe JC. The endothelin peptides ET-1, ET-2, ET-3 and sarafotoxin S6b are co-mitogenic with platelet-derived growth factor for vascular smooth muscle cells. Atherosclerosis. 1990;85:257–62.

66. Dong F, Zhang X, Wold LE, Ren Q, Zhang Z, Ren J. Endothelin-1 enhances oxidative stress, cell proliferation and reduces apoptosis in human umbilical vein endothelial cells: role of ETB receptor, NADPH oxidase and caveolin-1. Br J Pharmacol. 2005;145:323–33.
67. Kuhlmann CR, Most AK, Li F, et al. Endothelin-1-induced proliferation of human endothelial cells depends on activation of K+ channels and Ca+ influx. Acta Physiol Scand. 2005;183:161–9.
68. Chen S, Khan ZA, Barbin Y, Chakrabarti S. Pro-oxidant role of heme oxygenase in mediating glucose-induced endothelial cell damage. Free Radic Res. 2004;38:1301–10.
69. Farhangkhoee H, Khan ZA, Mukherjee S, et al. Heme oxygenase in diabetes-induced oxidative stress in the heart. J Mol Cell Cardiol. 2003;35:1439–48.
70. Flores C, Rojas S, Aguayo C, et al. Rapid stimulation of L-arginine transport by D-glucose involves p42/44(mapk) and nitric oxide in human umbilical vein endothelium. Circ Res. 2003;92:64–72.
71. Vasquez R, Farias M, Vega JL, et al. D-glucose stimulation of L-arginine transport and nitric oxide synthesis results from activation of mitogen-activated protein kinases p42/44 and Smad2 requiring functional type II TGF-beta receptors in human umbilical vein endothelium. J Cell Physiol. 2007;212:626–32.
72. Gelinas DS, Bernatchez PN, Rollin S, Bazan NG, Sirois MG. Immediate and delayed VEGF-mediated NO synthesis in endothelial cells: role of PI3K, PKC and PLC pathways. Br J Pharmacol. 2002;137:1021–30.
73. Dimmeler S, Fleming I, Fisslthaler B, Hermann C, Busse R, Zeiher AM. Activation of nitric oxide synthase in endothelial cells by Akt-dependent phosphorylation. Nature. 1999; 399:601–5.
74. Scotland RS, Morales-Ruiz M, Chen Y, et al. Functional reconstitution of endothelial nitric oxide synthase reveals the importance of serine 1179 in endothelium-dependent vasomotion. Circ Res. 2002;90:904–10.
75. Giugliano D, Marfella R, Coppola L, et al. Vascular effects of acute hyperglycemia in humans are reversed by L-arginine. Evidence for reduced availability of nitric oxide during hyperglycemia. Circulation. 1997;95:1783–90.
76. Cukiernik M, Mukherjee S, Downey D, Chakabarti S. Heme oxygenase in the retina in diabetes. Curr Eye Res. 2003;27:301–8.
77. Kinoshita JH, Nishimura C. The involvement of aldose reductase in diabetic complications. Diabetes Metab Rev. 1988;4:323–37.
78. Yabe-Nishimura C. Aldose reductase in glucose toxicity: a potential target for the prevention of diabetic complications. Pharmacol Rev. 1998;50:21–33.
79. Greene DA, Chakrabarti S, Lattimer SA, Sima AA. Role of sorbitol accumulation and myo-inositol depletion in paranodal swelling of large myelinated nerve fibers in the insulin-deficient spontaneously diabetic bio-breeding rat. Reversal by insulin replacement, an aldose reductase inhibitor, and myo-inositol. J Clin Invest. 1987;79:1479–85.
80. Chakrabarti S, Sima AA. The effect of myo-inositol treatment on basement membrane thickening in the BB/W-rat retina. Diabetes Res Clin Pract. 1992;16:13–7.
81. Trueblood N, Ramasamy R. Aldose reductase inhibition improves altered glucose metabolism of isolated diabetic rat hearts. Am J Physiol. 1998;275:H75–83.
82. Demaine AG. Polymorphisms of the aldose reductase gene and susceptibility to diabetic microvascular complications. Curr Med Chem. 2003;10:1389–98.
83. Sivenius K, Niskanen L, Voutilainen-Kaunisto R, Laakso M, Uusitupa M. Aldose reductase gene polymorphisms and susceptibility to microvascular complications in Type 2 diabetes. Diabet Med. 2004;21:1325–33.

84. Wang Y, Ng MC, Lee SC, et al. Phenotypic heterogeneity and associations of two aldose reductase gene polymorphisms with nephropathy and retinopathy in type 2 diabetes. Diabetes Care. 2003;26:2410–5.
85. Sorbinil Retinopathy Trial Research Group. A randomized trial of sorbinil, an aldose reductase inhibitor, in diabetic retinopathy. Arch Ophthalmol. 1990;108:1234–44.
86. Sun W, Oates PJ, Coutcher JB, Gerhardinger C, Lorenzi M. A selective aldose reductase inhibitor of a new structural class prevents or reverses early retinal abnormalities in experimental diabetic retinopathy. Diabetes. 2006;55:2757–62.
87. Brownlee M. Biochemistry and molecular cell biology of diabetic complications. Nature. 2001;414:813–20.
88. Du XL, Edelstein D, Rossetti L, et al. Hyperglycemia-induced mitochondrial superoxide overproduction activates the hexosamine pathway and induces plasminogen activator inhibitor-1 expression by increasing Sp1 glycosylation. Proc Natl Acad Sci USA. 2000;97:12222–6.
89. Chatham JC, Not LG, Fulop N, Marchase RB. Hexosamine biosynthesis and protein O-glycosylation: the first line of defense against stress, ischemia, and trauma. Shock. 2008;29:431–40.
90. Xin X, Khan ZA, Chen S, Chakrabarti S. Extracellular signal-regulated kinase (ERK) in glucose-induced and endothelin-mediated fibronectin synthesis. Lab Invest. 2004;84:1451–9.
91. Khan ZA, Barbin YP, Farhangkhoee H, Beier N, Scholz W, Chakrabarti S. Glucose-induced serum- and glucocorticoid-regulated kinase activation in oncofetal fibronectin expression. Biochem Biophys Res Commun. 2005;329:275–80.
92. Xin X, Khan ZA, Chen S, Chakrabarti S. Glucose-induced Akt1 activation mediates fibronectin synthesis in endothelial cells. Diabetologia. 2005;48:2428–36.
93. Ishii H, Koya D, King GL. Protein kinase C activation and its role in the development of vascular complications in diabetes mellitus. J Mol Med. 1998;76:21–31.
94. Koya D, King GL. Protein kinase C activation and the development of diabetic complications. Diabetes. 1998;47:859–66.
95. Nishizuka Y. Intracellular signaling by hydrolysis of phospholipids and activation of protein kinase C. Science. 1992;258:607–14.
96. Idris I, Gray S, Donnelly R. Protein kinase C activation: isozyme-specific effects on metabolism and cardiovascular complications in diabetes. Diabetologia. 2001;44:659–73.
97. Inoguchi T, Battan R, Handler E, Sportsman JR, Heath W, King GL. Preferential elevation of protein kinase C isoform beta II and diacylglycerol levels in the aorta and heart of diabetic rats: differential reversibility to glycemic control by islet cell transplantation. Proc Natl Acad Sci USA. 1992;89:11059–63.
98. Huang Q, Yuan Y. Interaction of PKC and NOS in signal transduction of microvascular hyperpermeability. Am J Physiol. 1997;273:H2442–51.
99. Khamaisi M, Dahan R, Hamed S, Abassi Z, Heyman SN, Raz I. Role of protein kinase C in the expression of endothelin converting enzyme-1. Endocrinology. 2009;150:1440–9.
100. Yokota T, Ma RC, Park JY, et al. Role of protein kinase C on the expression of platelet-derived growth factor and endothelin-1 in the retina of diabetic rats and cultured retinal capillary pericytes. Diabetes. 2003;52:838–45.
101. Park JY, Takahara N, Gabriele A, et al. Induction of endothelin-1 expression by glucose: an effect of protein kinase C activation. Diabetes. 2000;49:1239–48.
102. Pomero F, Allione A, Beltramo E, et al. Effects of protein kinase C inhibition and activation on proliferation and apoptosis of bovine retinal pericytes. Diabetologia. 2003;46:416–9.
103. Aiello LP, Bursell SE, Clermont A, et al. Vascular endothelial growth factor-induced retinal permeability is mediated by protein kinase C in vivo and suppressed by an orally effective beta-isoform-selective inhibitor. Diabetes. 1997;46:1473–80.

104. Cotter MA, Jack AM, Cameron NE. Effects of the protein kinase C beta inhibitor LY333531 on neural and vascular function in rats with streptozotocin-induced diabetes. Clin Sci (Lond). 2002;103:311–21.
105. Danis RP, Bingaman DP, Jirousek M, Yang Y. Inhibition of intraocular neovascularization caused by retinal ischemia in pigs by PKCbeta inhibition with LY333531. Invest Ophthalmol Vis Sci. 1998;39:171–9.
106. Ishii H, Jirousek MR, Koya D, et al. Amelioration of vascular dysfunctions in diabetic rats by an oral PKC beta inhibitor. Science. 1996;272:728–31.
107. Kowluru RA, Jirousek MR, Stramm L, Farid N, Engerman RL, Kern TS. Abnormalities of retinal metabolism in diabetes or experimental galactosemia: V. Relationship between protein kinase C and ATPases. Diabetes. 1998;47:464–9.
108. Aiello LP, Davis MD, Girach A, et al. Effect of ruboxistaurin on visual loss in patients with diabetic retinopathy. Ophthalmology. 2006;113:2221–30.
109. Awazu M, Ishikura K, Hida M, Hoshiya M. Mechanisms of mitogen-activated protein kinase activation in experimental diabetes. J Am Soc Nephrol. 1999;10:738–45.
110. Tomlinson DR. Mitogen-activated protein kinases as glucose transducers for diabetic complications. Diabetologia. 1999;42:1271–81.
111. Pearson G, Robinson F, Beers Gibson T, et al. Mitogen-activated protein (MAP) kinase pathways: regulation and physiological functions. Endocr Rev. 2001;22:153–83.
112. Liebmann C. Regulation of MAP kinase activity by peptide receptor signalling pathway: paradigms of multiplicity. Cell Signal. 2001;13:777–85.
113. Liu W, Schoenkerman A, Lowe Jr WL. Activation of members of the mitogen-activated protein kinase family by glucose in endothelial cells. Am J Physiol Endocrinol Metab. 2000;279:E782–90.
114. Hayashi M, Kim SW, Imanaka-Yoshida K, et al. Targeted deletion of BMK1/ERK5 in adult mice perturbs vascular integrity and leads to endothelial failure. J Clin Invest. 2004;113:1138–48.
115. Olson EN. Undermining the endothelium by ablation of MAPK-MEF2 signaling. J Clin Invest. 2004;113:1110–2.
116. Pap M, Cooper GM. Role of glycogen synthase kinase-3 in the phosphatidylinositol 3-Kinase/Akt cell survival pathway. J Biol Chem. 1998;273:19929–32.
117. Scheid MP, Woodgett JR. PKB/AKT: functional insights from genetic models. Nat Rev Mol Cell Biol. 2001;2:760–8.
118. Hammes HP, Du X, Edelstein D, et al. Benfotiamine blocks three major pathways of hyperglycemic damage and prevents experimental diabetic retinopathy. Nat Med. 2003;9:294–9.
119. Nishikawa T, Edelstein D, Du XL, et al. Normalizing mitochondrial superoxide production blocks three pathways of hyperglycaemic damage. Nature. 2000;404:787–90.
120. Wolff SP. Diabetes mellitus and free radicals. Free radicals, transition metals and oxidative stress in the aetiology of diabetes mellitus and complications. Br Med Bull. 1993;49: 642–52.
121. Warnholtz A, Nickenig G, Schulz E, et al. Increased NADH-oxidase-mediated superoxide production in the early stages of atherosclerosis: evidence for involvement of the renin-angiotensin system. Circulation. 1999;99:2027–33.
122. Zafari AM, Ushio-Fukai M, Akers M, et al. Role of NADH/NADPH oxidase-derived H2O2 in angiotensin II-induced vascular hypertrophy. Hypertension. 1998;32:488–95.
123. Li L, Sawamura T, Renier G. Glucose enhances endothelial LOX-1 expression: role for LOX-1 in glucose-induced human monocyte adhesion to endothelium. Diabetes. 2003;52:1843–50.

124. Farhangkhoee H, Khan ZA, Barbin Y, Chakrabarti S. Glucose-induced up-regulation of CD36 mediates oxidative stress and microvascular endothelial cell dysfunction. Diabetologia. 2005;48:1401–10.
125. Parthasarathy S, Wieland E, Steinberg D. A role for endothelial cell lipoxygenase in the oxidative modification of low density lipoprotein. Proc Natl Acad Sci USA. 1989;86: 1046–50.
126. Diffley JM, Wu M, Sohn M, Song W, Hammad SM, Lyons TJ. Apoptosis induction by oxidized glycated LDL in human retinal capillary pericytes is independent of activation of MAPK signaling pathways. Mol Vis. 2009;15:135–45.
127. Decker P, Muller S. Modulating poly (ADP-ribose) polymerase activity: potential for the prevention and therapy of pathogenic situations involving DNA damage and oxidative stress. Curr Pharm Biotechnol. 2002;3:275–83.
128. Kaur H, Chen S, Xin X, Chiu J, Khan ZA, Chakrabarti S. Diabetes-induced extracellular matrix protein expression is mediated by transcription coactivator p300. Diabetes. 2006;55:3104–11.
129. Obrosova IG, Pacher P, Szabo C, et al. Aldose reductase inhibition counteracts oxidative-nitrosative stress and poly(ADP-ribose) polymerase activation in tissue sites for diabetes complications. Diabetes. 2005;54:234–42.
130. Zheng L, Szabo C, Kern TS. Poly(ADP-ribose) polymerase is involved in the development of diabetic retinopathy via regulation of nuclear factor-kappaB. Diabetes. 2004;53: 2960–7.
131. Hassa PO, Haenni SS, Buerki C, et al. Acetylation of poly(ADP-ribose) polymerase-1 by p300/CREB-binding protein regulates coactivation of NF-kappaB-dependent transcription. J Biol Chem. 2005;280:40450–64.
132. Ota K, Kameoka M, Tanaka Y, Itaya A, Yoshihara K. Expression of histone acetyltransferases was down-regulated in poly(ADP-ribose) polymerase-1-deficient murine cells. Biochem Biophys Res Commun. 2003;310:312–7.
133. Vlassara H. Recent progress in advanced glycation end products and diabetic complications. Diabetes. 1997;46 Suppl 2:S19–25.
134. Vlassara H. The AGE-receptor in the pathogenesis of diabetic complications. Diabetes Metab Res Rev. 2001;17:436–43.
135. Bierhaus A, Hofmann MA, Ziegler R, Nawroth PP. AGEs and their interaction with AGE-receptors in vascular disease and diabetes mellitus. I. The AGE concept. Cardiovasc Res. 1998;37:586–600.
136. Schmidt AM, Hori O, Cao R, et al. RAGE: a novel cellular receptor for advanced glycation end products. Diabetes. 1996;45 Suppl 3:S77–80.
137. Schmidt AM, Hori O, Chen JX, et al. Advanced glycation endproducts interacting with their endothelial receptor induce expression of vascular cell adhesion molecule-1 (VCAM-1) in cultured human endothelial cells and in mice. A potential mechanism for the accelerated vasculopathy of diabetes. J Clin Invest. 1995;96:1395–403.
138. Stitt AW, He C, Vlassara H. Characterization of the advanced glycation end-product receptor complex in human vascular endothelial cells. Biochem Biophys Res Commun. 1999;256:549–56.
139. Stitt AW, Li YM, Gardiner TA, Bucala R, Archer DB, Vlassara H. Advanced glycation end products (AGEs) co-localize with AGE receptors in the retinal vasculature of diabetic and of AGE-infused rats. Am J Pathol. 1997;150:523–31.
140. Esposito C, Gerlach H, Brett J, Stern D, Vlassara H. Endothelial receptor-mediated binding of glucose-modified albumin is associated with increased monolayer permeability and modulation of cell surface coagulant properties. J Exp Med. 1989;170:1387–407.

141. Vasan S, Foiles PG, Founds HW. Therapeutic potential of AGE inhibitors and breakers of AGE protein cross-links. Expert Opin Investig Drugs. 2001;10:1977–87.
142. Yamagishi S, Yonekura H, Yamamoto Y, et al. Advanced glycation end products-driven angiogenesis in vitro. Induction of the growth and tube formation of human microvascular endothelial cells through autocrine vascular endothelial growth factor. J Biol Chem. 1997;272:8723–30.
143. Xu X, Li Z, Luo D, et al. Exogenous advanced glycosylation end products induce diabetes-like vascular dysfunction in normal rats: a factor in diabetic retinopathy. Graefes Arch Clin Exp Ophthalmol. 2003;241:56–62.
144. Kalfa TA, Gerritsen ME, Carlson EC, Binstock AJ, Tsilibary EC. Altered proliferation of retinal microvascular cells on glycated matrix. Invest Ophthalmol Vis Sci. 1995;36:2358–67.
145. Kern TS, Engerman RL. Pharmacological inhibition of diabetic retinopathy: aminoguanidine and aspirin. Diabetes. 2001;50:1636–42.
146. Bolton WK, Cattran DC, Williams ME, et al. Randomized trial of an inhibitor of formation of advanced glycation end products in diabetic nephropathy. Am J Nephrol. 2004;24:32–40.
147. Khan ZA, Chakrabarti S. Growth factors in proliferative diabetic retinopathy. Exp Diabesity Res. 2003;4:287–301.
148. Merimee TJ, Zapf J, Froesch ER. Insulin-like growth factors. Studies in diabetics with and without retinopathy. N Engl J Med. 1983;309:527–30.
149. Cassidy L, Barry P, Shaw C, Duffy J, Kennedy S. Platelet derived growth factor and fibroblast growth factor basic levels in the vitreous of patients with vitreoretinal disorders. Br J Ophthalmol. 1998;82:181–5.
150. Sivalingam A, Kenney J, Brown GC, Benson WE, Donoso L. Basic fibroblast growth factor levels in the vitreous of patients with proliferative diabetic retinopathy. Arch Ophthalmol. 1990;108:869–72.
151. Hirase K, Ikeda T, Sotozono C, Nishida K, Sawa H, Kinoshita S. Transforming growth factor beta2 in the vitreous in proliferative diabetic retinopathy. Arch Ophthalmol. 1998;116:738–41.
152. Aiello LP, Avery RL, Arrigg PG, et al. Vascular endothelial growth factor in ocular fluid of patients with diabetic retinopathy and other retinal disorders. N Engl J Med. 1994;331:1480–7.
153. Baeuerle PA. Pro-inflammatory signaling: last pieces in the NF-kappaB puzzle? Curr Biol. 1998;8:R19–22.
154. Chen S, Mukherjee S, Chakraborty C, Chakrabarti S. High glucose-induced, endothelin-dependent fibronectin synthesis is mediated via NF-kappa B and AP-1. Am J Physiol Cell Physiol. 2003;284:C263–72.
155. Quehenberger P, Bierhaus A, Fasching P, et al. Endothelin 1 transcription is controlled by nuclear factor-kappaB in AGE-stimulated cultured endothelial cells. Diabetes. 2000;49:1561–70.
156. Romeo G, Liu WH, Asnaghi V, Kern TS, Lorenzi M. Activation of nuclear factor-kappaB induced by diabetes and high glucose regulates a proapoptotic program in retinal pericytes. Diabetes. 2002;51:2241–8.
157. Harada C, Harada T, Mitamura Y, et al. Diverse NF-kappaB expression in epiretinal membranes after human diabetic retinopathy and proliferative vitreoretinopathy. Mol Vis. 2004;10:31–6.
158. Mitamura Y, Harada T, Harada C, et al. NF-kappaB in epiretinal membranes after human diabetic retinopathy. Diabetologia. 2003;46:699–703.
159. Zheng L, Howell SJ, Hatala DA, Huang K, Kern TS. Salicylate-based anti-inflammatory drugs inhibit the early lesion of diabetic retinopathy. Diabetes. 2007;56:337–45.
160. Glomb MA, Monnier VM. Mechanism of protein modification by glyoxal and glycolaldehyde, reactive intermediates of the Maillard reaction. J Biol Chem. 1995;270:10017–26.

161. Shaulian E, Karin M. AP-1 in cell proliferation and survival. Oncogene. 2001;20: 2390–400.
162. Chinenov Y, Kerppola TK. Close encounters of many kinds: Fos-Jun interactions that mediate transcription regulatory specificity. Oncogene. 2001;20:2438–52.
163. Barnes PJ. Anti-inflammatory actions of glucocorticoids: molecular mechanisms. Clin Sci (Lond). 1998;94:557–72.
164. Adcock IM, Ito K, Barnes PJ. Glucocorticoids: effects on gene transcription. Proc Am Thorac Soc. 2004;1:247–54.
165. Behl Y, Krothapalli P, Desta T, Roy S, Graves DT. FOXO1 plays an important role in enhanced microvascular cell apoptosis and microvascular cell loss in type 1 and type 2 diabetic rats. Diabetes. 2009;58:917–25.
166. Hoekman MF, Jacobs FM, Smidt MP, Burbach JP. Spatial and temporal expression of FoxO transcription factors in the developing and adult murine brain. Gene Expr Patterns. 2006;6:134–40.
167. Maiese K, Chong ZZ, Shang YC, Hou J. A "FOXO" in sight: targeting Foxo proteins from conception to cancer. Med Res Rev. 2009;29:395–418.
168. Tanaka J, Li Q, Banks AS, et al. Foxo1 links hyperglycemia to LDL oxidation and eNOS dysfunction in vascular endothelial cells. Diabetes. 2009;58:2344–54.
169. Qiu P, Ritchie RP, Gong XQ, Hamamori Y, Li L. Dynamic changes in chromatin acetylation and the expression of histone acetyltransferases and histone deacetylases regulate the SM22alpha transcription in response to Smad3-mediated TGFbeta1 signaling. Biochem Biophys Res Commun. 2006;348:351–8.
170. Kalkhoven E. CBP and p300: HATs for different occasions. Biochem Pharmacol. 2004;68:1145–55.
171. Goodman RH, Smolik S. CBP/p300 in cell growth, transformation, and development. Genes Dev. 2000;14:1553–77.
172. Chiu J, Xu BY, Chen S, Feng B, Chakrabarti S. Oxidative stress-induced, poly(ADP-ribose) polymerase-dependent upregulation of ET-1 expression in chronic diabetic complications. Can J Physiol Pharmacol. 2008;86:365–72.
173. Reddy MA, Villeneuve LM, Wang M, Lanting L, Natarajan R. Role of the lysine-specific demethylase 1 in the proinflammatory phenotype of vascular smooth muscle cells of diabetic mice. Circ Res. 2008;103:615–23.
174. Villeneuve LM, Reddy MA, Lanting LL, Wang M, Meng L, Natarajan R. Epigenetic histone H3 lysine 9 methylation in metabolic memory and inflammatory phenotype of vascular smooth muscle cells in diabetes. Proc Natl Acad Sci USA. 2008;105:9047–52.
175. Brasacchio D, Okabe J, Tikellis C, et al. Hyperglycemia induces a dynamic cooperativity of histone methylase and demethylase enzymes associated with gene-activating epigenetic marks that coexist on the lysine tail. Diabetes. 2009;58:1229–36.
176. El-Osta A, Brasacchio D, Yao D, et al. Transient high glucose causes persistent epigenetic changes and altered gene expression during subsequent normoglycemia. J Exp Med. 2008;205:2409–17.

14

IGFBP-3 as a Regulator of the Growth-Hormone/ Insulin-Like Growth Factor Pathway in Proliferative Retinopathies

Andreas Stahl, Ann Hellstrom, Chatarina Lofqvist, and Lois Smith

CONTENTS

Keywords Insulin-like growth factor • IGF • IGF-binding protein • IGFBP-3 • Diabetic retinopathy • Retinopathy of prematurity • ROP • Growth hormone • GH • Angiogenesis • Neovascularization

INTRODUCTION

Growth of retinal vessels is not only a crucial factor for retinal development but also one of the hallmarks of two of the most common causes of blindness in the industrialized world: retinopathy of prematurity (ROP) and proliferative diabetic retinopathy (PDR). Visual impairment in both diseases is causally linked to the growth of abnormal blood vessels in the retina. The current clinically established laser treatments for both conditions aim at destroying avascular areas of the affected retina to reduce the production of angiogenic mediators [1]. However, laser treatment is only partially effective and associated with the destruction of healthy retina and subsequent local visual field loss [2]. Over the recent years, considerable progress has been made in both understanding and treating proliferative retinopathies using medical instead of surgical or laser

From: *Ophthalmology Research: Visual Dysfunction in Diabetes*
Edited by: J. Tombran-Tink et al. (eds.), DOI 10.1007/978-1-60761-150-9_14

approaches. One medical treatment that has advanced furthest from basic science into clinical practice is the inhibition of vascular endothelial growth factor (VEGF). Anti-VEGF compounds were initially developed for treatment of wet age-related macular degeneration (AMD) but have recently also found their way into clinical trials for PDR (reviewed in [3]) and are considered for treatment of ROP [4–9].

VEGF has been extensively studied and is rightfully considered a "master switch" for angiogenesis [10]. It is unquestionably one of the major players in proliferative retinopathies and a valid target for anti-angioproliferative treatment approaches. However, both ROP as well as PDR have underlying pathomechanisms that are regulated by extensive and intricate metabolic pathways both locally in the retina as well as on a systemic level. It is therefore not only legitimate but rather essential to further investigate the underlying pathomechanisms of ROP and PDR to unveil angiogenic mediators that function upstream of VEGF expression. In proliferative retinopathies as well as in other angiogenesis-related diseases, VEGF can be viewed as possibly the most important mediator of a final common angiogenic pathway that is, however, activated through a variety of upstream mechanisms that can be very disease-specific [11]. Instead of targeting VEGF at the end of the angiogenic cascade, altering these disease-specific mediators upstream of VEGF might be a more effective approach to treating PDR and ROP. By summarizing our current knowledge about IGFBP-3 in regard to proliferative retinopathies, this chapter aims at evaluating the pathogenetic relevance as well as the potential therapeutic potential of one of the factors that might alter disease mechanisms upstream of VEGF expression in proliferative retinopathies.

THE GROWTH-HORMONE/INSULIN-LIKE GROWTH FACTOR PATHWAY IN PROLIFERATIVE RETINOPATHIES

Proliferative Diabetic Retinopathy (PDR)

Various systemic factors have been identified in diabetic patients that affect the severity of PDR: Obesity, smoking, and unstable control of blood glucose have all been found to be associated with increased severity of PDR. A potential role of growth hormone (GH) in PDR has been first suggested in the 1950s after anecdotal observations of attenuated diabetic retinopathy in women with postpartum hemorrhagic necrosis of the pituitary gland (Sheehan syndrome) [12]. Numerous studies thereafter have found that pituitary dysfunction can prevent or reverse proliferative retinopathy in diabetes patients [13–20]. Additionally, it was reported that GH replacement therapy for patients with GH deficiency can induce a diabetic-like retinopathy, which is attenuated after discontinuation of GH treatment [21].

These early observations about the role of GH in PDR have led to intense research into the downstream mediators of GH signaling. In this respect, insulin-like growth factor 1 (IGF-1) appears not only interesting as one of GH's prime downstream effectors but also because IGF-1 shares receptor-binding affinities with insulin, the disease-defining hormone in diabetes. Clinical studies have found increased levels of IGF-1 in serum and vitreous of patients with PDR [22–31]. However, a clear correlation between disease stage or progression and IGF-1 levels could not be confirmed in all studies [32–34]. These differing results may in part be attributed to differing methodologies for measuring IGF-1. Some studies did not distinguish between free IGF-1 and IGF-1 bound to binding proteins (IGFBPs; reviewed in [2]). The role of IGFBPs in regulating IGF bioavailability and action will be the focus of Section "IGFBP-3 as a Regulator of the Growth-Hormone/Insulin-Like Growth Factor Pathway" of this chapter.

Retinopathy of Prematurity (ROP)

Early investigations in humans have found that the severity of ROP is mainly determined by (1) postnatal oxygen exposure, (2) low gestational age/birth weight, and (3) slow postpartum weight gain [35–43]. The fact that prematurity is the most significant risk factor for ROP suggests that factors involved in growth and development are critical. Hellstrom et al. were the first to describe a direct link between low growth hormone levels and reduced retinal vascularization in children with congenital GH deficiency [44]. Consequent studies focused on one of the prime downstream mediators of GH function: IGF-1. IGF-1 is expressed in liver cells when they are exposed to GH stimulation [45, 46] and plays an important role in fetal growth and development during all stages of pregnancy but particularly in the third trimester [47]. The serum concentration of IGF-1, but not IGF-2, increases with gestational age and correlates with fetal size [48, 49]. IGF-1 levels rise significantly in the third trimester of pregnancy, but after birth decrease due to the loss of IGF-1 provided by the placenta [47]. Intriguingly, low levels of IGF-1 in preterm infants postpartum have been found to prevent normal retinal vascular growth [50] and correlated directly with the severity of clinical ROP [51–54].

The role of IGF-1 in ROP, however, becomes more complex when later disease stages are considered: While physiologic IGF-1 levels might be necessary during early retinal development to prevent ROP, IGF-1 might play a detrimental role during the proliferative stages of ROP. If during the course of postnatal retinal development in the preterm infant the retinal vascular development fails to keep up with the increased retinal demand for oxygen, the peripheral avascular parts of the developing retina will eventually respond to this oxygen shortage by expressing pathologically high levels of pro-angiogenic mediators like VEGF to boost retinal vessel growth. Due to this pro-angiogenic overstimulation retinal vessel growth becomes erratic and abnormal vessels begin to sprout from the retina into the vitreous. These disorganized neovascular tufts eventually lead to severe complications like intravitreal bleeding or retinal detachment caused by traction of the abnormal vessels on the underlying retina. In this second phase of ROP, IGF-1 can act as a permissive factor for retinal neovascularization amplifying VEGF-stimulated pathological vessel growth in the hypoxic retina. The detrimental role of IGF-1 during this phase of proliferative retinopathy is illustrated by the observation that inhibition of IGF-1 prevents hypoxia-induced retinal neovascularization despite high levels of intraocular VEGF [55]. Targeting IGF-1 in ROP infants therefore needs to be carefully timed and correlated to the clinical stage of the disease: During the early stages, when normal vascularization of the retina can still be achieved, IGF-1 levels should be monitored and increased to physiologic levels if needed. This first phase of ROP occurs from birth to approximately postmenstrual age 30–32 weeks. If by this time the retinal vasculature has not developed sufficiently to meet the demands of the maturing retina, high growth factor concentrations from the avascular parts of the retina will induce pathological neovascularization. This marks the second phase of ROP. During the second phase of ROP, IGF-1 supplementation can have detrimental effects by augmenting the growth of pathologic neovessels (reviewed in [50]).

Animal Models of Proliferative Retinopathies

Most of our understanding regarding the underlying mechanisms of proliferative retinopathies comes from the use of animal models of oxygen-induced retinopathy (OIR) that closely mimic the disease process of ROP in humans. In contrast to humans, many animals such as mice, rats, kittens, and beagle pups have incompletely vascularized retinas at birth and therefore resemble the immature retinal state of premature infants. The model that is most widely used to study disease mechanisms and possible interventions is a mouse model of OIR that was first described in 1994 [56]. In this model, neonatal mice are exposed to 75% oxygen from postnatal day 7–12. During this 5-day exposure to hyperoxia, vessel regression and the cessation of normal radial vessel growth occurs, mimicking the first phase of ROP. Other animal models also mimic this early phase of oxygen-induced vessel regression [57, 58]. The second phase of ROP that is characterized by abnormal vessel formation can also be studied in the OIR mouse model: When mice are returned to room air on postnatal day 12, the non-perfused parts of the retina become hypoxic and induce the expression of angiogenic growth factors. As a consequence, formation of abnormal retinal vascular tufts can be observed that closely resemble the erratic neovascularizations seen during the second phase of ROP in human preterm infants. Diabetic retinopathy shows a similar pattern with a first phase characterized by slow loss of retinal capillaries and a second phase of retinal neovascularization. The OIR model can therefore also be used as a tool to investigate some aspects of PDR. This is important as the currently established diabetic animal models do not develop proliferative retinopathy.

The OIR model has greatly promoted our understanding of the growth-hormone/insulin-like growth factor pathway in ROP. Early animal studies have found that normal retinal blood vessels grow more slowly in IGF-1 knockout mouse than in wild-type controls, a pattern very similar to that seen in premature babies with ROP [51]. Subsequent studies using the OIR model have found that mice with low IGF-1 levels and transgenic mice expressing a GH receptor antagonist are resistant to hypoxia-induced retinopathy [59]. Direct proof of the pro-angiogenic role of IGF-1 in the second phase of ROP was established using an IGF-1 receptor antagonist, which was found to suppress retinal neovascularization without altering retinal VEGF levels [55]. Additionally, mice with vascular endothelial cell-specific knockout of either the IGF-1 receptor or insulin receptor show a substantial reduction in retinal neovascularization compared to control mice [60]. Mechanistically, it was suggested that IGF-1 regulates retinal neovascularization at least in part through control of VEGF activation of p44/42 MAPK, establishing a hierarchical relationship between IGF-1 and VEGF receptors [51, 55].

As outlined earlier in this chapter, no good animal models for PDR exist to date. However, an animal study of normoglycemic/normoinsulinemic transgenic mice overexpressing IGF-1 through an insulin promoter at supraphysiological levels in the retina showed loss of pericytes and thickening of basement membrane of retinal capillaries [61]. In older transgenic mice overexpressing IGF-1, neovascularization of the retina and vitreous cavity was observed which was consistent with increased IGF-1 induction of VEGF expression in retinal cells [62]. These accumulated findings suggest that once proliferative neovascular (and therefore leaky) vessels occur in the retina, leaked serum IGF-1 may further promote the proliferation of retinal vessels through stimulation of VEGF. However, it has not been established that serum IGF-1 in the absence of leaky vessels causes proliferative disease. Although local production of IGF-1 in the retina appears to play only a minor role compared to the considerably higher levels of IGF-1 in the serum, local expression of other components of the GH/IGF-1 signaling pathway in the retina might have an impact on the response of retinal neovessels to IGF-1. This possibility will be discussed in the next chapter with regard to retinal expression of IGFBP-3 as locally regulating the GH/IGF-1 pathway.

IGFBP-3 AS A REGULATOR OF THE GROWTH-HORMONE/ INSULIN-LIKE GROWTH FACTOR PATHWAY

As outlined above, the growth-hormone/insulin-like growth factor pathway appears to be involved in both the development of PDR as well as ROP. This leads to the questions of how GH/IGF-1 signaling might be regulated both endogenously as well as by putative interventions using pharmacological approaches. The IGF-binding proteins (IGFBPs) have been found to play an important role in this respect by regulating both the actions as well as the bioavailability of IGF-1 [63]. Systemically, the vast majority of IGF-1 (up to 98%) is bound to one of the six IGFBPs. Within the IGFBP family, IGFBP-3 is by far the most abundant binding protein, with concentrations in the range of 100 nM, compared with the 2–15-nM concentrations of other binding proteins [64]. IGFBP-3 is bound to IGF-1 or IGF-2 in a ternary complex with a glycoprotein, the acid-labile subunit (ALS). This complexation of IGF-1 with IGFBP-3 and ALS leads to a greatly extended circulating half-life of IGF-1. By increasing IGF's serum half-life, IGFBP-3 might theoretically increase the biological effects of IGF-1 [65]. Once in the tissue, however, IGFBPs can either potentiate IGF signaling by releasing IGF-1 in proximity of its receptors or, conversely, hinder signaling by sequestering IGF-1 (reviewed in [66]). IGFBP-3 specifically has been found to have mainly inhibitory functions on IGF signaling. IGFBP-3 can inhibit IGF-1 effects by interfering with IGF signaling or by direct, IGF-independent effects (reviewed in [67]). In vitro, addition of IGFBP-3 to HUVECs stimulated with IGF-1 or VEGF reversed both IGF-1- and VEGF-induced proliferation and prevented the survival induced by these factors [68]. IGFBP-3 can directly bind to the retinoid X receptor-alpha independent of IGF. Binding to this receptor can modulate cell cycle and apoptosis through interference with TGF beta and other signaling pathways (reviewed in [66]).

In regard to proliferative retinopathies, IGFBP-3 has been found to be increased in the vitreous of diabetic rats and human diabetic patients. Interestingly, vitreal IGFBPs were elevated even before the onset of overt retinopathy. This was interpreted as vitreal IGFBPs being involved in early ocular events in the diabetic process as opposed to being the result of end-stage retinopathy [69]. Another study measuring serum free and total IGF-1 as well as IGFBP-3 levels in 56 insulin-treated diabetic patients and 52 healthy sex- and age-matched controls found lower serum levels both for IGF-1 and IGFBP-3 in diabetic patients. However, age-adjusted free IGF-1 levels in subjects with diabetic retinopathy were higher than those in subjects without diabetic retinopathy [32].

Similar to IGF-1, IGFBP-3 is not only present in the serum but also produced locally in the eye [70]. Retinal expression of IGFBP-3 was found to be highly elevated in rats that were exposed to hypoxia [71]. Another study investigated the retinal expression of several IGF-linked genes in greater detail using laser-capture microdissection [72]. This study could localize the hypoxia-induced surge in retinal IGFBP-3 to the neovascular tufts suggesting a direct role for IGFBP-3 during the course of proliferative retinopathy. It has not been investigated if IGFBP-3 alters IGF-1 signaling or has a direct, IGF-independent effect in this context. Considering the fact that IGFBP-3 can affect such divergent cellular functions as mobility, adhesion, apoptosis, survival, and the cell cycle, it would be of great interest for future studies to investigate the exact cellular pathways affected by hypoxia-induced local expression of IGFBP-3 in neovascular tufts. Especially in the light of IGFBP-3 having pro-angiogenic effects in some systems while inhibiting it in others [73], it remains open at this point if IGFBP-3 expression in neovascular tufts plays a role in inducing or rather limiting pathologic retinal neovascularization, although lower mRNA expression levels of IGFBP-3 are associated with more retinopathy [75].

In addition to the direct effects of IGFBP-3 on local angiogenesis, recent work from Chang et al. found that IGFBP-3 also has a critical role in promoting migration, tube formation, and differentiation of endothelial progenitor cells (EPCs) [74]. Recruitment of EPCs to neovascular tufts in the hypoxic retina may thus be another possible role for local IGFBP-3 in proliferative retinopathy. At this point it can only be speculated that increased EPC recruitment through retinal IGFBP-3 might lead to a more organized regrowth of normal vessels as opposed to the erratic growth observed in neovascular tuft formation. EPC recruitment might be one of the mechanisms by which IGFBP-3 promotes retinal repair after oxygen-induced vessel loss [75].

THERAPEUTIC CONSIDERATIONS FOR IGFBP-3 IN PROLIFERATIVE RETINOPATHIES

In children with ROP, serum levels of IGF-1 are inversely correlated with disease severity (see Section "Retinopathy of Prematurity (ROP)" of this chapter). Thus, increasing IGF-1 by exogenous administration might appear as a reasonable treatment option to improve ROP risk in premature babies with low IGF-1 levels. However, bolus injections of IGF-1 potentially can induce hypoglycemic episodes [76]. IGF-induced hypoglycemia, however, can be blocked by coadministering equimolar concentrations of IGFBP-3 together with IGF-1. This finding emphasizes the important regulatory role of IGFBP-3 on serum IGF-1 levels and systemic IGF-1 activity. It also stresses the importance of IGFBP-3 for therapeutic interventions involving the GH/IGF-1 pathway in human patients. The importance of IGFBP-3 substitution along with IGF-1 is further stressed by the fact that in premature infants of 30–35 weeks postmenstrual age, IGFBP-3 levels were found to be significantly diminished in infants with ROP compared to those without [75]. Further, in IGFBP-3-deficient mice, there is a dose-dependent increase in oxygen-induced retinal vessel loss. Wild-type mice treated with exogenous IGFBP-3 had a significant increase in vessel regrowth without any change in IGF-1 levels. This correlated with a 30% increase in EPCs in the retina at postnatal day 15, indicating that IGFBP-3 could be serving as a progenitor cell chemoattractant. These results suggest that IGFBP-3, acting independently of IGF-1, helps to prevent oxygen-induced vessel loss and to promote vascular regrowth after vascular destruction in vivo in a dose-dependent manner, resulting in less retinal neovascularization [75]. As a consequence, clinical trials aiming at correcting IGF-1 deficiency in premature infants use equimolar combinations of IGF-1 and IGFBP-3 [77].

In regard to diabetic retinopathy, there is a substantial body of work indicating that hyperglycemia is associated with reduced serum IGF-1 concentrations [32, 78]. Similar to ROP, the early stages of diabetic retinopathy are associated with low levels of systemic IGF-1. From a therapeutic point of view, it has been shown in clinical studies that restoring normal IGF-1 levels in insulin-treated patients using combined IGF-1/IGFBP-3 regimens results in a concomitant reduction in insulin requirement to maintain euglycemia [79, 80].

One other critical event during the course of diabetic retinopathy is an event known as "early worsening" of proliferative retinopathy. This term refers to an acute increase in retinal proliferative disease coinciding with the onset of exogenous insulin administration. This phenomenon is thought to be linked to an insulin-induced stimulation of the GH/IGF-1 axis. A recent case series with poorly controlled type 1 diabetic patients found that after glycemic control was improved by intensified insulin therapy, serum IGF-1 levels acutely increased and PDR progressed with development of macular edema and proliferation of new vessels [81]. Similarly, a prospective study with 103 pregnant women with type 1 diabetes found that progression of retinopathy during pregnancy was significantly associated with a pregnancy-related increase in IGF-1 levels [82].

It appears likely that the above-described increased serum levels of IGF-1 during "early worsening" of PDR are major contributors to increased retinal IGF-1 signaling. First, serum IGF-1 and IGFBP-3 levels are 10–100 times higher than those measured in the vitreous [26]. Second, patients with PDR show a significant positive correlation between serum and vitreous levels of IGF-1 and the increase in vitreous levels of IGF-1, IGF-2, and IGFBP-3 parallels the increase in vitreous of liver-derived serum proteins [25]. This correlation between serum and vitreal levels is likely due to a disease-associated increase in leakiness of the blood-retina barrier of patients with PDR [26, 83]. Measuring serum levels of IGF-1 and IGFBP-3 in diabetic patients can therefore give a good indication of the retina's exposure to these growth factors. From a therapeutic point of view, it can be speculated that exogenously administered IGFBP-3 could blunt the observed surge of serum IGF-1 by complexing free IGF-1 in the serum and thus preventing IGF-1 from accumulating in the retina. However, the safety of IGFBP-3 administration in PDR patients must be carefully evaluated especially in the context of pregnancy-induced IGF-1 increase.

CONCLUSION

The data on GH, IGF-1, and IGFBP-3 summarized in this chapter illustrate the close association of these three molecules with the development of proliferative retinopathies both in the setting of PDR as well as ROP. This chapter also suggests a number of possibilities to intervene medically in the development of retinopathy by targeting the GH/IGF-1 pathway. IGFBP-3 is one candidate for therapeutic interventions due to its role as a regulator of the GH/IGF-1 pathway. However, it must be emphasized that timing is critical to any intervention targeting proliferative retinopathies. Inhibition of IGF-1 early after birth in premature babies or during the early phases of diabetic retinopathy might prevent normal blood vessel development or increase the loss of established retinal vasculature. Instead of IGF inhibition, careful supplementation of IGF-1 might be needed during these early phases of retinopathy. In these cases, IGFBP-3 should be used as an adjunct to IGF-1 supplementation to regulate the bioavailability and activity of exogenously administered IGF-1 and to avoid IGF-induced hypoglycemic episodes. Once active proliferation in the retina has developed (stage II of ROP or PDR), further supplementation of IGF-1 might be deleterious to the retina. At these stages, IGF-1 acts as a permissive factor for proliferative retinopathy and inhibition of IGF-1 might be needed to limit retinal neovascularization. IGFBP-3 might play a role in this context through its inhibitory role on IGF-1 signaling. However, before IGFBP-3 can be suggested for clinical use during the proliferative stages of ROP or PDR, more work needs to be done deciphering the exact effects of IGFBP-3 both on IGF-1 signaling as well as on IGFBP-3's direct actions independent of IGF-1.

REFERENCES

1. Aiello LM. Perspectives on diabetic retinopathy. Am J Ophthalmol. 2003;136(1):122–35.
2. Wilkinson-Berka JL, Wraight C, Werther G. The role of growth hormone, insulin-like growth factor and somatostatin in diabetic retinopathy. Curr Med Chem. 2006;13(27):3307–17.
3. Jardeleza MS, Miller JW. Review of anti-VEGF therapy in proliferative diabetic retinopathy. Semin Ophthalmol. 2009;24(2):87–92.
4. Chung EJ et al. Combination of laser photocoagulation and intravitreal bevacizumab (avastin) for aggressive zone I retinopathy of prematurity. Graefes Arch Clin Exp Ophthalmol. 2007;245(11):1727–30.
5. Honda S et al. Acute contraction of the proliferative membrane after an intravitreal injection of bevacizumab for advanced retinopathy of prematurity. Graefes Arch Clin Exp Ophthalmol. 2008;246(7):1061–3.
6. Kusaka S et al. Efficacy of intravitreal injection of bevacizumab for severe retinopathy of prematurity: a pilot study. Br J Ophthalmol. 2008;92(11):1450–5.
7. Lalwani GA et al. Off-label use of intravitreal bevacizumab (avastin) for salvage treatment in progressive threshold retinopathy of prematurity. Retina. 2008;28(3 Suppl):S13–8.
8. Mintz-Hittner HA, Kuffel Jr RR. Intravitreal injection of bevacizumab (avastin) for treatment of stage 3 retinopathy of prematurity in zone I or posterior zone II. Retina. 2008;28(6): 831–8.
9. Quiroz-Mercado H et al. Antiangiogenic therapy with intravitreal bevacizumab for retinopathy of prematurity. Retina. 2008;28(3 Suppl):S19–25.
10. Ferrara N. Vascular endothelial growth factor. Arterioscler Thromb Vasc Biol. 2009;29(6): 789–91.
11. Stahl A et al. Rapamycin reduces VEGF expression in retinal pigment epithelium (RPE) and inhibits RPE-induced sprouting angiogenesis in vitro. FEBS Lett. 2008;582(20):3097–102.
12. Schimek RA. Hypophysectomy for diabetic retinopathy; a preliminary report. AMA Arch Ophthalmol. 1956;56(3):416–25.
13. Alzaid AA et al. The role of growth hormone in the development of diabetic retinopathy. Diabetes Care. 1994;17(6):531–4.
14. Kohner EM et al. Pituitary ablation in the treatment of diabetic retinopathy (a randomized trial). Trans Ophthalmol Soc U K. 1972;92:79–90.

15. Lundbaek K et al. Diabetes, diabetic angiopathy, and growth hormone. Lancet. 1970; 2(7664):131–3.
16. Merimee TJ et al. Diabetes mellitus and sexual ateliotic dwarfism: a comparative study. J Clin Invest. 1970;49(6):1096–102.
17. Poulsen JE. Diabetes and anterior pituitary insufficiency. Final course and postmortem study of a diabetic patient with Sheehan's syndrome. Diabetes. 1966;15(2):73–7.
18. Sharp PS et al. Long-term follow-up of patients who underwent yttrium-90 pituitary implantation for treatment of proliferative diabetic retinopathy. Diabetologia. 1987;30(4):199–207.
19. Wright AD et al. Serum growth hormone levels and the response of diabetic retinopathy to pituitary ablation. Br Med J. 1969;2(5653):346–8.
20. Wright AD et al. Serum growth hormone levels and size of pituitary tumour in untreated acromegaly. Br Med J. 1969;4(5683):582–4.
21. Hansen R, Koller EA, Malozowski S. Full remission of growth hormone (GH)-induced retinopathy after GH treatment discontinuation: long-term follow-up. J Clin Endocrinol Metab. 2000;85(7):2627.
22. Sonksen PH, Russell-Jones D, Jones RH. Growth hormone and diabetes mellitus. A review of sixty-three years of medical research and a glimpse into the future? Horm Res. 1993;40(1–3): 68–79.
23. Merimee TJ, Zapf J, Froesch ER. Insulin-like growth factors. Studies in diabetics with and without retinopathy. N Engl J Med. 1983;309(9):527–30.
24. Amiel SA et al. Effect of diabetes and its control on insulin-like growth factors in the young subject with type I diabetes. Diabetes. 1984;33(12):1175–9.
25. Grant M et al. Insulin-like growth factors in vitreous. Studies in control and diabetic subjects with neovascularization. Diabetes. 1986;35(4):416–20.
26. Pfeiffer A et al. Growth factor alterations in advanced diabetic retinopathy: a possible role of blood retina barrier breakdown. Diabetes. 1997;46 Suppl 2:S26–30.
27. Meyer-Schwickerath R et al. Vitreous levels of the insulin-like growth factors I and II, and the insulin-like growth factor binding proteins 2 and 3, increase in neovascular eye disease. Studies in nondiabetic and diabetic subjects. J Clin Invest. 1993;92(6):2620–5.
28. Hyer SL et al. A two-year follow-up study of serum insulinlike growth factor-I in diabetics with retinopathy. Metabolism. 1989;38(6):586–9.
29. Dills DG et al. Is insulinlike growth factor I associated with diabetic retinopathy? Diabetes. 1990;39(2):191–5.
30. Dills DG et al. Association of elevated IGF-I levels with increased retinopathy in late-onset diabetes. Diabetes. 1991;40(12):1725–30.
31. Boulton M et al. Intravitreal growth factors in proliferative diabetic retinopathy: correlation with neovascular activity and glycaemic management. Br J Ophthalmol. 1997;81(3):228–33.
32. Janssen JA et al. Free and total insulin-like growth factor I (IGF-I), IGF-binding protein-1 (IGFBP-1), and IGFBP-3 and their relationships to the presence of diabetic retinopathy and glomerular hyperfiltration in insulin-dependent diabetes mellitus. J Clin Endocrinol Metab. 1997;82(9):2809–15.
33. Wang Q et al. Does insulin-like growth factor I predict incidence and progression of diabetic retinopathy? Diabetes. 1995;44(2):161–4.
34. Frystyk J et al. The relationship between the circulating IGF system and the presence of retinopathy in Type 1 diabetic patients. Diabet Med. 2003;20(4):269–76.
35. Campbell K. Intensive oxygen therapy as a possible cause of retrolental fibroplasia; a clinical approach. Med J Aust. 1951;2(2):48–50.
36. Patz A, Hoeck LE, De La Cruz E. Studies on the effect of high oxygen administration in retrolental fibroplasia. I. Nursery observations. Am J Ophthalmol. 1952;35(9):1248–53.

37. Ashton N, Ward B, Serpell G. Role of oxygen in the genesis of retrolental fibroplasia; a preliminary report. Br J Ophthalmol. 1953;37(9):513–20.
38. Ashton N, Ward B, Serpell G. Effect of oxygen on developing retinal vessels with particular reference to the problem of retrolental fibroplasia. Br J Ophthalmol. 1954;38(7): 397–432.
39. Kinsey VE et al. PaO_2 levels and retrolental fibroplasia: a report of the cooperative study. Pediatrics. 1977;60(5):655–68.
40. Flynn JT. Acute proliferative retrolental fibroplasia: multivariate risk analysis. Trans Am Ophthalmol Soc. 1983;81:549–91.
41. Lutty GA et al. Proceedings of the third international symposium on retinopathy of prematurity: an update on ROP from the lab to the nursery (November 2003, Anaheim, California). Mol Vis. 2006;12:532–80.
42. Smith LE. Pathogenesis of retinopathy of prematurity. Semin Neonatol. 2003;8(6):469–73.
43. Tasman W et al. Retinopathy of prematurity: the life of a lifetime disease. Am J Ophthalmol. 2006;141(1):167–74.
44. Hellstrom A et al. Reduced retinal vascularization in children with growth hormone deficiency. J Clin Endocrinol Metab. 1999;84(2):795–8.
45. Le Roith D et al. The somatomedin hypothesis: 2001. Endocr Rev. 2001;22(1):53–74.
46. Velloso CP. Regulation of muscle mass by growth hormone and IGF-I. Br J Pharmacol. 2008;154(3):557–68.
47. Langford K, Nicolaides K, Miell JP. Maternal and fetal insulin-like growth factors and their binding proteins in the second and third trimesters of human pregnancy. Hum Reprod. 1998;13(5):1389–93.
48. Lassarre C et al. Serum insulin-like growth factors and insulin-like growth factor binding proteins in the human fetus. Relationships with growth in normal subjects and in subjects with intrauterine growth retardation. Pediatr Res. 1991;29(3):219–25.
49. Reece EA et al. The relation between human fetal growth and fetal blood levels of insulin-like growth factors I and II, their binding proteins, and receptors. Obstet Gynecol. 1994;84(1): 88–95.
50. Chen J, Smith LE. Retinopathy of prematurity. Angiogenesis. 2007;10(2):133–40.
51. Hellstrom A et al. Low IGF-I suppresses VEGF-survival signaling in retinal endothelial cells: direct correlation with clinical retinopathy of prematurity. Proc Natl Acad Sci USA. 2001;98(10):5804–8.
52. Hellstrom A et al. Postnatal serum insulin-like growth factor I deficiency is associated with retinopathy of prematurity and other complications of premature birth. Pediatrics. 2003;112(5):1016–20.
53. Lofqvist C et al. Postnatal head growth deficit among premature infants parallels retinopathy of prematurity and insulin-like growth factor-1 deficit. Pediatrics. 2006;117(6):1930–8.
54. Smith LE. Pathogenesis of retinopathy of prematurity. Growth Horm IGF Res. 2004; 14(Suppl A):S140–4.
55. Smith LE et al. Regulation of vascular endothelial growth factor-dependent retinal neovascularization by insulin-like growth factor-1 receptor. Nat Med. 1999;5(12):1390–5.
56. Smith LE et al. Oxygen-induced retinopathy in the mouse. Invest Ophthalmol Vis Sci. 1994;35(1):101–11.
57. Flower RW. Perinatal ocular physiology and ROP in the experimental animal model. Doc Ophthalmol. 1990;74(3):153–62.
58. Penn JS, Tolman BL, Henry MM. Oxygen-induced retinopathy in the rat: relationship of retinal nonperfusion to subsequent neovascularization. Invest Ophthalmol Vis Sci. 1994;35(9):3429–35.

59. Smith LE et al. Essential role of growth hormone in ischemia-induced retinal neovascularization. Science. 1997;276(5319):1706–9.
60. Kondo T et al. Knockout of insulin and IGF-1 receptors on vascular endothelial cells protects against retinal neovascularization. J Clin Invest. 2003;111(12):1835–42.
61. Ruberte J et al. Increased ocular levels of IGF-1 in transgenic mice lead to diabetes-like eye disease. J Clin Invest. 2004;113(8):1149–57.
62. Punglia RS et al. Regulation of vascular endothelial growth factor expression by insulin-like growth factor I. Diabetes. 1997;46(10):1619–26.
63. Baxter RC. Insulin-like growth factor binding proteins in the human circulation: a review. Horm Res. 1994;42(4–5):140–4.
64. Mohan S, Baylink DJ. Insulin-like growth factor system components and the coupling of bone formation to resorption. Horm Res. 1996;45 Suppl 1:59–62.
65. Mohan S, Baylink DJ. IGF-binding proteins are multifunctional and act via IGF-dependent and -independent mechanisms. J Endocrinol. 2002;175(1):19–31.
66. Firth SM, Baxter RC. Cellular actions of the insulin-like growth factor binding proteins. Endocr Rev. 2002;23(6):824–54.
67. Baxter RC. Insulin-like growth factor (IGF)-binding proteins: interactions with IGFs and intrinsic bioactivities. Am J Physiol Endocrinol Metab. 2000;278(6):E967–76.
68. Franklin SL, Ferry Jr RJ, Cohen P. Rapid insulin-like growth factor (IGF)-independent effects of IGF binding protein-3 on endothelial cell survival. J Clin Endocrinol Metab. 2003;88(2):900–7.
69. Waldbillig RJ et al. Vitreal insulin-like growth factor binding proteins (IGFBPs) are increased in human and animal diabetics. Curr Eye Res. 1994;13(7):539–46.
70. Burren CP et al. Localization of mRNAs for insulin-like growth factor-I (IGF-I), IGF-I receptor, and IGF binding proteins in rat eye. Invest Ophthalmol Vis Sci. 1996;37(7):1459–68.
71. Averbukh E et al. Gene expression of insulin-like growth factor-I, its receptor and binding proteins in retina under hypoxic conditions. Metabolism. 1998;47(11):1331–6.
72. Lofqvist C et al. Quantification and localization of the IGF/insulin system expression in retinal blood vessels and neurons during oxygen-induced retinopathy in mice. Invest Ophthalmol Vis Sci. 2009;50(4):1831–7.
73. Granata R et al. Dual effects of IGFBP-3 on endothelial cell apoptosis and survival: involvement of the sphingolipid signaling pathways. FASEB J. 2004;18(12):1456–8.
74. Chang KH et al. IGF binding protein-3 regulates hematopoietic stem cell and endothelial precursor cell function during vascular development. Proc Natl Acad Sci USA. 2007; 104(25):10595–600.
75. Lofqvist C et al. IGFBP-3 suppresses retinopathy through suppression of oxygen-induced vessel loss and promotion of vascular regrowth. Proc Natl Acad Sci USA. 2007;104(25): 10589–94.
76. Firth SM et al. Impaired blockade of insulin-like growth factor I (IGF-I)-induced hypoglycemia by IGF binding protein-3 analog with reduced ternary complex-forming ability. Endocrinology. 2002;143(5):1669–76.
77. Lofqvist C et al. A pharmacokinetic and dosing study of intravenous insulin-like growth factor-I and IGF-binding protein-3 complex to preterm infants. Pediatr Res. 2009;65:574–9.
78. Dunger DB, Cheetham TD, Crowne EC. Insulin-like growth factors (IGFs) and IGF-I treatment in the adolescent with insulin-dependent diabetes mellitus. Metabolism. 1995;44(10 Suppl 4):119–23.
79. Clemmons DR et al. The combination of insulin-like growth factor I and insulin-like growth factor-binding protein-3 reduces insulin requirements in insulin-dependent type 1 diabetes: evidence for in vivo biological activity. J Clin Endocrinol Metab. 2000;85(4):1518–24.

80. O'Connell T, Clemmons DR. IGF-I/IGF-binding protein-3 combination improves insulin resistance by GH-dependent and independent mechanisms. J Clin Endocrinol Metab. 2002;87(9):4356–60.
81. Chantelau E, Frystyk J. Progression of diabetic retinopathy during improved metabolic control may be treated with reduced insulin dosage and/or somatostatin analogue administration—a case report. Growth Horm IGF Res. 2005;15(2):130–5.
82. Lauszus FF et al. Increased serum IGF-I during pregnancy is associated with progression of diabetic retinopathy. Diabetes. 2003;52(3):852–6.
83. Spranger J et al. Systemic levels contribute significantly to increased intraocular IGF-I, IGF-II and IGF-BP3 [correction of IFG-BP3] in proliferative diabetic retinopathy. Horm Metab Res. 2000;32(5):196–200.

15
Neurotrophic Factors in Diabetic Retinopathy

Anne R. Murray and Jian-xing Ma

CONTENTS

Keywords Angiogenesis • Diabetic retinopathy • Neurotrophic factors • PEDF • VEGF

DIABETIC RETINOPATHY

The incidence of diabetes worldwide is staggering. Millions of people have been diagnosed with either Type 1 or Type 2 diabetes, and it is estimated that approximately 10% of diabetes cases are Type 1 [1], while approximately 90% of patients diagnosed with diabetes are Type 2. Type 2 diabetes currently affects more than 150 million people worldwide [1, 2], and it has been estimated that with an increasingly sedentary lifestyle and prevalence of obesity, the incidence of diabetes worldwide is expected to reach 366 million by the year 2020 [2, 3].

The maintenance of normal glucose levels is essential for the health of most organs. In fact, it has been shown that the incidence of issues such as peripheral neuropathy [4], oxidative stress [5], and vascular complications [4, 6, 7] increases greatly upon chronic exposure of elevated glucose levels. One severe diabetic complication involves the eye. Chronic exposure of the retina to elevated glucose levels leads to proliferative diabetic retinopathy (DR), a condition that is characterized by retinal inflammation, vascular leakage, abnormal blood vessel formation (neovascularization), and intraretinal hemorrhages [8]. Upon the progression of DR, the microvascular circulation in the retina fails, leading to ischemia (Fig. 1) [8]. If not properly monitored and regulated, the newly

From: *Ophthalmology Research: Visual Dysfunction in Diabetes*
Edited by: J. Tombran-Tink et al. (eds.), DOI 10.1007/978-1-60761-150-9_15

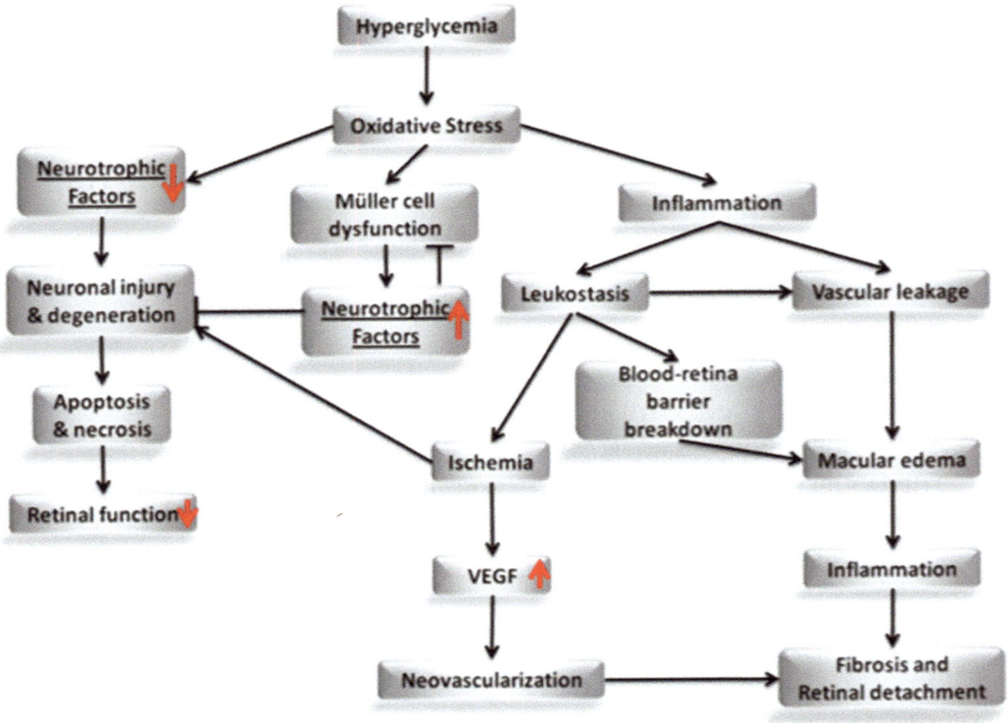

Fig. 1. The molecular pathway leading to decreased retinal function as well as to the neovascularization, fibrosis, and retinal detachment in diabetic retinopathy. Neurotrophic factors in the retina play essential roles in the development and progression of symptomatic DR. Upon oxidative stress signals in the retina, there is a decrease in most neurotrophic factors along with an increase in inflammatory factors and Müller cell dysfunction. The Müller cell then signals the release of several neurotrophic factors to aid in the survival of the retina. Upon the progression of DR and prolonged hyperglycemia, the retinal cells succumb to apoptosis and necrosis with a concomitant increase in vascular leakage and macular edema leading to vision loss.

formed retinal blood vessels will extend into the vitreous, which can lead to hemorrhage and retinal detachment. In addition to the ischemia and new vessel growth, another DR complication is the development of macular edema. Breakdown of the blood-retinal barrier (BRB) that maintains the retinal environment leads to leakage of macromolecules from the vessels into the retina and swelling of the central portion of the retina, the macula. This swelling will often progress and affect the patient's central vision. This combination of complications will often, if untreated, lead to irreversible vision loss and blindness.

Although proliferation of the retinal vasculature and macular edema are the devastating end points of proliferative DR, it has been suggested that at early stages of DR, there are changes in the retinal neurons and glia [8–10]. Experimental models have shown that changes in functional molecules and the viability of neurons in the retina occur immediately after the onset of diabetes [11, 12]. Prior to sight-threatening signs of abnormal angiogenesis, damages to the neurons in the inner [12] and outer retina [13] as well as glial cell activation [14] have been observed. The changes in these cells lead to retinal hypoxia, a damaging precursor to the angiogenesis, and further DR pathologies.

NEUROTROPHIC FACTORS

The retina is comprised of several cell types that each play a specific role in maintaining normal visual function. In order to ensure neuronal cell survival, several of these cells produce neurotrophic factors (NF). NFs have several functions in the neuron including neuronal cell development, synapse formation, synaptic plasticity, proper neuronal cell function, and the promotion of neuronal cell survival [15]. Several NFs promote these cellular functions via two classes of transmembrane receptor proteins, the tropomyosin receptor kinase (Trk) and neurotrophin receptor p75 (Fig. 2 and Table 1) [16]. Binding of the NF to the p75 receptor acts to signal cell death, while binding of a NF to the Trk family promotes signaling for cell survival and differentiation [16].

Several studies have shown that in the diabetic retina, even before the onset of DR and DR-associated retinal neovascularization, there is an increase in neuronal cell death along with glial changes and a reduction in the levels of several NFs [11, 17–19]. A caveat to this phenomenon is that although a decrease is observed in many of the retinal NFs, there is an increase in the pro-angiogenic NF VEGF [20–22]. The disruption in NF function observed in the pre-DR retina can be caused by several potential

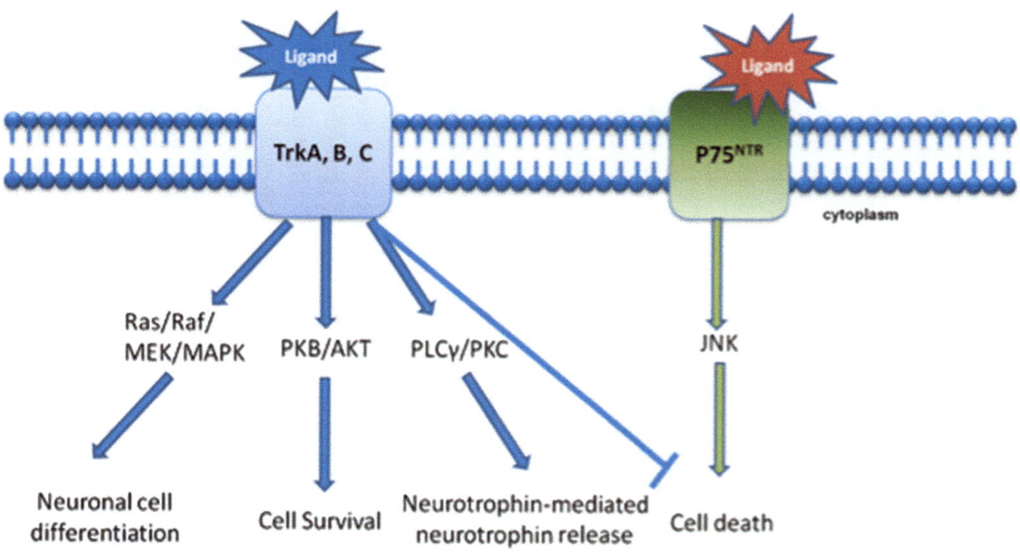

Fig. 2. Neurotrophic factors in the retina are responsible for several functions via two receptor families. Some neurotrophic factors, such as BDNF, can interact with two cell-surface receptors, the Trk and/or the p75 family. Upon binding to the Trk family of receptors, neurotrophic factor-associated intracellular signaling can occur through three major pathways, the Ras/Raf/MEK/MAPK, PKB/Akt, or PLCγ/PKC, to induce neuronal cell differentiation, cell survival, or neurotrophin-mediated neurotrophin release. Binding of a neurotrophic factor to the p75 receptor results in activation of the JNK signaling pathway and leads to the promotion of cell death.

Table 1. Neurotrophic factors involved in diabetic retinopathy

Neurotrophic factor	Secreted by	Additional function(s)	Known receptor(s)	References
Nerve growth factor (NGF)	Müller cells	Growth factor	$p75^{NGFR}$ (low affinity) and $NGFR^{TrkA}$	[16, 63, 109]
Glial cell-derived neurotrophic factor (GDNF)	Müller cells	Glial differentiation	Complex composed of GFRα1 and the transmembrane protein kinase Ret	[27, 110]
Ciliary neurotrophic factor (CNTF)	Müller cells	Protect retina against light damage; axonal regeneration of RGCs; neuronal differentiation factor; growth factor	Receptor complex: CNTFRα, LIFRβ+GP130 in Müller, RGCs, amacrine, horizontal, RPE, rods, and cones	[32, 37, 63, 77, 111]
Pigment-epithelium-derived factor (PEDF)	RPE, RCEC, and Müller cells	Retinal development; neuron differentiation; angiogenesis inhibitor; anti-inflammatory factor	PEDFR	[53, 112–115]
SERPINA3K	Unknown	Anti-fibrosis; angiogenesis inhibitor	Low-density lipoprotein receptor-like protein 6 (LRP6)	[116, 117]
Brain-derived neurotrophic factor (BDNF)	Retinal ganglion cells (RGCs) and Müller cells	Retinal development; synaptic modulator; hypertrophy of the retinal dopaminergic system in the retina; protect retina against light damage; angiogenesis	$Gp140^{TrkB}$ (signaling) and $p75^{NGFR}$ (low affinity)	[11, 16, 63, 64, 70, 77, 109, 118]
Fibroblast growth factor (FGF)	RPE	Retinal development; angiogenesis; growth factor; protect retina against light damage	FGFR1 and FGFR2	[37, 71, 77, 119]
Insulin	Pancreas	Growth factor	Insulin receptor	[37]
Insulin-like growth factor (IGF)	RPE	Retinal development; neurogenesis; angiogenesis	Insulin receptor	[78–81, 120]
Erythropoietin (EPO)	Neural cells	Angiogenesis	EpoR	[87, 89, 90]
Vascular endothelial growth factor (VEGF)	Retinal pigment epithelium (RPE) and Müller cells	Proliferation and migration; angiogenesis; neurogenesis; increasing axonal outgrowth; vascular permeability enhancer; apoptosis inhibitor	VEGFR-1, VEGFR-2 (receptor tyrosine kinase); neuropilins (NP) 1 and 2 (nonreceptor tyrosine kinase)	[92, 121, 122]

features (1) decreased NF synthesis, (2) disrupted transport of the NF in the neuronal cell, (3) modifications in the NF-associated signal transduction pathways, or (4) the ability of the cells that produce the NF is affected, including those cells that produce NF responsible for neuron survival [23].

Although the list of NFs associated with neural diseases is extensive, the roles that they play in DR have not been exhaustively studied. Several NFs are expressed in the retina, but some have not been confirmed to be expressed in retinal diseases including DR. Currently, there are several potential therapies employing the use of neurotrophic factors in neurological diseases in diseases such as DR. In fact, there are several ongoing studies that endeavor to develop practical therapies for DR using neurotrophic factors. The following sections provide a brief description of what is known about DR-associated neurotrophic factors.

NEUROTROPHINS AND OTHERS

Nerve Growth Factor

Nerve growth factor (NGF), a member of the neurotrophin gene family, has been widely studied in diabetic neuropathy [24]. However, a definitive characterization of this factor has not been examined in DR. The low-affinity NGF receptor, p75NGR, is expressed in both the Müller and retinal ganglion cells (RGCs) [12, 25]. Streptozotocin (STZ) injected rat retinas showed increased immunoreactivity of the receptor in the retina, including throughout the RGC and the outer nuclear layer [12]. Further examination revealed that upon the induction of diabetes, the Müller cells are the major source of the receptor upregulation [12].

Therapeutic intervention using NGF found that NGF prevented programmed cell death in both RGC and Müller cells in the diabetes-induced rat retina [12]. NGF's therapeutic potential will need to be examined further to determine its efficacy in treating patients with DR.

Glial-Cell-Derived Neurotrophic Factor

Glial-cell-derived neurotrophic factor (GDNF) was originally characterized as a neurotrophic differentiation factor in the central nervous system and retina [26]. GDNF and its receptor have been well documented in DR and have been shown to be secreted by the glial cells of the retina [27]. GDNF's role in maintaining neural cell survival is known, but GDNF has also been linked to proper glial cell development [28, 29]. Along with its roles in development and cell survival, GDNF may also modulate vascular permeability in the BRB via modulating the function of tight junctions [30].

The therapeutic potential of GDNF in the treatment of DR has not been studied.

Ciliary Neurotrophic Factor

Ciliary neurotrophic factor (CNTF) was first identified as a factor that supported the survival of ciliary neurons in the chick embryo [31]. It is a member of the interleukin-6 (IL-6) family of cytokines and binds to two common IL-6 family receptor components, gp130 and LIFRβ [32, 33]. In order to facilitate proper function, CNTF also requires the CNTFRα receptor subunit [34]. CNTF is primarily localized in Müller cells and is expressed in both the developing and mature retina in the rat [35, 36]. The CNTF receptor is located in the retinal Müller, horizontal, amacrine, and ganglion cells [35].

CNTF has several functions in the retina including, but not limited to, promoting the survival and axonal regeneration of RGCs, promoting green cone cell differentiation, and inhibiting rod cell differentiation [31, 32]. The majority of CNTF's functions are through the JAK/STAT intracellular signaling pathway [37], although it can also activate the ERK [38] and PI3-K/Akt pathways [39].

CNTF has been shown to aid in the survival of the retinal neurons in several retinal degenerative disorders [31, 40]. Intravitreal injection of recombinant CNTF into a retinal degeneration model led to a short-term rescue of photoreceptors [35, 40]. In another study, injection of an adenovirus expressing CNTF delayed photoreceptor degeneration in retinal degeneration (*rd/rd*) mice [41, 42]. Future studies are considering the use of an intravitreal implant that would apply a prolonged delivery of CNTF to the retina for longer neuronal protection [43].

ANTI-ANGIOGENIC NEUROTROPHIC FACTORS

Pigment-Epithelium-Derived Factor

Pigment-epithelium-derived factor (PEDF) is a member of the SERPIN gene family [44] and was first isolated from fetal retinal pigment epithelial cells [8, 45]. PEDF's actions were initially characterized as being primarily involved in neuronal differentiation [46]. However, as more information was gathered about PEDF, its role as an angiogenic inhibitor was revealed [47]. In fact, PEDF and another neurotrophic factor, vascular endothelial growth factor (VEGF), play reciprocal roles in the angiogenic process [8]. In models of oxygen-induced retinopathy (OIR) and DR, as the levels of the pro-angiogenic factor (VEGF) increase, the levels of PEDF decrease in the aqueous humor and vitreous of the eye [17, 18, 48–51]. This intricate balance between VEGF and PEDF levels is essential in maintaining the BRB integrity through prevention of vascular permeability [47, 52]. However, a reduced level of PEDF in the ischemic, nondiabetic eye has also been observed. This indicates that the reduced level of PEDF observed in DR is due to hypoxia rather than hyperglycemia [17]. In addition to its potent anti-angiogenic properties, recent findings have shown that PEDF is also an anti-inflammatory factor in the eye [53]. PEDF plays a role in inhibiting reactive oxygen species (ROS) as well as the subsequent VEGF increase that is seen in DR [47].

The effects of exogenous PEDF treatments on angiogenesis and other DR-associated symptoms have been studied. For instance, intraperitoneal administration of PEDF was shown to inhibit retinal neovascularization in a neonatal mouse exposed to hypoxic conditions [54]. A second study used an adenovirus expressing PEDF (AAV-PEDF). Intravitreal injection of AAV-PEDF inhibited both retinal and choroidal neovascularization in the mouse [8, 55]. In a third study, retinal vascular permeability and inflammatory factors were reduced in animal models of DR and OIR after intravitreal injection of PEDF [53].

SERPINA3K

SERPINA3K, a member of the SERPIN family, is a specific inhibitor of tissue kallikrein (a serine proteinase) and is often referred to as kallikrein-binding protein (KBP) [56, 57]. The kallikrein-kinin system was originally characterized to have functions in inflammation, local blood flow, and vasodilation regulation [58, 59]. As research continued on SERPINA3K, additional functions were uncovered, including its role as an anti-angiogenic factor [60].

In the STZ-induced diabetic rat model, the retinal levels of KBP are decreased, hinting at an essential role in the progression of DR [61]. In 2008, it was uncovered that SERPINA3K can function in a protective manner in both Müller and retinal neuronal cells against oxidative stress-induced damage, conditions seen in DR [62]. This protective effect occurs through blocking the intracellular calcium overload induced by oxidative stress [62].

THE DOUBLE-EDGED SWORDS: PRO-ANGIOGENIC NEUROTROPHIC FACTORS

As the knowledge increases about the anti-angiogenic neurotrophic factors in the retina, the relationship between neuronal cell protection and pro-angiogenic factors has broadened. Several pro-angiogenic factors, to be described below, have dual functions in the cell: promoting angiogenesis while promoting neuronal cell maintenance, differentiation, and development.

Brain-Derived Neurotrophic Factor

Brain-derived neurotrophic factor (BDNF) shares a similar structure to the most highly studied neurotrophic factor, NGF, as both are members of the neurotrophin gene family [63]. In the retinal tissue, BDNF targets (and is expressed in) RGCs and Müller glia [64] and has been shown to be important for the survival of RGCs and bipolar cells [11, 65, 66]. In addition, it has been shown that BDNF can prevent amacrine cell death [67, 68]. Upon the degeneration of dopaminergic amacrine cells in the retinas of STZ-induced diabetic rats, there is a reduction in the levels of BDNF in both RGCs and Müller cells [11]. BDNF's ability to bind to both the Trk and p75-type receptors facilitates its action in both retinal development and survival [69]. However, a recent study has suggested a novel pro-angiogenic role for BDNF in ischemic tissues [70].

The therapeutic potential of BDNF has been examined. Upon intraocular injection, BDNF prevented dopaminergic amacrine cell neurodegeneration [11]. In order to gather more information on BDNF and its usefulness in treating DR-associated pathological phenotypes, more studies remain to be conducted.

Fibroblast Growth Factors

Fibroblast growth factor (FGF) was first characterized as a growth and differentiation factor for mesoderm- and neuroectoderm-derived cells [71]. However, researchers have isolated two derivatives of the FGF family from the bovine retina, basic FGF (βFGF) and acidic FGF (αFGF) [71, 72]. βFGF is constitutively expressed by the RPE at considerably higher amounts than αFGF [71, 73]. During retinal ischemia and instances of proliferative DR, the retinal levels of βFGF are increased [22, 71]. In fact, it is speculated that during retinal hemorrhage, infiltrative macrophages in the vitreous may induce an enhanced secretion of βFGF [71, 74].

Although early studies on FGF had showed a link between its elevated expression and angiogenesis, now the primary function of FGF is thought to be neurotrophic and neuroprotective [22, 75]. Although βFGF has not been utilized for DR therapies, injections of βFGF into the eye of rats with either inherited retinal degeneration or ischemic injury led to a delay in the progression of degeneration [76, 77].

Insulin and Insulin-Like Growth Factor 1

Insulin and Insulin-like growth factor 1 (IGF-1) have been shown to prolong the survival of retinal neurons in culture as well as decrease apoptosis and stimulate cell proliferation, differentiation, and maturation [78–80]. In DR, increased levels of IGF-1 were observed in the vitreous of patients; IGF-2 levels do not increase [81, 82].

The use of IGF-1 has been examined as a possible therapeutic agent in the treatment of DR. Exogenous exposure of IGF-1 to cultured retinal neurons led to the enhanced survival of amacrine neurons [83]. Exposure of high levels of either insulin or IGF-2 led to the same effects [83]. Furthermore, upon depletion of these factors, there was an increase in amacrine apoptosis [83].

Erythropoietin

Erythropoietin (EPO) was initially described as a regulator of red blood cell production, or erythropoiesis, throughout the body [84, 85]. However, as the information about EPO broadened, it was found to be expressed in the retina [86]. In the retina, as well as in the brain, EPO is both a neurotrophic factor and an endothelial survival factor [22, 87]. EPO is elevated in the diabetic eye, and although it is neuroprotective in the retina, it has been shown in both in vitro and in vivo studies to stimulate angiogenesis [87]. EPO is regulated by hypoxia-inducible factor (HIF), and oxidative stress stimulates EPO production in the eye [88]. However, EPO's production is not solely dependent on the presence of oxidative stress because elevated levels of EPO were observed in cases of macular edema, a condition that is not solely dependent on hypoxic conditions [84, 86].

Intravitreal injection of EPO has been found to prevent apoptosis during early stages of DR [89]. In addition, suberythropoietic administration of EPO reduces the unnecessary side effects that can be associated with other potential EPO therapies, such as induction of angiogenesis, oxidative stress, and pericyte loss [87]. Another study explored the possibility of using siRNAs to EPO as a novel therapeutic agent for DR. Intravitreal injections of siRNA to EPO resulted in reduced levels of EPO and subsequent suppression of retinal neovascularization [90]. Although the results from the siRNA study are promising, methods to knock down EPO are risky due to its dual role as both an angiogenic stimulator and a neurotrophic factor in the retina.

Vascular Endothelial Growth Factor

VEGF is constitutively secreted by the retinal pigment epithelium [8, 91]. There are at least five different splice forms of VEGF, and each shows a differing amount of angiogenic activity [92, 93]. A combination of the isoforms can stimulate a higher degree of angiogenesis than a single isoform and provide prolonged efficacy in accelerating angiogenesis [94]. $VEGFA_{165}$ is the most commonly studied isoform and displays both neurotrophic and angiogenic properties [92].

Low levels of VEGF secretion are presumed to be responsible for its neurotrophic functions in the eye [9, 95]. However, VEGF is found prominently in the vitreous of patients with proliferative DR, pre-proliferative DR, and nondiabetics with choroidal neovascularization [9]. Under ischemic conditions and the appearance of new vessels, such as those observed in DR, VEGF levels increase [9, 96, 97]. In fact, VEGF expression is noted in Müller cells of the retina before any noticeable neovascularization has occurred in DR [9]. The induction of angiogenesis and vascular leakage that occur in DR are thought to occur when higher levels of VEGF are secreted due to the pathological conditions (e.g., ischemia) observed in DR [9]. VEGF increases vascular permeability [98] and thus has been suggested to play a role in the breakdown of the BRB, perhaps leading to diabetic macular edema [9, 99, 100].

Systemic anti-VEGF therapies have disadvantages when considered as possible therapies for patients with DR. Its dual roles as both a neurotrophic and a pro-angiogenic factor, though beneficial in some aspects, could prove detrimental in DR patients with systemic vascular problems [8]. Therefore, direct intraocular administration of a VEGF therapy is favorable. Currently, an aptamer consisting of a 28-base oligonucleotide that binds to the VEGF is in clinical trials toward the treatment of age-related macular degeneration, a condition that involves neovascularization of the choroid [101]. Another therapeutic potential is the use of ranibizumab, an antibody with high affinity to inhibit all VEGF isoforms. Clinical trials are underway to determine if this drug would be a useful therapy in the treatment of DR [102]. Regulating the expression of VEGF receptors (VEGFR-1 and 2) may be another therapeutic option. In fact, a drug that blocks VEGFR-2 has undergone initial tests as an angiogenesis inhibitor for the treatment of cancer [103], but it has not been tested as a treatment of DR.

NEUROTROPHIC FACTORS AND THE FUTURE OF DR RESEARCH

There are several additional NFs that are expressed in the retina that have yet to be studied in the diabetic retina. Although these factors may play a role in retinal development or have been shown to be modulated during retinal degeneration, they have not been studied in DR. These factors include, but are not limited to, neurturin [104, 105] and members of the neurotrophin family (NT-3, 4, 5, and 6) [106–108].

It is important to realize that characterization of retinal NFs in both the normal and DR retina would be a useful tool to further the development of therapies in the treatment and/or prevention of DR. As the incidence rates for diabetes increase and the world population's life expectancy increases, there is an ever-increasing desire to prolong quality of life. This includes delaying the progression or halting the onset of DR. A useful therapeutic tool, such as some that have been mentioned above, to promote neuronal survival through the manipulation of neurotrophic factors in the retina could be a powerful technique in order to halt or reverse the progression of the pathological manifestations of DR.

REFERENCES

1. Kolfschoten IG et al. Role and therapeutic potential of microRNAs in diabetes. Diabetes Obes Metab. 2009;11 Suppl 4:118–29.
2. Wild S et al. Global prevalence of diabetes: estimates for the year 2000 and projections for 2030. Diabetes Care. 2004;27(5):1047–53.
3. Pandey AK et al. MicroRNAs in diabetes: tiny players in big disease. Cell Physiol Biochem. 2009;23(4–6):221–32.
4. Al-Maskari F, El-Sadig M. Prevalence of risk factors for diabetic foot complications. BMC Fam Pract. 2007;8:59.
5. De Mattia G et al. Endothelial dysfunction and oxidative stress in type 1 and type 2 diabetic patients without clinical macrovascular complications. Diabetes Res Clin Pract. 2008;79(2):337–42.
6. Avogaro A et al. Incidence of coronary heart disease in type 2 diabetic men and women: impact of microvascular complications, treatment, and geographic location. Diabetes Care. 2007;30(5):1241–7.
7. Happich M et al. Cross-sectional analysis of adult diabetes type 1 and type 2 patients with diabetic microvascular complications from a German retrospective observational study. Curr Med Res Opin. 2007;23(6):1367–74.
8. Frank RN. Diabetic retinopathy. N Engl J Med. 2004;350(1):48–58.
9. Amin RH et al. Vascular endothelial growth factor is present in glial cells of the retina and optic nerve of human subjects with nonproliferative diabetic retinopathy. Invest Ophthalmol Vis Sci. 1997;38(1):36–47.
10. Lutty GA et al. Localization of vascular endothelial growth factor in human retina and choroid. Arch Ophthalmol. 1996;114(8):971–7.
11. Seki M et al. Involvement of brain-derived neurotrophic factor in early retinal neuropathy of streptozotocin-induced diabetes in rats: therapeutic potential of brain-derived neurotrophic factor for dopaminergic amacrine cells. Diabetes. 2004;53(9):2412–9.
12. Hammes HP, Federoff HJ, Brownlee M. Nerve growth factor prevents both neuroretinal programmed cell death and capillary pathology in experimental diabetes. Mol Med. 1995;1(5):527–34.
13. Park SH et al. Apoptotic death of photoreceptors in the streptozotocin-induced diabetic rat retina. Diabetologia. 2003;46(9):1260–8.
14. Asnaghi V et al. A role for the polyol pathway in the early neuroretinal apoptosis and glial changes induced by diabetes in the rat. Diabetes. 2003;52(2):506–11.
15. Reichardt LF. Neurotrophin-regulated signalling pathways. Philos Trans R Soc Lond B Biol Sci. 2006;361(1473):1545–64.
16. Bibel M, Barde YA. Neurotrophins: key regulators of cell fate and cell shape in the vertebrate nervous system. Genes Dev. 2000;14(23):2919–37.
17. Spranger J et al. Loss of the antiangiogenic pigment epithelium-derived factor in patients with angiogenic eye disease. Diabetes. 2001;50(12):2641–5.
18. Ogata N et al. Pigment epithelium-derived factor in the vitreous is low in diabetic retinopathy and high in rhegmatogenous retinal detachment. Am J Ophthalmol. 2001;132(3):378–82.
19. Feng Y et al. Vasoregression linked to neuronal damage in the rat with defect of polycystin-2. PLoS One. 2009;4(10):e7328.
20. Tanaka Y et al. Vascular endothelial growth factor in diabetic retinopathy. Lancet. 1997; 349(9064):1520.
21. Vinores SA et al. Upregulation of vascular endothelial growth factor in ischemic and nonischemic human and experimental retinal disease. Histol Histopathol. 1997;12(1):99–109.

22. Layton CJ, Becker S, Osborne NN. The effect of insulin and glucose levels on retinal glial cell activation and pigment epithelium-derived fibroblast growth factor-2. Mol Vis. 2006;12: 43–54.
23. Porte D, Sherwin RS, editors. Ellenberg and Rifkin's diabetes mellitus:theory and practice. 5th ed. Stamford: Appleton and Lange; 1997. p. 1027–8.
24. Pittenger G, Vinik A. Nerve growth factor and diabetic neuropathy. Exp Diabesity Res. 2003;4(4):271–85.
25. Carmignoto G et al. Expression of NGF receptor and NGF receptor mRNA in the developing and adult rat retina. Exp Neurol. 1991;111(3):302–11.
26. Lin LF et al. GDNF: a glial cell line-derived neurotrophic factor for midbrain dopaminergic neurons. Science. 1993;260(5111):1130–2.
27. Harada T et al. Neurotrophic factor receptors in epiretinal membranes after human diabetic retinopathy. Diabetes Care. 2002;25(6):1060–5.
28. Strelau J, Unsicker K. GDNF family members and their receptors: expression and functions in two oligodendroglial cell lines representing distinct stages of oligodendroglial development. Glia. 1999;26(4):291–301.
29. Nishikiori N et al. Glial cell line-derived neurotrophic factor in the vitreous of patients with proliferative diabetic retinopathy. Diabetes Care. 2005;28(10):2588.
30. Nishikiori N et al. Glial cell-derived cytokines attenuate the breakdown of vascular integrity in diabetic retinopathy. Diabetes. 2007;56(5):1333–40.
31. Li Y et al. CNTF induces regeneration of cone outer segments in a rat model of retinal degeneration. PLoS One. 2010;5(3):e9495.
32. Rhee KD, Yang XJ. Expression of cytokine signal transduction components in the postnatal mouse retina. Mol Vis. 2003;9:715–22.
33. Ip NY. The neurotrophins and neuropoietic cytokines: two families of growth factors acting on neural and hematopoietic cells. Ann N Y Acad Sci. 1998;840:97–106.
34. Ip NY et al. The alpha component of the CNTF receptor is required for signaling and defines potential CNTF targets in the adult and during development. Neuron. 1993;10(1): 89–102.
35. Liang FQ et al. AAV-mediated delivery of ciliary neurotrophic factor prolongs photoreceptor survival in the rhodopsin knockout mouse. Mol Ther. 2001;3(2):241–8.
36. Kirsch M et al. Evidence for multiple, local functions of ciliary neurotrophic factor (CNTF) in retinal development: expression of CNTF and its receptors and in vitro effects on target cells. J Neurochem. 1997;68(3):979–90.
37. Liu X et al. Suppressors of cytokine-signaling proteins induce insulin resistance in the retina and promote survival of retinal cells. Diabetes. 2008;57(6):1651–8.
38. Boulton TG, Stahl N, Yancopoulos GD. Ciliary neurotrophic factor/leukemia inhibitory factor/interleukin 6/oncostatin M family of cytokines induces tyrosine phosphorylation of a common set of proteins overlapping those induced by other cytokines and growth factors. J Biol Chem. 1994;269(15):11648–55.
39. Oh H et al. Activation of phosphatidylinositol 3-kinase through glycoprotein 130 induces protein kinase B and p70 S6 kinase phosphorylation in cardiac myocytes. J Biol Chem. 1998;273(16):9703–10.
40. LaVail MM et al. Protection of mouse photoreceptors by survival factors in retinal degenerations. Invest Ophthalmol Vis Sci. 1998;39(3):592–602.
41. Cayouette M et al. Intraocular gene transfer of ciliary neurotrophic factor prevents death and increases responsiveness of rod photoreceptors in the retinal degeneration slow mouse. J Neurosci. 1998;18(22):9282–93.

42. Cayouette M, Gravel C. Adenovirus-mediated gene transfer of ciliary neurotrophic factor can prevent photoreceptor degeneration in the retinal degeneration (rd) mouse. Hum Gene Ther. 1997;8(4):423–30.
43. Tao W et al. Encapsulated cell-based delivery of CNTF reduces photoreceptor degeneration in animal models of retinitis pigmentosa. Invest Ophthalmol Vis Sci. 2002;43(10):3292–8.
44. Steele FR, Chader GJ, Johnson LV, Tombran-Tink J. Pigment epithelium-derived factor: neurotrophic activity and identification as a member of the serine protease inhibitor gene family. Proc Natl Acad Sci USA. 1992;90:1526–30.
45. King GL, Suzuma K. Pigment-epithelium-derived factor – a key coordinator of retinal neuronal and vascular functions. N Engl J Med. 2000;342(5):349–51.
46. Tombran-Tink J, Chader GG, Johnson LV. PEDF: a pigment epithelium-derived factor with potent neuronal differentiative activity. Exp Eye Res. 1991;53(3):411–4.
47. Yamagishi S et al. Pigment epithelium-derived factor inhibits advanced glycation end product-induced retinal vascular hyperpermeability by blocking reactive oxygen species-mediated vascular endothelial growth factor expression. J Biol Chem. 2006;281(29): 20213–20.
48. Ogata N et al. Unbalanced vitreous levels of pigment epithelium-derived factor and vascular endothelial growth factor in diabetic retinopathy. Am J Ophthalmol. 2002;134(3):348–53.
49. Boehm BO et al. Proliferative diabetic retinopathy is associated with a low level of the natural ocular anti-angiogenic agent pigment epithelium-derived factor (PEDF) in aqueous humor. a pilot study. Horm Metab Res. 2003;35(6):382–6.
50. Boehm BO et al. Low content of the natural ocular anti-angiogenic agent pigment epithelium-derived factor (PEDF) in aqueous humor predicts progression of diabetic retinopathy. Diabetologia. 2003;46(3):394–400.
51. Gao G et al. Unbalanced expression of VEGF and PEDF in ischemia-induced retinal neovascularization. FEBS Lett. 2001;489(2–3):270–6.
52. Zhang SX et al. Pigment epithelium-derived factor downregulates vascular endothelial growth factor (VEGF) expression and inhibits VEGF-VEGF receptor 2 binding in diabetic retinopathy. J Mol Endocrinol. 2006;37(1):1–12.
53. Zhang SX et al. Pigment epithelium-derived factor (PEDF) is an endogenous antiinflammatory factor. FASEB J. 2006;20(2):323–5.
54. Stellmach V et al. Prevention of ischemia-induced retinopathy by the natural ocular antiangiogenic agent pigment epithelium-derived factor. Proc Natl Acad Sci U S A. 2001;98(5): 2593–7.
55. Mori K et al. Pigment epithelium-derived factor inhibits retinal and choroidal neovascularization. J Cell Physiol. 2001;188(2):253–63.
56. Chao J et al. Tissue kallikrein-binding protein is a serpin. I. Purification, characterization, and distribution in normotensive and spontaneously hypertensive rats. J Biol Chem. 1990; 265(27):16394–401.
57. Chao J et al. Identification of a new tissue-kallikrein-binding protein. Biochem J. 1986; 239(2):325–31.
58. Ma JX et al. Intramuscular delivery of rat kallikrein-binding protein gene reverses hypotension in transgenic mice expressing human tissue kallikrein. J Biol Chem. 1995;270(1):451–5.
59. Bhoola KD, Figueroa CD, Worthy K. Bioregulation of kinins: kallikreins, kininogens, and kininases. Pharmacol Rev. 1992;44(1):1–80.
60. Gao G et al. Kallikrein-binding protein inhibits retinal neovascularization and decreases vascular leakage. Diabetologia. 2003;46(5):689–98.
61. Hatcher HC et al. Kallikrein-binding protein levels are reduced in the retinas of streptozotocin-induced diabetic rats. Invest Ophthalmol Vis Sci. 1997;38(3):658–64.

62. Zhang B, Ma JX. SERPINA3K prevents oxidative stress induced necrotic cell death by inhibiting calcium overload. PLoS One. 2008;3(12):e4077.
63. Holtzman DM, Mobley WC. Neurotrophic factors and neurologic disease. West J Med. 1994;161(3):246–54.
64. Seki M et al. BDNF is upregulated by postnatal development and visual experience: quantitative and immunohistochemical analyses of BDNF in the rat retina. Invest Ophthalmol Vis Sci. 2003;44(7):3211–8.
65. Johnson JE et al. Brain-derived neurotrophic factor supports the survival of cultured rat retinal ganglion cells. J Neurosci. 1986;6(10):3031–8.
66. Kano T et al. Protective effect against ischemia and light damage of iris pigment epithelial cells transfected with the BDNF gene. Invest Ophthalmol Vis Sci. 2002;43(12): 3744–53.
67. Cusato K et al. Cell death in the inner nuclear layer of the retina is modulated by BDNF. Brain Res Dev Brain Res. 2002;139(2):325–30.
68. Kido N et al. Neuroprotective effects of brain-derived neurotrophic factor in eyes with NMDA-induced neuronal death. Brain Res. 2000;884(1–2):59–67.
69. Cellerino A et al. Brain-derived neurotrophic factor modulates the development of the dopaminergic network in the rodent retina. J Neurosci. 1998;18(9):3351–62.
70. Kermani P, Hempstead B. Brain-derived neurotrophic factor: a newly described mediator of angiogenesis. Trends Cardiovasc Med. 2007;17(4):140–3.
71. Sivalingam A et al. Basic fibroblast growth factor levels in the vitreous of patients with proliferative diabetic retinopathy. Arch Ophthalmol. 1990;108(6):869–72.
72. Courty J et al. Bovine retina contains three growth factor activities with different affinity to heparin: eye derived growth factor I, II, III. Biochimie. 1985;67(2):265–9.
73. Baird A et al. Retina- and eye-derived endothelial cell growth factors: partial molecular characterization and identity with acidic and basic fibroblast growth factors. Biochemistry. 1985;24(27):7855–60.
74. Baird A, Mormede P, Bohlen P. Immunoreactive fibroblast growth factor in cells of peritoneal exudate suggests its identity with macrophage-derived growth factor. Biochem Biophys Res Commun. 1985;126(1):358–64.
75. Ozaki H et al. Basic fibroblast growth factor is neither necessary nor sufficient for the development of retinal neovascularization. Am J Pathol. 1998;153(3):757–65.
76. Faktorovich EG et al. Photoreceptor degeneration in inherited retinal dystrophy delayed by basic fibroblast growth factor. Nature. 1990;347(6288):83–6.
77. Unoki K, LaVail MM. Protection of the rat retina from ischemic injury by brain-derived neurotrophic factor, ciliary neurotrophic factor, and basic fibroblast growth factor. Invest Ophthalmol Vis Sci. 1994;35(3):907–15.
78. Frade JM et al. Insulin-like growth factor-I stimulates neurogenesis in chick retina by regulating expression of the alpha 6 integrin subunit. Development. 1996;122(8): 2497–506.
79. Kermer P et al. Insulin-like growth factor-I protects axotomized rat retinal ganglion cells from secondary death via PI3-K-dependent Akt phosphorylation and inhibition of caspase-3 In vivo. J Neurosci. 2000;20(2):2–8.
80. Hernandez-Sanchez C et al. Autocrine/paracrine role of insulin-related growth factors in neurogenesis: local expression and effects on cell proliferation and differentiation in retina. Proc Natl Acad Sci U S A. 1995;92(21):9834–8.
81. Grant M et al. Insulin-like growth factors in vitreous. Studies in control and diabetic subjects with neovascularization. Diabetes. 1986;35(4):416–20.
82. Merimee TJ. Diabetic retinopathy. A synthesis of perspectives. N Engl J Med. 1990;322(14): 978–83.

83. Politi LE et al. Insulin-like growth factor-I is a potential trophic factor for amacrine cells. J Neurochem. 2001;76(4):1199–211.
84. Garcia-Ramirez M, Hernandez C, Simo R. Expression of erythropoietin and its receptor in the human retina: a comparative study of diabetic and nondiabetic subjects. Diabetes Care. 2008;31(6):1189–94.
85. Fisher JW. Erythropoietin: physiology and pharmacology update. Exp Biol Med (Maywood). 2003;228(1):1–14.
86. Hernandez C et al. Erythropoietin is expressed in the human retina and it is highly elevated in the vitreous fluid of patients with diabetic macular edema. Diabetes Care. 2006;29(9): 2028–33.
87. Wang Q et al. Low-dose erythropoietin inhibits oxidative stress and early vascular changes in the experimental diabetic retina. Diabetologia. 2010;53(6):1227–38.
88. Grimm C et al. HIF-1-induced erythropoietin in the hypoxic retina protects against light-induced retinal degeneration. Nat Med. 2002;8(7):718–24.
89. Zhang J et al. Intravitreal injection of erythropoietin protects both retinal vascular and neuronal cells in early diabetes. Invest Ophthalmol Vis Sci. 2008;49(2):732–42.
90. Chen J et al. Suppression of retinal neovascularization by erythropoietin siRNA in a mouse model of proliferative retinopathy. Invest Ophthalmol Vis Sci. 2009;50(3):1329–35.
91. Blaauwgeers HG et al. Polarized vascular endothelial growth factor secretion by human retinal pigment epithelium and localization of vascular endothelial growth factor receptors on the inner choriocapillaris. Evidence for a trophic paracrine relation. Am J Pathol. 1999;155(2):421–8.
92. Sakowski SA et al. Neuroprotection using gene therapy to induce vascular endothelial growth factor-A expression. Gene Ther. 2009;16(11):1292–9.
93. Falk T, Zhang S, Sherman SJ. Vascular endothelial growth factor B (VEGF-B) is up-regulated and exogenous VEGF-B is neuroprotective in a culture model of Parkinson's disease. Mol Neurodegener. 2009;4:49.
94. Whitlock PR et al. Adenovirus-mediated transfer of a minigene expressing multiple isoforms of VEGF is more effective at inducing angiogenesis than comparable vectors expressing individual VEGF cDNAs. Mol Ther. 2004;9(1):67–75.
95. Adamis AP et al. Synthesis and secretion of vascular permeability factor/vascular endothelial growth factor by human retinal pigment epithelial cells. Biochem Biophys Res Commun. 1993;193(2):631–8.
96. Aiello LP et al. Hypoxic regulation of vascular endothelial growth factor in retinal cells. Arch Ophthalmol. 1995;113(12):1538–44.
97. Shweiki D et al. Vascular endothelial growth factor induced by hypoxia may mediate hypoxia-initiated angiogenesis. Nature. 1992;359(6398):843–5.
98. Dvorak HF et al. Vascular permeability factor/vascular endothelial growth factor, microvascular hyperpermeability, and angiogenesis. Am J Pathol. 1995;146(5):1029–39.
99. Vinores SA et al. Blood-ocular barrier breakdown in eyes with ocular melanoma. A potential role for vascular endothelial growth factor/vascular permeability factor. Am J Pathol. 1995;147(5):1289–97.
100. Murata T et al. The relation between expression of vascular endothelial growth factor and breakdown of the blood-retinal barrier in diabetic rat retinas. Lab Invest. 1996;74(4): 819–25.
101. Barakat MR, Kaiser PK. VEGF inhibitors for the treatment of neovascular age-related macular degeneration. Expert Opin Investig Drugs. 2009;18(5):637–46.
102. Rodriguez-Fontal M et al. Ranibizumab for diabetic retinopathy. Curr Diabetes Rev. 2009;5(1):47–51.

103. Rosen LS. Clinical experience with angiogenesis signaling inhibitors: focus on vascular endothelial growth factor (VEGF) blockers. Cancer Control. 2002;9(2 Suppl):36–44.
104. Kotzbauer PT et al. Neurturin, a relative of glial-cell-line-derived neurotrophic factor. Nature. 1996;384(6608):467–70.
105. Jomary C et al. Expression patterns of neurturin and its receptor components in developing and degenerative mouse retina. Invest Ophthalmol Vis Sci. 1999;40(3):568–74.
106. Oku H et al. Gene expression of neurotrophins and their high-affinity Trk receptors in cultured human Muller cells. Ophthalmic Res. 2002;34(1):38–42.
107. Lavail MM et al. Sustained delivery of NT-3 from lens fiber cells in transgenic mice reveals specificity of neuroprotection in retinal degenerations. J Comp Neurol. 2008;511(6):724–35.
108. Liu X et al. Regulation of neonatal development of retinal ganglion cell dendrites by neurotrophin-3 overexpression. J Comp Neurol. 2009;514(5):449–58.

16
The Role of CTGF in Diabetic Retinopathy

R.J. van Geest, E.J. Kuiper, I. Klaassen,
C.J.F. van Noorden, and R.O. Schlingemann

Contents

Keywords CTGF • VEGF • TGF-β • Preclinical diabetic retinopathy • Proliferative diabetic retinopathy • Basal lamina (thickening) • Basement membrane • ECM (remodeling) • Wound healing • AGEs • Angiofibrotic switch

INTRODUCTION

Diabetic retinopathy (DR) is a leading cause of ocular morbidity [1]. The pathogenesis of DR is only partly understood. Before development of clinical signs, it involves a complex sequence of events, a phase called preclinical DR (PCDR), finally leading to retinal vascular occlusion and ischemia. This causes the clinical manifestations of the disease: vision-threatening vascular leakage and macular edema, and preretinal neovascularization [2, 3]. The latter condition, proliferative DR (PDR), is essentially a wound healing-like response.

In PCDR, the capillary basal lamina (BL) thickens, along with pericyte and endothelial cell apoptosis, and diffusely increased vascular permeability [2, 4]. In experimental rodent models of DR, retinal VEGF and VEGF-receptor (VEGF-R) mRNA are increased in this stage, suggesting an early role of VEGF in DR, possibly as a result of high glucose levels, advanced glycation end products (AGEs), and/or other factors altered by

From: *Ophthalmology Research: Visual Dysfunction in Diabetes*
Edited by: J. Tombran-Tink et al. (eds.), DOI 10.1007/978-1-60761-150-9_16

the diabetic milieu [2, 3]. Although the exact sequence of events and the relative importance of the early changes are poorly understood, capillary BL thickening is a hallmark of early DR and may be causal in endothelial and pericyte dysfunction. BL thickening is the result of extracellular matrix (ECM) remodeling, resulting in increased deposition of BL components such as collagen type IV, laminin, and fibronectin.

In addition, tissue fibrosis plays a role in PDR. Uncontrolled retinal neovascularization is followed by fibrosis, scarring, tractional retinal detachment, and blindness. PDR patients with established neovascularization and imminent fibrosis find themselves in a situation associated with poor prognosis, despite aggressive laser treatment or surgical procedures. The major mediator of vascular leakage and angiogenesis in PDR is VEGF-A, which is overexpressed in ischemic retina [2, 3, 5, 6]. However, little is known about growth factors that are involved in the subsequent fibrotic phase in PDR.

Connective tissue growth factor (CTGF) is a candidate for contributing to the fibrotic responses observed in both PCDR and PDR. CTGF acts as a mitogen for fibroblasts and induces increased ECM production [7–11]. CTGF functions as a downstream mediator of transforming growth factor (TGF)-β signaling in certain cell types and seems essential for effectuation of the profibrotic actions of TGF-β, such as ECM production [12].

In long-standing diabetes, structural and functional ECM alterations, including BL thickening, lead to microvascular diabetic complications, such as DR, nephropathy, cardiomyopathy, peripheral vascular disease, cerebrovascular disorders, and atherosclerosis [13, 14]. CTGF is involved in these diabetic microvascular complications [15–17]. In diabetic nephropathy, CTGF is strongly overexpressed in the kidney glomerulus [16, 18], and its levels in urine and plasma correlate with progression of the disease [19, 20].

Similarly to its involvement in fibrosis in diabetic nephropathy [7, 9, 10, 21], CTGF may have a causal role in capillary BL thickening in PCDR and in fibrosis in PDR. This chapter discusses the roles of CTGF in the pathogenesis of DR in relation to ECM remodeling and wound healing mechanisms, and explores whether CTGF is be a novel therapeutic target in the clinical management of early as well as late stages of DR.

ECM REMODELING AND WOUND HEALING MECHANISMS IN DIABETIC RETINOPATHY

ECM Remodeling in PCDR

Normal constituents of the retinal capillary BL are collagen type IV, which is the predominant component, laminin, and fibronectin. The latter is mainly located at pericyte-endothelial cell contacts [22, 23]. Changes in the BL in diabetes are considered to result from a disturbed balance between synthesis and degradation of these matrix components [4, 24, 25]. The thickening BL may consist of matrix proteins that are normally found in the ECM, or proteins that are not present in the BL under physiological conditions, or both [4, 22, 26–28].

ECM alterations including BL thickening are hallmarks in all target organs affected by diabetes [29] and are directly related to loss of function of these organs [30]. It is considered to represent a pathological response to prolonged hyperglycemia, which is the major factor associated with the onset of microvascular complications, as has been

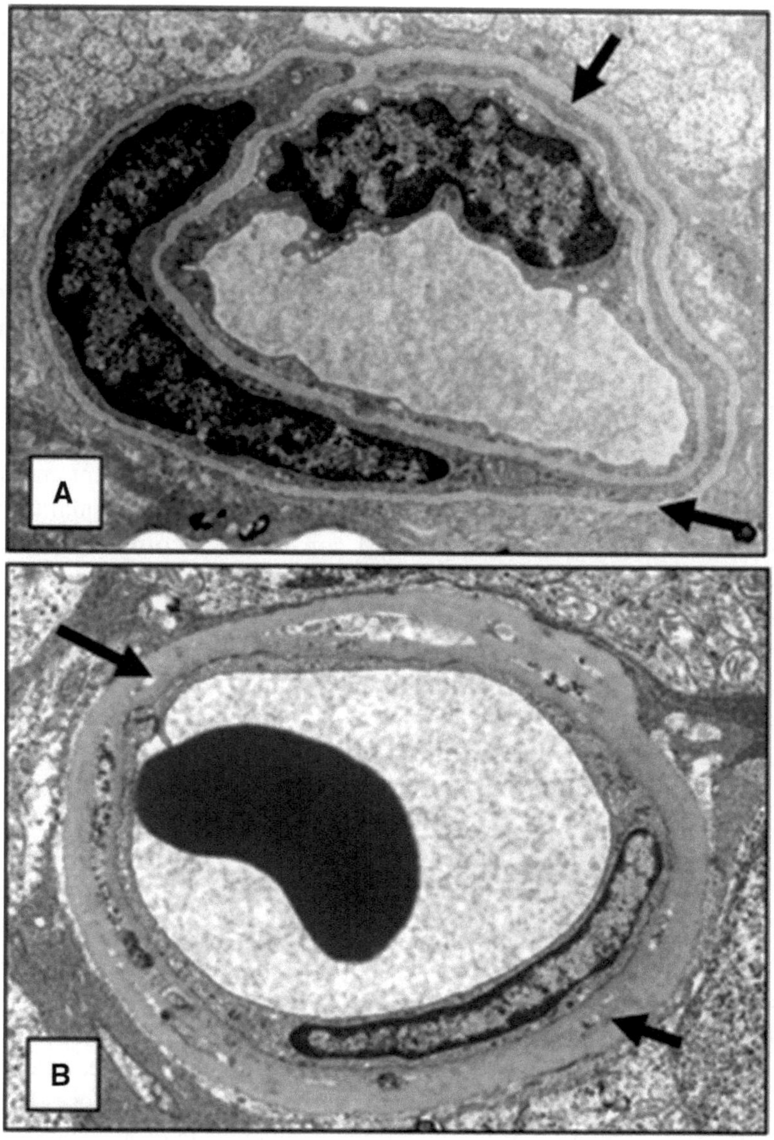

Fig. 1. Vascular basal lamina (BL) thickening in diabetic retinal capillaries. Electron microscopy of rat retina shows that retinal capillary BL (*arrows*) is susceptible to thickening after 12 months diabetes (compare nondiabetic (**A**) with diabetic (**B**)). (Reprinted by permission from Macmillan Publishers Ltd: Eye [34]Copyright (2009)).

shown in prospective studies for both type I and type II diabetes (DCCT and UKPDS, respectively) [31, 32].

The early thickening of the retinal capillary BL in diabetes was recognized already 60 years ago [33]. In the following decades, numerous electron microscopic studies have demonstrated increased thickness of the BL in diabetic humans and animals (Fig. 1) [34–36]. The only clear structural retinal change after 1–3 years of diabetes was capillary BL thickening in a canine model of diabetes, which eventually displays all early features of diabetic retinal microvascular damage – BL thickening, loss of pericytes, the formation of microaneurysms, and capillary closure [37]. In this model, widespread loss of pericytes was noted only after 4 years of diabetes. BL thickening is likely to be instrumental in progression of early DR [38]. In an experimental model of diabetes in galactose-fed rats, downregulation of synthesis of the BL component fibronectin with the use of antisense oligos not only downregulated retinal BL thickening at least partly but also reduced other more advanced features of PCDR, such as apoptosis of pericytes and endothelial cells, as well as the development of acellular capillaries. In another study, concomitant downregulation of fibronectin, laminin, and collagen type IV expression with the use of antisense oligos injected in eyes of rats with streptozotocin-induced diabetes reduced vascular leakage [23]. These findings indicate that BL thickening is a crucial step in the progression of DR.

Thickening of the BL, often incorrectly referred to as "basement membrane," is characterized by accumulation of ECM components, as well as a qualitative change in ECM composition [30]. Structurally, ECM consists of a complex network of collagens, elastins, structural glycoproteins, and proteoglycan–hyaluronans, and differs quantitatively and qualitatively in the various tissues. ECM provides mechanical support to cells in tissues, is involved in differentiation of cells, and regulates interactions between (vascular) cells and the ECM itself [39]. Under physiological circumstances, ECM continuously undergoes remodeling by synthesis and degradation, with a balanced turnover. ECM remodeling is required for maintaining the normal structure and function of tissues [40]. Multiple growth factors are involved in the induction of ECM synthesis, such as TGF-β [41], CTGF [42], insulin-like growth factor I (IGF-I), fibroblast growth factor (FGF), epidermal growth factor (EGF), and platelet-derived growth factor (PDGF) [30]. ECM degradation and remodeling is regulated by proteases such as the matrix metalloproteinases (MMPs) [43] and serine proteases [44], as well as their respective inhibitors, the tissue inhibitors of metalloproteinases (TIMPs) [45] and plasminogen activator inhibitor-I (PAI-1) [43].

Different metabolic mechanisms underlie the hyperglycemia-induced changes in expression of growth factors and ECM turnover [35, 46]. Hyperglycemia affects various biochemical pathways, among which increased formation of glucose-derived AGEs [24, 47]. In DR, AGE formation has a causal role in changes in growth factor expression and ECM turnover [25, 48]. AGE formation also increases synthesis of BL components, most likely via upregulation of TGF-β signaling and its downstream effectors, including CTGF [49].

Wound Healing Mechanisms in PDR

The early features of retinal microvascular diabetic damage such as BL thickening eventually lead to retinal vascular occlusion and ischemia. In response, new vessels develop from the preexisting retinal vasculature, which hallmarks the progression of clinical DR from nonproliferative to proliferative disease. VEGF is the major mediator of this hypoxia-driven neovascularization [2, 3, 5, 6] and may act in concert with other angiogenic factors, such as angiopoietin-2 (Ang-2) [50]. The initial step in retinal neovascularization is degradation of the BL and ECM components surrounding the vascular cells, followed by invasion, migration, and proliferation of endothelial cells, and finally the formation of new vascular tubes [51]. Remodeling of the ECM in angiogenesis is exerted by MMPs, which are induced by angiogenic stimuli such as VEGF and Ang-2 [52].

Neovascularization and the switch to subsequent fibrosis in PDR can be considered as a wound healing-like response [53]. Fibrosis is the deposition and cross-linking of collagen in the terminal phase of the normal wound healing response [54, 55], which has mainly been studied in the skin. Wound healing in the skin is initiated by tissue injury [56–58], which involves vascular damage, hemorrhage, and activation of the clotting system. The subsequent response can be divided into three phases: an inflammatory phase, a proliferative phase, and a maturation phase [56].

During the inflammatory phase, angiogenic and profibrotic cytokines and growth factors are released from activated cells, such as platelets and macrophages. In the proliferation phase, fibroblasts contribute to the synthesis of the ECM [59], and endothelial cells form "sprouts" and new capillaries. Sprouting angiogenesis is initiated by the presence of a fibrin matrix and growth factors at the wound healing edge [60]. Besides ECM components, fibroblasts also produce growth factors and various enzymes such as proteases which are of importance for reepithelialization and angiogenesis. During the wound healing response, the ECM itself serves as a reservoir for growth factors, thereby regulating their activity and presentation to receptors. In the proliferation phase, formation of ECM, angiogenesis, and reepithelialization take place [56].

In the maturation phase, angiogenesis ceases whereas the production of ECM continues [56]. Under normal conditions, after this switch from angiogenesis to fibrosis, ECM production ceases when sufficient quantities of collagen have been synthesized [56, 61–63]. Then, remodeling of the newly formed ECM reduces the wound thickness and increases the strength of the regenerating tissue. This breakdown of collagen is tightly regulated by a balance between proteases such as MMPs and their endogenous inhibitors such as TIMPs [64, 65].

Most features of the wound healing response in human skin can also be recognized in pathological wound healing responses characterizing various disease states in other organs. These pathological conditions have in common that tissue-specific wound healing responses are initiated, but that the wound healing process is not properly terminated, leading to pathological fibrosis [54, 66]. This is a situation in which normal scarring progresses to excessive production, limited degradation, altered deposition, and/or contraction of the ECM, probably due to an imbalance between pro- and antifibrotic factors causing a profibrotic state.

Several eye conditions lead to blindness by the involvement of wound healing-like responses culminating in scarring or excessive fibrosis (see Section on "CTGF in the Eye"). Although the initial wound healing response may have a functional meaning in restoring ocular integrity, it also results in loss of visual function and is therefore deemed to be pathological [67, 68].

CTGF STRUCTURE AND FUNCTION

CTGF is a member of the CCN family of growth factors, named after the first three members identified, *C*yr61 (CCN1), *C*TGF (CCN2), and *N*ov (CCN3), but also includes CCN4 (WISP-1), CCN5 (WISP-2), and CCN6 (WISP-3) as well [69–71]. CTGF exhibits a unique domain structure, made up of five modules including a signal peptide,

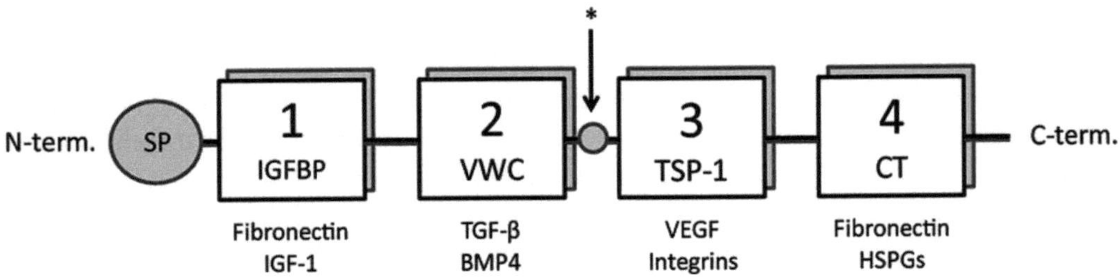

Fig. 2. Modular structure of the CTGF protein. CTGF consists of an N-terminal secretory signaling peptide (SP) and four distinct domains, through which CTGF binds extracellular ligands like VEGF, TGF-β, and fibronectin, and cell surface proteins like integrins and heparin-sulfate proteoglycans. (*Asterisks*) Hinge region. CTGF can be cleaved by proteases, such as MMPs, in between the domains. Cleavage products can accumulate in biological fluids and may serve as clinical markers.

encoded by five exons (Fig. 2) [71]. CTGF exerts its biological activities by interactions with ECM components, such as fibronectin, extracellular signaling molecules, and cell surface proteins, such as integrins, through its various interaction domains [70, 72–76]. Most likely, CTGF also indirectly regulates signaling by modulating the activity of other growth factors [77, 78]. For instance, binding of CTGF and VEGF suppresses VEGF-induced angiogenesis, and cleavage of CTGF by MMPs recovers the angiogenic activity of VEGF [79].

The biological functions of CTGF are diverse and cell and context dependent. CTGF was first discovered in conditioned media of endothelial cells as a molecule affecting the activity of fibroblasts [80]. CTGF is induced during wound healing [81], is overexpressed in fibrosis [82, 83], and acts as an essential downstream mediator for most of the profibrotic activity of TGF-β, in particular in stimulation of ECM production [66], and fibroblast proliferation [84–86]. The synergy between CTGF and TGF-β1 may be explained by binding of the unique TGF-β response element of CTGF, which enhances receptor binding and signaling activity of TGF-β (Fig. 2). For example, skin fibrosis in newborn mice was persistent only after coinjection of both TGF-β1 and CTGF, and not after injection of TGF-β1 or CTGF alone [87, 88]. In humans, CTGF is upregulated in diseases that are characterized by pathological fibrosis including renal diseases of various etiology, liver, lung, cardiovascular diseases, and in the eye.

Biological functions of CTGF include induction of angiogenesis, chondrogenesis, osteogenesis, and control of cell proliferation and differentiation, migration, adhesion, apoptosis, and survival of fibroblasts [10, 89], but the exact function of CTGF in normal tissues is not known yet; CTGF is expressed in the placenta during embryo implantation [90] and during the development of ovarian follicles [91]. Recently, a role CTGF was suggested in (nonfibrotic) tissue repair in the eye, as it was required for reepithelialization in human cornea [92].

CTGF IN THE EYE

CTGF in Ocular Fibrosis

It has been suggested that CTGF functions in the eye primarily as a profibrotic growth factor. In the human eye, CTGF has been identified in various diseases complicated by fibrosis, both in the anterior and posterior segments [93–100]. Several major eye conditions lead to blindness due to scarring or pathological fibrosis [101] as a consequence of tissue-specific wound healing responses. In subretinal neovascularization as well as in PDR and other ischemic retinopathies, these responses are driven by neovascularization, like in skin wound healing. In other conditions such as proliferative vitreoretinopathy (PVR), these responses are mainly avascular. There is accumulating evidence that CTGF is an important pathogenic factor in these conditions. For instance, in the vitreous of patients with PVR, CTGF is present in higher levels as compared to nonproliferative retinal diseases [101, 102], in correlation with TGF-β [103]. In human PVR membranes, CTGF has been identified as well [104–106].

CTGF in Ocular Angiogenesis

CTGF has been suggested to play a role in ocular angiogenesis. In the rat eye, corneal micropocket implants containing murine CTGF induced neovascularization [7]. Moreover, CTGF and VEGF colocalized in vascular cells in human choroidal neovascular membranes, and levels of CTGF were increased in the vitreous of patients with active PDR [107]. However, VEGF-induced angiogenesis was inhibited by combined exogenous administration of CTGF and VEGF in the back of mice, as well as in a mouse model of hindlimb ischemia, as a result of binding of VEGF by CTGF [108, 109]. When CTGF is upregulated by VEGF [11], it can reduce the bioavailability of VEGF through direct binding. The involvement of CTGF in angiogenesis in ocular disease in general and in DR in particular is also questionable because of findings in human PDR and in distinct angiogenesis models applied to CTGF transgenic mice [101, 110, 111]. In human PDR, CTGF levels consistently correlated with degree of fibrosis and not with angiogenesis activity [101, 111]. In studies in transgenic mice lacking the CTGF gene, vascular outgrowth from metatarsals of 17-day-old $CTGF^{-/-}$ embryos, cultured in the presence or absence of VEGF, did not differ significantly from outgrowth of wild-type or heterozygous $CTGF^{+/-}$ metatarsals [110]. These data indicate that CTGF is not required for (VEGF-induced) angiogenesis in this model. Secondly, the effect of CTGF gene deletion was investigated in two ocular angiogenesis models. In the oxygen-induced retinopathy model [112], in which retinal hypoxia-induced VEGF overexpression causes preretinal angiogenesis, differences between $CTGF^{+/+}$ and $CTGF^{+/-}$ mice were not observed. In another ocular angiogenesis model, choroidal neovascularization was induced in $CTGF^{+/+}$ and $CTGF^{+/-}$ mice by laser burns [113, 114], but statistical differences between $CTGF^{+/+}$ and $CTGF^{+/-}$ mice were not found [110]. Taken together, these data suggest that CTGF is a dispensable factor in the complex interplay of hypoxic signaling and VEGF- or wound healing-driven ocular angiogenesis.

CTGF IN DIABETIC RETINOPATHY

In various organs other than the eye, CTGF (in combination with TGF-β) is considered to cause ECM accumulation and fibrosis as a consequence of diabetic pathology [30]. This is based on experimental diabetic models, where CTGF mRNA and protein were found to be upregulated in kidney, heart, and liver [115]. TGF-β1 is generally considered to be the main profibrotic factor in diabetic nephropathy [75], with CTGF as an important downstream mediator. Diabetes-induced thickening of glomerular BL in mouse kidney, analogous to BL thickening of retinal capillaries, was shown to be diminished in CTGF-deficient mice [115]. CTGF expression is not only induced by TGF-β but also by high glucose levels, AGEs, RAAS, TNF-α, mechanical stress, and CTGF itself [15–17, 49, 116–118]. There is increasing evidence confirming this role of CTGF in diabetic nephropathy. In diabetic patients, glomerular CTGF mRNA levels were upregulated, both in patients with microalbuminuria as well as in overt nephropathy [18]. Moreover, CTGF mRNA levels correlated with the degree of albuminuria [119]. In a baboon model of type I diabetes, expression levels of tubular CTGF protein after 5 years predicted albuminuria after 10 years [120]. Accordingly, in human diabetic patients, CTGF levels in urine [19] and plasma [20] correlated with progression of diabetic nephropathy.

The role of CTGF in the development of DR was less clear. However, recent evidence suggests that CTGF is involved in both the early stages and in the late proliferative stage of DR.

CTGF in BL Thickening in PCDR

In the light of its known role in matrix remodeling in other diabetic microvascular complications, CTGF is a candidate causal factor in diabetic BL thickening in the human retina. We studied CTGF expression in a series of 36 diabetic patients and 18 nondiabetic controls [121]. Immunohistochemical staining with a highly-specific antibody against CTGF revealed a distinct and specific cellular cytoplasmic staining in the retina, suggesting local cellular expression of the CTGF protein. In the normal human retina, CTGF staining was present in paravascular microglia. However, in the retina of diabetic subjects, microglial staining was significantly decreased whereas expression of CTGF in microvascular pericytes was significantly increased. Therefore, two main patterns of CTGF expression can be distinguished: either predominant staining of microglia or predominant staining of pericytes. The predominant pericyte staining correlated almost exclusively with the presence of diabetes. The constitutive expression in paravascular microglia in the normal retina suggests a role in retinal microvascular physiology. In the light of known functions of CTGF in other cells and tissues, it is tempting to speculate that microglia-derived CTGF is involved in retinal matrix or vascular BL homeostasis in normal conditions. However, Abu El-Asrar et al. [104] did not find immunostaining of CTGF in the nondiabetic retina, whereas the diabetic retina showed CTGF staining in ganglion cells, cells in the inner nuclear layer, and in cells identified as microglia, in agreement with the study by Kuiper et al. The difference in the two studies may be explained by the different antibodies used. It was also investigated whether altered CTGF expression in diabetes was associated with established DR [121].

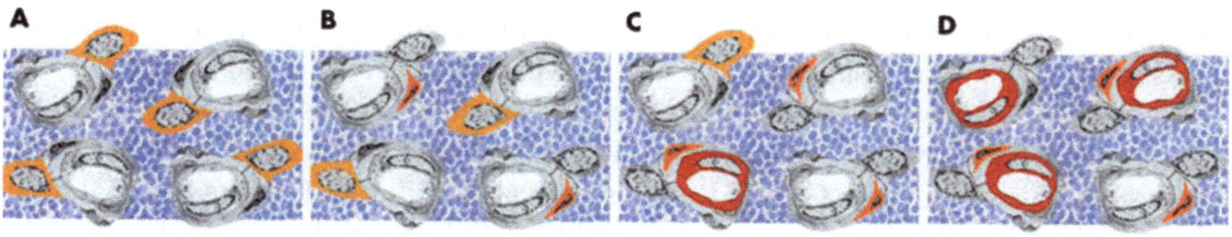

Fig. 3. Model of CTGF expression patterns during the development of DR. Progressive degrees of nonproliferative DR are indicated by an increase in PAL-E-positive endothelial cells (*red*). (**A**) Control subject without PAL-E staining, showing CTGF-positive microglia only (*yellow*). (**B–D**) Diabetic subjects with or without PAL-E staining, showing decreased CTGF-positive microglia (*yellow*) and increased CTGF-positive pericytes (*orange*). (Reproduced from [121] with permission from BMJ Publishing Group Ltd.).

Staining with the use of the endothelium-specific monoclonal antibody PAL-E recognizing plasmalemma vesicle-associated protein (PLVAP), a marker associated with local vascular leakage [122, 123], revealed no correlation with CTGF staining patterns in pericytes or microglia. In fact, CTGF seemed to be evenly distributed in diabetes, irrespective of PAL-E staining (Fig. 3). Apparently, CTGF expression patterns in pericytes of the diabetic retina are not related to clinical DR, but rather are associated with preclinical changes in the retina in diabetes. Increased pericyte CTGF expression may be related to BL thickening and/or pericyte apoptosis, both important early events in PCDR.

AGEs and CTGF in BL Thickening in PCDR

One of the proposed mechanisms of BL thickening in PCDR is the formation of AGEs. Treatment of diabetic rats with the AGE-inhibitor aminoguanidine markedly reduced AGE formation in the retinal vasculature, but also protected against retinal capillary BL thickening [46]. AGEs can also induce synthesis of ECM in diabetic rat kidney [117]. A similar induction of ECM synthesis is mediated by CTGF, both in diabetic kidney [115] and retina [124]. In the diabetic rat kidney, AGEs induce expression of fibronectin and collagen type IV, possibly partly through CTGF [125, 126]. Furthermore, AGEs induced CTGF expression in cultured retinal vascular cells [125].

Therefore, it seems likely that AGE-induced BL thickening in the retina is mediated by CTGF. We recently investigated the levels of CTGF and ECM-related molecules in both the STZ-induced diabetic rat retina, treated with or without aminoguanidine, and in the retina of mice infused with AGEs [49]. In rats, STZ treatment resulted in a significant increase in carboxy-methyl-lysine (CML) plasma levels, a marker for AGE formation, at 6 and 12 weeks of diabetes. Aminoguanidine treatment had no effect on CML levels at 6 weeks, but decreased CML levels by 25% after 12 weeks. At this time point, retinal CTGF mRNA levels were elevated twofold in diabetic rats compared to nondiabetic controls, but treatment with aminoguanidine almost completely prevented this increase. Similarly, CTGF protein levels were increased in the retina of diabetic rats, and aminoguanidine prevented this effect. Other ECM components, such as collagen type IV and TIMP-1, also showed elevated mRNA levels after 6 or 12 weeks of diabetes, which were significantly reduced by aminoguanidine treatment. TGF-β and fibronectin levels in the retina were unaffected at 6 and 12 weeks of diabetes in this model.

Retinal mRNA analysis in mice that received exogenous AGEs daily for 7 consecutive days revealed a twofold increase in CTGF levels as compared with control mice. Expression levels of Cyr61 (CCN1) were also elevated in the AGE-treated animals, but other CCN family members were not affected [49]. Taken together, these data present evidence that AGEs are both necessary and sufficient to cause increased levels of CTGF in the diabetic retina, concomitantly with ECM-related molecules [49]. Therefore, CTGF, and possibly Cyr61 as well, may have a role in thickening of the BL.

Another crucial feature of PCDR is loss of retinal capillary pericytes. Pericytes maintain capillary structure and integrity and regulate homeostasis of the endothelium. Cultured rat retinal pericytes exposed to AGEs expressed increased levels of CTGF [125]. In these cells, AGEs induced anoikis, a form of apoptosis caused by loss of cell–matrix interactions. Likewise, overexpression of CTGF promoted detachment and anoikis of retinal pericytes. The authors suggested that accumulation of CTGF in the retinal capillaries at the onset of diabetes may alter vascular structure and organization and have a role in pericyte apoptosis in PCDR.

Role of VEGF in BL Thickening

VEGF, a potent vascular permeability and angiogenic factor in PDR, is also increased early in PCDR [6, 127, 128]. Neutralizing VEGF with an antibody partly prevented diabetes-induced BL thickening in the retina of obese type 2 diabetic mice [129]. To test whether VEGF itself is capable to induce expression of genes that contribute to BL thickening in PCDR, we investigated the effect of VEGF injected in the vitreous of rat eyes on the retinal expression of CTGF, other CCN family members, TGF-β, and ECM-related molecules [26]. Adult Wistar rats were injected intravitreously with recombinant rat $VEGF_{164}$ in one eye and with solvent only in the contralateral eye. Retinal gene expression and protein levels were examined at various time points afterwards. At 24 h after injection, CTGF mRNA expression showed a 2.3-fold increase. TGF-β1 mRNA, but not TGF-β2 mRNA, was also induced significantly at 24 h after injection. Of the ECM-related molecules examined, fibronectin and TIMP-1 were significantly upregulated at 24 h. TIMP-2, collagen type IV, and laminin B1 mRNA levels were unaffected by VEGF. At the protein level, CTGF and fibronectin were clearly increased at 48 h after injection in VEGF-injected eyes. TGF-β and fibronectin immunostaining in retinal sections was more intense in the microvasculature in VEGF-injected eyes as compared to PBS-injected and noninjected eyes. VEGF stimulation in bovine retinal endothelial cells (BRECs) resulted in an early increase of CTGF, TGF-β1, TGF-β2, and fibronectin expression. At 24 h, TIMP-1 mRNA was significantly increased. In bovine retinal pericytes (BRPCs), fibronectin, collagen type IV, and TIMP-1 mRNA levels were significantly upregulated at 24 h after VEGF stimulation. CTGF, TGF-β1, and TGF-β2 expression was not affected by VEGF in BRPCs.

Overall, VEGF was able to induce expression of genes related to ECM remodeling in the rat retina. The specificity of this response was demonstrated by the fact that induction of expression of ECM-related genes was selective and that the expression profile correlated to changes in protein levels. In vitro, comparable gene expression profiles

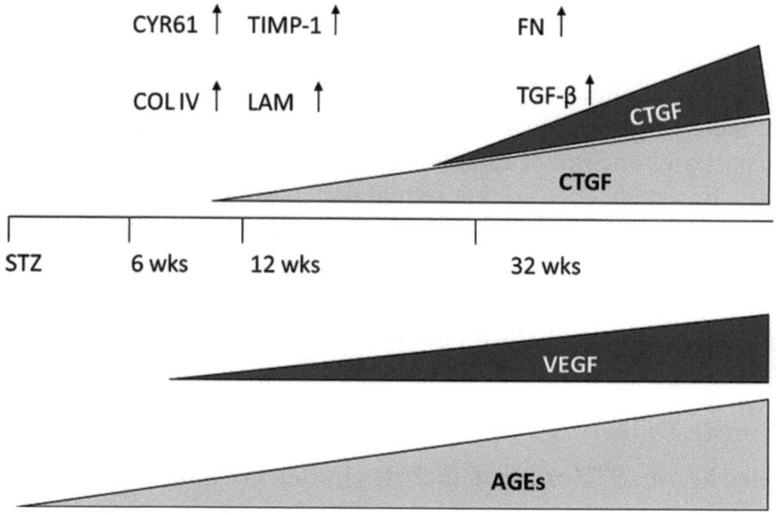

Fig. 4. Hypothetical model of diabetes-induced BL thickening. The model was developed on the basis of data obtained in both the VEGF-induced retinopathy model and the STZ-induced diabetes study, and what is know from the literature [42]. During diabetes, levels of AGEs and VEGF increase, and ECM molecules are induced at different time points after the onset of diabetes. Both AGEs and VEGF contribute to the induction of CTGF expression.

were found in retinal endothelial cells and pericytes, suggesting that the retinal vasculature plays an important role in the altered gene expression profile found in rat retina. Thus, early expression of VEGF in PCDR may contribute directly, and/or via CTGF, to BL thickening and further development of DR. Based on the VEGF-induced retinopathy model and the STZ-induced diabetes study, we developed a model of the expression of profibrotic genes involved in diabetes-induced BL thickening (Fig. 4).

TGF-β and CTGF in BL Thickening

TGF-β plays a causal role in BL thickening in mouse brain capillaries [130] and in the diabetic kidney [131, 132]. However, evidence for such a role in DR is scarce or indirect. Recently, it was shown that two drugs that are effective in the suppression of experimental DR had in common that upregulation of expression of members of the TGF-β pathway was suppressed, suggesting that TGF-β signaling plays a major role in the early pathogenesis of DR [133]. More specifically, retinal vessels in diabetic rats showed both increased TGF-β activity and increased CTGF mRNA expression [133].

To further identify the possible role of TGF-β in BL thickening in DR, its downstream effects were characterized in cultured retinal vascular cells [134]. BRECs and BRPCs were incubated with both low and high concentrations of TGF-β1, and expression levels of ECM-related molecules downstream of TGF-β were analyzed. In BRECs, only high concentrations of TGF-β induced mRNA expression of specific downstream TGF-β effector genes, including fibronectin, but not of CTGF. In BRPCs, both low and high concentrations of TGF-β induced expression of fibronectin and CTGF. Specific inhibition of the TGF-β receptor ALK5 significantly decreased expression levels of fibronectin in both cell types. CTGF expression was decreased with TGF-β inhibition in BRPCs only. Fibronectin protein was present in higher levels in BRPCs. These results show that TGF-β has differential effects on ECM-related gene expression in BRECs and BRPCs. Pericytes are more responsive to TGF-β, and CTGF expression seemed to be regulated by TGF-β in pericytes and not in endothelial cells.

In summary, this study showed that retinal pericytes in particular have the essential characteristics to allow for a role of TGF-β in BL thickening in PCDR. Pericytes are of mesenchymal origin like fibroblasts, which may explain their TGF-β-dependent CTGF regulation. These results suggest that in retinal endothelial cells, CTGF expression is regulated by other pathways and factors, acting independently of TGF-β, such as VEGF, AGEs, and/or high glucose levels [26].

BL Thickening in Diabetic CTGF-Knockout Mice

As indicated above, STZ-induced diabetes in rodents is associated with a twofold increase in CTGF gene expression in total retina, which can be attenuated by treatment with the ACE-inhibitor perindopril or aminoguanidine, respectively [49, 135]. We studied the effects of STZ-induced diabetes on retinal capillary BL thickness in transgenic CTGF$^{+/-}$ mice [136] and wild-type mice (CTGF$^{+/+}$) [124]. BL thickness was calculated by quantitative analysis of electron microscopic (EM) images of transversally sectioned capillaries in and around the inner nuclear layer of the retina. In the retinal capillaries, a significant increase in particularly the endothelial cell BL was detected in diabetic CTGF$^{+/+}$ mice as compared to control CTGF$^{+/+}$ mice, using two independent quantitative methods in EM images (Fig. 5). This preferential thickening of the endothelial BL and pericyte BL in diabetic mice had been observed previously [137].

In this study, the CTGF$^{+/-}$ and CTGF$^{+/+}$ mice were in a similar diabetic state with respect to blood glucose levels. However, there was a clear genotype effect on CTGF expression in the CTGF$^{+/-}$ mice. Approximately 50% lower CTGF protein expression levels in plasma and urine were found in control animals lacking one CTGF allele. Retinal CTGF levels were not analyzed in this study. However, renal CTGF mRNA levels in diabetic CTGF$^{+/-}$ mice were 50% of those in diabetic CTGF$^{+/+}$ mice. This suggests that retinal CTGF protein levels may also have been lower and prevented the diabetes-induced BL thickening of the retinal capillaries. Renal TGF-β1 mRNA levels were significantly increased due to diabetes, irrespective of the CTGF genotype. Similarly to the retinal vessels, a genotypic effect on the BL of glomeruli was found in diabetic mouse kidney [115].

Taken together, the data of this study indicate that CTGF is necessary for BL thickening in diabetes. This provides important direct evidence for an essential role of CTGF in diabetic retinal BL thickening. In concert with the supportive indirect evidence for such a role as described above, these data identify CTGF as a possible therapeutic target to prevent early changes in PCDR. This may be clinically relevant, as experimental animal studies have shown that prevention of BL thickening can ameliorate the subsequent development of other preclinical changes in DR [38].

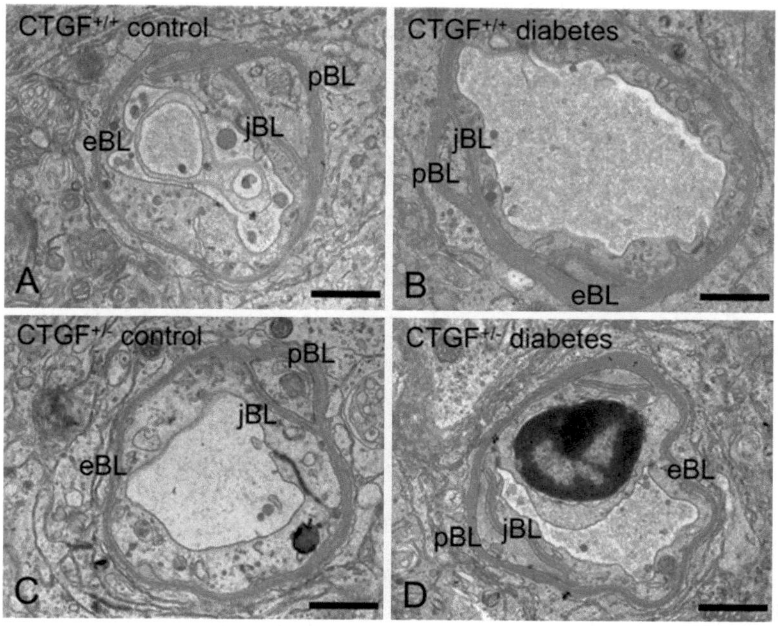

Fig. 5. Examples of retinal capillaries analyzed for BL thickness. Distinct regions of the BL are identified as endothelial BL (eBL), pericyte BL (pBL), and joint endothelial cell and pericyte BL (jBL). Note the diabetes-induced BL thickening in diabetic CTGF$^{+/+}$ mice (**B**) as compared with control CTGF$^{+/+}$ mice (**A**) and the absence of this effect in diabetic CTGF$^{+/-}$ mice (**D**) compared with control CTGF$^{+/-}$ mice (**C**). Bar = 1 μm. (Reproduced from: Journal of Histochemistry and Cytochemistry. Online by Kuiper EJ et al. Copyright 2008 by Histochemical Society Inc. Reproduced with permission of Histochemical Society Inc in the format Trade book via Copyright Clearance Center).

CTGF in PDR

In PDR, CTGF was found in fibrovascular membranes, predominantly localized in myofibroblasts [104, 107], with a significant correlation between the number of α-SMA-positive myofibroblasts and the number of myofibroblasts expressing CTGF [104]. Myofibroblasts are activated matrix-producing fibroblasts, associated with (persistent) fibrosis [59]. Furthermore, CTGF was detected in endothelial cells in these membranes [104]. In the vitreous of a small series of patients with active PDR, levels of the N-terminal CTGF fragment were increased as compared to nondiabetic patients and patients with quiescent PDR [107]. Vitreous levels of full-length CTGF were similar in all groups, whereas the C-terminal fragment was not detectable. N-terminal CTGF levels were also higher in diabetic patients with vitreous hemorrhage than in nondiabetic patients with vitreous hemorrhage, who had similar N-CTGF levels as nondiabetic controls. This finding suggests that local synthesis of CTGF plays a role in PDR. On the basis of the association between CTGF levels and PDR, these authors concluded that CTGF has a role in angiogenesis. However, we showed that elevated CTGF levels are associated with degree of fibrosis and not with angiogenic activity in vitreoretinal conditions, including PDR, in a series of vitreous samples of 119 patients (Fig. 6) [101].

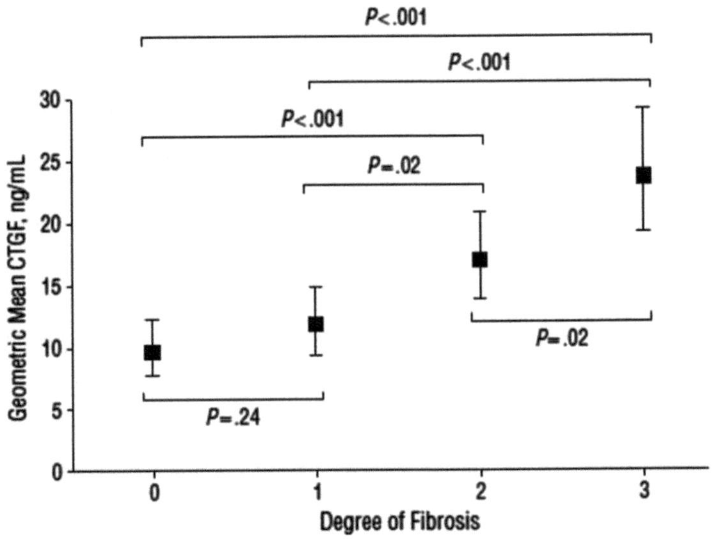

Fig. 6. Geometric mean of CTGF levels in relation to degree of fibrosis. Fibrosis was graded as 0 when no fibrosis was present, 1 with only a few preretinal membranes present, 2 with some proliferative membranes/PVR grade a/b, or 3 with abundant proliferative membranes/PVR grade c/d. Error bars represent the 95% confidence intervals. (Reproduced from [101] Copyright © (2006) American Medical Association. All rights reserved).

In addition, the degree of fibrosis was best predicted by CTGF levels. Possibly, TGF-β has a role in regulating CTGF levels intravitreally and thereby fibrosis in DR. An earlier study has shown that TGF-β2 was associated with fibrotic proliferation in the vitreous of patients with PDR [138]. Furthermore, vitreous levels of both TGF-β2 and CTGF in patients with PDR were significantly higher than in those with nonproliferative diseases, with a correlation between the levels of TGF-β2 and CTGF [103].

Role of CTGF and VEGF in the "Angiofibrotic Switch" in PDR

In PDR, neovascularization progresses to a fibrotic phase. VEGF is considered to be the primary angiogenesis factor in this process [2, 6]. In vitreoretinal disorders (including PDR), N-terminal CTGF levels in the vitreous are elevated [107] and are strongly correlated with the degree of fibrosis [101]. Therefore, it was proposed that CTGF is a causal factor of fibrosis and scarring in PDR.

In vitreous of PDR and PVR patients, Kita et al. [139] found no significant correlation between the levels of CTGF and VEGF, even though concentrations of CTGF and VEGF were both significantly higher compared to those in vitreous from patients with nonproliferative diseases. With regard to a possible role of CTGF in retinal neovascularization, it was concluded that CTGF may have no direct effect on retinal neovascularization, but possibly works indirectly by modulation of VEGF levels.

We investigated the correlation between VEGF and CTGF levels and the degree of fibrosis and neovascularization in the vitreous of a series of 68 patients with PDR and other vitreoretinal disorders (macular hole or macular pucker) [111]. Neovascularization

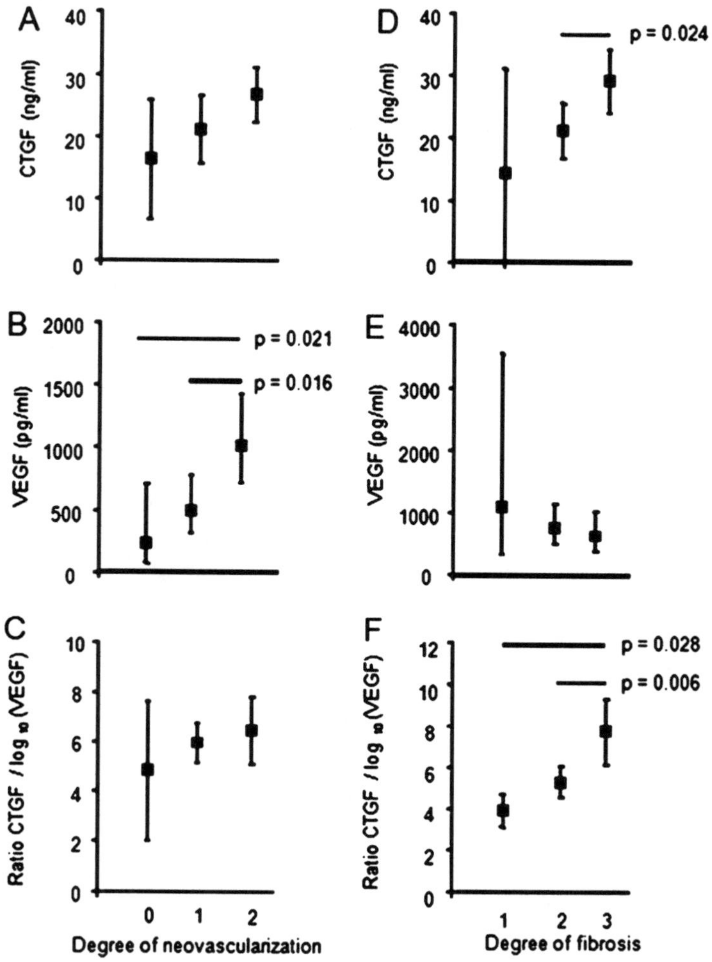

Fig. 7. Mean levels of CTFG (**A**, **D**), geometric mean levels of VEGF (**B**, **D**), and mean ratio CTGF/log10(VEGF) (**c**, **f**) in relation with degree of neovascularization (**A–C**) and degree of fibrosis (**D–F**) in the vitreous of 32 PDR patients. Vertical bars represent 95% confidence intervals. Significant differences between groups are indicated. (From [101]).

and fibrosis in various degrees occurred almost exclusively in PDR patients, in which vitreous CTGF levels were significantly associated with the degree of fibrosis and with VEGF levels, but not with neovascularization. On the other hand, VEGF levels were associated only with neovascularization, in agreement with the widely accepted role of VEGF as the major angiogenic factor in PDR (Fig. 7). As the ratio of CTGF and VEGF levels was the strongest predictor of the degree of fibrosis, the results suggested that the balance of VEGF and CTGF levels in the vitreous determines progression of fibrovascular proliferation in PDR.

These findings led to the following concept of regulation of angiogenesis and fibrosis in ocular disease and in wound healing in general: angiogenesis in the vitreous is driven by VEGF, which upregulates the profibrotic factor CTGF in various cell types in the newly formed neovascular membranes. The elevated CTGF levels do not significantly

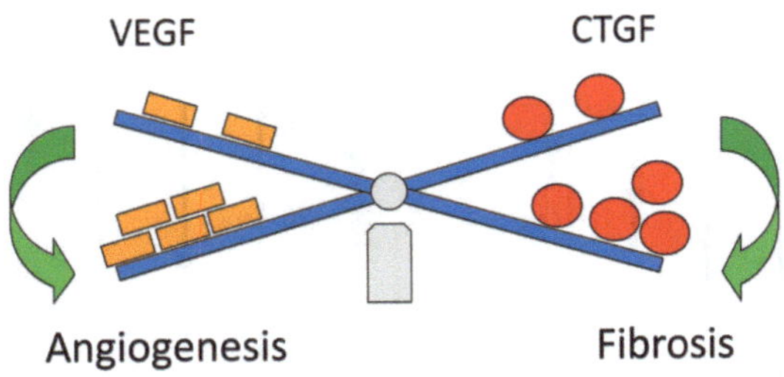

Fig. 8. Hypothesis of the angiofibrotic switch in PDR. Angiogenesis in the vitreous is driven by VEGF, which upregulates the profibrotic factor CTGF. Increasing levels of CTGF inactivate VEGF, and when the balance between these two factors shifts to a certain threshold ratio, the angiofibrotic switch occurs: angiogenesis ceases, and fibrosis driven by excess of CTGF leads to scarring and blindness.

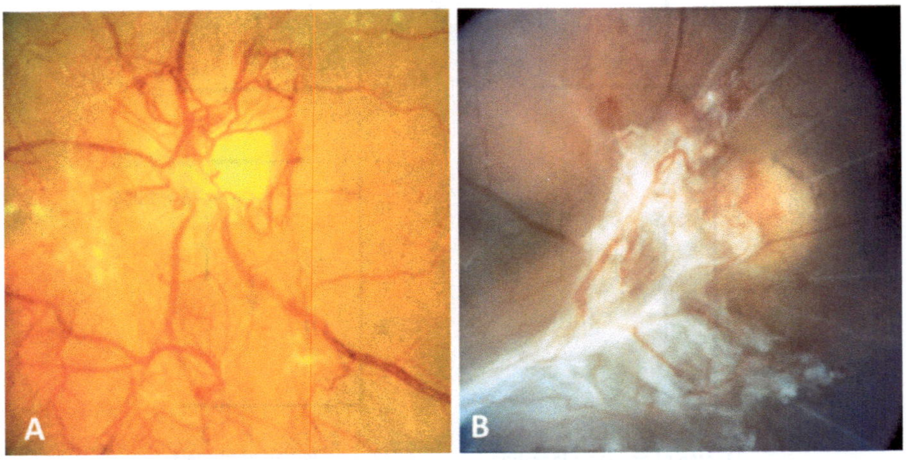

Fig. 9. Fundus photographs of a patient with PDR and new vessels along the lower vascular arcade, before (**A**) and at 8 months after (**B**) an injection with bevacizumab followed by pan-retinal photocoagulation. Note the increase in fibrosis after combined anti-VEGF and laser treatment (**B**).

contribute to ocular angiogenesis. In contrast, increased levels of CTGF sequester VEGF, and when the balance between these two factors shifts to a certain threshold ratio, the angiofibrotic switch occurs: angiogenesis ceases, and fibrosis driven by excess CTGF leads to scarring and blindness (Fig. 8).

This concept predicts that a sharp decline in VEGF levels in a patient with active neovascularization due to PDR inhibits angiogenesis, causes the angiofibrotic switch, and temporarily increases fibrosis. This is supported by clinical observations in patients with active neovascularization treated with intravitreal inhibitors of VEGF, such as bevacizumab and ranibizumab, and/or pan-retinal laser photocoagulation, which destroys large areas of retina and markedly reduces intraocular VEGF levels [5]. Regression of neovascularization and the predicted temporary increase in fibrosis was observed in a non-systematic survey of a small series of patients (Fig. 9) [111]. Others have also reported exacerbation and subsequent contraction of fibrous tissue leading to tractional retinal detachment in patients who received anti-VEGF treatment for active PDR [140, 141]. PDR patients treated with bevacizumab also showed a remarkable inhibition of angiogenesis, suggesting that the elevated CTGF levels, which remain in the vitreous after VEGF inhibition, are not able to maintain the angiogenic response, providing further evidence that CTGF has no proangiogenic role in PDR.

As indicated, it appears from the available evidence that CTGF, in a critical balance with VEGF, drives the angiofibrotic switch and subsequent fibrosis in PDR. This indicates that CTGF-targeted therapy, in particular in combination with anti-VEGF agents, is a possible novel option to prevent sight-threatening fibrosis in PDR and other ocular diseases that are associated with neovascularization and fibrosis.

CONCLUSIONS

We conclude that in DR, CTGF has a role in two important stages of the disease. Early in the pathogenesis, CTGF contributes to thickening of the retinal capillary BL, which is a crucial step in the progression of DR. In this stage, CTGF interacts with AGEs, and growth factors such as VEGF and TGF-β are involved as well. Designing treatment strategies against TGF-β, a major inducer of fibrosis in many diabetic complications, is unfavorable, as this growth factor also exhibits (beneficial) immunosuppressive and anti-inflammatory activity. Targeting CTGF, a major regulator of profibrotic TGF-β action, may therefore be a much more suitable option in the preclinical stage of the disease.

In a later stage of the disease, the switch from neovascularization to a fibrotic phase in PDR is driven by CTGF, in a critical balance with VEGF. This indicates that CTGF-targeted therapy, in particular in combination with anti-VEGF agents, is a possible novel option to prevent sight-threatening fibrosis in PDR and other ocular diseases that are associated with neovascularization and fibrosis.

On the basis of these and other data, CTGF has been suggested to be useful as a biomarker for a wide range of fibrotic disorders [142]. A simple tool as ELISAs can be used to monitor the expression of CTGF and VEGF proteins in the vitreous of PDR patients, which enables the monitoring of disease progression and drug efficacy in clinical trials.

REFERENCES

1. Frank RN. Diabetic retinopathy. N Engl J Med. 2004;350(1):48–58.
2. Aiello LP, Wong JS. Role of vascular endothelial growth factor in diabetic vascular complications. Kidney Int Suppl. 2000;77:S113–9.
3. Schlingemann RO, van Hinsbergh VW. Role of vascular permeability factor/vascular endothelial growth factor in eye disease. Br J Ophthalmol. 1997;81(6):501–12.
4. Spirin KS, Saghizadeh M, Lewin SL, Zardi L, Kenney MC, Ljubimov AV. Basement membrane and growth factor gene expression in normal and diabetic human retinas. Curr Eye Res. 1999;18(6):490–9.
5. Aiello LP, Avery RL, Arrigg PG, et al. Vascular endothelial growth factor in ocular fluid of patients with diabetic retinopathy and other retinal disorders. N Engl J Med. 1994;331(22):1480–7.

6. Witmer AN, Vrensen GFJM, Van Noorden CJF, Schlingemann RO. Vascular endothelial growth factors and angiogenesis in eye disease. Prog Retin Eye Res. 2003;22(1):1–29.
7. Babic AM, Chen CC, Lau LF. Fisp12/mouse connective tissue growth factor mediates endothelial cell adhesion and migration through integrin alphavbeta3, promotes endothelial cell survival, and induces angiogenesis in vivo. Mol Cell Biol. 1999;19(4):2958–66.
8. Fan WH, Pech M, Karnovsky MJ. Connective tissue growth factor (CTGF) stimulates vascular smooth muscle cell growth and migration in vitro. Eur J Cell Biol. 2000;79(12): 915–23.
9. Moussad EE, Brigstock DR. Connective tissue growth factor: what's in a name? Mol Genet Metab. 2000;71(1–2):276–92.
10. Shimo T, Nakanishi T, Nishida T, et al. Connective tissue growth factor induces the proliferation, migration, and tube formation of vascular endothelial cells in vitro, and angiogenesis in vivo. J Biochem. 1999;126(1):137–45.
11. Suzuma K, Naruse K, Suzuma I, et al. Vascular endothelial growth factor induces expression of connective tissue growth factor via KDR, Flt1, and phosphatidylinositol 3-kinase-akt-dependent pathways in retinal vascular cells. J Biol Chem. 2000;275(52): 40725–31.
12. Duncan MR, Frazier KS, Abramson S, et al. Connective tissue growth factor mediates transforming growth factor beta-induced collagen synthesis: downregulation by cAMP. FASEB J. 1999;13(13):1774–86.
13. Khan ZA, Chakrabarti S. Growth factors in proliferative diabetic retinopathy. Exp Diabesity Res. 2003;4(4):287–301.
14. Zimmet P, Alberti KGMM, Shaw J. Global and societal implications of the diabetes epidemic. Nature. 2001;414(6865):782–7.
15. Murphy M, Godson C, Cannon S, et al. Suppression subtractive hybridization identifies high glucose levels as a stimulus for expression of connective tissue growth factor and other genes in human mesangial cells. J Biol Chem. 1999;274(9):5830–4.
16. Riser BL, Denichilo M, Cortes P, et al. Regulation of connective tissue growth factor activity in cultured rat mesangial cells and its expression in experimental diabetic glomerulosclerosis. J Am Soc Nephrol. 2000;11(1):25–38.
17. Twigg SM, Chen MM, Joly AH, et al. Advanced glycosylation end products up-regulate connective tissue growth factor (insulin-like growth factor-binding protein-related protein 2) in human fibroblasts: a potential mechanism for expansion of extracellular matrix in diabetes mellitus. Endocrinology. 2001;142(5):1760–9.
18. Umezono T, Toyoda M, Kato M, et al. Glomerular expression of CTGF, TGF-beta 1 and type IV collagen in diabetic nephropathy. J Nephrol. 2006;19(6):751–7.
19. Nguyen TQ, Tarnow L, Andersen S, et al. Urinary connective tissue growth factor excretion correlates with clinical markers of renal disease in a large population of type 1 diabetic patients with diabetic nephropathy. Diabetes Care. 2006;29(1):83–8.
20. Roestenberg P, Van Nieuwenhoven FA, Wieten L, et al. Connective tissue growth factor is increased in plasma of type 1 diabetic patients with nephropathy. Diabetes Care. 2004;27(5):1164–70.
21. Goldschmeding R, Aten J, Ito Y, Blom I, Rabelink T, Weening JJ. Connective tissue growth factor: just another factor in renal fibrosis? Nephrol Dial Transplant. 2000;15(3):296–9.
22. Lorenzi M, Gerhardinger C. Early cellular and molecular changes induced by diabetes in the retina. Diabetologia. 2001;44(7):791–804.
23. Oshitari T, Polewski P, Chadda M, Li A, Sato T, Roy S. Effect of combined antisense oligonucleotides against high-glucose- and diabetes-induced overexpression of extracellular matrix components and increased vascular permeability. Diabetes. 2006;55(1):86–92.

24. Nishikawa T, Edelstein D, Du XL, et al. Normalizing mitochondrial superoxide production blocks three pathways of hyperglycaemic damage. Nature. 2000;404(6779):787–90.
25. Roy S, Maiello M, Lorenzi M. Increased expression of basement-membrane collagen in human diabetic-retinopathy. J Clin Invest. 1994;93(1):438–42.
26. Kuiper EJ, Hughes JM, Van Geest RJ, et al. Effect of VEGF-A on expression of profibrotic growth factor and extracellular matrix genes in the retina. Invest Ophthalmol Vis Sci. 2007;48(9):4267–76.
27. Ljubimov AV, Burgeson RE, Butkowski RJ, et al. Basement membrane abnormalities in human eyes with diabetic retinopathy. J Histochem Cytochem. 1996;44(12):1469–79.
28. Roy S, Sala R, Cagliero E, Lorenzi M. Overexpression of fibronectin induced by diabetes or high glucose: phenomenon with a memory. Proc Natl Acad Sci USA. 1990;87(1):404–8.
29. Brownlee M, Spiro RG. Biochemistry of the basement membrane in diabetes mellitus. Adv Exp Med Biol. 1979;124:141–56.
30. Ban CR, Twigg SM. Fibrosis in diabetes complications: pathogenic mechanisms and circulating and urinary markers. Vasc Health Risk Manag. 2008;4(3):575–96.
31. The Diabetes Control and Complications Trial Research Group. The effect of intensive treatment of diabetes on the development and progression of long-term complications in insulin-dependent diabetes mellitus. N Engl J Med. 1993;329(14):977–86.
32. UK Prospective Diabetes Study (UKPDS) Group. Intensive blood-glucose control with sulphonylureas or insulin compared with conventional treatment and risk of complications in patients with type 2 diabetes (UKPDS 33). Lancet. 1998;352(9131):837–53.
33. Friedenwald J, Day R. The vascular lesions of diabetic retinopathy. Bull Johns Hopkins Hosp. 1950;86(4):253–4.
34. Curtis TM, Gardiner TA, Stitt AW. Microvascular lesions of diabetic retinopathy: clues towards understanding pathogenesis? Eye (Lond). 2009;23(7):1496–508.
35. Mansour SZ, Hatchell DL, Chandler D, Saloupis P, Hatchell MC. Reduction of basement membrane thickening in diabetic cat retina by sulindac. Invest Ophthalmol Vis Sci. 1990;31(3):457–63.
36. Stitt AW, Anderson HR, Gardiner TA, Archer DB. Diabetic retinopathy: quantitative variation in capillary basement membrane thickening in arterial or venous environments. Br J Ophthalmol. 1994;78(2):133–7.
37. Gardiner TA, Stitt AW, Anderson HR, Archer DB. Selective loss of vascular smooth muscle cells in the retinal microcirculation of diabetic dogs. Br J Ophthalmol. 1994;78(1):54–60.
38. Roy S, Sato T, Paryani G, Kao R. Downregulation of fibronectin overexpression reduces basement membrane thickening and vascular lesions in retinas of galactose-fed rats. Diabetes. 2003;52(5):1229–34.
39. Hayden MR, Sowers JR, Tyagi SC. The central role of vascular extracellular matrix and basement membrane remodeling in metabolic syndrome and type 2 diabetes: the matrix preloaded. Cardiovasc Diabetol. 2005;4(1):9.
40. Tyagi SC, Kumar SG, Banks J, Fortson W. Co-expression of tissue inhibitor and matrix metalloproteinase in myocardium. J Mol Cell Cardiol. 1995;27(10):2177–89.
41. Riser BL, Cortes P, Yee J, et al. Mechanical strain- and high glucose-induced alterations in mesangial cell collagen metabolism: role of TGF-beta. J Am Soc Nephrol. 1998;9(5):827–36.
42. Twigg SM, Cooper ME. The time has come to target connective tissue growth factor in diabetic complications. Diabetologia. 2004;47(6):965–8.
43. McLennan SV, Fisher E, Martell SY, et al. Effects of glucose on matrix metalloproteinase and plasmin activities in mesangial cells: possible role in diabetic nephropathy. Kidney Int Suppl. 2000;77:S81–7.

44. Geiger M, Binder BR. Plasminogen activation in diabetes mellitus. Kinetics of plasmin formation with tissue plasminogen activator and plasminogen from individual diabetic donors and with in vitro glucosylated plasminogen. Enzyme. 1988;40(2–3):149–57.
45. Gomez DE, Alonso DF, Yoshiji H, Thorgeirsson UP. Tissue inhibitors of metalloproteinases: structure, regulation and biological functions. Eur J Cell Biol. 1997;74(2):111–22.
46. Gardiner TA, Anderson HR, Stitt AW. Inhibition of advanced glycation end-products protects against retinal capillary basement membrane expansion during long-term diabetes. J Pathol. 2003;201(2):328–33.
47. Goldin A, Beckman JA, Schmidt AM, Creager MA. Advanced glycation end products: sparking the development of diabetic vascular injury. Circulation. 2006;114(6):597–605.
48. Brownlee M, Cerami A, Vlassara H. Advanced glycosylation end products in tissue and the biochemical basis of diabetic complications. N Engl J Med. 1988;318(20):1315–21.
49. Hughes JM, Kuiper EJ, Klaassen I, et al. Advanced glycation end products cause increased CCN family and extracellular matrix gene expression in the diabetic rodent retina. Diabetologia. 2007;50(5):1089–98.
50. Feng Y, Wang Y, Pfister F, Hillebrands JL, Deutsch U, Hammes HP. Decreased hypoxia-induced neovascularization in angiopoietin-2 heterozygous knockout mouse through reduced MMP activity. Cell Physiol Biochem. 2009;23(4–6):277–84.
51. van Hinsbergh VW, Koolwijk P. Endothelial sprouting and angiogenesis: matrix metalloproteinases in the lead. Cardiovasc Res. 2008;78(2):203–12.
52. Das A, Fanslow W, Cerretti D, Warren E, Talarico N, McGuire P. Angiopoietin/Tek interactions regulate mmp-9 expression and retinal neovascularization. Lab Invest. 2003;83(11):1637–45.
53. Schlingemann RO, Witmer AN. Treatment of retinal diseases with VEGF antagonists. Prog Brain Res. 2009;175:253–67.
54. Gabbiani G. The myofibroblast in wound healing and fibrocontractive diseases. J Pathol. 2003;200(4):500–3.
55. van der Slot-Verhoeven AJ, van Dura EA, Attema J, et al. The type of collagen cross-link determines the reversibility of experimental skin fibrosis. Biochim Biophys Acta. 2005;1740(1):60–7.
56. Baum CL, Arpey CJ. Normal cutaneous wound healing: clinical correlation with cellular and molecular events. Dermatol Surg. 2005;31(6):674–86.
57. Blom IE, van Dijk AJ, Wieten L, et al. In vitro evidence for differential involvement of CTGF, TGFbeta, and PDGF-BB in mesangial response to injury. Nephrol Dial Transplant. 2001;16(6):1139–48.
58. Franklin TJ. Therapeutic approaches to organ fibrosis. Int J Biochem Cell Biol. 1997;29(1):79–89.
59. Chen Y, Shi-Wen X, van BJ, et al. Matrix contraction by dermal fibroblasts requires transforming growth factor-beta/activin-linked kinase 5, heparan sulfate-containing proteoglycans, and MEK/ERK: insights into pathological scarring in chronic fibrotic disease. Am J Pathol. 2005;167(6):1699–711.
60. Kaijzel EL, Koolwijk P, van Erck MG, van Hinsbergh VW, de Maat MP. Molecular weight fibrinogen variants determine angiogenesis rate in a fibrin matrix in vitro and in vivo. J Thromb Haemost. 2006;4(9):1975–81.
61. Buck M, Houglum K, Chojkier M. Tumor necrosis factor-alpha inhibits collagen alpha1(I) gene expression and wound healing in a murine model of cachexia. Am J Pathol. 1996;149(1):195–204.
62. Gailit J, Clark RA. Wound repair in the context of extracellular matrix. Curr Opin Cell Biol. 1994;6(5):717–25.

63. Granstein RD, Murphy GF, Margolis RJ, Byrne MH, Amento EP. Gamma-interferon inhibits collagen synthesis in vivo in the mouse. J Clin Invest. 1987;79(4):1254–8.
64. Madlener M, Parks WC, Werner S. Matrix metalloproteinases (MMPs) and their physiological inhibitors (TIMPs) are differentially expressed during excisional skin wound repair. Exp Cell Res. 1998;242(1):201–10.
65. Soo C, Shaw WW, Zhang X, Longaker MT, Howard EW, Ting K. Differential expression of matrix metalloproteinases and their tissue-derived inhibitors in cutaneous wound repair. Plast Reconstr Surg. 2000;105(2):638–47.
66. Leask A, Abraham DJ. TGF-beta signaling and the fibrotic response. FASEB J. 2004;18(7): 816–27.
67. Pastor JC, de la Rua ER, Martin F. Proliferative vitreoretinopathy: risk factors and pathobiology. Prog Retin Eye Res. 2002;21(1):127–44.
68. Schlingemann RO. Role of growth factors and the wound healing response in age-related macular degeneration. Graefes Arch Clin Exp Ophthalmol. 2004;242(1):91–101.
69. Brigstock DR, Goldschmeding R, Katsube KI, et al. Proposal for a unified CCN nomenclature. Mol Pathol. 2003;56(2):127–8.
70. Lau LF, Lam SC. The CCN family of angiogenic regulators: the integrin connection. Exp Cell Res. 1999;248(1):44–57.
71. Perbal B. CCN proteins: multifunctional signalling regulators. Lancet. 2004;363(9402): 62–4.
72. Brigstock DR. The connective tissue growth factor/cysteine-rich 61/nephroblastoma overexpressed (CCN) family. Endocr Rev. 1999;20(2):189–206.
73. Gao R, Brigstock DR. Connective tissue growth factor (CCN2) induces adhesion of rat activated hepatic stellate cells by binding of its C-terminal domain to integrin alpha(v) beta(3) and heparan sulfate proteoglycan. J Biol Chem. 2004;279(10):8848–55.
74. Hoshijima M, Hattori T, Inoue M, et al. CT domain of CCN2/CTGF directly interacts with fibronectin and enhances cell adhesion of chondrocytes through integrin alpha5beta1. FEBS Lett. 2006;580(5):1376–82.
75. Nguyen TQ, Goldschmeding R. Bone morphogenetic protein-7 and connective tissue growth factor: novel targets for treatment of renal fibrosis? Pharm Res. 2008;25(10): 2416–26.
76. Pi L, Ding X, Jorgensen M, et al. Connective tissue growth factor with a novel fibronectin binding site promotes cell adhesion and migration during rat oval cell activation. Hepatology. 2008;47(3):996–1004.
77. Abreu JG, Ketpura NI, Reversade B, De Robertis EM. Connective-tissue growth factor (CTGF) modulates cell signalling by BMP and TGF-beta. Nat Cell Biol. 2002;4(8): 599–604.
78. Lam S, van der Geest RN, Verhagen NA, et al. Connective tissue growth factor and IGF-I are produced by human renal fibroblasts and cooperate in the induction of collagen production by high glucose. Diabetes. 2003;52(12):2975–83.
79. Hashimoto G, Inoki I, Fujii Y, Aoki T, Ikeda E, Okada Y. Matrix metalloproteinases cleave connective tissue growth factor and reactivate angiogenic activity of vascular endothelial growth factor 165. J Biol Chem. 2002;277(39):36288–95.
80. Bradham DM, Igarashi A, Potter RL, Grotendorst GR. Connective tissue growth factor: a cysteine-rich mitogen secreted by human vascular endothelial cells is related to the SRC-induced immediate early gene product CEF-10. J Cell Biol. 1991;114(6):1285–94.
81. Liu LD, Shi HJ, Jiang L, et al. The repairing effect of a recombinant human connective-tissue growth factor in a burn-wounded rhesus-monkey (Macaca mulatta) model. Biotechnol Appl Biochem. 2007;47(2):105–12.

82. Igarashi A, Okochi H, Bradham DM, Grotendorst GR. Regulation of connective tissue growth factor gene expression in human skin fibroblasts and during wound repair. Mol Biol Cell. 1993;4(6):637–45.
83. Leask A, Abraham DJ. The role of connective tissue growth factor, a multifunctional matricellular protein, in fibroblast biology. Biochem Cell Biol. 2003;81(6):355–63.
84. Grotendorst GR, Rahmanie H, Duncan MR. Combinatorial signaling pathways determine fibroblast proliferation and myofibroblast differentiation. FASEB J. 2004;18(3):469–79.
85. Grotendorst GR, Duncan MR. Individual domains of connective tissue growth factor regulate fibroblast proliferation and myofibroblast differentiation. FASEB J. 2005;19(7):729–38.
86. Uchio K, Graham M, Dean NM, Rosenbaum J, Desmouliere A. Down-regulation of connective tissue growth factor and type I collagen mRNA expression by connective tissue growth factor antisense oligonucleotide during experimental liver fibrosis. Wound Repair Regen. 2004;12(1):60–6.
87. Leask A, Sa S, Holmes A, Shiwen X, Black CM, Abraham DJ. The control of ccn2 (ctgf) gene expression in normal and scleroderma fibroblasts. Mol Pathol. 2001;54(3):180–3.
88. Mori T, Kawara S, Shinozaki M, et al. Role and interaction of connective tissue growth factor with transforming growth factor-beta in persistent fibrosis: a mouse fibrosis model. J Cell Physiol. 1999;181(1):153–9.
89. Leask A, Abraham DJ. All in the CCN family: essential matricellular signaling modulators emerge from the bunker. J Cell Sci. 2006;119(23):4803–10.
90. Surveyor GA, Wilson AK, Brigstock DR. Localization of connective tissue growth factor during the period of embryo implantation in the mouse. Biol Reprod. 1998;59(5):1207–13.
91. Slee RB, Hillier SG, Largue P, Harlow CR, Miele G, Clinton M. Differentiation-dependent expression of connective tissue growth factor and lysyl oxidase messenger ribonucleic acids in rat granulosa cells. Endocrinology. 2001;142(3):1082–9.
92. Secker GA, Shortt AJ, Sampson E, Schwarz QP, Schultz GS, Daniels JT. TGFbeta stimulated re-epithelialisation is regulated by CTGF and Ras/MEK/ERK signalling. Exp Cell Res. 2008;314(1):131–42.
93. Esson DW, Neelakantan A, Iyer SA, et al. Expression of connective tissue growth factor after glaucoma filtration surgery in a rabbit model. Invest Ophthalmol Vis Sci. 2004;45(2):485–91.
94. Ho SL, Dogar GF, Wang J, et al. Elevated aqueous humour tissue inhibitor of matrix metalloproteinase-1 and connective tissue growth factor in pseudoexfoliation syndrome. Br J Ophthalmol. 2005;89(2):169–73.
95. Khaw PT, Occleston NL, Schultz G, Grierson I, Sherwood MB, Larkin G. Activation and suppression of fibroblast function. Eye. 1994;8(2):188–95.
96. Nagai N, Klimava A, Lee WH, Izumi-Nagai K, Handa JT. CTGF is increased in basal deposits and regulates matrix production through the ERK (p42/p44mapk) MAPK and the p38 MAPK signaling pathways. Invest Ophthalmol Vis Sci. 2009;50(4):1903–10.
97. Neumann C, Yu A, Welge-Lussen U, Lutjen-Drecoll E, Birke M. The effect of TGF-beta2 on elastin, type VI collagen, and components of the proteolytic degradation system in human optic nerve astrocytes. Invest Ophthalmol Vis Sci. 2008;49(4):1464–72.
98. Razzaque MS, Foster CS, Ahmed AR. Role of connective tissue growth factor in the pathogenesis of conjunctival scarring in ocular cicatricial pemphigoid. Invest Ophthalmol Vis Sci. 2003;44(5):1998–2003.
99. van SG, Aspiotis M, Blalock TD, Grotendorst G, Schultz G. Connective tissue growth factor in pterygium: simultaneous presence with vascular endothelial growth factor – possible

contributing factor to conjunctival scarring. Graefes Arch Clin Exp Ophthalmol. 2003;241(2):135–9.
100. Yamanaka O, Saika S, Ohnishi Y, Kim-Mitsuyama S, Kamaraju AK, Ikeda K. Inhibition of p38MAP kinase suppresses fibrogenic reaction in conjunctiva in mice. Mol Vis. 2007;13:1730–9.
101. Kuiper EJ, de Smet MD, van Meurs JC, et al. Association of connective tissue growth factor with fibrosis in vitreoretinal disorders in the human eye. Arch Ophthalmol. 2006;124(10):1457–62.
102. He S, Chen Y, Khankan R, et al. Connective tissue growth factor as a mediator of intraocular fibrosis. Invest Ophthalmol Vis Sci. 2008;49(9):4078–88.
103. Kita T, Hata Y, Miura M, Kawahara S, Nakao S, Ishibashi T. Functional characteristics of connective tissue growth factor on vitreoretinal cells. Diabetes. 2007;56(5):1421–8.
104. Abu El-Asrar AM, Van den Steen PE, Al-Amro SA, Missotten L, Opdenakker G, Geboes K. Expression of angiogenic and fibrogenic factors in proliferative vitreoretinal disorders. Int Ophthalmol. 2007;27(1):11–22.
105. Cui JZ, Chiu A, Maberley D, Ma P, Samad A, Matsubara JA. Stage specificity of novel growth factor expression during development of proliferative vitreoretinopathy. Eye. 2007;21(2):200–8.
106. Hinton DR, He S, Jin ML, Barron E, Ryan SJ. Novel growth factors involved in the pathogenesis of proliferative vitreoretinopathy. Eye. 2002;16(4):422–8.
107. Hinton DR, Spee C, He S, et al. Accumulation of NH2-terminal fragment of connective tissue growth factor in the vitreous of patients with proliferative diabetic retinopathy. Diabetes Care. 2004;27(3):758–64.
108. Inoki I, Shiomi T, Hashimoto G, et al. Connective tissue growth factor binds vascular endothelial growth factor (VEGF) and inhibits VEGF-induced angiogenesis. FASEB J. 2002;16(2):219–21.
109. Jang HS, Kim HJ, Kim JM, et al. A novel ex vivo angiogenesis assay based on electroporation-mediated delivery of naked plasmid DNA to skeletal muscle. Mol Ther. 2004;9(3):464–74.
110. Kuiper EJ, Roestenberg P, Ehlken C, et al. Angiogenesis is not impaired in connective tissue growth factor (CTGF) knock-out mice. J Histochem Cytochem. 2007;55(11):1139–47.
111. Kuiper EJ, Van Nieuwenhoven FA, de Smet MD, et al. The angio-fibrotic switch of VEGF and CTGF in proliferative diabetic retinopathy. PLoS One. 2008;3(7):e2675.
112. Agostini H, Boden K, Unsold A, et al. A single local injection of recombinant VEGF receptor 2 but not of Tie2 inhibits retinal neovascularization in the mouse. Curr Eye Res. 2005;30(4):249–57.
113. Lambert V, Munaut C, Noel A, et al. Influence of plasminogen activator inhibitor type 1 on choroidal neovascularization. FASEB J. 2001;15(6):1021–7.
114. Lambert V, Munaut C, Carmeliet P, et al. Dose-dependent modulation of choroidal neovascularization by plasminogen activator inhibitor type I: implications for clinical trials. Invest Ophthalmol Vis Sci. 2003;44(6):2791–7.
115. Roestenberg P, Van Nieuwenhoven FA, Joles JA, et al. Temporal expression profile and distribution pattern indicate a role of connective tissue growth factor (CTGF/CCN-2) in diabetic nephropathy in mice. Am J Physiol Renal Physiol. 2006;290(6):F1344–54.
116. Cooker LA, Peterson D, Rambow J, et al. TNF-alpha, but not IFN-gamma, regulates CCN2 (CTGF), collagen type I, and proliferation in mesangial cells: possible roles in the progression of renal fibrosis. Am J Physiol Renal Physiol. 2007;293(1):F157–65.
117. Twigg SM, Cao Z, McLennan SV, et al. Renal connective tissue growth factor induction in experimental diabetes is prevented by aminoguanidine. Endocrinology. 2002;143(12):4907–15.

118. Wahab NA, Harper K, Mason RM. Expression of extracellular matrix molecules in human mesangial cells in response to prolonged hyperglycaemia. Biochem J. 1996;316(3): 985–92.
119. Adler SG, Kang SW, Feld S, et al. Glomerular mRNAs in human type 1 diabetes: biochemical evidence for microalbuminuria as a manifestation of diabetic nephropathy. Kidney Int. 2001;60(6):2330–6.
120. Thomson SE, McLennan SV, Kirwan PD, et al. Renal connective tissue growth factor correlates with glomerular basement membrane thickness and prospective albuminuria in a non-human primate model of diabetes: possible predictive marker for incipient diabetic nephropathy. J Diabetes Complications. 2008;22(4):284–94.
121. Kuiper EJ, Witmer AN, Klaassen I, Oliver N, Goldschmeding R, Schlingemann RO. Differential expression of connective tissue growth factor in microglia and pericytes in the human diabetic retina. Br J Ophthalmol. 2004;88(8):1082–7.
122. Schlingemann RO, Dingjan GM, Emeis JJ, Blok J, Warnaar SO, Ruiter DJ. Monoclonal antibody PAL-E specific for endothelium. Lab Invest. 1985;52(1):71–6.
123. Schlingemann RO, Hofman P, Vrensen GF, Blaauwgeers HG. Increased expression of endothelial antigen PAL-E in human diabetic retinopathy correlates with microvascular leakage. Diabetologia. 1999;42(5):596–602.
124. Kuiper EJ, van Zijderveld R, Roestenberg P, et al. Connective tissue growth factor is necessary for retinal capillary basal lamina thickening in diabetic mice. J Histochem Cytochem. 2008;56(8):785–92.
125. Liu H, Yang R, Tinner B, Choudhry A, Schutze N, Chaqour B. Cysteine-rich protein 61 and connective tissue growth factor induce deadhesion and anoikis of retinal pericytes. Endocrinology. 2008;149(4):1666–77.
126. Zhou G, Li C, Cai L. Advanced glycation end-products induce connective tissue growth factor-mediated renal fibrosis predominantly through transforming growth factor beta-independent pathway. Am J Pathol. 2004;165(6):2033–43.
127. Boulton M, Foreman D, Williams G, McLeod D. VEGF localisation in diabetic retinopathy. Br J Ophthalmol. 1998;82(5):561–8.
128. Mathews MK, Merges C, McLeod DS, Lutty GA. Vascular endothelial growth factor and vascular permeability changes in human diabetic retinopathy. Invest Ophthalmol Vis Sci. 1997;38(13):2729–41.
129. Flyvbjerg A, Dagnaes-Hansen F, De Vriese AS, Schrijvers BF, Tilton RG, Rasch R. Amelioration of long-term renal changes in obese type 2 diabetic mice by a neutralizing vascular endothelial growth factor antibody. Diabetes. 2002;51(10):3090–4.
130. Wyss-Coray T. Alzheimer's disease-like cerebrovascular pathology in transforming growth factor-beta 1 transgenic mice and functional metabolic correlates. Ann N Y Acad Sci. 2000;903:317–23.
131. Fujimoto M, Maezawa Y, Yokote K, et al. Mice lacking Smad3 are protected against streptozotocin-induced diabetic glomerulopathy. Biochem Biophys Res Commun. 2003; 305(4):1002–7.
132. Wolf G. From the periphery of the glomerular capillary wall toward the center of disease – podocyte injury comes of age in diabetic nephropathy. Diabetes. 2005;54(6):1626–34.
133. Gerhardinger C, Dagher Z, Sebastiani P, Park YS, Lorenzi M. The transforming growth factor-beta pathway is a common target of drugs that prevent experimental diabetic retinopathy. Diabetes. 2009;58(7):1659–67.
134. van Geest RJ, Klaassen I, vogels IME, van noorden CJF, Schlingemann RO. Differential TGF-β signaling in retinal vascular cells: a role in diabetic retinopathy. Invest Ophthalmol vis sci? 2010;51(4): 1857–65.

135. Tikellis C, Cooper ME, Twigg SM, Burns WC, Tolcos M. Connective tissue growth factor is up-regulated in the diabetic retina: amelioration by angiotensin-converting enzyme inhibition. Endocrinology. 2004;145(2):860–6.
136. Ivkovic S, Yoon BS, Popoff SN, et al. Connective tissue growth factor coordinates chondrogenesis and angiogenesis during skeletal development. Development. 2003;130(12): 2779–91.
137. Fischer F, Gartner J. Morphometric analysis of basal laminae in rats with long-term streptozotocin diabetes L. II. Retinal capillaries. Exp Eye Res. 1983;37(1):55–64.
138. Hirase K, Ikeda T, Sotozono C, Nishida K, Sawa H, Kinoshita S. Transforming growth factor beta2 in the vitreous in proliferative diabetic retinopathy. Arch Ophthalmol. 1998;116(6):738–41.
139. Kita T, Hata Y, Kano K, et al. Transforming growth factor-beta2 and connective tissue growth factor in proliferative vitreoretinal diseases: possible involvement of hyalocytes and therapeutic potential of Rho kinase inhibitor. Diabetes. 2007;56(1):231–8.
140. Arevalo JF, Maia M, Flynn Jr HW, et al. Tractional retinal detachment following intravitreal bevacizumab (Avastin) in patients with severe proliferative diabetic retinopathy. Br J Ophthalmol. 2008;92(2):213–6.
141. Moradian S, Ahmadieh H, Malihi M, Soheilian M, Dehghan MH, Azarmina M. Intravitreal bevacizumab in active progressive proliferative diabetic retinopathy. Graefes Arch Clin Exp Ophthalmol. 2008;246(12):1699–705.
142. Leask A. Trial by CCN2: a standardized test for fibroproliferative disease? J Cell Commun Signal. 2009;3(1):87–8.

Part IV
How Can Vision Loss Be Limited: Experimental Therapies

17
Ranibizumab and Other VEGF Antagonists for Diabetic Macular Edema

Ben J. Kim, Diana V. Do, and Quan Dong Nguyen

CONTENTS

Keywords Tractional retinal detachment • Neovascularization • Microaneurysms • Vascular endothelial growth factor • Early Treatment of Diabetic Retinopathy Study • Combination laser treatment

INTRODUCTION

Vision loss from diabetic retinopathy has a tremendous impact on society as it is the most prevalent cause of vision loss in the working-age population of developed countries. By 2025, it is expected that there will be more than 300 million people worldwide with diabetes [1]. While severe loss can be caused by vitreous hemorrhage or tractional retinal detachment, diabetic macular edema (DME) is the most common cause of moderate vision loss [2]. The prevalence of DME is estimated at 10–25% of the diabetic population, with this percentage being higher in those patients with more severe retinopathy [3, 4]. While patients can strive to optimize glycemic control, blood pressure, and weight

From: *Ophthalmology Research: Visual Dysfunction in Diabetes*
Edited by: J. Tombran-Tink et al. (eds.), DOI 10.1007/978-1-60761-150-9_17

loss, these efforts alone often do not cause significant improvement of DME. Thus, the development of treatments is critical to relieve the burden that DME causes on society. This chapter will discuss the recent development of ranibizumab and other therapy directed against vascular endothelial growth factor (VEGF) for the treatment of DME.

PATHOGENESIS OF DME AND CURRENT STANDARD OF CARE

DME occurs when plasma leaks out of vessels and accumulates within the retinal tissue of the macula. Hyperglycemia leads to impairment of pericytes and subsequent loss of the normal integrity of the retinal vasculature [5] and the blood-retina barrier [6]. As visualized by fluorescein angiography (Fig. 1A, B), this plasma leakage can be focal leakage from microaneurysms or more diffuse leakage. In the latter case, there are no structural abnormalities seen on angiography, and nonfocal leakage from retinal capillaries occurs. For both focal and diffuse edema, the interstitial fluid distorts the structure of the macula, leading to visual dysfunction. This fluid accumulation can be imaged by optical coherence tomography (OCT) (Fig. 1C). Normal vision can be restored if the leakage ceases and the excess fluid resorbs. However, the chronic presence of fluid can lead to permanent damage to neurons, thus limiting the potential for vision restoration. Treatments for DME are aimed at reducing leakage by physically changing the vasculature or by reducing permeability through molecular signaling pathways.

While the pathogenesis of DME at the molecular level is not completely understood, VEGF is thought to have a significant role. VEGF contributes to the pathogenesis of numerous retinal diseases that involve abnormal vessels, including choroidal neovascularization from age-related macular degeneration (AMD). There has been an explosion of basic science knowledge about this protein that ultimately has led to therapeutic development. VEGF is a 45-kDa glycoprotein that is a potent stimulator of vascular permeability and neovascularization [7, 8]. VEGF is a family of growth factors, and the most studied member is VEGF-A, which was known simply as VEGF before the discovery of other members. Different exon splicing pathways of the human VEGF-A gene can lead to the production of at least nine isoforms of VEGF-A [9]. In particular, the VEGF 165 isoform is thought to have a key part in disease development [10, 11]. The expression of VEGF is associated with DME, as it has been found that vitreous levels of VEGF are significantly higher in patients with DME as compared to patients without diabetes [12]. Hypoxia contributes to the development of DME [13], and VEGF expression is stimulated by hypoxia as it is regulated by the transcription factor hypoxia-inducible factor-1 [14]. Hyperglycemia also stimulates VEGF expression [15]. Once there is increased VEGF production in the setting of poorly controlled diabetes, the VEGF can affect the tight junction complexes that are a significant component of the blood-retina barrier. Specifically, it has been shown that VEGF may cause vascular permeability by down-regulating the tight junctional protein occludin in retinal endothelial cells [16]. Using a primate model, Ozaki et al. found that intravitreal injection of a sustained release VEGF pellet causes breakdown of the blood-retinal barrier. This finding was accompanied by significant leakage of fluorescein from vessels and macular edema [17]. Together, these studies suggested that VEGF antagonists may serve as a treatment for DME.

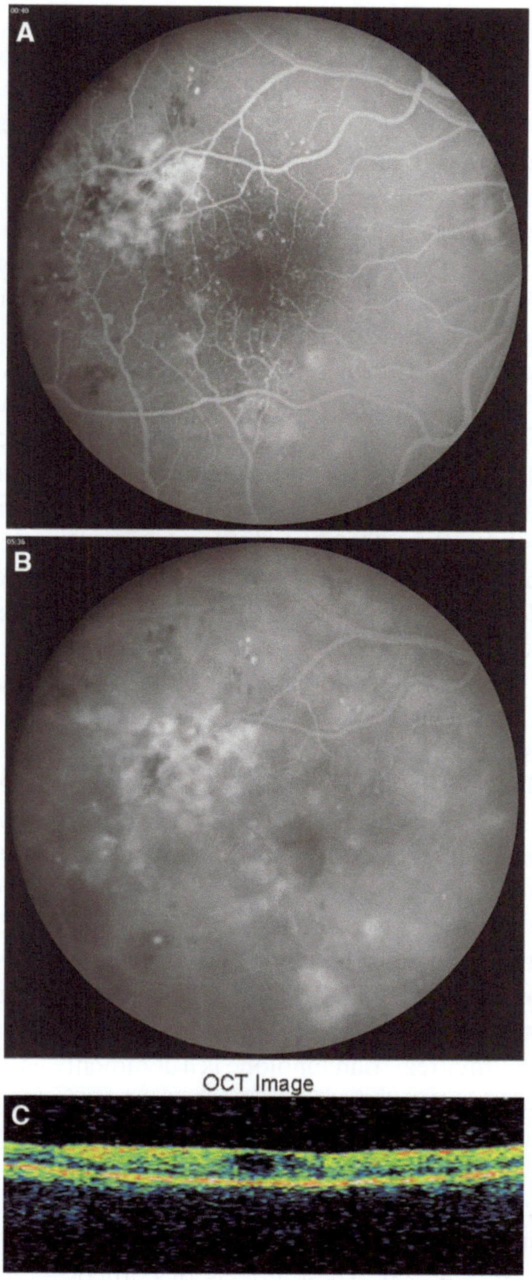

Fig. 1. Early transit (**A**) and late transit (**B**) images from the fluorescein angiogram of a 55-year-old patient with severe nonproliferative diabetic retinopathy and diabetic macular edema (DME). Staining from prior focal laser scars is seen superotemporally. There are multiple microaneurysms and enlargement of the foveal avascular zone. Leakage is seen in the late frame. An optical coherence tomography image (**C**) from the same patient demonstrates intraretinal fluid.

The first study examining this possibility involved an orally active nonselective blocker of VEGF receptors called PKC412 [18]. PKC412 is a kinase inhibitor that blocks VEGF receptors 1 and 2, platelet-derived growth factor, the receptors for stem cell factor, and several isoforms of protein kinase C [19, 20]. In a dose-dependent manner, PKC412 reduced macular thickening and modestly improved visual acuity in patients with DME. However, the therapeutic dose also caused liver toxicity, and thus the study pointed to a need for a more selective and locally administered medication. Before anti-VEGF therapy for DME is discussed further, it is important to discuss other treatments in use to understand the potential impact that anti-VEGF therapy may have on patient care.

Currently, the standard of care for the treatment of DME is focal/grid laser. With this treatment, laser is applied focally to leaking microaneurysms or applied in a grid pattern to areas of diffuse leakage. Laser was the first evidence-based treatment developed for DME as delineated by the Early Treatment of Diabetic Retinopathy Study (ETDRS) [21]. The ETDRS demonstrated that laser treatment reduces the risk of moderate vision loss by 50% (30–15%) at 3 years. As one considers other treatments, it is worthwhile to consider that the ETDRS showed reduction in vision loss with as much as 3 years of follow-up. This demonstration of the long-term benefit of laser treatment cannot be ignored. Nevertheless, the prevailing thought has been that laser treatment effectively reduces the risk of vision loss, but laser is not effective at improving vision. Yet it must be noted that many of the patients in the ETDRS study had good visual acuities, and 85% of the patients had vision better than 20/40 [21]. This stems from the fact that the study was not designed originally to demonstrate visual improvement from laser. To investigate this question of vision improvement from laser treatment, a subset of 114 eyes in the ETDRS was examined. These eyes had definite center thickening on photographs, visual acuity worse than 20/32, and mild to moderate nonproliferative retinopathy at baseline. It was found that laser treatment led to a median change from baseline visual acuity of +4 letters at 2 years and that 29% improved ten or more letters [22]. Thus, laser treatment reduces the risk of vision loss and, in some cases, can lead to vision improvement. Since the ETDRS, laser treatment has been the gold standard to which new treatments must be compared.

In addition to laser, intravitreal triamcinolone is a commonly used treatment for DME. The rationale and evidence supporting its use is important to touch upon. It is known that inflammation can lead to vascular permeability through leukocyte-mediated mechanisms [23]. These mechanisms involve the increased expression of leukocyte adhesion proteins such as intercellular adhesion molecule-1 (ICAM-1) and CD18. Corticosteroids can reduce inflammatory signals that lead to vascular permeability as well as reduce the expression of VEGF [24, 25]. These effects of corticosteroids form the basis for the use of intravitreal triamcinolone as a treatment for DME. Gillies et al. conducted a randomized clinical trial in which 69 eyes were treated with 4 mg of intravitreal triamcinolone or a placebo injection of subconjunctival saline [26]. All eyes entered into the study were considered to have DME that persisted or recurred despite previous laser treatment. Eyes received additional injections every 6 months as needed. After 2 years, it was found that 56% of the treated eyes gained five or more letters compared to 26% of the placebo eyes. The most concerning side effects were glaucoma and cataract. An increase of intraocular pressure by 5 mmHg or more was seen in 68% of the treated eyes vs. only 10% of the untreated eyes. Cataract surgery was performed in 54% of the treated eyes and none of the untreated eyes. This study showed the potential benefits of intravitreal triamcinolone, but it did not directly compare this treatment to laser.

A major randomized clinical trial of the Diabetic Retinopathy Clinical Research Network (DRCR) recently compared laser and intravitreal triamcinolone for the treatment of DME [22]. Eight hundred and forty eyes received either focal/grid photocoagulation, 1 mg of intravitreal triamcinolone, or 4 mg of intravitreal triamcinolone. Additional treatments were given every 4 months for persistent or new edema. This study found that after 2 years of follow-up, laser was superior to triamcinolone in preventing vision loss and caused less complications. However, the differences in vision were modest. The average change in visual acuity at 2 years was 1 ± 17 letters for the laser group, -2 ± 18 letters for the 1-mg triamcinolone group, and -3 ± 22 letters for the 4-mg triamcinolone group. This statistically significant difference was not caused by steroid-induced cataract; an analysis of patients that were pseudophakic or without clinically relevant lens changes did not show a benefit of triamcinolone over laser. Greater than or equal to 15 letters of improvement was seen in 18% of the laser group, 14% of the 1-mg triamcinolone group, and 17% of the 4-mg triamcinolone group. Overall, this study emphasized that laser treatment remains the gold standard of treatment for DME. When considering these treatments, potential drawbacks are that triamcinolone is limited by complications of glaucoma and cataract, and both treatments lead to a relatively modest amount of vision improvement. These issues set the stage for ranibizumab treatment. The use of ranibizumab for DME can avoid the potential complications of triamcinolone while potentially providing significant vision improvement.

RANIBIZUMAB FOR DME

Ranibizumab is the Fab fragment of a humanized monoclonal antibody that binds all isoforms of VEGF-A, thereby inhibiting its signaling pathway. To produce ranibizumab, the portion of an anti-VEGF murine monoclonal antibody that binds VEGF was mass produced and then altered by affinity maturation [27, 28] (Fig. 2A). Ranibizumab was originally developed for the treatment of AMD. The MARINA and ANCHOR clinical trials demonstrated that after 2 years of follow-up, ranibizumab prevents moderate visual loss in approximately 95% of patients [29, 30]. Moderate vision gain was seen in approximately one-third of patients. While demonstrating the efficacy of ranibizumab for AMD, these studies also established the safety profile of this treatment. The most severe complications are endophthalmitis and retinal detachment, which have a risk of only 1.3% and less than 1.0%, respectively [29]. With the proven safety and efficacy of ranibizumab for neovascular AMD, it was logical to consider whether ranibizumab could improve DME.

Nguyen et al. first studied ranibizumab for DME in a nonrandomized clinical trial involving ten patients [31]. The investigation has been referred to as the READ-1 study (*R*anibizumab for *E*dema of the m*A*cula in *D*iabetes). In this study, each patient received 0.5 mg of ranibizumab at baseline and at 1, 2, 4, and 6 months. One week after the first injection, the median and mean reductions in foveal thickness were 88 and 130 μm, respectively. At the study's primary end point of 7 months, the schedule of ranibizumab

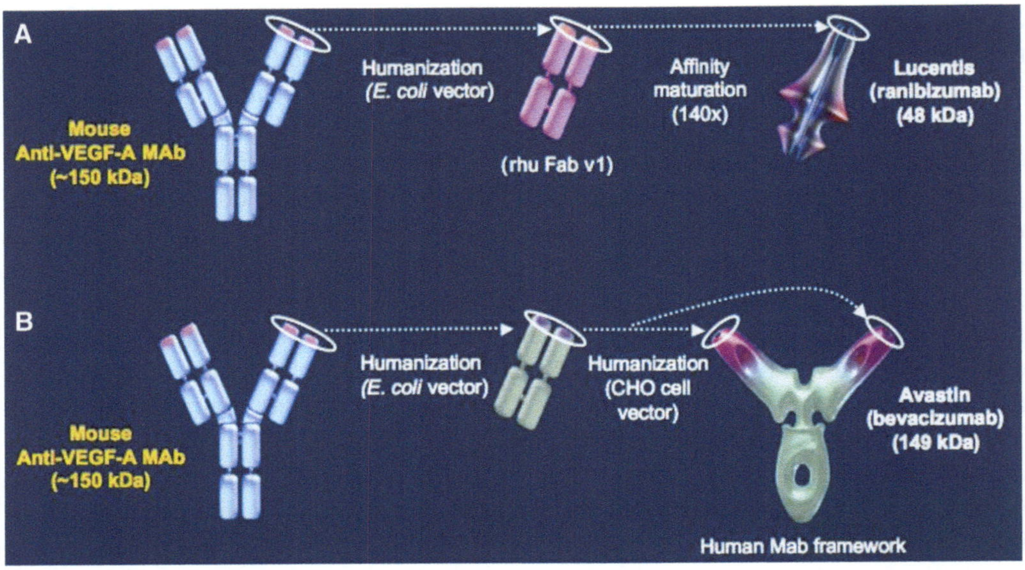

Fig. 2. Ranibizumab and bevacizumab were both created from an anti-VEGF murine monoclonal antibody (Anti-VEGF-A MAb). To create ranibizumab (**A**), the portion that binds VEGF was inserted into a human FAb framework and mass produced using an *Escherichia coli* vector to produce rhuFAb version 1. Through affinity maturation, rhuFAb V1 is modified to increase its binding ability by approximately 140-fold. The final product is ranibizumab. To create bevacizumab (**B**), the portion of the murine antibody that binds to VEGF was inserted into a different humanized Fab variant. This antibody was then mass produced in Chinese hamster ovary (CHO) cells to produce bevacizumab (courtesy of Genentech, Inc.).

injections led to median and mean reductions in foveal thickness of 261 and 246 μm, respectively. This was a mean reduction in excess foveal thickening of 85%. While READ-1 was not a masked, placebo-controlled trial, the visual acuity at 7 months also improved from baseline with a median and mean of 11 and 12.3 letters gained, respectively. Another study by Chun et al. evaluated five patients treated with 0.3 mg of ranibizumab and another five patients treated with 0.5 mg of ranibizumab [32]. The patients received injections at baseline, 1 and 2 months. At 3 months, both doses were associated with an improvement in visual acuity and a decrease in central retinal thickness. The 0.3-mg group gained an average of 12 ± 20 letters, and the 0.5-mg group gained a mean of 7.8 ± 8.1 letters. For central retinal thickness at 3 months, the 0.3-mg group had an average decrease of 45.3 ± 196.3 μm, and the 0.5-mg group had an average decrease of 197.8 ± 85.9 μm. Because of the small sample size, no conclusions can be drawn about any potential differences of the dosing regimens. Interestingly, for both groups, the mean amount of improvement in visual acuity fell after 3 months as the patients were followed out to 6 months, while the mean central retinal thickness continued to improve from month 3 to 6. With regard to the safety of ranibizumab injections for these patients, Nguyen et al. found no adverse systemic or ocular side effects. Chun et al. also did not see adverse systemic side effects, although five of the patients did have intraocular inflammation that resolved within several weeks. The ranibizumab used in

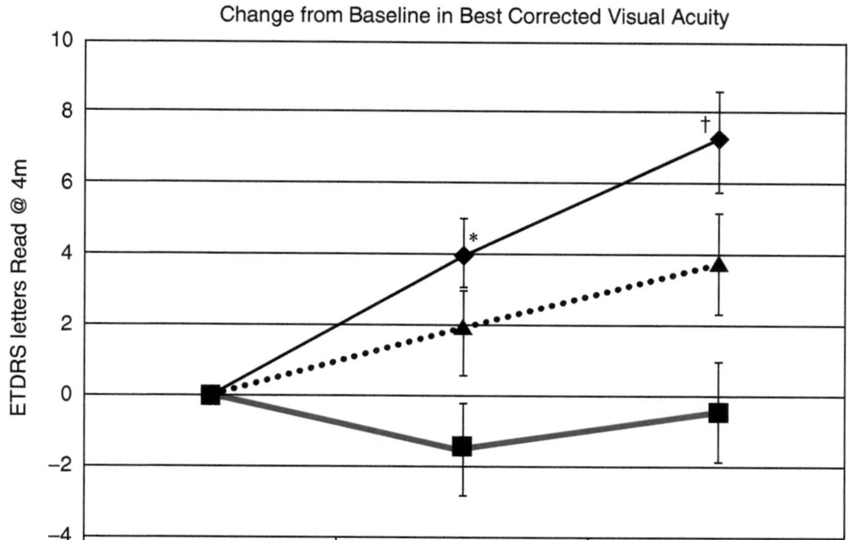

	Baseline	Month 3	Month 6
RBZ	0.00	3.93	7.24
Laser	0.00	−1.48	−0.43
RBZ+Laser	0.00	1.93	3.80

Fig. 3. Changes in visual acuity from baseline in patients with DME treated in the READ-2 study with ranibizumab, focal/grid laser, or a combination of ranibizumab and laser. The mean (±standard error of the mean) change from baseline in number of letters read at 4 m at 3 and 6 months was significantly greater for ranibizumab alone vs. focal/grid laser alone. The combination group was not significantly different from the other two groups at either time point. *$P=0.01$; †$P=0.0003$ by one-way analysis of variance and Bonferroni post hoc analysis. *RBZ* ranibizumab; *ETDRS* Early Treatment Diabetic Retinopathy Study [33].

the study by Chun et al. has been reformulated since then, with the ranibizumab used in the READ-1 study and subsequent studies in DME and AMD having not induced any reported inflammation. Thus, these studies suggested that ranibizumab is safe and can play a key role in DME. However, the study also raised a question regarding the optimal dosing schedule for ranibizumab and pointed to the need for a larger, double-masked, randomized, controlled trial.

The READ-2 study took this next step. READ-2 is a prospective, controlled, multicenter trial involving 126 patients with DME to be conducted over 36 months, with the primary end point at 6 months [33]. Subjects were randomized 1:1:1 into three groups: (group 1) 0.5 mg of ranibizumab at baseline and months 1, 3, and 5; (group 2) focal/grid laser photocoagulation at baseline and month 3 if needed; and (group 3) a combination of 0.5 mg of ranibizumab and focal/grid laser at baseline and month 3. Group 2 thus provided a comparison to the current standard of care. Group 3 was designed because it has been hypothesized that, in cases of extensive retinal thickening, laser may be more effective if ranibizumab is first administered to reduce the retinal thickening. In turn, the more effective laser treatment may then enable less frequent administration of ranibizumab.

The primary end point for the READ-2 study was the change from baseline visual acuity at 6 months [33]. Study subjects who received ranibizumab only (group 1) had the most improvement in visual acuity at 6 months with a mean gain of 7.24 letters (Fig. 3). This was statistically significant from those that received laser only (group 2), as this group had a mean loss of 0.43 letters. Those that received both ranibizumab and laser (group 3) had a mean gain of 3.80 letters, and this was not statistically significant from group 1 or 2. The mean reduction in excess foveal thickness was 50, 33, and 45% in groups 1, 2, and 3, respectively. Thus, the OCT measurements showed similar trends between the groups. However, it should be noted that the laser only group had no improvement in visual acuity despite a 33% mean reduction in excess foveal thickness. It is well known that a reduction in macular edema after laser photocoagulation is not always accompanied by an improvement of visual acuity [34].

The randomized, controlled READ-2 clinical trial demonstrated that ranibizumab may be superior to focal/grid laser. Nevertheless, this conclusion should be considered carefully as the follow-up for the study was only 6 months and only included up to two laser treatments. Patients in the READ-2 study are being followed until month 36. It is certainly possible that longer follow-up may yield different results in the future, which was exemplified by the previously mentioned DRCR study comparing focal/grid laser with intravitreal triamcinolone acetonide [22]. In this study, 4 mg of triamcinolone led to a greater improvement in mean visual acuity than the laser treated group after 4 months. But as discussed earlier, the laser group had a superior visual acuity outcome compared to triamcinolone at 2 years. The differences in visual acuity shifted in favor of the laser group with longer follow-up. While the results of the READ-2 study are noteworthy, it will be important to see the results of more extensive follow-up. The larger RISE and RIDE phase III trials sponsored by Genentech (South San Francisco, CA) evaluate monthly injections of ranibizumab (0.3 and 0.5 mg) with a 2-year primary end point; patients in the control arm received sham injections. Rescue therapy with laser photocoagulation began at month 3 of these studies. The primary outcome measure for these studies is the proportion of patients who gain 15 letters in BCVA compared to baseline. These two studies have completed recruitment and should have completed data collection for the primary outcome by October 2012.

Yet another consideration highlighted by the READ-2 study is the optimal dosing regimen of ranibizumab for DME. While it was a study with only ten patients, the READ-1 trial had a more aggressive regimen of injections (baseline, 1, 2, 4, and 6 months) compared to the READ-2 study (baseline, 1, 3, and 5 months) [31]. This resulted in an 85% reduction of excess foveal thickness as compared to the 50% achieved by the READ-2 study. The RISE and RIDE trials may help determine if monthly dosing of ranibizumab for DME is more effective. Ultimately, it will be important to devise an effective anti-VEGF treatment that is longer lasting.

PEGAPTANIB FOR DME

Ranibizumab is one of the several anti-VEGF treatments that have been studied for the treatment of DME. The first anti-VEGF medication that was delivered by intravitreal injection was pegaptanib, an aptamer that specifically inhibits the VEGF 165 isoform

[35]. Like ranibizumab, pegaptanib was originally developed for the treatment of neovascular AMD and subsequently studied for DME. A phase II study with 172 patients evaluated the use of pegaptanib (0.3, 1, and 3 mg) for DME compared to sham injections [11]. Injections were given at baseline, week 6, 12, and then additional injections or laser treatments were given as needed during the next 18 weeks. The 0.3-mg pegaptanib group had the best vision results. 0.3 mg of pegaptanib led to a better mean visual acuity change of +4.7 letters as compared to the sham group with −0.4 letters ($P = 0.04$). Additionally, less patients within the 0.3-mg pegaptanib arm received laser as compared to sham treatment (25 vs. 48%, $P = 0.04$). These results contribute to the conclusion that VEGF plays a critical role in the pathogenesis of DME. The weaknesses of this study are that it did not directly compare pegaptanib to laser treatment alone and had a limited follow-up period.

BEVACIZUMAB FOR DME

Bevacizumab is another anti-VEGF medication that is being extensively studied for DME. Similar to ranibizumab and also produced by Genentech (South San Francisco, CA), bevacizumab is a humanized monoclonal antibody that inhibits all isoforms of VEGF-A. It is a whole antibody instead of only a Fab fragment (Fig. 17.2B). Bevacizumab is currently approved by the Food and Drug Administration (FDA) for metastatic colorectal cancer, breast cancer, and non-small cell lung cancer [36]. Although there have been no large-scale, randomized, ophthalmic clinical trials involving bevacizumab, many retina specialists are using this medication as an off-label treatment for neovascular AMD [37]. The primary motivation for the use of bevacizumab instead of ranibizumab is the significantly lower cost of bevacizumab. It is reasonable to consider that bevacizumab may also be used widely for DME if clinical trials demonstrate efficacy for ranibizumab.

The DRCR has completed a phase II clinical trial evaluating bevacizumab for DME [36]. One hundred and twenty-one patients were randomized to one of five groups: (A) laser at baseline, (B) 1.25 mg of bevacizumab at baseline and 6 weeks, (C) 2.5 mg of bevacizumab at baseline and 6 weeks, (D) 1.25 mg of bevacizumab at baseline and sham injection at 6 weeks, and (E) 1.25 mg of bevacizumab at baseline and 6 weeks with laser at 3 weeks. Both doses of bevacizumab caused reduction in central retinal thickness, and within the limits of the study, there was no clear difference between the two doses. The similar efficacy of these doses has also been found by others [38]. Defining a significant response as exceeding an 11% reduction in thickness compared to baseline, about half of the eyes treated with bevacizumab had a significant response of retinal thickness. While those eyes treated with bevacizumab had a greater reduction in thickness as compared to laser at 3 weeks, there was no significant difference seen with longer follow-up out to 12 weeks. The improvement in retinal thickness seemed to plateau or decrease between the 3- and 6-week visits, suggesting that subsequent injections should be sooner than 6 weeks. For visual acuity, groups B and C compared with laser, each had a significant difference of about 1-line greater improvement at 12 weeks. Within the short-term follow-up of the study, the combination treatment of bevacizumab and laser did not show any additional benefit compared to the other groups. While there was one case of endophthalmitis, there were no complications that could clearly be attributed to the medication. Overall, the study showed the potential efficacy of bevacizumab and emphasized the need for a phase III trial to study both efficacy and safety.

A report by Kook et al. examined a population of 126 patients with chronic, diffuse DME followed for 6–12 months after treatment with bevacizumab [39]. In this study, "chronic" was defined as the presence of DME for more than 12 months. "Diffuse" edema was defined as thickening that included the fovea and extended to the arcades. All of the patients had received at least one previous treatment that included: focal laser (62%), vitrectomy with internal limiting membrane peeling (11%), and intravitreal triamcinolone (41%). Eleven percent had more than one focal laser treatment, and about 9% had more than one triamcinolone injection. Thirty-eight percent of the patients never had focal laser treatment because the clinician believed that the edema was too severe to respond to laser or that the source of leakage was too close to the fovea. None of the patients had received treatment within 6 months of the first bevacizumab injection. Additional bevacizumab injections were given as frequently as every 4 weeks if there was improvement from the prior injection or if there was significant recurrence of edema after injections were stopped. For the 59 (47%) patients that completed 12 months of follow-up, the mean number of injections was 2.7. While there was no significant improvement of visual acuity at 6 months, there was a significant improvement of +5.1 letters for the 47% of patients that completed 12 months of follow-up. Significant improvements of central retinal thickness compared to baseline (463 μm) were seen at both 6 months (374 μm) and 12 months (357 μm). While this study population was heterogeneous and lacked a control group, the results suggest that bevacizumab may still be beneficial in recalcitrant cases of DME. Importantly, the study raises the question of what subtypes of DME may be resistant to a certain therapy.

Soheilian et al. recently reported the results of a randomized trial comparing bevacizumab alone, bevacizumab with triamcinolone, and macular laser treatment for DME [40]. In this study, 150 eyes were randomized to one of these arms. The primary outcome was visual acuity at 24 weeks, but patients were followed out to 36 weeks. The bevacizumab dose was the commonly used amount of 1.25 mg, but it should be noted that the triamcinolone dose was only 2 mg. Instead of the more typical dose of 4 mg of triamcinolone, a 2-mg dose was chosen to minimize side effects. For all groups, retreatment was given every 12 weeks if the vision was not better than 20/40, and there was persistent clinically significant macular edema. Only one injection was given in 72% of the patients in the bevacizumab alone group. In this group, visual acuity improved significantly from baseline at all follow-up visits up; at 24 weeks, there was a change of -0.23 ± 0.22 logarithm of the minimum angle of resolution (log MAR). The bevacizumab with triamcinolone group and the laser alone group did not have a significant change in vision at 24 weeks compared to baseline. The percentage of patients with a >2 Snellen lines improvement at 36 weeks was 37, 25, and 14.8% of the bevacizumab alone, bevacizumab with triamcinolone, and laser alone groups, respectively. The central macular thickness decreased significantly in all groups only at the sixth week visit, and there was no significant difference among the groups. The authors suggested that bevacizumab alone may be a better primary treatment than laser, although they acknowledge that longer follow-up is needed to demonstrate a lasting benefit over laser. For this

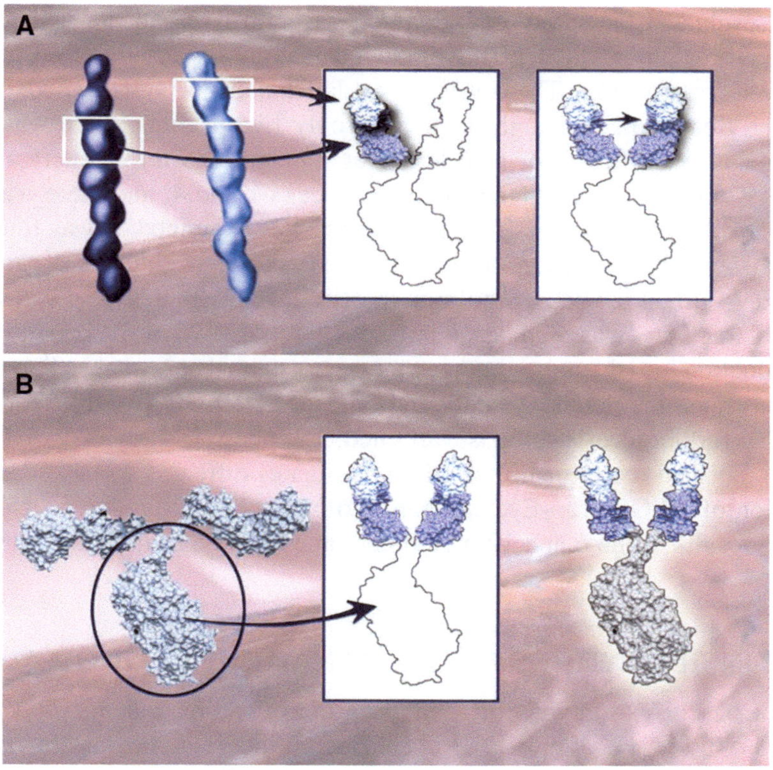

Fig. 4. VEGF Trap-Eye is a fusion protein consisting of all human amino acid sequences. As shown here, the key domain (**A**) from VEGF receptors 1 and 2 have been fused (**B**) with the Fc portion of human IgG. This protein can penetrate the layers of the retina and binds with high affinity to all VEGF-A isoforms and placental growth factor more tightly than the native receptors (courtesy of Regeneron Pharmaceuticals, Inc.)

study, it is notable that only one injection was given to 72% of the bevacizumab alone group. This finding again raises the question as to what is the optimal dosing regimen of bevacizumab for DME.

VEGF TRAP-EYE FOR DME

VEGF Trap-Eye is another potential treatment on the horizon. It is a 115-kDa recombinant fusion protein designed such that the VEGF-binding domains of human VEGF receptors 1 and 2 are fused to the Fc domain of IgG_1 [41] (Fig. 4A, B). In contrast to ranibizumab, VEGF Trap-Eye has a longer half-life and binds all VEGF-A isoforms as well as placental growth factor. VEGF Trap-Eye has a binding constant of approximately 0.5 pM K_d, and this is about 140 times that of ranibizumab [42, 43]. It is estimated that VEGF Trap-Eye has significant intravitreal activity for up to 10 weeks [42]. Thus, the medication has the potential to be given less frequently than ranibizumab, while perhaps being more efficacious.

Do et al. evaluated the safety of VEGF Trap-Eye in five patients with DME [44]. A single intravitreal injection of 4.0 mg of the medication was administered, and patients were followed for 6 weeks. There was no ocular toxicity or systemic adverse events related to the treatment. Although there were only five patients, there was a median improvement in visual acuity of nine letters at 1 month and three letters at 6 weeks. The gain in visual acuity was highest between weeks 1 and 4, and there was less of a gain after 6 weeks. When excess foveal thickness was examined, it was found that all five patients showed a reduction. The median excess foveal thickness was 69 μm at 1 month and 74 μm at 6 weeks. Similar to the visual acuity trend, the greatest effect on excess foveal thickness was seen between weeks 1 and 4. Two of the patients were able to have a reduction into the normal range that was sustained at 6 weeks. This small pilot study demonstrated the potential safety and efficacy of VEGF Trap-Eye for DME and suggested that further investigation is warranted. The phase II study of VEGF Trap-Eye in DME has finished recruitment; the trial investigates different doses and intervals of administration of VEGF Trap-Eye compared to laser photocoagulation. It is expected that detailed results of the 6 and 12 month outcomes will be available in late 2010 and early 2011.

OTHER CONSIDERATIONS IN THE MANAGEMENT OF DME

Treatment based on subtypes of DME is a consideration that may become more relevant in the future. Focal/grid laser is considered the gold standard for any type of DME. However, as exemplified in the above-mentioned study by Kook et al., some retina specialists think that laser treatment is less effective when there is extensive or diffuse edema [39]. A criticism of the DRCR study that compared laser with triamcinolone [22] is that the study does not compare subtypes of DME. There is potential to categorize DME more specifically based on the constellation of angiographic findings, clinical exam, duration, and OCT measurements. Perhaps there are cases of DME that are more responsive to one particular treatment over another. As these considerations move forward, it will be important to define what exactly the DME subtypes are such that effective comparisons can be made; currently, there are no established clinical trial definitions of DME subtypes. This is emphasized in a report by Browning et al. [45]. While there are many papers using the terms "focal" and "diffuse," there are varying definitions for these terms. Browning et al. point out the need to arrive at a consensus on how to categorize DME subtypes. If DME is to be divided into subtypes, then the ability to grade the DME needs to be reproducible between clinicians and reading centers for clinical trials. Until such definitions are determined, one should be careful when interpreting conclusions about subtypes of DME and suggesting that certain therapeutic approaches may be more appropriate for certain types of DME.

In addition to subtypes based on angiography or clinical exam, future treatment of DME could also be stratified by biomarkers. What is it that causes one patient to have an astonishing improvement from ranibizumab while another patient's response is only modest? One possibility is that biomarkers could indicate what level of response a patient may have to a treatment or whether a different treatment should be considered. There are numerous potential cytokines at play in DME. As discussed above, ICAM-1 is thought to have an important role in leukocyte-mediated vascular permeability [23]. A recent report by Funatsu et al. suggests that vitreous levels of ICAM-1 and VEGF correlate independently with increased vascular permeability and the severity of DME [46]. Previous reports have also implicated interleukin-6 (IL-6). IL-6 is a proinflammatory cytokine with multiple functions. It can be involved in the pathogenesis of uveitis [47], is associated with breakdown of the blood-retina barrier, and can lead to VEGF expression [48]. After analyzing aqueous humor samples obtained from 54 diabetic patients during cataract surgery, Funatsu et al. also reported that aqueous levels of VEGF and IL-6 correlate with the severity of DME [12]. In addition to VEGF, it is possible that the profile of other proteins within a patient's vitreous at a given point in time may affect the severity of DME and the response to treatment. Analysis of biomarkers may have a role in the management of DME, especially as treatments with different mechanisms of action become established. Nevertheless, care should be taken when interpreting such studies as an elevated cytokine level does not necessarily prove that there is a role for it in the pathophysiology of DME. If the receptor for the cytokine is downregulated or if a soluble, inhibitory receptor is present, then the measured cytokine level may not have the expected effect [49].

Another potentially important consideration is the balance between VEGF angiogenic and antiangiogenic isoforms. Through differential splicing, an antiangiogenic VEGF-A isoform called VEGF165b can be produced. VEGF165b has a different C-terminal amino acid sequence from angiogenic forms of VEGF [50]. It inhibits angiogenesis by binding to, but not activating, VEGF receptor 2. While studying colonic carcinoma cells, Varey et al. found that bevacizumab inhibited the growth of cells predominantly expressing VEGF 165, while those cells predominantly expressing VEGF 165b were resistant to treatment with bevacizumab [51]. Perrin et al. have found that under normal conditions, the eye expresses VEGF165b and other potentially antiangiogenic isoforms of VEGF [52]. They have suggested that a shift in the balance of antiangiogenic and angiogenic isoforms of VEGF occurs in diabetic retinopathy. One would expect that patients with DME would predominantly have the angiogenic isoforms of VEGF but still might have some expression of VEGF165b. As discussed by Perrin et al., it is not known whether current anti-VEGF treatments also target VEGF165b, potentially limiting their own efficacy. Therefore, the levels of angiogenic vs. antiangiogenic VEGF isoforms could serve as biomarkers that would predict the response to anti-VEGF treatment.

COMBINATION TREATMENT FOR DME

While ranibizumab and triamcinolone have been compared to laser treatment, it is possible that combination laser treatment may be superior to any of these individual treatments. As discussed with the READ-2 trial [33], laser may be more effective and provide longer lasting benefit after an agent has been given to temporarily reduce the macular edema. When the combination of ranibizumab and laser was studied by the READ-2 trial at 6 months, the improvement in BCVA was not statistically different from the ranibizumab alone group or the laser alone group. However, as the follow-up period was short at the primary end point in the READ-2 study, it is worthwhile to further investigate combination treatments that attack DME with complimentary mechanisms. The DRCR has completed enrollment for a trial comparing combination treatments.

The mentioned DRCR protocol compares four groups: (A) sham injections plus laser, (B) 0.5 mg of ranibizumab followed by laser, (C) 0.5 mg of ranibizumab followed by deferred laser, and (D) 4 mg of intravitreal triamcinolone followed by laser [53]. For groups A, B, and D, the laser treatment occurs 3–10 days after the injection. For group C, there is no laser during the first 24 weeks. After 24 weeks, patients within this group receive laser treatment if there has been no improvement from the last two injections and there is macular edema for which laser would be indicated. The primary outcome is the visual acuity after 1 year of follow-up. With this study design, the trial may provide a more definitive answer regarding the potential benefit of combination therapy for DME.

DME AND QUALITY OF LIFE

On a separate note from the details of VEGF mechanisms and the pathogenesis of DME, clinicians must continuously listen to the visual needs of each individual patient. Recommendations based on clinical trial data are based primarily on visual acuity outcomes. Visual acuity measurements are not necessarily always the most comprehensive means of quantifying how DME may affect a patient's daily visual needs and emotional well-being. One must ask if intensive treatments are actually making an improvement on the patient's visual needs and not just the patient's visual acuity measurements. Such visual needs may include the patient's ability to read, to pick an item off the shelf at a grocery store, to interact socially with others, and to perform well at work. It is not surprising that past reports have shown an association between diabetic retinopathy and psychosocial well-being [54–56]. The National Eye Institute 25-Item Visual Function Questionnaire (NEI-VFQ-25) is a questionnaire that can assess the impact that an eye disease can have on quality of life [57]. The NEI-VFQ-25 has been validated and used for different eye diseases [58–63]. Recently, Bressler et al. have shown that treatment of neovascular AMD patients with ranibizumab positively affects the NEI-VFQ-25 scores at 24 months [64]. Such data supports the use of ranibizumab for neovascular AMD patients and demonstrates how qualify-of-life measurements can be used within clinical trials. The NEI-VFQ-25 has utility as a measurement of central visual function in patients with diabetes [60, 63]. However, there is limited literature on how visual function is specifically affected by DME. With a group of 33 patients, Hariprasad has shown that patients with DME can have NEI-VFQ-25 scores similar to patients with AMD [65]. Lamoureux has used the vision-specific functioning scale (VF-11) to show that patients with proliferative diabetic retinopathy (PDR) and vision-threatening diabetic retinopathy (VTDR) have difficulty with vision-specific daily activities [66]. In this study, VTDR was defined as severe nonproliferative retinopathy, PDR, or macular edema within 500 μm of the foveal center, or focal laser scars at the macula. Of the 357 study participants, only 5% had macular edema, and this was determined by photographs and not by clinical exam. There is a need for further studies demonstrating the relationship between DME and vision-related quality of life. As future clinical trials are developed for DME, it will be important to determine if new treatments positively affect a patient's quality of life. Lastly, considerations of quality of life should include low-vision referrals as part of the management regimen. The sometimes overlooked benefits that a patient may have from an evaluation by a low-vision specialist should be recognized. While improved anti-VEGF treatments are on the forefront, a low-vision referral for the patient with significantly decreased vision from refractory DME can be helpful and improve their quality of life.

CONCLUSIONS

There is ample evidence that VEGF plays a critical role in the pathogenesis of DME. Recent clinical trials, such as the READ-2 study and early studies with VEGF Trap-Eye, have demonstrated that anti-VEGF therapy can be effective for DME [33]. Importantly, evidence suggests that such treatment may be more effective than the current gold standard of focal/grid laser photocoagulation. As anti-VEGF therapy for DME becomes more established, one can expect that ranibizumab and bevacizumab may be used by practitioners for DME; currently, there is no clinical trial demonstrating that one medication is inferior to the other. The optimal dosing schedule for these treatments is unclear, but additional information will be forthcoming to help resolve this issue. Depending on the results of further clinical trials, the use of these anti-VEGF treatments in combination with laser or other therapies is a possible trend that will emerge. There are many reasons to be optimistic about these new treatment regimens for DME. Nevertheless, one limitation of current anti-VEGF therapies is the requirement of frequent dosing. If a safe and long-lasting anti-VEGF therapy is developed, then it would be especially effective in reducing the societal burden of DME.

REFERENCES

1. WHO. Fact sheet no. 138. Geneva: WHO; 2002.
2. Klein R. Retinopathy in a population-based study. Trans Am Ophthalmol Soc. 1992;90: 561–94.
3. Klein R, Klein BE, Moss SE, Cruickshanks KJ. The Wisconsin Epidemiologic Study of Diabetic Retinopathy: XVII. The 14-year incidence and progression of diabetic retinopathy and associated risk factors in type 1 diabetes. Ophthalmology. 1998;105(10):1801–15.
4. Hardy RA, Crawford JB. Retina. In: Vaughn D, Asbury T, Riordan-Eva P, editors. General ophthalmology. 15th ed. Stamford: Appleton & Lange; 1999. p. 178–99.
5. Moore J, Bagley S, Ireland G, McLeod D, Boulton ME. Three dimensional analysis of microaneurysms in the human diabetic retina. J Anat. 1999;194(Pt 1):89–100.
6. Antcliff RJ, Marshall J. The pathogenesis of edema in diabetic maculopathy. Semin Ophthalmol. 1999;14(4):223–32.
7. Aiello LP, Bursell SE, Clermont A, et al. Vascular endothelial growth factor-induced retinal permeability is mediated by protein kinase C in vivo and suppressed by an orally effective beta-isoform-selective inhibitor. Diabetes. 1997;46(9):1473–80.
8. Senger DR, Galli SJ, Dvorak AM, Perruzzi CA, Harvey VS, Dvorak HF. Tumor cells secrete a vascular permeability factor that promotes accumulation of ascites fluid. Science. 1983;219(4587):983–5.
9. Takahashi H, Shibuya M. The vascular endothelial growth factor (VEGF)/VEGF receptor system and its role under physiological and pathological conditions. Clin Sci (Lond). 2005;109(3):227–41.
10. Tolentino MJ, Miller JW, Gragoudas ES, et al. Intravitreous injections of vascular endothelial growth factor produce retinal ischemia and microangiopathy in an adult primate. Ophthalmology. 1996;103(11):1820–8.

11. Cunningham Jr ET, Adamis AP, Altaweel M, et al. A phase II randomized double-masked trial of pegaptanib, an anti-vascular endothelial growth factor aptamer, for diabetic macular edema. Ophthalmology. 2005;112(10):1747–57.
12. Funatsu H, Yamashita H, Ikeda T, Mimura T, Eguchi S, Hori S. Vitreous levels of interleukin-6 and vascular endothelial growth factor are related to diabetic macular edema. Ophthalmology. 2003;110(9):1690–6.
13. Nguyen QD, Shah SM, Van Anden E, Sung JU, Vitale S, Campochiaro PA. Supplemental oxygen improves diabetic macular edema: a pilot study. Invest Ophthalmol Vis Sci. 2004; 45(2):617–24.
14. Forsythe JA, Jiang BH, Iyer NV, et al. Activation of vascular endothelial growth factor gene transcription by hypoxia-inducible factor 1. Mol Cell Biol. 1996;16(9):4604–13.
15. Lu M, Kuroki M, Amano S, et al. Advanced glycation end products increase retinal vascular endothelial growth factor expression. J Clin Invest. 1998;101(6):1219–24.
16. Antonetti DA, Barber AJ, Khin S, Lieth E, Tarbell JM, Gardner TW. Vascular permeability in experimental diabetes is associated with reduced endothelial occludin content: vascular endothelial growth factor decreases occludin in retinal endothelial cells. Penn State Retina Research Group. Diabetes. 1998;47(12):1953–9.
17. Ozaki H, Hayashi H, Vinores SA, Moromizato Y, Campochiaro PA, Oshima K. Intravitreal sustained release of VEGF causes retinal neovascularization in rabbits and breakdown of the blood-retinal barrier in rabbits and primates. Exp Eye Res. 1997;64(4):505–17.
18. Campochiaro PA. Reduction of diabetic macular edema by oral administration of the kinase inhibitor PKC412. Invest Ophthalmol Vis Sci. 2004;45(3):922–31.
19. Fabbro D, Buchdunger E, Wood J, et al. Inhibitors of protein kinases: CGP 41251, a protein kinase inhibitor with potential as an anticancer agent. Pharmacol Ther. 1999;82(2–3): 293–301.
20. Fabbro D, Ruetz S, Bodis S, et al. PKC412–a protein kinase inhibitor with a broad therapeutic potential. Anticancer Drug Des. 2000;15(1):17–28.
21. Early Treatment Diabetic Retinopathy Study Research Group. Photocoagulation for diabetic macular edema Early Treatment Diabetic Retinopathy Study report number 1. Arch Ophthalmol. 1985;103(12):1796–806.
22. Diabetic Retinopathy Clinical Research Network. A randomized trial comparing intravitreal triamcinolone acetonide and focal/grid photocoagulation for diabetic macular edema. Ophthalmology. 2008;115(9):1447–9, 1449e1–10.
23. Joussen AM, Poulaki V, Le ML, et al. A central role for inflammation in the pathogenesis of diabetic retinopathy. FASEB J. 2004;18(12):1450–2.
24. Nauck M, Karakiulakis G, Perruchoud AP, Papakonstantinou E, Roth M. Corticosteroids inhibit the expression of the vascular endothelial growth factor gene in human vascular smooth muscle cells. Eur J Pharmacol. 1998;341(2–3):309–15.
25. Nauck M, Roth M, Tamm M, et al. Induction of vascular endothelial growth factor by platelet-activating factor and platelet-derived growth factor is downregulated by corticosteroids. Am J Respir Cell Mol Biol. 1997;16(4):398–406.
26. Gillies MC, Sutter FK, Simpson JM, Larsson J, Ali H, Zhu M. Intravitreal triamcinolone for refractory diabetic macular edema: two-year results of a double-masked, placebo-controlled, randomized clinical trial. Ophthalmology. 2006;113(9):1533–8.
27. Presta LG, Chen H, O'Connor SJ, et al. Humanization of an anti-vascular endothelial growth factor monoclonal antibody for the therapy of solid tumors and other disorders. Cancer Res. 1997;57(20):4593–9.
28. Chen Y, Wiesmann C, Fuh G, et al. Selection and analysis of an optimized anti-VEGF antibody: crystal structure of an affinity-matured Fab in complex with antigen. J Mol Biol. 1999;293(4):865–81.

29. Rosenfeld PJ, Brown DM, Heier JS, et al. Ranibizumab for neovascular age-related macular degeneration. N Engl J Med. 2006;355(14):1419–31.
30. Brown DM, Kaiser PK, Michels M, et al. Ranibizumab versus verteporfin for neovascular age-related macular degeneration. N Engl J Med. 2006;355(14):1432–44.
31. Nguyen QD, Tatlipinar S, Shah SM, et al. Vascular endothelial growth factor is a critical stimulus for diabetic macular edema. Am J Ophthalmol. 2006;142(6):961–9.
32. Chun DW, Heier JS, Topping TM, Duker JS, Bankert JM. A pilot study of multiple intravitreal injections of ranibizumab in patients with center-involving clinically significant diabetic macular edema. Ophthalmology. 2006;113(10):1706–12.
33. Nguyen Q, Shah SM, Heier JS, Do DV, Lim J, Boyer D, et al. Primary end point (six months) results of the ranibizumab for edema of the Macula in Diabetes (READ-2) Study. Ophthalmology. 2009;116(11):2175–81.
34. Schmid KE, Neumaier-Ammerer B, Stolba U, Binder S. Effect of grid laser photocoagulation in diffuse diabetic macular edema in correlation to glycosylated haemoglobin (HbA1c). Graefes Arch Clin Exp Ophthalmol. 2006;244(11):1446–52.
35. Gragoudas ES, Adamis AP, Cunningham Jr ET, Feinsod M, Guyer DR. Pegaptanib for neovascular age-related macular degeneration. N Engl J Med. 2004;351(27):2805–16.
36. Scott IU, Edwards AR, Beck RW, et al. A phase II randomized clinical trial of intravitreal bevacizumab for diabetic macular edema. Ophthalmology. 2007;114(10):1860–7.
37. Mittra RA, Savino PJ, editors. ASRS 2006 preferences and trends membership survey. Chico: American Society of Retina Specialists; 2007.
38. Arevalo JF, Sanchez JG, Fromow-Guerra J, et al. Comparison of two doses of primary intravitreal bevacizumab (Avastin) for diffuse diabetic macular edema: results from the Pan-American Collaborative Retina Study Group (PACORES) at 12-month follow-up. Graefes Arch Clin Exp Ophthalmol. 2009;247(6):735–43.
39. Kook D, Wolf A, Kreutzer T, et al. Long-term effect of intravitreal bevacizumab (avastin) in patients with chronic diffuse diabetic macular edema. Retina. 2008;28(8):1053–60.
40. Soheilian M, Ramezani A, Obudi A, et al. Randomized trial of intravitreal bevacizumab alone or combined with triamcinolone versus macular photocoagulation in diabetic macular edema. Ophthalmology. 2009;116(6):1142–50.
41. Holash J, Davis S, Papadopoulos N, et al. VEGF-Trap: a VEGF blocker with potent antitumor effects. Proc Natl Acad Sci USA. 2002;99(17):11393–8.
42. Stewart MW, Rosenfeld PJ. Predicted biological activity of intravitreal VEGF Trap. Br J Ophthalmol. 2008;92(5):667–8.
43. Kaiser PK. Vascular endothelial growth factor Trap-Eye for diabetic macular oedema. Br J Ophthalmol. 2009;93(2):135–6.
44. Do DV, Nguyen QD, Shah SM, et al. An exploratory study of the safety, tolerability and bioactivity of a single intravitreal injection of vascular endothelial growth factor Trap-Eye in patients with diabetic macular oedema. Br J Ophthalmol. 2009;93(2):144–9.
45. Browning DJ, Altaweel MM, Bressler NM, Bressler SB, Scott IU. Diabetic macular edema: what is focal and what is diffuse? Am J Ophthalmol. 2008;146(5):649–55, 655e641–6.
46. Funatsu H, Noma H, Mimura T, Eguchi S, Hori S. Association of vitreous inflammatory factors with diabetic macular edema. Ophthalmology. 2009;116(1):73–9.
47. Hoekzema R, Verhagen C, van Haren M, Kijlstra A. Endotoxin-induced uveitis in the rat. The significance of intraocular interleukin-6. Invest Ophthalmol Vis Sci. 1992;33(3):532–9.
48. Cohen T, Nahari D, Cerem LW, Neufeld G, Levi BZ. Interleukin 6 induces the expression of vascular endothelial growth factor. J Biol Chem. 1996;271(2):736–41.
49. Gardner TW, Antonetti DA. Novel potential mechanisms for diabetic macular edema: leveraging new investigational approaches. Curr Diab Rep. 2008;8(4):263–9.

50. Woolard J, Wang WY, Bevan HS, et al. VEGF165b, an inhibitory vascular endothelial growth factor splice variant: mechanism of action, in vivo effect on angiogenesis and endogenous protein expression. Cancer Res. 2004;64(21):7822–35.
51. Varey AH, Rennel ES, Qiu Y, et al. VEGF 165 b, an antiangiogenic VEGF-A isoform, binds and inhibits bevacizumab treatment in experimental colorectal carcinoma: balance of pro- and antiangiogenic VEGF-A isoforms has implications for therapy. Br J Cancer. 2008;98(8):1366–79.
52. Perrin RM, Konopatskaya O, Qiu Y, Harper S, Bates DO, Churchill AJ. Diabetic retinopathy is associated with a switch in splicing from anti- to pro-angiogenic isoforms of vascular endothelial growth factor. Diabetologia. 2005;48(11):2422–7.
53. Elman MJ, Aiello LP, Beck RW, et al. Randomized trial evaluating ranibizumab plus prompt or deferred laser or triamcinolone plus prompt laser for diabetic macular edema. Ophthalmology. 2010;117(6):1064–77. www.drcr.net.
54. Bernbaum M, Albert SG, Duckro PN. Psychosocial profiles in patients with visual impairment due to diabetic retinopathy. Diabetes Care. 1988;11(7):551–7.
55. Bernbaum M, Albert SG, Duckro PN, Merkel W. Personal and family stress in individuals with diabetes and vision loss. J Clin Psychol. 1993;49(5):670–7.
56. Wulsin LR, Jacobson AM, Rand LI. Psychosocial aspects of diabetic retinopathy. Diabetes Care. 1987;10(3):367–73.
57. Mangione CM, Lee PP, Gutierrez PR, Spritzer K, Berry S, Hays RD. Development of the 25-item National Eye Institute Visual Function Questionnaire. Arch Ophthalmol. 2001;119(7):1050–8.
58. Miskala PH, Bressler NM, Meinert CL. Relative contributions of reduced vision and general health to NEI-VFQ scores in patients with neovascular age-related macular degeneration. Arch Ophthalmol. 2004;122(5):758–66.
59. Clemons TE, Chew EY, Bressler SB, McBee W. National Eye Institute Visual Function Questionnaire in the Age-Related Eye Disease Study (AREDS): AREDS report no. 10. Arch Ophthalmol. 2003;121(2):211–7.
60. Klein R, Moss SE, Klein BE, Gutierrez P, Mangione CM. The NEI-VFQ-25 in people with long-term type 1 diabetes mellitus: the Wisconsin Epidemiologic Study of Diabetic Retinopathy. Arch Ophthalmol. 2001;119(5):733–40.
61. Jampel HD, Schwartz A, Pollack I, Abrams D, Weiss H, Miller R. Glaucoma patients' assessment of their visual function and quality of life. J Glaucoma. 2002;11(2):154–63.
62. Deramo VA, Cox TA, Syed AB, Lee PP, Fekrat S. Vision-related quality of life in people with central retinal vein occlusion using the 25-item National Eye Institute Visual Function Questionnaire. Arch Ophthalmol. 2003;121(9):1297–302.
63. Cusick M, SanGiovanni JP, Chew EY, et al. Central visual function and the NEI-VFQ-25 near and distance activities subscale scores in people with type 1 and 2 diabetes. Am J Ophthalmol. 2005;139(6):1042–50.
64. Bressler NM, Chang TS, Suner IJ, et al. Vision-related function after ranibizumab treatment by better- or worse-seeing eye: clinical trial results from MARINA and ANCHOR. Ophthalmology. 2010;117(4):747–56e744.
65. Hariprasad SM, Mieler WF, Grassi M, Green JL, Jager RD, Miller L. Vision-related quality of life in patients with diabetic macular oedema. Br J Ophthalmol. 2008;92(1):89–92.
66. Lamoureux EL, Tai ES, Thumboo J, et al. Impact of diabetic retinopathy on vision-specific function. Ophthalmology. 2010;117(4):757–65.

18
Neurodegeneration, Neuropeptides, and Diabetic Retinopathy

Cristina Hernández, Marta Villarroel, and Rafael Simó

Contents

Keywords Adrenomedullin • Angiotensin II • Brain-derived neurotrophic factor • Ciliary neurotrophic factor • Erythropoietin • Glial cell line-derived neurotrophic factor • Neuroprotectin D1 • Neurodegeneration • Neuropeptides • Pigment epithelial-derived factor • Renin-angiotensin system • Somatostatin

From: *Ophthalmology Research: Visual Dysfunction in Diabetes*
Edited by: J. Tombran-Tink et al. (eds.), DOI 10.1007/978-1-60761-150-9_18

INTRODUCTION

Diabetic retinopathy (DR) is the leading cause of blindness in working-age individuals in developed countries [1]. The tight control of blood glucose levels and blood pressure is essential in preventing or arresting DR development. However, the therapeutic objectives are difficult to achieve, and in consequence, DR appears in a high proportion of patients. When DR appears, laser photocoagulation remains as the main tool in the therapeutic armamentarium. The objective of laser photocoagulation is not to improve visual acuity but to stabilize DR, thus preventing severe visual loss. When laser photocoagulation is indicated in time, the risk of blindness is reduced by 90% in the following 5 years, and the loss of visual acuity is reduced in 50% in those patients with macular edema [2]. However, timely indication is often passed, and therefore, the effectiveness of laser photocoagulation in current clinical practice is significantly lower. In addition, laser photocoagulation destroys a part of the healthy retina, and in consequence, side effects such as loss in visual acuity, impairment of both dark adaptation and color vision, and visual field loss may appear. Vitreoretinal surgery could be indicated in advanced stages of DR (i.e., hemovitreous, retinal detachment). However, this therapeutic option requires a skillful team of ophthalmologists, is expensive, and fails in more than 30% of cases. With this scenario, it seems clear that new treatments based on greater physiopathological knowledge of DR are needed.

DR has been classically considered to be a microcirculatory disease of the retina due to the deleterious metabolic effects of hyperglycemia per se and the metabolic pathways triggered by hyperglycemia (polyol pathway, hexosamine pathway, DAG-PKC pathway, advanced glycation end products [AGEs], and oxidative stress). However, before any microcirculatory abnormalities can be detected in ophthalmoscopic examination, retinal neurodegeneration is already present. In other words, retinal neurodegeneration is an early event in the pathogenesis of DR which antedates and participates in the microcirculatory abnormalities that occur in DR [3, 4]. Therefore, the study of the mechanisms that lead to neurodegeneration will be essential for identifying new therapeutic targets in the early stages of DR.

NEURODEGENERATION AS AN EARLY EVENT IN THE PATHOGENESIS OF DR

The concept of neurodegeneration as an early event in the pathogenesis of DR was first introduced by Barber et al. [5]. These authors observed that 1 month after inducing diabetes in rats by using streptozotocin, there was a high rate of apoptosis (TUNEL-positive cells) in the neuroretina without a significant apoptosis in endothelial cells. In the same paper, the authors found a higher rate of apoptosis in the neuroretina from diabetic donors in comparison with nondiabetic donors, even in the case of a diabetic donor without microvascular abnormalities. These findings have been further confirmed in experimental models. In addition, it has been demonstrated that, apart from apoptosis, another of the features of retinal neurodegeneration is glial activation [3–7]. Our research group has been able to demonstrate that both apoptosis and glial activation occur in the retina of diabetic patients and precede microvascular abnormalities [8, 9] (Fig. 1). This is important because the experimental model in which these findings had

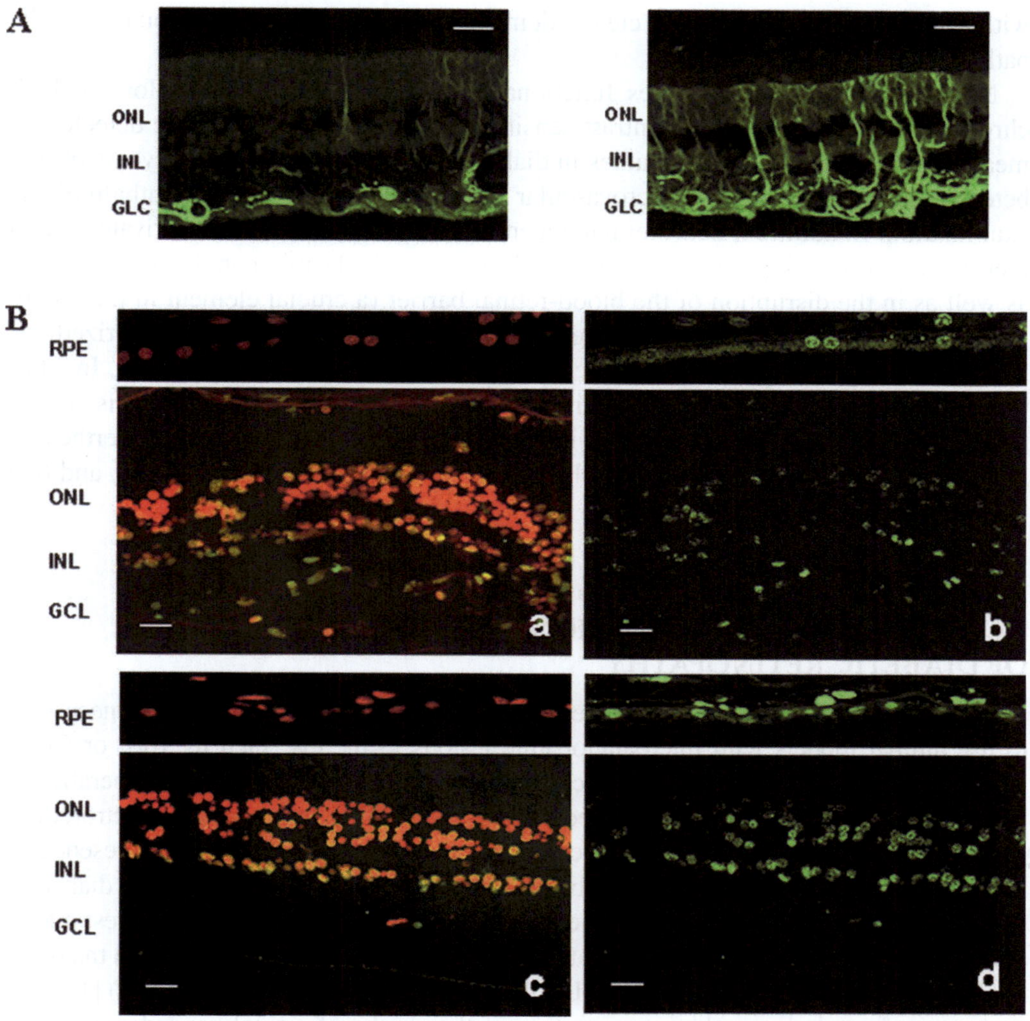

Fig. 1. Comparison of the two key elements of neurodegeneration (glial activation and apoptosis) between a representative case of diabetic patient without DR and a nondiabetic subject. As can be seen, neurodegeneration is higher in the retina from the diabetic donor. (**A**) Glial activation in the human retina. Glial fibrillar acidic protein (GFAP) immunofluorescence (*green*) from a nondiabetic donor (*left panel*) and a diabetic donor (*right panel*). (**B**) Apoptosis in the human retina. *Upper panel*: nondiabetic donor (*a*: propidium iodide, *b*: TUNEL immunofluorescence). *Low panel*: diabetic donor (*c*: propidium iodide, *d*: TUNEL immunofluorescence). *RPE* retinal pigment epithelium; *ONL* outer nuclear layer; *INL* inner nuclear layer; *GCL* ganglion cell layer. The bar represents 20 μm.

been observed was the rat with streptozotocin-induced diabetes (STZ-DM). Streptozotocin is a potent neurotoxic agent and is able to produce neural degeneration. Therefore, neurodegeneration (apoptosis + glial activation) observed in rats with STZ-DM could be due to STZ itself rather than the metabolic pathways related to diabetes. However, our observation that these changes also occur in the retina of diabetic patients and the further demonstration that they are also present in retinal explants cultured with a media with a high content of AGEs [10] clearly demonstrate that neurodegeneration is a crucial pathogenic factor of DR.

Neuroretinal damage produces functional abnormalities such as the loss of both chromatic discrimination and contrast sensitivity. These alterations can be detected by means of electrophysiological studies in diabetic patients with less than 2 years of diabetes duration, that is before microvascular lesions can be detected in ophthalmologic examination. In addition, neuroretinal degeneration will initiate and/or activate several metabolic and signaling pathways which will participate in the microangiopathic process, as well as in the disruption of the blood-retinal barrier (a crucial element in the pathogenesis of DR). Nevertheless, these metabolic pathways remain to be characterized.

The mechanisms involved in DR neurodegeneration are poorly understood. In addition, it is unknown which of the two primordial pathological elements (apoptosis or glial activation) is the first to appear and is, in consequence, the primary event. Nevertheless, it seems that these diabetes-induced changes occur in the early stages of DR, and that they are closely related.

IN VIVO EXPERIMENTAL MODELS TO STUDY RETINAL NEURODEGENERATION IN THE SETTING OF DIABETIC RETINOPATHY

Since neurodegeneration is an early event in the pathogenesis of DR, it is not necessary to use animal models with microangiopathic lesions in the eye such us Torii or Goto-Kakizaki rats. The experimental model currently used to study retinal neurodegeneration in DR is the rat with streptozotocin-induced diabetes (STZ-DM). In this model, electroretinographic abnormalities are present 2 weeks after inducing diabetes, and the presence of neural apoptosis and glial reaction can be clearly detected 1 month after starting diabetes.

Retinal ganglion cells (RGCs) are the earliest cells affected and with the highest rate of apoptosis [11]. However, an elevated rate of apoptosis has been also observed in the outer nuclear layer (photoreceptors) [12] and in the retinal pigment epithelium (RPE) [13].

As commented above, the interpretation of the results of retinal neurodegeneration in STZ-DM rats is hampered by the neurotoxic effect of STZ. It is worthy of mention that pathological changes to the brain after intraventricular injection of STZ are very similar to the neurodegeneration reported in DR [14]. Therefore, it may be advisable to use murine models with a spontaneous development of diabetes or at least experimental models in which diabetes has not been induced by a neurotoxic drug.

Mice have been much less used than rats as experimental models for the study of DR and retinal neurodegeneration. This is because they are more resistant to the STZ effect (mice need 3–5 doses of STZ to induce diabetes, whereas in rats, one dose is sufficient), have lower eyecups, and present a lower degree of lesions in comparison with rats. This relative protection to developing pathological lesions related to diabetes can be partly attributed to lower activity of aldose reductase (polyol pathway) in comparison with rats [7]. Nevertheless, because of its great potential for genetic manipulation, the mouse offers a unique opportunity to study the molecular pathways involved in disease development. Among mice, C57BL/KsJ-*db/db* is the model that best reproduces the neurodegenerative features observed in patients with DR (Fig. 2). C57BL/KsJ-*db/db* mice carry a mutation in the leptin receptor gene, and they are a model for obesity-induced

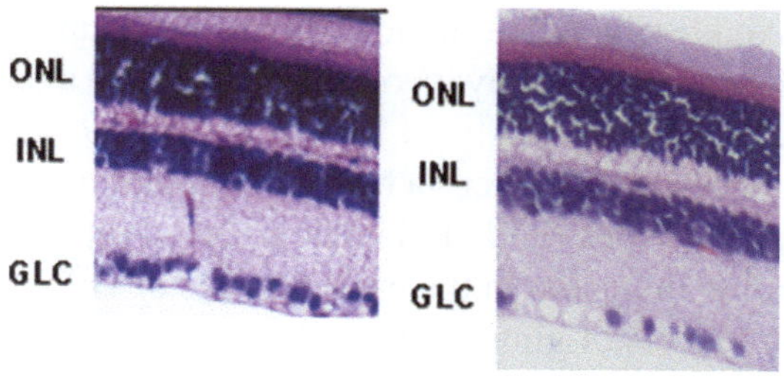

Fig. 2. Comparison of neurodegenerative features in the retina between C57BL/KsJ-*db/db* (*left panel*) and nondiabetic wild-type mice (*right panel*). *ONL* outer nuclear layer; *INL* inner nuclear layer; *GCL* ganglion cell layer.

type 2 diabetes. They develop hyperglycemia starting at ~8 weeks of age as a result of excessive food consumption. It is noteworthy that they present an abundant expression of aldose reductase in the retina (this is an important differential trait from other mouse models) [15]. Therefore, C57BL/KsJ-*db/db* seems appropriate for investigating the underlying mechanisms of retinal neurodegeneration associated with diabetes and for testing new drugs.

NEUROPEPTIDES INVOLVED IN THE PATHOGENESIS OF DR

The final result of retinal neurodegeneration is the loss of neurotransmitters such as dopamine, adrenaline, noradrenaline, acetylcholine, and several neuropeptides, which may play a critical role in the development of visual deficits in diabetes. However, rather than focusing on these deficits, it may be more interesting from both the pathophysiological and therapeutic point of views to go over the main factors accounting for this deleterious effect.

The main neurotoxic metabolite involved in diabetic retinal neurodegeneration is glutamate. In addition, there is emerging information on the neurotoxicity due to angiotensin II in the setting of the renin-angiotensin system (RAS) overexpression that exists in DR. The role of other neurotoxic factors has still to be elucidated. Among the neuroprotective factors, pigment epithelial-derived factor (PEDF), somatostatin (SST), erythropoietin (Epo), neuroprotectin D1 (NPD1), brain-derived neurotrophic factor (BDNF), glial cell line-derived neurotrophic factor (GDNF), ciliary neurotrophic factor (CNTF), and adrenomedullin (AM) have been the most extensively studied. Nevertheless, it should be noted that it is the balance between the neurotoxic and neuroprotective factors that will determine the fate of the retinal neurons.

GLUTAMATE

Glutamate is the major excitatory neurotrasmitter in the retina and is involved in neurotransmission from photoreceptors to bipolar cells and from bipolar cells to ganglion cells. However, an elevated glutamate level (which results in excessive stimulation) is implicated in the so-called "excitotoxicity" which leads to neurodegeneration. The excitoxicity of glutamate is the result of overactivation of *N*-methyl-d-aspartame (NMDA) receptors, which have been found overexpressed in DM-STZ rat [16]. There are at least two mechanisms involved in glutamate-induced apoptosis: a caspase-3-dependent pathway and a caspase-independent pathway involving calpain and mitochondrial apoptosis-inducing factor (AIF).

Elevated levels of glutamate in the retina have been found in experimental models of diabetes, as well as in the vitreous fluid of diabetic patients with PDR. However, there is no information concerning this issue in the earlier stages of DR.

The cause of the high levels of glutamate in DR has been related to a dysfunction of macroglia in metabolizing glutamate [17]. The reason for this dysfunction seems to be related to an impairment in the glutamate transporter of Müller cells due to diabetes-induced oxidative stress [18]. In addition, two enzymatic abnormalities in glutamate metabolism have been found in the diabetic retina: transamination to alpha-ketoglutarate and amination to glutamine. The reduced flux through these pathways may be associated with the accumulation of glutamate [19].

ANGIOTENSIN II

The blockade of the RAS with a converting enzyme (ACE) inhibitor or by using angiotensin II type 1 (AT_1) receptor blockers (ARBs) is one of the most used strategies for hypertension treatment in diabetic patients. Apart from the kidney, the RAS system is expressed in the eye. In the retina, RAS components are largely found and synthesized in two sites: neurons and glia cells in the inner retina and in blood vessels [20]. The finding of renin and angiotensin in glia and neurons suggests a role for these molecules in neuromodulation.

There is growing evidence that RAS activation in the eye plays an important role in the pathogenesis of DR [20]. Therefore, apart form lowering blood pressure, the blockade of the RAS could also be beneficial "per se" in reducing the development and progression of DR. In fact, recent evidence supports the concept that RAS blockade in normotensive patients has beneficial effects in the incidence and progression of DR [21–23].

The major components of RAS have been identified in ocular tissues, and they are overexpressed in the diabetic retina. Angiotensin II binds and activates two primary receptors, AT1-R and AT2-R. In adult humans, activation of the AT1-R dominates the pathological states. AT1-R activation by angiotensin II produced by the retina stimulates several pathways involved in the pathogenesis of DR such as inflammation, oxidative stress, cell proliferation, pericyte migration, remodeling of extracellular matrix by increasing matrix metalloproteinases, angiogenesis, and fibrosis [20]. In addition, AT1-R activation by angiotensin II promotes leukostasis (the inappropriate adherence of leukocytes to the retinal capillaries) and neurodegeneration [20, 24]. Apart from reducing microvascular disease, there is growing evidence pointing to neuroprotection as a relevant mechanism involved in the beneficial effects of ARBs in DR. In this regard, it has been recently reported that candesartan (the ARB with the best diffusion across the blood–brain barrier) has a neuroprotective effect after brain focal ischemia [25]. In addition, telmisartan and valsartan inhibit the synaptophysin degradation that exists in the retina of a murine model of DR [26]. Moreover, valsartan is able to prolong the survival of astrocytes and reduce glial activation in the retina of rats with hypoxia-induced retinopathy [27]. Furthermore, mitochondrial oxidative stress associated with retinal neurodegeneration has been improved using losartan in a model of spontaneously hypertensive rats [28]. Taken together, it seems that neuroprotection is a relevant mechanism involved in the beneficial effects of ARBs in DR.

PIGMENT EPITHELIAL-DERIVED FACTOR

PEDF is a 50-kDa protein encoded by a single gene that is preserved across phyla from fish to mammals. It shares homology with the serine proteinase inhibitor (Serpin) family, but lacks proteinase activity. PEDF was first purified from human RPE cells, and it was described as a neurotrophic factor with neuroprotective properties [29]. In this regard, it should be noted that intraocular gene transfer of PEDF significantly increases neuroretinal cells survival after ischemia-reperfusion injury and excessive light exposure. In addition, PEDF protects neurons from glutamate-mediated neurodegeneration.

Apart from its neurotrophic and neuroprotective properties, there is growing evidence that PEDF is among the most important natural inhibitors of angiogenesis and that it is the main factor accounting for the antiangiogenic activity of vitreous fluid where it is found in abundant quantities [30]. PEDF is responsible for the avascularity of the cornea and vitreous, and under hypoxic conditions, its secretion is decreased. In addition, elevated glucose downregulates PEDF expression in RPE cells. In 2006, the human Transport Secretion Protein-2.2 (TTS-2.2)/independent phospholipase A2 (PLA2) ξ, a novel lipase critical for triglyceride metabolism (also known in mice as adipose triglyceride lipase [ATGL], desnutrin, and patatin-like phospholipase domain-containing protein [PNPLP2]), has been identified as a specific receptor for PEDF (PEDF-R) in the retina [31]. In addition, it has been suggested that antiangiogenic and neurotrophic activities reside in separate regions of the molecule, thus suggesting that more than one receptor exists [32].

Therefore, there are enough arguments to propose PEDF as a serious new candidate for diabetic retinopathy treatment. PEDF can successfully be delivered to the eye by viral vectors. As an alternative to viral-mediated gene transfer, transplantation of autologous cells transfected with plasmids encoding for PEDF delivers therapeutic doses of PEDF to the eye. Another mechanism for delivering PEDF to the eye is to exploit its endogenous availability or production. It seems likely that much of the endogenous PEDF in the eye is bound to extracellular matrix molecules and thus may not be active. Drugs that release PEDF from these matrix molecules could increase free PEDF to therapeutic levels. In addition, levels of PEDF mRNA and secreted protein could be increased by either dexamethasone or retinoic acid [33]. Therefore, new strategies for diabetic retinopathy treatment based on PEDF activation are warranted.

SOMATOSTATIN

SST is a peptide that was originally identified as the hypothalamic factor responsible for the inhibition of the release of the growth hormone (GH) from the anterior pituitary. Subsequent studies have shown that SST has a much broader spectrum of inhibitory actions and that it is much more widely distributed in the body, occurring not only in

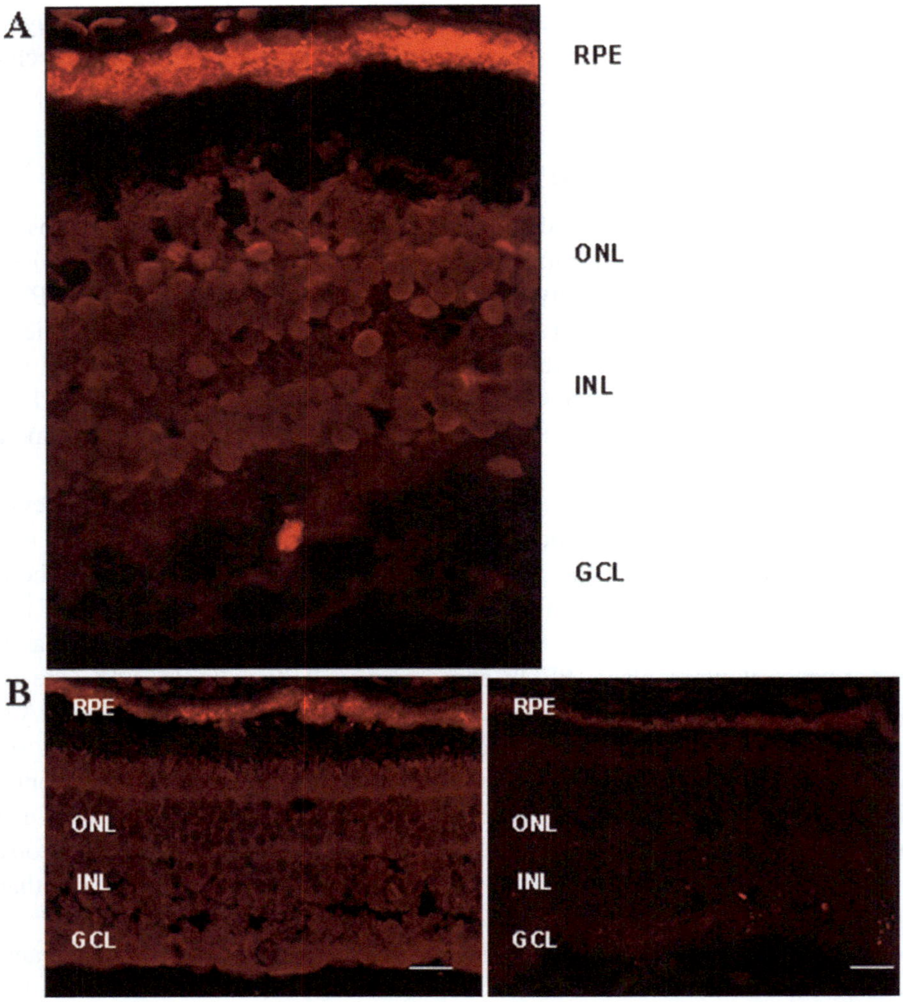

Fig. 3. (**A**) SST immunofluorescence (*red*) in the human retina showing a higher expression in RPE than in the neuroretina from human eye donors. (**B**) Higher content of SST in the retina (RPE and neuroretina) of a nondiabetic subject (*left panel*) than in a diabetic donor (*right panel*). *RPE* retinal pigment epithelium; *ONL* outer nuclear layer; *INL* inner nuclear layer; *GCL* ganglion cell layer. The bar represents 20 μm.

many regions of the central nervous system but also in many tissues of the digestive tract and in the retina [34]. SST mediates its multiple biologic effects via specific plasma membrane receptors that belong to the family of G-protein-coupled receptors having seven transmembrane domains. So far, five SST receptor subtypes (SSTRs) have been identified (SSTRs 1–5).

Neuroretina and, in particular, the amacrine cells have been classically described as the main source of SST in the retina. However, we have found that SST expression and content is higher in RPE than in the neuroretina from human eye donors [8] (Fig. 3A). Therefore, RPE rather than neuroretina is the main source of SST, at least in humans.

The amount of SST produced by the human retina is significant as can be deduced by the strikingly high levels found in the vitreous fluid [35, 36]. Apart from SST, SSTRs are also expressed in the retina, with SSTR1 and SSTR2 being the most widely expressed [34, 37, 38]. The production of both SST and its receptors simultaneously suggests an autocrine action in the human retina.

The main functions of SST for retinal homeostasis are the following: (1) SST acts as a neuromodulator through multiple pathways, including intracellular Ca^{2+} signaling, nitric oxide function, and glutamate release from the photoreceptors. In addition, a loss of SST immunoreactivity occurs after degeneration of the ganglion cells. Therefore, the neuroretinal damage that occurs in DR might be the reason for the decreased SST levels detected in the vitreous fluid of these patients. In fact, we have recently found that low SST expression and production is an early event in DR and is associated with retinal neurodegeneration (apoptosis and glial activation) [8]. (2) SST is a potent angiostatic factor. SST may reduce endothelial cell proliferation and neovascularization by multiple mechanisms, including the inhibition of postreceptor signaling events of peptide growth factors such as IGF-I, VEGF, epidermal growth factor (EGF), and PDGF [39]. (3) SST has been involved in the transport of water and ions. Various ion/water transport systems are located on the apical side of the RPE, adjacent to the subretinal space, and indeed, a high expression of SST-2 has been shown in this apical membrane of the RPE [37].

In DR, there is a downregulation of SST (Fig. 3B) that is associated with retinal neurodegeneration [8]. The lower expression of SST in RPE and neuroretina is associated with a dramatic decrease of intravitreal SST levels in both PDR [35, 36] and DME [40]. As a result, the physiological role of SST in preventing both neovascularization and fluid accumulation within the retina could be reduced, and consequently, the development of PDR and DME is favored. In addition, the loss of neuromodulator activity could also contribute to neuroretinal damage. For all these reasons, intravitreal injection of SST analogues or gene therapy has been proposed as a new therapeutic approach in DR [41].

ERYTHROPOIETIN

Erythropoietin (Epo) was first described as a glycoprotein produced exclusively in fetal liver and adult kidney that acts as a major regulator of erythropoiesis. However, Epo expression has also been found in the human brain and in the human retina [42, 43]. In recent years, we have demonstrated that not only Epo but also its receptor (EpoR) are expressed in the adult human retina (Fig. 4) [44]. Epo and EpoR mRNAs are significantly higher in RPE than in the neuroretina [44]. In addition, intravitreal levels of Epo are ~3.5-fold higher that those found in plasma [43]. The role of Epo in the retina remains to be elucidated, but it seems that it has a potent neuroprotective effect [45, 46].

Epo is upregulated in DR [43, 44, 47, 48]. Epo overexpression has been found in both the RPE and neuroretina of diabetic eyes [43, 44]. This is in agreement with the elevated concentrations of Epo found in the vitreous fluid of diabetic patients (~30-fold higher than plasma and ~10-fold higher than in nondiabetic subjects) [43]. Hypoxia is a major stimulus for both systemic and intraocular Epo production. In fact, high intravitreous levels of Epo have recently been reported in ischemic retinal diseases such as

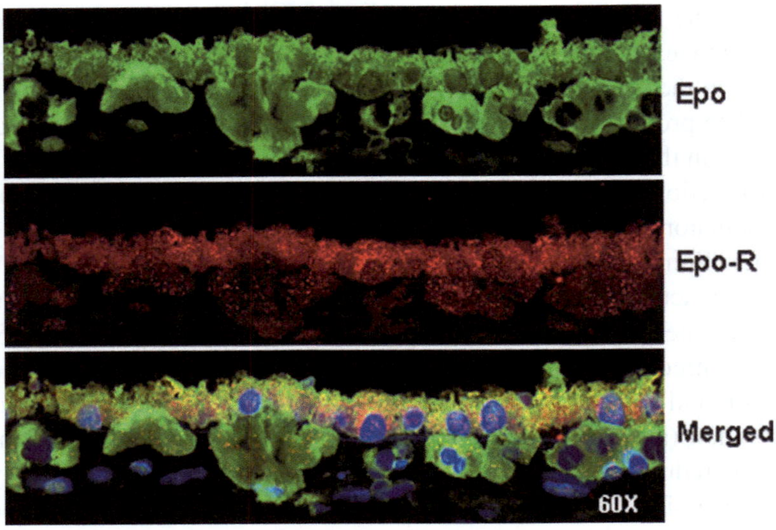

Fig. 4 Epo (*green*) and Epo receptor (*red*) immunofluorescence in the retinal pigment epithelium of human retina. In the merged image (*lower panel*), the nuclei have been stained using DAPI (*blue*)

PDR [43, 47–49]. In addition, it has been reported that Epo has an angiogenic potential equivalent to VEGF [48, 50]. Therefore, Epo could be an important factor involved in stimulating retinal angiogenesis in PDR. However, intravitreal levels of Epo have been found at a similar range in PDR to that in DME (a condition in which hypoxia is not a predominant event). In addition, intravitreal Epo levels are not elevated in nondiabetic patients with macular edema secondary to retinal vein occlusion [51]. Finally, a higher expression of Epo has been detected in the retinas from diabetic donors at early stages of DR in comparison with nondiabetic donors, and this overexpression is unrelated to mRNA expression of hypoxic inducible factors (HIF-1α and HIF-1β) [44]. Therefore, stimulating agents other than hypoxia/ischemia are involved in the upregulation of Epo that exists in the diabetic eye.

The reason why Epo is increased in DR remains to be elucidated, but the bulk of the available information points to a protective effect rather than a pathogenic effect, at least in the early stages of DR. In addition, Epo is a potent physiological stimulus for the mobilization of endothelial progenitor cells (EPCs), and therefore, it could play a relevant role in regulating the traffic of circulating EPCs toward injured retinal sites [52]. In this regard, the increase of intraocular synthesis of Epo that occurs in DR can be contemplated as a compensatory mechanism to restore the damage induced by the diabetic milieu. In fact, exogenous Epo administration by intravitreal injection in early diabetes may prevent retinal cell death and protect the BRB function in STZ-DM rats [53]. Nevertheless, in advanced stages, the elevated levels of Epo could potentiate the effects of VEGF, thus contributing to neovascularization and, in consequence, worsening PDR [52, 54].

The potential advantages of Epo or EpoR agonists in the treatment of DR include neuroprotection, vessel stability, and enhanced recruitment of EPCs to the pathological area. However, as mentioned above, timing is critical since if Epo is given at later hypoxic stages, the severity of DR could even increase. However, in the case of the eye, disease progression is easy to follow without invasive investigation and allows timing of the administration of drugs to be carefully monitored, hopefully resulting in better clinical outcomes.

DOCOSAHEXAENOIC ACID AND NEUROPROTECTIN D1

Delivery of fatty acids such as docosahexaenoic acid (DHA) to the photoreceptors is important for visual function [55]. DHA is an essential omega-3 fatty acid that cannot be synthesized by neural tissue but is required as structural protein by the membranes of neurons and photoreceptors. DHA is synthesized from its precursor, linolenic acid, in the liver and transported in the blood bound to plasma lipoprotein where it is taken up in a concentration-dependent manner. Apart from the RPE's functional integrity, DHA is the precursor of NPD1, a docosatriene that is required for the functional integrity of RPE. NPD1 protects RPE cells from oxidative stress, has an antiapoptotic effect, and inhibits the expression of IL-β-stimulated expression of COX-2 [56, 57]. Therefore, NPD1 can be postulated as a retinal neuroprotective factor.

BRAIN-DERIVED NEUROTROPHIC FACTOR

BDNF is a neurotrophin expressed in RGCs, Müller cells, and amacrine cells (both cholinergic and dopaminergic) in the retina [58]. BDNF expression is upregulated by noradrenaline [59] and is important for the survival of RGCs and amacrine cells [60]. In addition, BDNF acts as a synaptic modulator and is essential for the development of the dopaminergic network in the rodent retina [61]. Dopaminergic amacrine cell degeneration is accompanied by a reduction in BDNF levels in the retina of STZ-DM rats, and BDNF intravitreal administration can rescue these cells from neurodegeneration [62]. Furthermore, induction of BDNF expression by adrenergic agonists may provide a therapeutic approach to retinal neurodegenerative disorders including DR.

GLIAL CELL LINE-DERIVED NEUROTROPHIC FACTOR

GDNF is a 20-kDa glycosylated homodimer belonging to the TGF-β superfamily that has been recognized for its ability to increase the survival of dopaminergic cells in animal models of Parkinson's disease [63].

GDNF signals directly through the cell surface receptors (GFR-α1 and GFR-α2) and indirectly through the transmembrane Ret receptor, tyrosine kinase [64]. Both receptors have been identified on embryonic chick RGCs, as well as on amacrine and horizontal cells [65]. GFR-α2 overexpression has also been found in the epiretinal membranes of patients with PDR [66]. In addition, high levels of GFR-α2 have been detected in the vitreous fluid of PDR patients [67]. Finally, several experimental studies support the concept that GDNF exerts a neuroprotective effect in the retina.

CILIARY NEUROTROPHIC FACTOR

CNTF was first identified as a survival factor in studies involving ciliary ganglion neurons in the chick eye. CNTF is a member of the IL-6 family of cytokines and acts through a heterodimeric receptor complex composed of CNTF receptor α plus two signal-transducing transmembrane subunits, leukemia inhibitory factor receptor β (LIFR), and glycoprotein gp130 (gp-130) [68]. The CNTF-α receptor is located on Müller glial membranes [69] and practically on all retinal layers [70]. CNTF is effective in retarding retinal degeneration in several experimental models of retinitis pigmentosa, amyotrophic lateral sclerosis, and in Huntington's disease. CNTF administered as eyedrops prevents retinal neurodegeneration in STZ-DM rats [71].

ADRENOMEDULLIN

Adrenomedullin (AM) is a multifunctional protein with neuroprotective actions [72]. Administration of AM is neuroprotective in cerebral ischemia through an increase in astrocyte survival which is attributed to the inhibition of oxidative stress signaling pathways [73]. Recently, it has been demonstrated that the *AM* gene is one of those retinal genes differentially expressed in the neuroprotection conferred by hypoxic preconditioning [74] and, therefore, could be a new therapeutic target in retinal ischemic diseases such as DR.

CONCLUDING REMARKS AND THERAPEUTIC IMPLICATIONS

Neurodegeneration is an early event in the pathogenesis of DR and, apart from its own deleterious effects, participates in the microcirculatory abnormalities that occur in DR. Whereas the role of neuropathy is essential at early stages of DR, in advanced stages of DR, microangiopathy will be the main protagonist from the pathophysiological point of view.

The two capital findings of retinal neurodegeneration are apoptosis and glial activation. Although the bulk of the information on this issue has been drawn from experimental models, it has also been demonstrated in the human diabetic retina. The experimental model currently used for studying retinal neurodegeneration is the STZ-DM rat. However, STZ has neurotoxic effects, thus hampering our ability to elucidate whether the neurotoxic effects are due to the diabetic milieu or to STZ. In this regard, the use of genetically modified mice with spontaneous diabetes such as C57BL/KsJ-*db/db* seems to be more appropriate.

Elevated levels of glutamate play an essential role in the neurodegenerative process that occurs in the diabetic retina, and recent evidence suggests that overexpression of the RAS system is also an important contributing factor. Among the neuroprotective factors, PEDF, SST, and Epo seem to play a critical role, but the effect of other neurotrophic factors such as NPD1, BDNF, GDNF, CNTF, and AM should also be taken into account. In fact, the balance between neurotoxic and neuroprotective factors rather than the levels of neurotoxic factors alone is determinant for the presence or not of retinal neurodegeneration in the diabetic eye.

Intravitreal injection permits neurotrophic drugs to effectively reach the retina and overcome the potential adverse effects related to systemic administration. However, this is an invasive procedure, with the potential for blinding sequelae such as endophthalmitis and retinal detachment. Although the incidence of these serious complications is low, cumulative risk exposure may be significant for patients requiring serial treatment over many years. Therefore, attempts have been made to formulate alternative delivery vehicles for these drugs. Gene therapy and stem cell therapy are new therapeutic strategies that permit us to reduce the frequency of injections, thus reducing local side effects. The use of eyedrops is another potential route of delivery for neurotrophic factors that is currently being explored. However, clinical trials addressed to the evaluation of both the effectiveness and safety of these new treatments in arresting or preventing DR are needed.

Finally, it should be underlined that at present, the milestones in DR treatment are the optimization of blood glucose levels, lowering of blood pressure, and regular fundoscopic screening. Therefore, while we are awaiting the results of clinical research on the use of neuroprotective agents, competent strategies targeting prevention are still required to overcome this disease which is one of the major causes of blindness in the Western world.

REFERENCES

1. Congdom N, Friedman DS, Lietman T. Important causes of visual impairment in the world today. JAMA. 2006;290:2057–60.
2. Mohamed Q, Gillies MC, Wong TY. Management of diabetic retinopathy: a systematic review. JAMA. 2007;298:902–16.
3. Barber AJ. A new view of diabetic retinopathy: a neurodegenerative disease of the eye. Prog Neuropsychopharmacol Biol Psychiatry. 2003;27:283–90.
4. Antonetti DA, Barber AJ, Bronson SK, et al. JDRF Diabetic Retinopathy Center Group. Diabetic retinopathy: seeing beyond glucose-induced microvascular disease. Diabetes. 2006;55:2401–11.
5. Barber AJ, Lieth E, Khin SA, Antonetti DA, Buchanan AG, Gardner TW. Neural apoptosis in the retina during experimental and human diabetes. Early onset and effect of insulin. J Clin Invest. 1988;102:783–91.
6. Lorenzi M, Gerhardinger C. Early cellular and molecular changes induced by diabetes in the retina. Diabetologia. 2001;44:791–804.
7. Asnaghi V, Gerhardinger C, Hoehn T, Adeboje A, Lorenzi M. A role for the polyol pathway in the early neuroretinal apoptosis and glial changes induced by diabetes in the rat. Diabetes. 2003;52:506–11.
8. Carrasco E, Hernandez C, Miralles A, Huguet P, Farrés J, Simó R. Lower somatostatin expression is an early event in diabetic retinopathy and is associated with retinal neurodegeneration. Diabetes Care. 2007;30:2902–8.
9. Carrasco E, Hernández C, de Torres I, Farrés J, Simó R. Lowered cortistatin expression is an early event in the human diabetic retina and is associated with apoptosis and glial activation. Mol Vis. 2008;14:1496–502.
10. Lecleire-Collet A, Tessier LH, Massin P, et al. Advanced glycation end products can induce glial reaction and neuronal degeneration in retinal explants. Br J Ophthalmol. 2005;89:1631–3.
11. Kern TS, Barber AJ. Retinal ganglion cells in diabetes. J Physiol. 2008;586:4401–8.
12. Park SH, Park JW, Park SJ, et al. Apoptotic death of photoreceptors in the streptozotocin-induced diabetic rat retina. Diabetologia. 2003;46:1260–8.
13. Aizu Y, Oyanagi K, Hu J, Nakagawa H. Degeneration of retinal neuronal processes and pigment epithelium in the early stage of the streptozotocin-diabetic rats. Neuropathology. 2002;22:161–70.

14. Shoham S, Bejar C, Kovalev E, Schorer-Apelbaum D, Weinstock M. Ladostigil prevents gliosis, oxidative-nitrative stress and memory deficits induced by intracerebroventricular injection of streptozotocin in rats. Neuropharmacology. 2007;52:836–43.
15. Cheung AK, Fung MK, Lo AC, et al. Aldose reductase deficiency prevents diabetes-induced blood-retinal barrier breakdown, apoptosis, and glial reactivation in the retina of db/db mice. Diabetes. 2005;54:3119–25.
16. Ng YK, Zeng XX, Ling EA. Expression of glutamate receptors and calcium-binding proteins in the retina of streptozotocin-induced diabetic rats. Brain Res. 2004;1018:66–72.
17. Lieth E, Barber AJ, Xu B, et al. Glial reactivity and impaired glutamate metabolism in short-term experimental diabetic retinopathy. Penn State Retina Research Group. Diabetes. 1998;47:815–20.
18. Li Q, Puro DG. Diabetes-induced dysfunction of the glutamate transporter in retinal Müller cells. Invest Ophthalmol Vis Sci. 2002;43:3109–16.
19. Lieth E, LaNoue KF, Antonetti DA, Ratz M. Diabetes reduces glutamate oxidation and glutamine synthesis in the retina. The Penn State Retina Research Group. Exp Eye Res. 2000;70:723–30.
20. Wilkinson-Berka JL. Angiotensin and diabetic retinopathy. Int J Biochem Cell Biol. 2006; 38:752–65.
21. Chaturvedi N, Porta M, Klein R, et al. DIRECT Programme Study Group. Effect of candesartan on prevention (DIRECT-Prevent 1) and progression (DIRECT-Protect 1) of retinopathy in type 1 diabetes: randomised, placebo-controlled trials. Lancet. 2008;372:1394–402.
22. Sjølie AK, Klein R, Porta M, et al. DIRECT Programme Study Group. Effect of candesartan on progression and regression of retinopathy in type 2 diabetes (DIRECT-Protect 2): a randomised placebo-controlled trial. Lancet. 2008;372:1385–93.
23. Mauer M, Zinman B, Gardiner R, et al. Renal and retinal effects of enalapril and losartan in type 1 diabetes. N Engl J Med. 2009;361:40–51.
24. Chen P, Scicli GM, Guo M, et al. Role of angiotensin II in retinal leukostasis in the diabetic rat. Exp Eye Res. 2006;83:1041–51.
25. Krikov M, Thone-Reineke C, Müller S, Villringer A, Unger T. Candesartan but not ramipril pretreatment improves outcome after stroke and stimulates neurotrophin BNDF/TrkB system in rats. J Hypertens. 2008;26:544–52.
26. Kurihara T, Ozawa Y, Nagai N, et al. Angiotensin II type 1 receptor signaling contributes to synaptophysin degradation and neuronal dysfunction in the diabetic retina. Diabetes. 2008;57:2191–8.
27. Downie LE, Pianta MJ, Vingrys AJ, Wilkinson-Berka JL, Fletcher EL. AT1 receptor inhibition prevents astrocyte degeneration and restores vascular growth in oxygen-induced retinopathy. Glia. 2008;56:1076–90.
28. Silva KC, Rosales MA, Biswas SK, Lopes de Faria JB, Lopes de Faria JM. Diabetic retinal neurodegeneration is associated with mitochondrial oxidative stress and is improved by angiotensin receptor blocker in a model that combines hypertension and diabetes. Diabetes. 2009;58:1382–90.
29. Barnstable CJ, Tombran-Tink J. Neuroprotective and antiangiogenic actions of PEDF in the eye: molecular targets and therapeutic potential. Prog Retin Eye Res. 2004;23:561–77.
30. Dawson DW, Volpert OV, Gillis P, et al. Pigment epithelium derived factor: a potent inhibitor of angiogenesis. Science. 1999;285:245–8.
31. Notari L, Baladron V, Aroca-Aguilar JD, et al. Identification of a lipase-linked cell-membrane receptor for pigment epithelium-derived factor (PEDF). J Biol Chem. 2006;281:38022–37.
32. Filleur S, Nelius T, de Riese W, Kennedy RC. Characterization of PEDF: a multi-functional serpin family protein. J Cell Biochem. 2009;106:769–75.

33. Tombran-Tink J, Lara N, Apricio SE, et al. Retinoic acid and dexamethasone regulate the expression of PEDF in retinal and endothelial cells. Exp Eye Res. 2004;78:945–55.
34. Cervia D, Casini G, Bagnoli P. Physiology and pathology of somatostatin in the mammalian retina: a current view. Mol Cell Endocrinol. 2008;286:112–22.
35. Simó R, Lecube A, Sararols L, et al. Deficit of somatostatin-like immunoreactivity in the vitreous fluid of diabetic patients: possible role in the development of proliferative diabetic retinopathy. Diabetes Care. 2002;25:2282–6.
36. Hernández C, Carrasco E, Casamitjana R, Deulofeu R, García-Arumí J, Simó R. Somatostatin molecular variants in the vitreous fluid: a comparative study between diabetic patients with proliferative diabetic retinopathy and nondiabetic control subjects. Diabetes Care. 2005;28:1941–7.
37. Lambooij AC, Kuijpers RW, van Lichtenauer-Kaligis EG, Kliffen M, Baarsma GS, van Hagen PM, et al. Somatostatin receptor 2A expression in choroidal neovascularization secondary to age-related macular degeneration. Invest Ophthalmol Vis Sci. 2000;41:2329–35.
38. Klisovic DD, O'Dorisio MS, Katz SE, et al. Somatostatin receptor gene expression in human ocular tissues: RT-PCR and immunohistochemical study. Invest Ophthalmol Vis Sci. 2001;42:2193–201.
39. Davis MI, Wilson SH, Grant MB. The therapeutic problem of proliferative diabetic retinopathy: targeting somatostatin receptors. Horm Metab Res. 2001;33:295–9.
40. Simó R, Carrasco E, Fonollosa A, García-Arumí J, Casamitjana R, Hernández C. Deficit of somatostatin in the vitreous fluid of patients with diabetic macular edema. Diabetes Care. 2007;30:725–7.
41. Hernández C, Simó R. Strategies for blocking angiogenesis in diabetic retinopathy by intravitreal therapy. From basic science to clinical practice. Expert Opin Investig Drugs. 2007;16:1209–26.
42. Marti HH. Erythropoietin and the hypoxic brain. J Exp Biol. 2004;207:3233–42.
43. Hernández C, Fonollosa A, García-Ramírez M, et al. Erythropoietin is expressed in the human retina and it is highly elevated in the vitreous fluid of patients with diabetic macular edema. Diabetes Care. 2006;29:2028–33.
44. García-Ramírez M, Hernández C, Simó R. Expression of erythropoietin and its receptor in the human retina: a comparative study of diabetic and nondiabetic subjects. Diabetes Care. 2008;31:1189–94.
45. Jelkmann W. Effects of erythropoietin on brain function. Curr Pharm Biotechnol. 2005; 6:65–79.
46. Becerra SP, Amaral J. Erythropoietin: an endogenous retinal survival factor. N Engl J Med. 2002;347:1968–70.
47. Katsura Y, Okano T, Matsuno K, et al. Erythropoietin is highly elevated in vitreous fluid of patients with proliferative diabetic retinopathy. Diabetes Care. 2005;28:2252–4.
48. Watanabe D, Suzuma K, Matsui S, et al. Erythropoietin as a retinal angiogenic factor in proliferative diabetic retinopathy. N Engl J Med. 2005;353:782–92.
49. Inomata Y, Hirata A, Takahashi E, Kawaji T, Fukushima M, Tanihara H. Elevated erythropoietin in vitreous with ischemic retinal diseases. Neuroreport. 2004;15:877–9.
50. Jaquet K, Krause K, Tawakol-Khodai M, Geidel S, Kuck KH. Erythropoietin and VEGF exhibit equal angiogenic potential. Microvasc Res. 2002;64:326–33.
51. García-Arumí J, Fonollosa A, Macià C, et al. Vitreous levels of erythropoietin in patients with macular oedema secondary to retinal vein occlusions: a comparative study with diabetic macular oedema. Eye. 2009;23:1066–71.
52. Chen J, Connor KM, Aderman CM, Smith LE. Erythropoietin deficiency decreases vascular stability in mice. J Clin Invest. 2008;118:526–33.

53. Zhang J, Wu Y, Jin Y, et al. Intravitreal injection of erythropoietin protects both retinal vascular and neuronal cells in early diabetes. Invest Ophthalmol Vis Sci. 2008;49:732–42.
54. Grant MB, Boulton ME, Ljubimov AV. Erythropoietin: when liability becomes asset in neurovascular repair. J Clin Invest. 2008;118:467–70.
55. Bazan NG, Gordon WC, Rodriguez de Turco EB. Docosahexaenoic acid uptake and metabolism in photoreceptors: retinal conservation by an efficient retinal pigment epithelial cell-mediated recycling process. Neurobiol Essent Fatty Acids. 1992;318:295–306.
56. Mukherjee PK, Marcheselli VL, Serhan CN, Bazan NG. Neuroprotectin D1: a docosahexaenoic acid-derived docosatriene protects human retinal pigment epithelial cells from oxidative stress. Proc Natl Acad Sci USA. 2004;101:8491–6.
57. Bazan NG. Neurotrophins induce neuroprotective signaling in the retinal pigment epithelial cell by activating the synthesis of the anti-inflammatory and antiapoptotic neuroprotectin-1. Adv Exp Med Biol. 2008;613:39–44.
58. Seki M, Nawa H, Fukuchi T, Abe H, Takei N. BDNF is upregulated by postnatal development and visual experience: quantitative and immunohistochemical analyses of BDNF in the rat retina. Invest Ophthalmol Vis Sci. 2003;44:3211–8.
59. Seki M, Tanaka T, Sakai Y, et al. Müller cells as a source of brain-derived neurotrophic factor in the retina: noradrenaline upregulates brain-derived neurotrophic factor levels in cultured rat Müller cells. Neurochem Res. 2005;30:1163–70.
60. Kido N, Tanihara H, Honjo M, et al. Neuroprotective effects of brain-derived neurotrophic factor in eyes with NMDA-induced neuronal death. Brain Res. 2000;884:59–67.
61. Cellerino A, Pinzón-Duarte G, Carroll P, Kohler K. Brain-derived neurotrophic factor modulates the development of the dopaminergic network in the rodent retina. J Neurosci. 1998;18:3351–62.
62. Seki M, Tanaka T, Nawa H, et al. Involvement of brain-derived neurotrophic factor in early retinal neuropathy of streptozotocin-induced diabetes in rats: therapeutic potential of brain-derived neurotrophic factor for dopaminergic amacrine cells. Diabetes. 2004;53:2412–9.
63. Lin LF, Doherty DH, Lile JD, Bektesh S, Collins F. GDNF: a glial cell line-derived neurotrophic factor for midbrain dopaminergic neurons. Science. 1993;260:1130–2.
64. Sariola H, Saarma M. Novel functions and signalling pathways for GDNF. J Cell Sci. 2003;116:3855–62.
65. Karlsson M, Lindqvist N, Mayordomo R, Hallböök F. Overlapping and specific patterns of GDNF, c-ret and GFR alpha mRNA expression in the developing chicken retina. Mech Dev. 2002;114:161–5.
66. Harada T, Harada C, Mitamura Y, et al. Neurotrophic factor receptors in epiretinal membranes after human diabetic retinopathy. Diabetes Care. 2002;25:1060–5.
67. Nishikiori N, Mitamura Y, Tashimo A, et al. Glial cell line-derived neurotrophic factor in the vitreous of patients with proliferative diabetic retinopathy. Diabetes Care. 2005;28:2588.
68. Stahl N, Yancopoulos GD. The tripartite CNTF receptor complex: activation and signaling involves components shared with other cytokines. J Neurobiol. 1994;25:1454–66.
69. Peterson WM, Wang Q, Tzekova R, Wiegand SJ. Ciliary neurotrophic factor and stress stimuli activate the Jak-STAT pathway in retinal neurons and glia. J Neurosci. 2000;20:4081–90.
70. Beltran WA, Rohrer H, Aguirre GD. Immunolocalization of ciliary neurotrophic factor receptor alpha (CNTFR alpha) in mammalian photoreceptor cells. Mol Vis. 2005;11:232–44.
71. Aizu Y, Katayama H, Takahama S, Hu J, Nakagawa H, Oyanagi K. Topical instillation of ciliary neurotrophic factor inhibits retinal degeneration in streptozotocin-induced diabetic rats. Neuroreport. 2003;14:2067–71.

72. Miyashita K, Itoh H, Arai H, et al. The neuroprotective and vasculo-neuro-regenerative roles of adrenomedullin in ischemic brain and its therapeutic potential. Endocrinology. 2006;147:1642–53.
73. Xia CF, Yin H, Borlongan CV, Chao J, Chao L. Adrenomedullin gene delivery protects against cerebral ischemic injury by promoting astrocyte migration and survival. Hum Gene Ther. 2004;15:1243–54.
74. Thiersch M, Raffelsberger W, Frigg R, et al. Analysis of the retinal gene expression profile after hypoxic preconditioning identifies candidate genes for neuroprotection. BMC Genomics. 2008;9:73.

19
Glial Cell–Derived Cytokines and Vascular Integrity in Diabetic Retinopathy

Shuichiro Inatomi, Hiroshi Ohguro, Nami Nishikiori, and Norimasa Sawada

CONTENTS

Keywords Blood retinal barrier • Inflammatory cytokines • Tight junctions • Retinoic acid

INTRODUCTION

Normal functions and environments of the retina are preferentially performed under homeostatic conditions which are exclusively maintained by the blood retinal barrier (BRB) [1, 2]. The BRB is composed of the inner BRB and the outer BRB. Endothelial cells of the retinal capillaries form the inner BRB, and pigment epithelial cells form the outer BRB. The structure of the inner BRB is considered to be analogous to that of the blood–brain barrier (BBB). The capillary endothelial cells of the BRB (hereafter, BRB is used to indicate the inner BRB) have highly impermeable tight junctions between endothelial cells composing the biological barrier, the most important cellular apparatus for the regulation of the paracellular passage [3]. In addition, the retinal capillaries are surrounded by end-feet of glial cells, similar to the BBB (Fig. 1A). It is believed that the glial cells have been supposed to enhance the barrier function of the BRB whose permeability is known to be regulated by glial cell–derived cytokines [4–6]. Thus, the retinal endothelial and glial cells form a functional unit of the biological barrier of the BRB to maintain retinal

From: *Ophthalmology Research: Visual Dysfunction in Diabetes*
Edited by: J. Tombran-Tink et al. (eds.), DOI 10.1007/978-1-60761-150-9_19

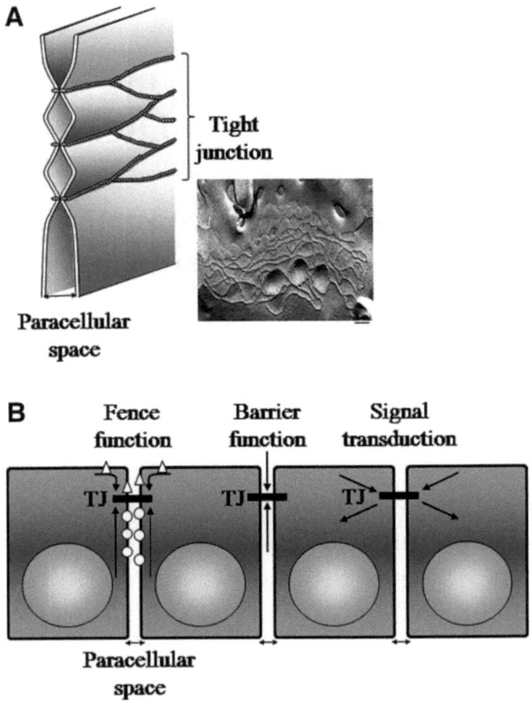

Fig. 1. Schematic representation of TJ. (**A**) *Left panel*, in this structural model of TJ, there are a number of intercrossing TJ strands (depicted as *small dots*) and three so-called kissing points of TJ. *Right panel*, freeze-fracture replica of a TJ. The TJ consists of an anastomosing network of strands that form irregular interstrand compartments and is comprised of a large number of protein components, including membrane proteins such as occludin and claudins, as well as cytoplasmic scaffolding proteins such as ZO-1. Scale bar, 50 nm. (**B**) In polarized cells, TJs are positioned at the boundary of the apical and basolateral plasma membrane domains to maintain cell polarity by forming a fence. TJs also seal cells together to generate the primary barrier and prevent diffusion of solutes through the paracellular pathway. In addition, a certain type of TJ protein such as occludin is a signaling molecule that has functions in receiving environmental cues and transmitting signals inside the cells.

homeostatic conditions, and this is often destroyed under pathological conditions, such as diabetic retinopathy [7, 8], uveitis [9], and other ocular inflammation and ischemia [10].

In diabetic retinopathy, microvascular complications such as macular edema and retinal neovascularization cause adult blindness of patients with diabetes mellitus [11]. During the pathological progression of diabetic retinopathy, leukocyte binding to the retinal vascular endothelium detected as an initial event results in early BRB breakdown, capillary nonperfusion, and endothelial cell death [12–15]. Also possible are molecular events within the initial stage of the diabetic retinopathy, an increase of vascular permeability caused by the breakdown of BRB, and upregulation of several cytokines and intracellular adhesion molecules. These pathological events merge to contribute to the development of retinal ischemia, diabetic macular edema, and neovascularization. In fact, during this pathological progression of the diabetic retinopathy, several intracellular adhesion molecules, including sICAM-1 and sVCAM-1 [16, 17], and inflammatory cytokines, including TNF-α and IL-1β [18, 19], VEGF [17, 20], GDNF [21, 22], and IL-6 [23], and others, are induced by high levels of glucose in vitro and in vivo, and high concentrations of these mediators are detected in vitreous or plasma specimens from patients with diabetic retinopathy. Based upon the release profiles of these mediators from pericytes, it was speculated that TNF-α and IL-1β are initially released and trigger the release of intercellular adhesion molecule-1 (ICAM-1) and sVCAM-1, which affect leukostasis, and VEGF, GDNF, and IL-6, which induce vascular permeability during the initial stage of the diabetic retinopathy [24].

In this chapter, to get a better understanding of the pathophysiological roles of glial cell–derived cytokines in the diabetic retinopathy, we focus on the structural and functional aspects of the BRB and its modulation by cytokines derived from glial cells under pathological conditions at an early phase of diabetic retinopathy. In addition, we also describe the possible treatment and prevention of the retinopathy with retinoic acid analogue that affects the glial cell–derived cytokines.

STRUCTURAL AND FUNCTIONAL ASPECTS OF THE BLOOD-RETINAL BARRIER (BRB)

The BRB Functional Unit Composed of Glial and Endothelial Cells

The endothelial cells of the BRB form near continuous sheets because of impermeable tight junctions (Fig. 1B). They also show no fenestration and few pinocytic vesicles. These distinctive features of the BRB-forming endothelial cells from capillary endothelial cells in other tissues maintain unique microenvironment essential for functions of retinal cells. Thus the tight junction of the BRB-forming endothelial cells is a substantial barrier that strictly regulates the paracellular pathways between the cells. Another unique feature of the endothelial cells that form the BRB and the BBB is that the capillaries forming the barrier are almost all ensheathed by vascular feet of astrocytes [25]. The anatomical relationship between the endothelial cells and astrocytes has prompted some of the researchers to explore a functional relationship between these cells. In fact, the characteristics of endothelial cells of the BBB are induced using chick-quail transplantation.

The transplantation of astrocytes into the avascular space of the anterior eye chamber showed that the capillaries that invaded the chamber were similar in characteristics to the BBB-forming endothelial cells. In vitro, a combination of an astrocyte-conditioned medium and cAMP made a tight junction resistant to the paracellular passage [26–28]. Since paracellular passage between the endothelial cells, mostly provided by the structural organization of tight junctions, astrocytes are strongly suggested to secrete some mediators that regulate paracellular passage, in terms of regulation of the tight junctions between the endothelial cells [29]. These findings strongly suggest an important insight into the permeability of the BBB between endothelial cells and astrocytes. In other words, it is feasible that astrocytes regulate the barrier function of the BRB, in terms of impermeability, in a paracrine manner.

Tight Junctions Between Endothelial Cells Are Substantial Barrier of the BRB

Tight junctions, the most apical component of intercellular junctional complexes, separate the apex from the basolateral cell surface domains to establish cell polarity (performing the function of a fence) [26–29] (Fig. 2). Tight junctions also possess a barrier function, inhibiting the flow of solutes and water through the paracellular space [26–29].

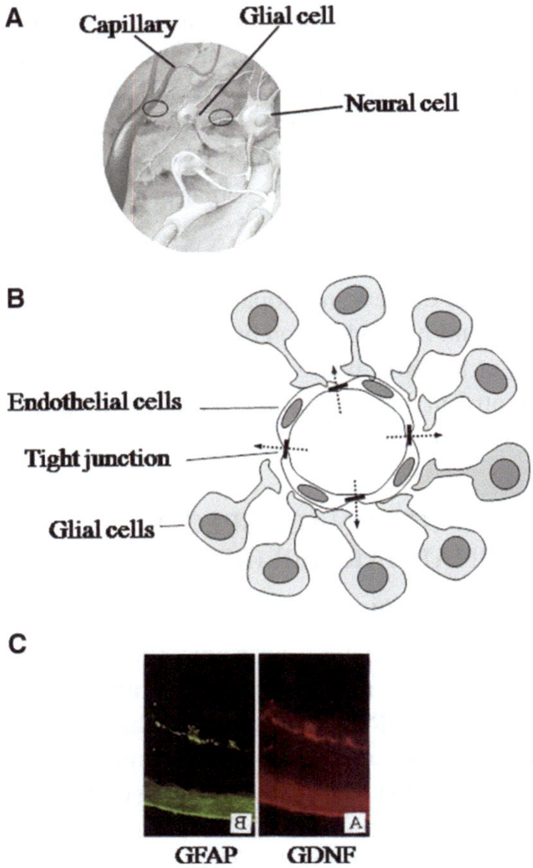

Fig. 2. Glial cell as a main component of BRB. (**A**) Schematic presentation of BRB. Note that cytoplasm of glial cell associates both with neural cell and capillary endothelium (*open circles*). (**B**) BRB is a biological unit comprised of specialized endothelial cells firmly connected by intercellular TJs and the endothelium-surrounding glial cells. Glial cell–derived cytokines such as VEGF and GDNF closely associate with the vascular integrity, which is regulated by modulating the TJ function of capillary endothelium in a paracrine manner. (**C**) BRB-forming glial cell expresses GDNF in the murine retina. Glial cell is highlighted by *red* in the retina, which is stained with anti-GFAP, a specific marker for glial cell in central nervous system and retina (**a**, *left panel*). GDNF expression shows similar distribution in *green*, suggesting that glial cell expresses GDNF protein (**B**, *right panel*).

They form a particular netlike meshwork of fibrils created by the integral membrane proteins, occludin and claudin, and members of the Ig superfamilies JAM and CAR [30]. Several peripheral membrane proteins related to tight junctions, such as ZO-1, ZO-2, ZO-3, 7H6 antigen, cingulin, symplekin, Rab3B, Ras target AF-6, and ASIP, an atypical protein kinase C-interacting protein, have been reported [3, 25, 31]. Recently, a new integral membrane protein tricellulin was also identified at tricellular contacts, which consist of three epithelial cells and have a barrier function [32].

ZO-1 and ZO-2 can independently determine whether and where claudins are polymerized [33]. Thus tight junctions are considered to be a large complex composed of at least 40 known proteins. Within the tight junction proteins, claudins with 20–27 kDa are the most indispensable proteins because they are solely capable of forming tight junction strands. The claudin family consists of 24 members, and, in general, more than two claudin members are expressed in epithelial and endothelial cells. Claudins are tetraspam proteins with a cytoplasmic N-terminus, two extracellular loops, and a C-terminus [30]. They have a PDZ (PSD-95/Dlg/ZO-1) binding motif at their C-terminus which is tethered to a PDZ domain of scaffold proteins such as ZO-1 and ZO-2. Since claudin family is solely able to form tight junction strands, endothelial permeability, in terms of the barrier function, depends on claudin expression.

In the endothelial cell forming of the BBB and/or the BRB, expression of claudin-1, claudin-3, claudin-5, and claudin-12 was identified by immunostaining and Western blot analyses [31]. Despite four isoforms of claudin being expressed in the BBB, claudin-5-deficient mice died in the first day after birth [34]. Furthermore, claudin-5 is shown to be indispensable for the BBB because claudin-5 functions as a barrier against small molecules. Expression of claudin-5 is regulated by a transcription factor SOX-18 in endothelial cells [35]. Recently VE-cadherin has been shown to upregulate claudin-5 expression by inhibition of transcriptional factor Fox01 [36]. Claudin-5 is phosphorylated at threonine 207 by PKA [37, 38] and Rho-A [39]. Regarding claudin-3, it has been reported that the canonical Wnt signal upregulates claudin-3 expression in cultured mouse brain microvascular endothelial cells, although the signal is very low after birth [40]. On the other hand, the regulation and functions of the other isoforms of the claudin family expressed in the BBB are yet unknown.

MAJOR CYTOKINES DERIVED FROM GLIAL CELLS AFFECTING TIGHT JUNCTIONS OF THE BRB

As major cytokines involved in the pathogenesis of the diabetic retinopathy, TNF-α, IL-1β, VEGF, GDNF, and IL-6 have been identified and characterized as described below. These cytokines were identified within vitreous specimens and their concentrations were significantly elevated in diabetic retinopathy [16–23].

TNF-α

TNF-α, a multifunctional proinflammatory cytokine belonging to the tumor necrosis factor (TNF) superfamily, is mainly secreted by macrophages and binds and functions through its specific receptors TNFRSF1A/TNFR1 and TNFRSF1B/TNFBR. Functionally, TNF-α is involved in the regulation of a wide spectrum of biological processes, including cell proliferation, differentiation, apoptosis, lipid metabolism, and coagulation. TNF-α has also been implicated in a variety of diseases, including autoimmune diseases, insulin resistance, and cancer [41, 42]. In diabetic retinopathy, TNF-α is identified as playing a role in promoting angiogenesis by altering endothelial cell morphology and stimulates mesenchymal cells to generate extracellular matrix proteins [43–45]. The susceptibility to diabetic retinopathy has been associated with the TNF-α gene polymorphism and expression of HLA-DR3 and HLA-DR-4 phenotypes [45]. As a possible contribution of TNF-α in the pathogenesis of the diabetic retinopathy, TNF-α induces adhesion of leukocytes to vascular endothelium by mediating increased production of adhesion molecules, such as ICAM-1 and platelet endothelial adhesion molecule-1 (PECAM-1) [12–16]. TNF-α is also known to affect the tight junctions between epithelium cells, thus increasing the flow of solutes across the epithelium [46].

IL-1β

IL-1β is a member of the interleukin 1 cytokine family. IL-1β and eight other interleukin 1 family genes form a cytokine gene cluster on chromosome 2. This is produced by activated macrophages as a proprotein, which becomes active through the proteolytic process by caspase 1 (CASP1/ICE). IL-1β is a pivotal mediator of the inflammatory response and is involved in a variety of cellular activities, including cell proliferation, differentiation, and apoptosis [47]. Similar to TNF-α, IL-1β also induces ICAM-1- and PECAM-1-induced leukostasis during the initial stage of the diabetic retinopathy [12–16]. IL-1β, in addition to acting directly, induces VEGF [48], TNF-α [49], and PEG2, and PEG2, in turn, can induce VEGF [50], emphasizing the complex interaction. Thus, TNF-α and IL-1β can increase vascular endothelial permeability.

VEGF

Vascular endothelial growth factor (VEGF) is a hypoxia-induced angiogenic and vasopermeability factor which is mainly involved in the pathogenesis of diabetic retinopathy by playing a role of leukocyte-mediated breakdown of the BRB and retinal neovascularization [51–55]. Based upon an experimental diabetes rat model, retinal VEGF levels increase with associated upregulation of ICAM-1 in retinal endothelia cells and its ligands, the β2-integrins, on the surfaces of peripheral blood neutrophils [56]. These molecular events result in an increased adhesion of leukocytes, predominantly neutrophils, and a concomitant increase in retinal vascular permeability. In experimental models, the intravitreal injection of VEGF in fact induced the retinal vascular changes including retinal leukostasis and concomitant BRB breakdown [57, 58]. In turn, these changes were abolished by the addition of inhibitors of VEGF, ICAM-1, or β2-integrin [59, 60]. In terms of the effects of VEGF on tight junctions of the BRB, in diabetes, several types of advanced glycation end-derivatives (AGEs), which are formed by a nonenzymatic reaction under hyperglycemic conditions, increase the expression of VEGF, and hypoxia induces VEGF expression. These conditions result in the disruption of the BRB, in diabetic retinopathy, because VEGF affects the expression of claudin-5 [61] and occludin [62].

GDNF

Glial cell line–derived neurotrophic factor (GDNF) was originally identified as a neurotrophic differentiation factor for dopaminergic neurons in the central nervous system and retina, and much has subsequently been learned about the neuroprotective effects of GDNF [63]. In a series of studies, we demonstrated that BRB-forming capillary endothelial cells express GDNF family receptor α1, a receptor for GDNF, and that GDNF enhances the barrier function of tight junctions in cultured endothelial cells [64]. We also demonstrated that glial cells in the retina show constitutive expression of GDNF, suggesting that retinal glia potentially regulates the permeability of the BRB [65]. In addition, AGEs increase the expression of VEGF while simultaneously decreasing GDNF expression from glial cells [4]. Additionally, they induce apoptosis in pericytes in diabetic retinopathy. These findings suggest that AGE-mediated phenotypic alterations of glial cells in hyperglycemia result in an increase of the vascular permeability of the BRB in vitro and lead causally to BRB breakdown in the diabetic retina [4].

APKAP12

A-kinase anchor protein 12 (APKAP12) is a putative tumor suppressor linked with protein A, and protein kinase C serves as a scaffolding protein in signal transduction. Src-suppressed C-kinase substrate (SSeCKS), the rodent ortholog of human AKAP12, is identified to be important for mouse brain homeostasis by regulating BBB formation [66]. Recently, VEGF has been reported to be downregulated by A-kinase anchor protein 12 (APKAP12), which in turn causes upregulation of angiopoietin-1 in glia cells [67]. Thus, it is suggested that APKAP12 may be involved in the BRB formation through antiangiogenesis and barriergenesis during the retinal development, and its defect can lead to a loss of tight junction components resulting in BRB dysfunctions.

IL-6

IL-6 is a cytokine that functions in inflammation and the maturation of B cells. IL-6 is primarily produced at sites of acute and chronic inflammation, where it is secreted into the serum and induces a transcriptional inflammatory response through the IL-6 receptor alpha. The functioning of IL-6 is implicated in a wide variety of inflammation-associated disease states, such as diabetes mellitus and systemic juvenile rheumatoid arthritis [47]. Similar to TNF-α, intravitreal injection of IL-6 has been reported to induce an ocular inflammation by breaking the BRB [68].

A POSSIBLE TREATMENT OF THE RETINOPATHY WITH RETINOIC ACID ANALOGUES

Retinoic acid (RA) is an established signaling molecule that is involved in a variety of neuronal functions, such as the development, regeneration, and maintenance of the nervous system [69, 70]. Such RA signaling is thought be assessed by binding to a transcription factors comprising the heterodimer of the RA receptor (RAR) and retinoic X receptor (RXR). In each receptors, three genes (α, β, and γ) are present, and together, the heterodimeric pair binds to a DNA sequence termed as a retinoic acid–response element (RARE). In addition to ligand binding, phosphorylation of the receptors and recruitment of coactivator or cosuppressors are required for the induction or suppression of gene transcription [71]. At present, more than 500 genes have been identified as RA-responsive [72].

Thang et al. reported that RA also plays a pivotal role in the induction of GDNF expression and its responsiveness in rat superior cervical ganglia [73]. This allows us to speculate that RA may also enhance GDNF expression in the retina and affect the barrier function of TJ in the BRB resulting in suppression of the vascular permeability. Consistent with this hypothesis, real-time PCR, semiquantitative RT-PCR, and ELISA demonstrated significant upregulation of GDNF and downregulation of VEGF by all-trans RA (atRA), a RAR pan-agonist and Am580 (4(5,6,7,8-tetrahydro-5,5,8,8-tetramethyl-2-naphtamido) benzoic acid) in glial cells. In contrast, such effects were not observed by 9-*cis*-RA, an RXR agonist, or RAR or RXR antagonists. In addition, RARα agonists enhanced the expression of glial fibrillary acidic protein (GFAP), an intermediate filament protein that is thought to be specific for glial cell in central nervous system and glial cells in the retina (Fig. 2).

We recently demonstrated that GDNF secreted from glia cells plays an important role in the regulation of vascular permeability of the BRB and the BBB in a biological unit comprised of capillary endothelial cells and glial cells [5, 6]. As shown in Fig. 3, recombinant GDNF and RARα stimulants significantly enhanced the TER and inhibited the flux through endothelial cells, which indicates enhancement of the permeability of the BRB. Furthermore, these effects were affected by the addition of GDNF-specific siRNA, which selectively silenced the constitutive expressed GDNF in glial cells. Upon systemic administration of RARα stimulants to a mouse model with diabetic retinopathy, vascular leakage of the mouse retina was significantly reduced (Fig. 4). Taken together, this RARα-mediated enhancement of the barrier function of the BRB is sufficient for significant reductions of vascular leakage and angiogenesis in the diabetic retina, suggesting that RARα significantly antagonizes the loss of TJ integrity induced under diabetes. As expected, upon administration with RARα stimulants, the expression levels of endothelial TJ proteins such as claudin-5, a major determinant of vascular permeability; occludin; and ZO-1 were markedly increased, indicating that RARα stimulants regulate barrier functions through modulation of expression of a number of TJ-associated genes [21]. Thus, it is very likely that RAs upregulate expression of GDNF in glial cells and GDNF then induces the TJ-associated gene-expression alterations in endothelial cells.

Regarding possible molecular mechanisms of RA-dependent upregulation of GDNF, it has been reported that RARα transcriptionally may stimulate GDNF expression through the p300/CREB-binding protein (CBP)–signal transducer and the activator of the transcription 3 (STAT3) pathway [21]. Consistently, as indicated in Fig. 4, we found that the treatments with atRA and Am580 remarkably increased the levels of p300/CBP, STAT3, smad1, Notch, Hes-1, and Hes-5 mRNA in glial cells. To confirm this possible mechanism responsible for the RAs-mediated GDNF upregulation, a ~1.8 kb putative promoter fragment including the transcription initiation codon was isolated and made into a deletion mutant (~1.2 kb) that lacked putative p300-binding motifs for promoter assay. atRA and Am580 significantly enhanced the promoter activity of GDNF, whereas a deletion mutant showed a marked decrease of the promoter activities. Furthermore, p300 was selectively recruited to the GDNF promoter after treatments with RAs, indicating that the expression of GDNF is exclusively regulated through the recruitment of an RARα-driven *trans*-acting coactivator to the ~1.8 kb 5′-flanking fragment of the GDNF promoter.

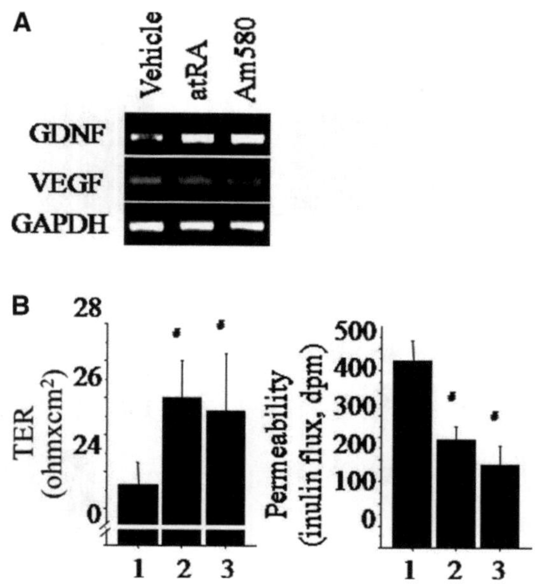

Fig. 3. Glial cell–derived cytokines regulate the vascular permeability in vitro. (**A**) Semi-quantitative RT-PCR analysis showing that expression of GDNF and VEGF is modulated in human astrocytes after treatments with 100 nM atRA and 10 nM Am580. RAs such as atRA and RARα stimulants Am580 upregulate GDNF mRNA expression and conversely decrease VEGF. (**B**) atRA and Am580-mediated gene-expression alteration is sufficient to promote endothelial barrier function. Primary cultures of bovine brain microvascular endothelial cell were grown to confluence on transwell semipermeable membranes (pore size, 0.4 μm). In our coculture experiments, glial cells cultured in the lower chamber of the transwell were treated with 100 nM atRA or 10 nM Am580 for 8 h and cocultured with endothelial cells that were grown to confluence on transwell membranes in the upper chamber. Transendothelial electrical resistance (TER) was measured using an EVOM voltohmmeter, and electric resistance was expressed in standard units of Ω cm^2. Paracellular tracer flux was measured by applying [^{14}C]-mannitol at 1×10^5 dpm/well and [^{14}C]-inulin at 5×10^5 dpm/well onto an endothelial monolayer in the apical compartment, and the samples were collected from the basolateral compartment in a time-dependent manner. Radioactivity of [^{14}C] was counted by scintillation counter. Group 1: cells treated with vehicle only; Group 2: cells treated with atRA; Group 3: cells treated with Am580. #: $p<0.05$, vs. cells treated with vehicle.

CONCLUSION

In this chapter, we described the BRB under physiological and diabetic conditions. Three conclusions reached are as follows: (1) The BRB is composed of glia and endothelial cells. The relationship between these cells is deeply functional as well as anatomical. (2) The barrier function of endothelial tight junctions, in terms of permeability of the BRB, is predominantly regulated by cytokines derived from glial cells. This fact clearly shows that glial cells are a promising therapeutic target of diabetic retinopathy, even at an

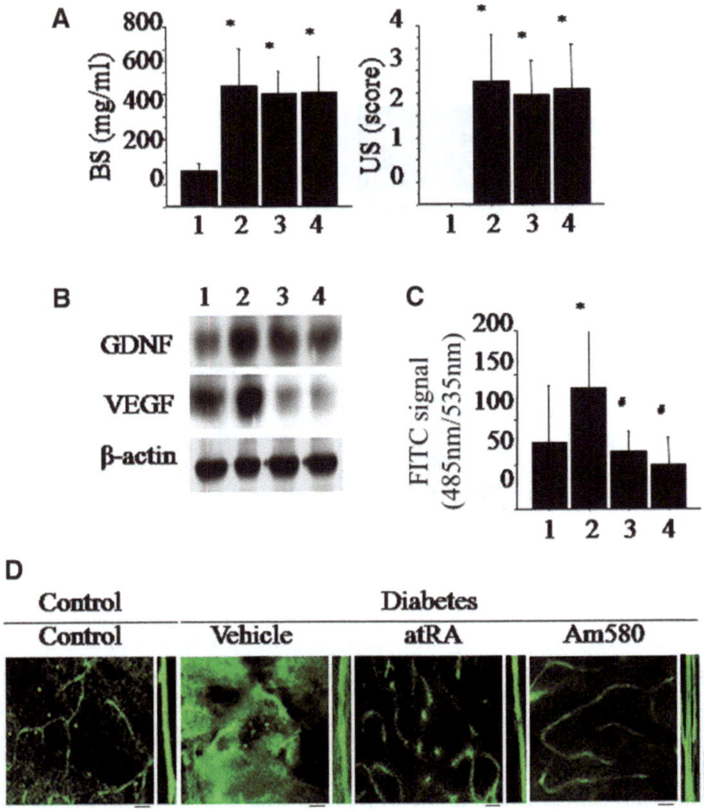

Fig. 4. RARα-mediated phenotypic transformation of glial cells antagonizes the loss of TJ integrity induced under diabetes. C57BL/6 male mice (5 weeks old) were intraperitoneally injected with 40 mg/kg streptozotocin for 5 consecutive days. Fourteen weeks after the verification of diabetes, mice were treated with 1.0 mg/kg atRA every day or 3.75 mg/kg Am580 every other day for 1 week. To examine the leakage of retinal vessels, we injected 50 mg/kg fluorescein isothiocyanate (FITC) dextran dissolved in saline into mice via the vena cava, and the mice were sacrificed and the bilateral eyes enucleated 5 min after the FITC injection under general anesthesia. FITC concentration was measured using right eye. Left eyes were flat mounted, and the FITC dextran–perfused retinas were analyzed by laser-scanning confocal analysis. To provide a quantitative control, the FITC concentration in cardiac blood of each mouse was calculated. (**A**) Blood sugar (BS) and urinal sugar (US) were increased in diabetic mice. US was assessed as follows: score 0, negative (–); score 1, slightly positive (±); score 2, weakly positive (+); score 3, moderately positive (++); and score 4, strongly positive (+++). Note that RAs did not affect these parameters, indicating evidence that RA is not a drug for diabetes. (**B**) Western blot analysis to demonstrate the increase of GDNF and decrease of VEGF expression in the mouse eye by the treatment of RAs. (**C**, **D**) FITC leakage from diabetic retina was assessed by quantification of FITC (**C**) and laser-scanning confocal microscope (**D**). FITC leakage is clearly observed in diabetic mice; however, phenotypic alterations mediated by RARα were sufficient for inhibiting the vascular leakage to maintain vascular integrity in the retinal microenvironment. Scale bars, 100 μm. Group 1: control animals; Group 2: diabetic mice without the treatment; Group 3: atRA-treated diabetic mice; and Group 4: Am580-treated diabetic mice. *: $p<0.05$, vs. control animals; #: $p<0.05$, vs. animals treated without RAs.

early phase of the disease. (3) RAs are promising candidates to prevent the progression of diabetic eye diseases, including retinopathy. It is worthy of mention that RAs down-regulate VEGF expression in glial cells.

Lastly, we hope that many researchers focus their attention on the regulation of tight junctions of the BRB from the viewpoints mentioned above. In particular, since tight junctions of the BRB-forming endothelial cells are regulated by cytokines in a paracrine manner, mechanisms of cytokine secretion from glial cells should be elucidated to develop a new rational therapy of diabetic retinopathy.

REFERENCES

1. Kim JH, Kim JH, Park JA, et al. Blood-neural barrier: intercellular communication at glio-vascular interface. J Biochem Mol Biol. 2006;39:339–45.
2. Cunha-Vaz JG. The blood-retinal barriers system. Basic concepts and clinical evaluation. Exp Eye Res. 2004;78:715–21.
3. Sawada N, Murata M, Kikuchi K, et al. Tight junctions and human diseases. Med Electron Microsc. 2003;36:147–56.
4. Miyajima H, Osanai M, Chiba H, et al. Glyceroaldehyde-derived advanced glycation end-products preferentially induce VEGF expression and reduce GDNF expression in human astrocytes. Biochem Biophys Res Commun. 2005;330:361–6.
5. Igarashi Y, Utsumi H, Chiba H, et al. Glial cell line-derived neurotrophic factor induces barrier function of endothelial cells forming the blood-brain barrier. Biochem Biophys Res Commun. 1999;261:108–12.
6. Igarashi Y, Chiba H, Sawada N, et al. Expression of receptors for glial cell line-derived neurotrophic factor (GDNF) and neurturin in the inner blood–retinal barrier of rats. Cell Struct Funct. 2000;25:237–41.
7. Felinski EA, Antonetti DA. Glucocorticoid regulation of endothelial cell tight junction gene expression: novel treatments for diabetic retinopathy. Curr Eye Res. 2005;30:949–57.
8. Kaur C, Foulds WS, Ling EA. Blood-retinal barrier in hypoxic ischaemic conditions: basic concepts, clinical features and management. Prog Retin Eye Res. 2008;27:622–47.
9. Luna JD, Chan CC, Derevjanik NL, et al. Blood-retinal barrier (BRB) breakdown in experimental autoimmune uveoretinitis: comparison with vascular endothelial growth factor, tumor necrosis factor α, and interleukin-1β-mediated breakdown. J Neurosci Res. 1997;49:268–80.
10. Kaur C, Foulds WS, Ling EA. Blood-retinal barrier in hypoxic ischaemic conditions: basic concepts, clinical features and management. Prog Retin Eye Res. 2008;27:622–47.
11. Frank RN. Diabetic retinopathy. N Engl J Med. 2004;350:48–58.
12. Miyamoto K, Khosrof S, Bursell SE, et al. Vascular endothelial growth factor (VEGF)-induced retinal vascular permeability is mediated by intercellular adhesion molecule-1 (ICAM-1). Am J Pathol. 2000;156:1733–9.
13. Joussen AM, Poulaki V, Qin W, et al. Retinal vascular endothelial growth factor induces intercellular adhesion molecule-1 and endothelial nitric oxide synthase expression and initiates early diabetic retinal leukocyte adhesion in vivo. Am J Pathol. 2002;160:501–9.
14. Qaum T, Xu Q, Joussen AM, et al. VEGF-initiated blood-retinal barrier breakdown in early diabetes. Invest Ophthalmol Vis Sci. 2001;42:2408–13.
15. Ishida S, Usui T, Yamashiro K, et al. VEGF164 is proinflammatory in the diabetic retina. Invest Ophthalmol Vis Sci. 2003;44:2155–62.
16. Barile GR, Chang SS, Park LS, Reppucci VS, Schiff WM, Schmidt AM. Soluble cellular adhesion molecules in proliferative vitreoretinopathy and proliferative diabetic retinopathy. Curr Eye Res. 1999;19:219–27.

17. Hernández C, Burgos R, Cantón A, Garcia-Arumi J, Segura RM, Simó R. Vitreous levels of vascular cell adhesion molecule and vascular endothelial growth factor in patients with proliferative diabetic retinopathy: a case-control study. Diabetes Care. 2001;24:516–21.
18. Demircan N, Safran BG, Soylu M, Ozcan AA, Sizmaz S. Determination of vitreous interleukin-1 (IL-1) and tumor necrosis factor (TNF) levels in proliferative diabetic retinopathy. Eye. 2006;20: 1366–9.
19. Doganay S, Everekliogl u C, Er H, et al. Comparison of serum NO, TNF-α, IL-1β, sIL-2R, IL-6 and IL-8 levels with grades of retinopathy in patients with diabetes mellitus. Eye. 2002;16:163–70.
20. Patel JI, Tombran-Tink J, Hykin PG, Gregor ZJ, Cree IA. Vitreous and aqueous concentrations of proangiogenic, antiangiogenic factors and other cytokines in diabetic retinopathy patients with macular edema: Implications for structural differences in macular profiles. Exp Eye Res. 2006;82:798–806.
21. Nishikiori N, Osanai M, Chiba H, et al. Glial cell-derived cytokines attenuates the breakdown of vascular integrity in diabetic retinopathy. Diabetes. 2007;56:1333–40.
22. Nishikiori N, Osanai M, Chiba H, et al. Glial cell line-derived neurotrophic factor in the vitreous of patients with proliferative diabetic retinopathy. Diabetes Care. 2005;28:2588.
23. Yuuki T, Kanda T, Kimura Y, et al. Inflammatory cytokines in vitreous fluid and serum of patients with diabetic vitreoretinopathy. J Diabetes Complications. 2001;15:257–9.
24. Nebme A, Edelman J. Dexamethasone inhibits high glucose-, TNF-α, and IL-1β-induced secretion of inflammatory and angiogenic mediators from retinal microvascular pericytes. Invest Ophthalmol Vis Sci. 2008;49:2030–8.
25. Abott NJ, Rönnbäck L, Hansson E. Astrocyte-endothelial interactions at the blood-brain barrier. Nat Rev Neurosci. 2006;7:41–53.
26. Balda MS, Matter K. Tight junctions at a glance. J Cell Sci. 2008;15:3677–82.
27. Tsukita S, Furuse M, Itoh M. Structural and signalling molecules come together at tight junctions. Curr Opin Cell Biol. 1999;11:628–33.
28. Tsukita S, Furuse M, Itoh M. Multifunctional strands in tight junctions. Nat Rev Mol Cell Biol. 2001;2:285–93.
29. Matter K, Balda MS. Signalling to and from tight junctions. Nat Rev Mol Cell Biol. 2003;4:225–36.
30. Chiba H, Osanai M, Murata M, Kojima T, Sawada N. Transmembrane proteins of tight junctions. Biochim Biophys Acta. 2008;1778:588–600.
31. Schneeberger EE, Lynch RD. The tight junction: a multifunctional complex. Am J Physiol Cell Physiol. 2004;286:1213–28.
32. Ikenouchi J, Furuse M, Furuse K, Sasaki H, Tsukita S, Tsukita S. Tricellulin constitutes a novel barrier at tricellular contacts of epithelial cells. J Cell Biol. 2005;171:939–45.
33. Umeda K, Ikenouchi J, Katahira-Tayama S, et al. ZO-1 and ZO-2 independently determine where claudins are polymerized in tight-junction strand formation. Cell. 2006;25(126): 741–54.
34. Nitta T, Hata M, Gotoh S, et al. Size-selective loosening of the blood-brain barrier in claudin-5-deficient mice. J Cell Biol. 2003;161:653–60.
35. Fontijn RD, Volger OL, Fledderus JO, Reijerkerk A, de Vries HE, Horrevoets AJ. SOX-18 controls endothelial-specific claudin-5 gene expression and barrier function. Am J Physiol Heart Circ Physiol. 2008;294:891–900.
36. Taddei A, Giampietro C, Conti A, et al. Endothelial adherens junctions control tight junction by VE-cadherin-mediated upregulation of claudin-5. Nat Cell Biol. 2008;10:923–34.
37. Ishizaki T, Chiba H, Kojima T, et al. Cyclic AMP induces phosphorylation of claudin-5 immunoprecipitates and expression of claudin-5 gene in blood-brain-barrier endothelial cells via protein kinase A-dependent and -independent pathways. Exp Cell Res. 2003;290:275–88.

38. Soma T, Chiba H, Kato-Mori Y, et al. Thr(207) of claudin-5 is involved in size-selective loosening of the endothelial barrier by cyclic AMP. Exp Cell Res. 2004;300:202–12.
39. Yamamoto M, Ramirez SH, Sato S, et al. Phosphorylation of claudin-5 and occludin by rho kinase in brain endothelial cells. Am J Pathol. 2008;172:521–33.
40. Liebner S, Corada M, Bangsow T, Gerhardt H, Dejana E, et al. Wnt/β-catenin signaling controls development of the blood-brain barrier. J Cell Biol. 2008;183:409–17.
41. Sun M, Fink PJ. A new class of reverse signaling costimulators belongs to the TNF family. J Immunol. 2007;179:4307–12.
42. Vinay DS, Kwon BS. TNF superfamily: costimulation and clinical applications. Cell Biol Int. 2009;33:453–65.
43. Limb GA, Chignell AH, Green W, LeRoy F, Dumonde DC. Distribution of TNF α and its reactive vascular adhesion molecules in fibrovascular membranes of proliferative diabetic retinopathy. Br J Ophthalmol. 1996;80:168–73.
44. Limb GA, Soomro H, Janikoun S, Hollifield RD, Shilling J. Evidence for control of tumour necrosis factor-alpha (TNF-α) activity by TNF receptors in patients with proliferative diabetic retinopathy. Clin Exp Immunol. 1999;115:409–14.
45. Hawrami K, Hitman GA, Rema M, et al. An association in non-insulin-dependent diabetes mellitus subjects between susceptibility to retinopathy and tumor necrosis factor polymorphism. Hum Immunol. 1996;46:49–54.
46. Mullin JM, Snock KV. Effect of tumor necrosis factor on epithelial tight junctions and transepithelial permeability. Cancer Res. 1990;50:2172–6.
47. Akira S, Hirano T, Taga T, Kishimoto T. Biology of multifunctional cytokines: IL 6 and related molecules (IL 1 and TNF). FASEB J. 1990;4:2860–7.
48. Li J, Perrella MA, Tsai JC, et al. Induction of vascular endothelial growth factor gene expression by interleukin-1b in rat aortic smooth muscle cells. J Biol Chem. 1995;270:308–12.
49. Camussi G, Albano E, Tetta C, Bussolino F. The molecular action of tumor necrosis factor-α. Eur J Biochem. 1991;202:3–14.
50. Harada S, Nagy JA, Sullivan KA, et al. Induction of vascular endothelial growth factor expression by prostaglandin E2 and E1 in osteoblasts. J Clin Invest. 1994;93:2490–6.
51. Shweiki D, Itin A, Soffer D, Keshet E. Vascular endothelial growth factor induced by hypoxia may mediate hypoxia-initiated angiogenesis. Nature. 1992;359:843–5.
52. Aiello LP, Avery RL, Arrigg PG, et al. Vascular endothelial growth factor in ocular fluid of patients with diabetic retinopathy and other retinal disorders. N Engl J Med. 1994;331:1480–7.
53. IkedaE, AchenMG, BreierG, RisauW. Hypoxia-induced transcriptional activation and increased mRNA stability of vascular endothelial growth factor in C6 glioma cells. J Biol Chem. 1995;270: 19761–6.
54. Senger DR, Galli SJ, Dvorak AM, Perruzzi CA, Harvey VS, Dvorak HF. Tumor cells secrete a vascular permeability factor that promotes accumulation of ascites fluid. Science. 1983;219:983–5.
55. Dvorak HF, Brown LF, Detmar M, Dvorak AM. Vascular permeability factor/vascular endothelial growth factor, microvascular hyperpermeability, and angiogenesis. Am J Pathol. 1995;146:1029–39.
56. Barouch FC, Miyamoto K, Allport JR, et al. Integrin-mediated neutrophil adhesion and retinal leukostasis in diabetes. Invest Ophthalmol Vis Sci. 2000;41:1153–8.
57. Pierce EA, Avery RL, Foley ED, Aiello LP, Smith LE. Vascular endothelial growth factor/ vascular permeability factor expression in a mouse model of retinal neovascularization. Proc Natl Acad Sci USA. 1995;92:905–9.
58. Murata T, Nakagawa K, Khalil A, Ishibashi T, Inomata H, Sueishi K. The relation between expression of vascular endothelial growth factor and breakdown of the blood-retinal barrier in diabetic rat retinas. Lab Invest. 1996;74:819–25.

59. Ng EW, Shima DT, Calias P, Cunningham Jr ET, Guyer DR, Adamis AP. Pegaptanib, a targeted anti-VEGF aptamer for ocular vascular disease. Nat Rev Drug Discov. 2006;5:123–32.
60. Comer GM, Ciulla TA. Pharmacotherapy for diabetic retinopathy. Curr Opin Ophthalmol. 2004;15:508–18.
61. Rodewald M, Herr D, Fraser HM, Hack G, Kreienberg R, Wulff C. Regulation of tight junction proteins occludin and claudin 5 in the primate ovary during the ovulatory cycle and after inhibition of vascular endothelial growth factor. Mol Hum Reprod. 2007;13:781–9.
62. Antonetti DA, Barber AJ, Khin S, Lieth E, Tarbell JM, Gardner TW. Vascular permeability in experimental diabetes is associated with reduced endothelial occludin content: vascular endothelial growth factor decreases occludin in retinal endothelial cells. Penn State Retina Research Group. Diabetes. 1998;47:1953–9.
63. Lin LF, Dohery DH, Lile JD, Bektesh S, Collins F. GDNF: a glial cell line-derived neurotrophic factor for midbrain dopaminergic neurons. Science. 1993;260:1130–2.
64. Henderson CE, Phillips HS, Pollock RA, et al. GDNF: a potent survival factor for motoneurons present in peripheral nerve and muscle. Science. 1994;266:1062–4.
65. Utsumi H, Chiba H, Kamimura Y, et al. Expression of GFRalpha-1, receptor for GDNF, in rat brain capillary during postnatal development of the BBB. Am J Physiol Cell Physiol. 2000;279:361–8.
66. Lee SW, Kim WJ, Choi YK, et al. SSeCKS regulates angiogenesis and tight junction formation in blood-brain barrier. Nat Med. 2003;9:828–9.
67. Choi YK, Kim JH, Kim WJ, et al. AKAP12 regulates human blood-retinal barrier formation by downregulation of hypoxia-inducible factor-1α. J Neurosci. 2007;27:4472–81.
68. Bamforth SD, Lightman S, Greenwood J. The effect of TNF-α and IL-6 on the permeability of the rat blood-retinal barrier in vivo. Acta Neuropathol. 1996;91:624–32.
69. Maden M. Retinoic acid in the development, regeneration and maintenance of the nervous system. Nat Rev Neurosci. 2007;8:755–65.
70. Blomhoff R, Blomhoff HK. Overview of retinoid metabolism and function. J Neurobiol. 2006;66:606–30.
71. Bastien J, Rochette-Egly C. Nuclear retinoid receptors and the transcription of retinoid-target genes. Gene. 2004;328:1–16.
72. Balmer JE, Blomhoff R. Gene expression regulation by retinoic acid. J Lipid Res. 2002;43:1773–808.
73. Thang SH, Kobayashi M, Matsuoka I. Regulation of glial cell line-derived neurotrophic factor responsiveness in developing rat sympathetic neurons by retinoic acid and bone morphogenetic protein-2. J Neurosci. 2000;20:2917–25.

20
Impact of Islet Cell Transplantation on Diabetic Retinopathy in Type 1 Diabetes

Iain S. Begg, Garth L. Warnock, and David M. Thompson

CONTENTS

From: *Ophthalmology Research: Visual Dysfunction in Diabetes*
Edited by: J. Tombran-Tink et al. (eds.), DOI 10.1007/978-1-60761-150-9_20

Keywords Wisconsin Epidemiologic Study of Diabetic Retinopathy • Diabetes Control and Complications Trial • Intensive insulin therapy • Diabetes Quality of Life Measure • Metabolic memory • Pancreas transplantation • Islet cell transplantation

INTRODUCTION

Diabetes Mellitus is a risk factor for other diseases, often termed complications, which impose a large and expanding healthcare problem. Type 1 diabetes accounts for about 10% of all cases of diabetes. Currently, there is a global increase in incidence of 3% per year [1], and it is predicted that the incidence will be 40% higher in 2010 than in 1998 [2]. It can be confronted today because of increased emphasis on intensive control of glycemia and control of blood pressure and lipids. The morbidity of diabetes includes blindness from retinopathy (the leading cause of new cases of legal blindness in North America in people in the age group 20–74 years [3]), end-stage renal disease, cardiovascular disease, and lower extremity amputations. In any individual, these complications are markers of the severity of the disease. In type 1 diabetes, amputation and poor visual acuity are significantly associated with mortality [4]. Epidemiology contributes to the etiology of type 1 and type 2 diabetes by refining the diagnosis and identifying and quantifying the risk factors for diabetic complications. A landmark clinical trial and observational follow-up of the efficacy of medical treatment, the Diabetes Control and Complications Trial (DCCT) [5]/Epidemiology of Diabetes Intervention and Complications (EDIC) Study [6], extended the findings of the Wisconsin Epidemiologic Study of Diabetic Retinopathy (WESDR) [7] to show beyond all doubt that diabetic microvascular and macrovascular complications can be diminished or modified through the use of intensive insulin therapy to lower HbA1c concentrations. The study also established a causal role of hyperglycemia in the development and progression of the complications. In some centers, annual incidence rates and prevalence of the complications of retinopathy are now decreasing following translation of the DCCT findings into clinical practice. Despite the benefits of improved blood glucose control in (1) reducing the incidence and progression of retinopathy, renal disease and neuropathy, (2) diminishing the risk of cardiovascular disease, and (3) increasing life expectancy, many patients fail to achieve their glycemic targets. Those who maintain good blood glucose control endure for a lifetime the burden of a rigid lifestyle to avoid hypoglycemia. This chapter focuses on the relationship between hyperglycemia and retinal microvascular disease and the benefits and adverse effects of intensive insulin therapy to provide the rationale for pancreas islet cell transplantation, an evolving treatment aimed at the lowest HbA1c that can be safely achieved to minimize long-term diabetic complications.

WHAT IS THE ASSOCIATION BETWEEN GLYCEMIA AND THE ONSET AND PROGRESSION OF RETINOPATHY, MACULAR EDEMA, AND PROLIFERATIVE RETINOPATHY IN TYPE 1 DIABETES?

Data from the WESDR, initiated in 1979–1980, identified important independent predictors of the incidence and progression of retinopathy and decreased survival [7, 8]. The study examined a sample selected from 10,135 people consisting of (1) people with diabetes developing before age 30 and taking insulin (defined as the equivalent of type 1 diabetes) and (2) people with diabetes developing after age 30 either taking insulin or not taking insulin stratified by duration of disease (defined as the equivalent of type 2 diabetes). A total of 995 persons with type 1 diabetes participated at baseline and at least one of the four follow-up examinations (including fundus photography) at 4, 10, 14, and 25 years, or died before the first follow-up examination [9–14]. This study population underwent examinations that followed a similar protocol, structured interview about medications, and objective masked recording of retinopathy using seven-standard field stereoscopic fundus photographs with a validated grading protocol modified from the Early Treatment Diabetic Retinopathy Study (ETDRS) adaptation of the modified Airlie House classification of diabetic retinopathy [15, 16] and evaluated for inter- and intraobserver errors.

With respect to the overall 25-year incidence of any retinopathy (97%), rates of progression of retinopathy (83%), progression to proliferative retinopathy (42%), improvement in retinopathy (18%), incidence of macular edema (29%), and incidence of clinically significant macular edema (17%), the strongest most consistent risk factor relationships throughout the study were with glycemia [10, 14]. In those multivariate models that included accurate baseline retinopathy severity level and baseline HbA1c, the duration of diabetes was not an independent risk factor for progression [17, 18]. The finding of a stepwise relationship between increasing HbA1c levels and increase in progression of retinopathy suggested a possible benefit in the reduction of risk of progression of retinopathy by lowering blood glucose at any level of hyperglycemia found within the population, at any time during the course of diabetes and at any level of severity before the onset of proliferative retinopathy [19].

Preliminary reports of small randomized controlled clinical trials of glycemic control recorded that after 8–12 months of follow-up, the frequency of deterioration of background retinopathy was greater in the group treated with an insulin pump than in the conventionally treated group [20, 21], especially in patients with the best glycemic control [20]. However, after 2 years of treatment, improvement of retinopathy was more frequent among patients treated with continuous subcutaneous insulin infusion (CSII) than among patients treated conventionally, although the difference was marginal [22–24]. The results of these studies and the findings of the WESDR prompted the need for larger longer controlled clinical trials of treatment in type 1 diabetes to provide unequivocal evidence as to whether or not intensive glycemic control aimed at lower levels of glycemia would reduce the development and progression of retinopathy. The Stockholm Diabetes Intervention Study (SDIS) evaluated the effect of intensified (mean HbA1c 7.1%) compared with standard treatment (mean HbA1c 8.5%) in people with type 1 diabetes. After 7.5 years of follow-up, intensified therapy reduced the risk of progression of nonproliferative retinopathy and retarded the development of "serious retinopathy" (proliferative retinopathy or macular edema requiring immediate photocoagulation) by an absolute amount of 25% [25]. The results of this trial and five other studies of more than 2 years duration were combined in a meta-analysis that confirmed that intensive therapy reduced progression of diabetic retinopathy in people with type 1 diabetes [26].

WHAT ARE THE BENEFITS AND RISKS OF REDUCING BLOOD GLUCOSE?

The DCCT was designed to assess whether an intensive treatment regimen aimed at achieving blood glucose values as close to the nondiabetic range as possible would affect the rates of onset and progression, or regression, of early retinal, renal, and neurological complications over time in insulin-dependent diabetes mellitus when compared with conventional treatment [5]. The study was performed before other potential confounding factors such as antihypertensives, blockers of the renin-angiotensin system, and lipid-lowering agents came into common use. It was a multicenter randomized prospective study (1983–1989) which involved 1,441 patients in good general health, aged 13–39 years and no severe complications and medical conditions, who were randomly assigned to receive either conventional or intensive insulin treatment [6]. The primary prevention group ($n=726$) had duration of diabetes less than 5 years, while the secondary prevention group ($n=715$) had duration of diabetes 1–15 years.

The conventional therapy group injected insulin without daily adjustments, once or twice daily and tested either urine or blood glucose daily, and received education about exercise and diet. The goals were (1) absence of symptoms of hyperglycemia, (2) absence of ketonuria, (3) maintenance of normal growth development and ideal body weight, and (4) freedom from frequent or severe hypoglycemia. The intensive therapy group received treatment with a three or more times daily insulin regimen either by injection or insulin pump with doses adjusted on the basis of self-monitored blood glucose measurements (four or more per day) and diet and exercise under the direction of an expert team. The regimen was adjusted by telephone contact and examinations were performed monthly [27]. It was determined that hemoglobin A1c (HbA1c) could be used as a surrogate marker for glycemia [28]. The targets' were preprandial blood glucose 3.9–6.7 mmol/L and postprandial blood glucose level lower than 10 mmol/L, weekly blood glucose 3 a.m. measurement higher than 3.6 mmol/L, and HbA1c values within the nondiabetic range (<6.05%) [29]. The most important primary outcome measures in the primary prevention cohort were persistent development of any retinopathy (at least one microaneurysm in either eye) at two consecutive visits scheduled at 6-month intervals, and in the secondary prevention cohort, sustained (at least two consecutive 6-month visits) three-step progression of diabetic retinopathy based on scores in both eyes.

At enrolment, the primary prevention group had no photographic evidence of retinopathy, visual acuity of 20/25 or better in each eye, and urinary albumin excretion less than 40 mg/24 h. The secondary prevention group had presence of very mild to moderate nonproliferative diabetic retinopathy (NPDR) in at least one eye and visual acuity of 20/32 or better in each eye [5].

Stereoscopic color fundus photographs of the seven-standard fields were taken every 6 months and graded in masked fashion at the University of Wisconsin Fundus Photograph Reading Center using the protocol of the ETDRS. Grades of the various lesions were used to construct an interim ETDRS score and a final score [15, 16]. Observations were performed for a mean of 6.5 years (range 3–9 years) after randomization. The study was completed by 99% of patients, and the assigned treatment was received 97% of the time [30].

Over the 9-year period of the study of both the primary and secondary prevention groups, the average difference in HbA1c between the two groups was statistically different, nearly 2% [29]. The average within-subject mean HbA1c was 9.1% in the conventional group vs. 7.2% in the intensive group. With regard to the distribution of HbA1c, 31% had a mean HbA1c between 8.5 and 9.49% in the conventional group vs. 5% of the intensive group. Conversely, among those in the intensive group, 50% had a mean HbA1c between 6.5 and 7.49% vs. 8% of the conventional group. Almost exactly 23% of intensive and conventional group subjects had a mean HbA1c between 7.5 and 8.49%.

ON AVERAGE, 3 YEARS WAS REQUIRED TO DEMONSTRATE THE BENEFICIAL EFFECT OF INTENSIVE TREATMENT

There was initial ("early") worsening of retinopathy (13.1% of subjects in the intensive insulin group and 7.6% in the conventional treatment group) in the first year of treatment [31] (except in the group with no retinopathy) similar to reports in the early feasibility studies [20–24]; then after 3 years, the rate of sustained progression was lower and the beneficial effects of intensive therapy increased over time for all retinopathy groups except moderate NPDR (43/<43), which took longer to demonstrate a beneficial effect. After 3 years, the magnitude of progression was also less as measured by the number of steps on the severity scale. These differences increased with longer follow-up and were associated with higher rates of recovery from progression of three or more steps on the scale compared to conventional therapy [32].

THE EARLIER IN THE COURSE OF DIABETES THAT INTENSIVE THERAPY IS INITIATED, EVEN BEFORE THE ONSET OF RETINOPATHY, THE GREATER THE LONG-TERM BENEFITS

In the primary prevention group on intensive insulin therapy, the 9-year cumulative incidence of developing at least one microaneurysm in persons with no diabetic retinopathy at baseline was 70% in persons with 2.5 or fewer years of duration of diabetes and 62% in persons with more than 2.5 years duration of diabetes at baseline. The 9-year cumulative incidence of sustained three-step progression in persons with diabetes duration of 2.5 years or less without retinopathy at baseline was 7% compared with 20% when the duration of diabetes was greater than 2.5 years. In the secondary prevention group on intensive insulin therapy, the 9-year cumulative incidence of sustained three-step progression in eyes with baseline level of severity 20/<20 to 35/<35 was lower compared to eyes with retinopathy severity level 43/<43 (11.5–18.2 vs. 43.8%) [33].

RISK REDUCTION IN THE PRIMARY PREVENTION COHORT

The incidence of diabetic retinopathy was reduced by 27% by intensive treatment over 9 years [33]. The adjusted mean risk of retinopathy sustained progression by three or more steps was reduced by 76% [29].

RISK REDUCTION IN THE SECONDARY PREVENTION COHORT

Intensive therapy reduced the mean risk of sustained progression by three or more steps by 65% during the entire study. Progression to severe NPDR or worse was reduced by 47%. The need for laser treatment of macular edema or proliferative retinopathy was reduced by 59%. The incidence of clinically significant macular edema in the intensive therapy group decreased but not statistically significantly [33].

THERE WAS NO GLYCEMIC THRESHOLD REGARDING PROGRESSION OF RETINOPATHY

There was significant reduction in the risk of retinopathy in an exponential relationship along the entire range of HbA1c in the study [34]. Although the magnitude of the absolute risk reduction declined with continuing proportional reductions in HbA1c, there were still meaningful further reductions in risk as HbA1c was reduced toward the normal range [35]. Each 10% reduction in HbA1c resulted in (a) 35% risk reduction in sustained onset, (b) 39% reduction in progression of three or more steps of severity, and (c) 37% reduction for development of severe NPDR or proliferative diabetic retinopathy (PDR) [34]. A simple exponential regression model showed that, in the combined groups, small differences in any given value of the HbA1c (assumed held constant over time) correspond to large differences in the cumulative incidence of sustained retinopathy progression over a period of many years [35]. The short-term (within-day) variability in blood glucose around a patient's mean value had no influence (independent of conventional therapy or intensive therapy) on the development or the progression of retinopathy [36]. However, glucose variability should be reduced as much as possible to limit hypoglycemia unawareness and severe hypoglycemia and maintain quality of life. Another study using DCCT data found that glycemic instability (SD of glucose profile set samples for each visit) had little influence on the HbA1c value of a patient [37]. Longer-term variability in HbA1c adds to the mean value in predicting microvascular complications. A 1% absolute increase in HbA1c SD results in at least a doubling in retinopathy [38].

A re-examination of previously presented DCCT findings [34] and additional analysis of DCCT data [39] show that virtually all (96%) of the beneficial effect of intensive vs. conventional therapy on progression of retinopathy and other outcomes is explained by the reductions in the mean HbA1c levels. The total glycemic exposure (HbA1c and duration of diabetes) explains only ~11% of the variation in retinopathy risk in the complete cohort. Subjects within the intensive and conventional treatment groups with similar HbA1c level over time have similar risks of retinopathy progression especially after adjusting for factors in which they differ [39].

THE RISK OF HYPOGLYCEMIA INCREASED CONTINUOUSLY BUT NOT PROPORTIONALLY AS THE GOAL OF NORMOGLYCEMIA WAS APPROACHED

Severe hypoglycemia was three times more common in the intensive therapy group compared with the conventional therapy group [29]. The rate of severe hypoglycemic episodes requiring treatment was 62/100 patients years in the intensive insulin treatment group compared with 19/100 patients years in the conventional arm of the study [40, 41]. This risk persisted over the duration of the study and was inversely correlated with the HbA1c. The risk of severe hypoglycemia within the intensive group increased exponentially as the HbA1c was reduced. Although the risk of severe hypoglycemia continues to increase at lower HbA1c values with intensive therapy, the risk gradient flattens substantially [29]. Among all risk factors for hypoglycemia, the dominant predictor was history of prior episodes of hypoglycemia [40, 42]. Among patients with HbA1c 6.0%, 21.3 events were predicted per 100 patient years.

DIABETIC KETOACIDOSIS (DKA)

In the DCCT, the risk of diabetic ketoacidosis (DKA) was similar between intensive and conventional treatment groups (1.8–2/100 patient years), despite lower HbA1c levels achieved in the intensive group [41]. Among the intensive treatment group, rates were higher for patients using CSII compared with those on multiple injections (3.09 vs. 1.39 per 100 patient years) [41]. In a meta-analysis evaluating the effect of intensive treatment on the risk of DKA using data from 14 randomized trials, the overall risk of DKA was greater for patients treated with intensive vs. conventional therapy largely due to the effect of CSII [43]. In the EURODIAB study, 8.6% of type 1 diabetes participants had been admitted to hospital for treatment of DKA in the previous 12 months [44].

EFFORTS TO NORMALIZE BLOOD GLUCOSE ARE ASSOCIATED WITH WEIGHT GAIN IN PEOPLE WITH TYPE 1 DIABETES

In the DCCT, the incidence of becoming overweight, defined as body mass index (BMI) ≥27.8 kg/m^2 for men and BMI ≥27.3 kg/m^2 for women during the median 6.5 years of follow-up was 41.5% in the intensive therapy group compared to only 26.9% in the conventional therapy group [41]. The rate of weight gain decreases with time up to 9 years [45]. Weight gain includes an increase in fat mass. The strongest predictors of weight gain were higher baseline HbA1c concentration and larger decrements in HbA1c during intensive therapy from baseline to 1 year. After adjusting for baseline HbA1c, weight, insulin dose (U/kg), and stimulated C-peptide, weight gain of experimental subjects still remained significantly greater than that of standard subjects [46]. The greater weight gain in people with severe hypoglycemia suggests that overeating is a causal factor. Insulin-induced weight gain and heightened risk of obesity, if undesirable, could diminish long-term compliance with intensive therapy and, if continued, could become a risk factor for cardiovascular disease.

CONNECTING PEPTIDE (C-PEPTIDE) RESPONDERS HAVE LESS RISK OF PROGRESSION OF RETINOPATHY

In the DCCT, 303 of 855 patients with type 1 diabetes of duration 1–5 years were C-peptide responders (C-peptide levels 0.20–0.50 pmol/mL) after ingestion of a mixed meal [47]. They were randomly assigned to receive either intensive or conventional treatment. Responders with C-peptide levels >0.50 pmol/mL were excluded from enrollment. Responders receiving intensive therapy maintained a higher stimulated C-peptide level and a lower likelihood of becoming nonresponders than did responders receiving conventional therapy. Among intensive therapy patients, responders had a lower HbA1c value, reduced risk for retinopathy progression, and a lower risk for severe hypoglycemia compared with nonresponders [47, 48]. The risk of losing C-peptide responses to stimulation was reduced by 57% by intensive treatment. The characteristic decline in β cell function was prolonged to the sixth year after initiation of intensive therapy, about 2 years beyond conventional therapy. Interestingly, no difference in the development of complications was seen between the previous responders and nonresponders in the conventional treatment group. Intensively treated nonresponders had the highest rate of severe hypoglycemia (17.3 episodes per 100 patient years). In the intensive therapy group, the adjusted odds for retinopathy were 3.2-fold higher for those with undetectable C-peptide than for those in the sustained C-peptide group. In those receiving conventional treatment, the odds of retinopathy were no different among C-peptide groups [48]. These findings support early introduction of intensive therapy to sustain endogenous insulin secretion which, in turn, is associated with better metabolic control and lower risk for hypoglycemia and progression of retinopathy. The weaker benefit of sustained C-peptide secretion in the conventional group compared with the intensive therapy group on microvascular complications suggests that glycemic control is potentially a more important factor in imparting the benefit of continuing β cell function than the direct effect of C-peptide secretion itself.

THE VALIDITY OF GENERALIZING THE DCCT RESULTS TO PATIENTS WITH INSULIN-DEPENDENT DIABETES MELLITUS IN THE GENERAL POPULATION WAS CONFIRMED

The DCCT cohort was generally similar to IDDM patients in the WESDR in terms of demography, rates of progression in the conventionally treated group, and association between HbA1c levels and progression of retinopathy [49].

WHAT ARE THE LONG-TERM EFFECTS OF INTENSIVE INSULIN THERAPY ON MICRO- AND MACROVASCULAR DISEASE?

The Epidemiology of Diabetes Interventions and Complications (EDIC) study was a multicenter, longitudinal, observational study designed to use the well-characterized DCCT cohort of >1,400 patients to determine the long-term effects of former separation of glycemic levels on micro- and macrovascular outcomes [30]. At the end of the DCCT, patients in the conventional treatment group were offered intensive therapy, and the care of all patients was transferred to their own physicians. To assess whether the benefits of intensive therapy persist, the EDIC study compared the effects of former intensive and conventional therapy on the occurrence and severity of retinopathy, nephropathy, neuropathy, and cardiovascular disease after the end of the DCCT. Retinopathy was evaluated on the basis of centrally graded fundus photographs in 1,208 subjects during the fourth year after the DCCT ended [50] and 1,211 subjects at year 10 [51].

EFFECTS OF IMPROVED CONTROL ON RETINOPATHY WERE SUSTAINED IN THE LONG-TERM

The proportion of patients who had worsening of retinopathy including PDR, macular edema, and the need for laser treatment was lower in the intensive therapy group than in the conventional therapy group in the 10 years following DCCT closeout [51]. At entry to the DCCT, the mean HbA1c level in each treatment group was 9.0%. Following 6.5 years of DCCT follow-up, the mean HbA1c levels were 7.3 and 9.0% in the intensive and conventional therapy groups, respectively. One year after the DCCT closeout, the HbA1c values had converged. Over 10 years in the EDIC study, the mean HbA1c levels in the two groups were almost the same (8.0% in the conventional therapy group vs. 7.98% in the intensive therapy group). Despite similar HbA1c levels over the period of observation following the DCCT closeout, the ongoing risk of worsening of retinopathy remained significantly reduced in the intensive compared with the conventional group ("metabolic memory"). The decrease in HbA1c from approximately 9% to about 8% resulted in only small reductions in the progression of retinopathy in the conventional group. After 4 years of follow-up in the EDIC study, 49% of the patients in the conventional treatment group had progression of retinopathy of three or more steps from the DCCT baseline as compared with 18% of the patients in the intensive therapy group [50]. After 10 years, 60.6% in the conventional group had progressed vs. 35.8% in the intensive therapy group. An important benefit was the continued significant reductions at 4 and 10 years of the EDIC follow-up in the adjusted odds of severe NPDR or worse, clinically significant macular edema, and photocoagulation. However, the difference in retinopathy between the two groups was becoming less. The odds reductions at 10 years were less than that observed at 4 years, except for those for photocoagulation. Metabolic memory waned faster in patients with more severe retinopathy than in those with milder retinopathy. As the severity of retinopathy increased at the DCCT closeout, the relative benefits of intensive therapy decreased. The risk of further progression of diabetic retinopathy increased significantly with higher HbA1c levels at DCCT baseline (19% increase in risk per 1% increase in HbA1c level), higher mean blood pressure at DCCT closeout (11% increase in risk per 5 mm increase in mean BP), and hyperlipidemia at DCCT closeout (70% increase in risk for those with hyperlipidemia vs. those without). Eighty-nine percent of the prolonged effect of DCCT intensive therapy treatment on further retinopathy progression was explained by the differences in the DCCT mean HbA1c levels, whereas the EDIC mean HbA1c levels explained only 1.6% of the prolonged intensive effect.

INTENSIVE DIABETES THERAPY HAS LONG-TERM BENEFICIAL EFFECTS ON THE RISK OF CARDIOVASCULAR DISEASE AND MORTALITY

A statistically significant benefit of intensive insulin therapy on cardiovascular disease in the DCCT/EDIC study was observed at year 11 of the EDIC study beyond the end of the DCCT. Patients who had been randomized to the intensive arm had a 42% reduction in any cardiovascular disease outcomes and a 57% reduction in the risk of nonfatal myocardial infarction, stroke, and cardiovascular death compared with those who had received conventional treatment [52]. As is the case with microvascular complications, it may be that a period of intensive diabetes management plays a greater role in lessening cardiovascular risk before atherosclerosis is well developed.

QUALITY OF LIFE MEASURE

The Diabetes Quality of Life Measure (DQOL) was designed to evaluate the relative burden of an intensive regimen among patients with type 1 diabetes enrolled in the DCCT [53]. During the DCCT, DQOL scores, functional health status, and psychological distress were similar between treatment groups. However, among intensively treated patients, symptoms of psychiatric distress rose with the frequency of hypoglycemic episodes [54]. Following islet cell transplantation, the risk of short-term clinical consequences found with intensive or conventional insulin therapy (DKA, hypoglycemia, poor glycemic control) is abolished. Quality of life appears to improve initially after islet cell transplantation due primarily to a reduced fear of hypoglycemia but declines with the loss of insulin independence [55].

"METABOLIC MEMORY": A PHENOMENON PRODUCING A LONG-TERM BENEFICIAL INFLUENCE OF EARLY METABOLIC CONTROL ON CLINICAL OUTCOMES

In the DCCT/EDIC studies, "metabolic memory" was used to describe the persistent benefit of reduced progression of retinopathy which outlasted by at least 10 years the period of good control of blood glucose on DCCT intensive insulin therapy [50].

An early glycemic environment is remembered in the target organs prone to hyperglycemic damage. The benefit is an apparent discrepancy between the level of hyperglycemia and the incidence and severity of diabetic complications. The duration of the metabolic memory can be determined by quantifying the continuing differences in the long-term clinical effects between the original treatment groups. It is not known how long the risk of progression will be delayed by the memory effect. AGE (Advanced Glycation End Products) is implicated in diabetic complications and metabolic memory [56, 57]. In the DCCT/EDIC studies, AGE formation measured in skin biopsy 1 year before DCCT closeout and 10 years later, was significantly lower in the intensive therapy subjects compared with conventional therapy subjects [58], and levels of AGEs were significantly associated with the 10-year incidence or progression of retinopathy [59].

The induction, maintenance, and "switching off" of the "metabolic memory" (a term coined by Cahill [60]) were recently reviewed by Ceriello et al. [61]. with evidence gathered in part from reviews on the pathobiology [62], reactive species [63], mitochondrial DNA mutators [64], endothelial dysfunction [65], and AGE [66]. In the hypothesis, intracellular hyperglycemia induces overproduction of superoxide, a ROS, at the mitochondrial level as a possible cause of the metabolic memory of hyperglycemic stress after glucose normalization. Overproduction of ROS is the first and key event in the activation of all other pathways involved in the pathogenesis of diabetic complications, such as the polyol pathway flux, increased AGE formation, activation of protein kinase C, and increased hexosamine pathway flux. Mitochondrial proteins are glycated in hyperglycemia, and this effect induces mitochondria to overproduce superoxide anion, a condition that does not depend on glycemic levels. Superoxide and similar reactive species target nucleic acid, proteins, and lipids/lipoproteins with a long half-life over a prolonged time. Binding of AGEs to the receptor for AGE called "RAGE" results in intracellular ROS generation, which promotes the expression of RAGE themselves. Mitochondrial DNA may influence gene expression and, at the same time, may contribute to an overgeneration of free radicals at the mitochondrial level. These self-maintaining conditions, leading to a persistent oxidative stress generation independent of the actual glycemic levels, may contribute to the appearance of the metabolic memory. It was shown in endothelial cells that reducing intracellular production of free radicals, particularly at the mitochondrial level, was capable of "switching off" the metabolic memory [67]. Additional explanations such as genetic factors may be needed to explain susceptibility to severe complications [68].

RECENT DECREASE OF ANNUAL INCIDENCE AND PREVALENCE OF RETINOPATHY

Several recent cohort studies report decreases in the cumulative incidence of progression of retinopathy [14], proliferative retinopathy [14, 69, 70], macular edema [10, 70], and new blindness in a diabetic population [71] with increasing calendar year of diagnosis. A meta-analysis, which reviewed rates of progression of diabetic retinopathy to proliferative retinopathy and/or severe visual loss in a combined population of persons with type 1 and type 2 diabetes, concluded that rates of these outcomes were lower since 1985 [72]. Estimates of the annual rates of progression of diabetic retinopathy, incidence of PDR, improvement of retinopathy, and incidence of macular edema and clinically significant macular edema for the four periods of the WESDR 25-year incidence study reveal that in contrast to the first 10 years of follow-up when incidence rates were constant, annualized incidence and progression rates decreased over the past 10–15 years [14]. Also, there was a lower prevalence and incidence of PDR in persons who were more recently diagnosed with diabetes. Coincident with these declines, there was gradual improvement in glycemic control. This data suggested the efficacy of improved glycemic control in reducing the complications of diabetic retinopathy, or else the reason for the decline in prevalence was explained by death, leading to survival of the healthiest. Also, there have been recent reports of lower prevalence of diabetic retinopathy [73], and proliferative retinopathy [74], compared with previous reports of prevalences in the WESDR baseline. However, a lower cumulative incidence of proliferative retinopathy was not found in the Pittsburgh Study [75]. The environment of the population could have changed because of an upward trend in the initiation of intensive blood glucose control by physicians as well as improved adherence to treatment by patients because of translation of the results of the DCCT. A greater percentage of people than before are monitoring blood glucose four times per day and injecting insulin three or more times daily. Also taking place is the aggressive management of both renal and cardiovascular disease with improved control of blood pressure and lipids and use of renin-angiotensin system blockers. Early blockade of the renin-angiotensin system with either an angiotensin-converting enzyme (ACE) inhibitor or an angiotensin system blocker (ARB) did not slow nephropathy progression but slowed the progression of retinopathy independently of changes in blood pressure [76].

NEED FOR A MORE PHYSIOLOGIC GLYCEMIC CONTROL REGIMEN

Does it matter by what means a reduction in glycemia is obtained? Risk gradients for progression of diabetic retinopathy as a function of mean HbA1c are similar for both intensive therapy and conventional therapy, which suggests that a similar benefit in the reduction of complications could be obtained with another less intensive, more effective regimen [34]. Diabetes management is an integral part of daily living and requires extensive education and lifestyle changes. It is difficult to sustain near-normal blood glucose continuously even with frequent corrective adjustments of insulin dosage. Intensive therapy regimen is extremely demanding and requires substantial effort and resources by the patients and the healthcare team to try to reach their goal of normal glycemia. People with type 1 diabetes have the burden of hypoglycemia, ketoacidosis, injections three to four times per day, insulin pump manipulations, self-monitoring blood glucose four times per day, adjusting insulin dose based on glucose level, meal size and composition, and activity level. In the EDIC 4-year follow-up examination, less than half of the patients in each group were performing self-monitoring of blood glucose four or more times per day [50]. It is the increased likelihood of severe hypoglycemia that reduces the capacity to increase administration of insulin to reach glycemic targets [77]. Both hypoglycemia and weight gain deter many patients and physicians from aiming at stringent control [78]. Insulin underuse may be adopted by patients attempting to limit weight gain [79, 80], as well as for manipulation, or because of recklessness, error, or fatigue in the day-to-day effort of managing diabetes. Failure to adhere to insulin treatment, expressed as a prescribed vs. dispensed index, was shown to occur in 28% of a cohort of adolescents and young adults with type 1 diabetes attending a teaching clinic and was directly associated with failure to take insulin, poor glycemic control, and acute hospital admissions for DKA [81]. Suboptimal adherence with diabetes treatment increases healthcare costs [82]. The UK Hypoglycemic Study Group [83] reported an incidence of severe hypoglycemia of 110 episodes per 100 patient years in patients with type 1 diabetes who were necessarily treated with insulin for <5 years and an incidence of 320 episodes per 100 patient years in those with type 1 diabetes for >15 years. The prospective population-based study of Donnelly et al. [84]. indicated that the incidence of any and of severe hypoglycemia was about 4,300 and 115 per 100 patient years, respectively in type 1 diabetes.

Insulin analogues (Insulin Lispro, Aspart, Glulisine, Glargine, and Detemir) were developed to provide greater efficacy, safety, and convenience compared with conventional human insulin [85]. These are synthetic insulins which have undergone small changes in amino acid sequence relative to human insulin with more rapid absorption resulting in pharmacokinetic and pharmacodynamic profiles that mimic the action of endogenous insulin more closely. Their effect, to reduce postprandial hyperglycemia, and frequency of nocturnal hypoglycemia help improve adherence with therapy in type 1 diabetes. Detemir has ability to limit weight gain [86]. The combination of both long- and short-acting insulin analogues leads to significant minor reductions in both HbA1c and nocturnal hypoglycemia in adults compared with NPH insulin and unmodified human insulin managed with a multiple injection regimen [87].

The limitations of long-acting insulin has driven the popularity of CSII but with the drawback of subcutaneous administration, carbohydrate counting, and the continued need for frequent adjustments of infusion rates based on intermittent self-monitoring of blood glucose and training in the management. Trials of modern insulin pumps report decreases of HbA1c 0.6–0.4% compared with multiple daily insulin injections with no increase in hypoglycemia [88].

Continuous glucose monitoring devices which measure interstitial glucose every 5 min for 72 h show glycemic excursions, permit immediate adjustment of insulin, and are able to alert patients to a falling glucose level. Alarm system specificity may not yet be sufficient for reliable use [89]. A meta-analysis of clinical trials comparing continuous glucose monitoring system with self-blood glucose monitoring in type 1 diabetes showed that the use of continuous monitoring did not significantly reduce HbA1c levels [90].

In the DCCT in which subjects were carefully selected for adherence with care, and received management from experts with access to full resources, 44% of patients in the intensive therapy group achieved HbA1c less than or equal to 6.05 mol/L on one or more occasions during the study. Fewer than 5% in the intensively treated group was able to maintain their HbA1c level at less than or equal to 6.05% over the course of the study [29]. At the upper limit of the nondiabetic range of 6.0%, it is estimated that there is still some risk of retinopathy progression (risk 0.52/100 patient years) [34]. Over a lifetime, even a small increment or decrement of mean HbA1c could substantially increase or decrease, respectively, the risk of progression to more severe complications.

Most countries' diabetes associations recommend achieving and monitoring an HbA1c level lower than 7.0%. European guidelines recommend HbA1c <6.5% [91]. The ADA recommends <7.0% and less than 6.0% for the individual patient [92] in whom it can be safely achieved. The Canadian Diabetes Association Clinical Practice Guidelines recommend <7.0% [93].

EFFECT OF INTENSIVE INSULIN THERAPY ON HYPOGLYCEMIA COUNTERREGULATION

The major limitation of near-normoglycemia is the increased risk of hypoglycemia due to lower blood glucose levels and/or defective hypoglycemia counterregulation. The pathophysiology of glucose counterregulation was recently reviewed by Cryer [94]. Glucagon release which is reduced in early years of diabetes history [95] cannot be improved in response to hypoglycemia with islet cell transplantation into liver [96]. Thereafter, catecholamine secretion becomes the most important hypoglycemia counterregulatory hormone [95, 97], until its secretion is impaired by hypoglycemia episodes. However, hypoglycemia counterregulation is partly restored by prevention of severe hypoglycemia and decreases in the incidences of mild hypoglycemia [98, 99]. Tight metabolic control is compatible with mostly intact hormonal hypoglycemia counterregulation for up to 7 years provided there is a low incidence of hypoglycemia [100]. Hormonal counterregulation becomes progressively impaired; even meticulous prevention of insulin-induced hypoglycemia cannot totally prevent this development in type 1 diabetic patients after a diabetes duration of more than 10 years [101].

β CELL FUNCTION

Diabetes is defined by a sole defect—the loss of β cell function below a level that is adequate to maintain euglycemia. C-peptide is co-secreted with insulin in equimolar concentration by the β cell as a by-product of the enzymatic cleavage of proinsulin to insulin. Measurement of C-peptide under standardized conditions provides a sensitive and clinically valid assessment of β cell function (endogenous insulin secretion) [102]. The diagnosis of type 1 diabetes can be enhanced with fasting C-peptide below 0.3 mmol/L or glucagon-stimulated C-peptide under 0.6 nmol/L or high concentration of islet cell antibodies, although with less sensitivity and specificity [102]. Continuing C-peptide insulin secretion is important in avoiding hypoglycemia. Stimulated C-peptide levels >0.2 pmol/mL at diagnosis is a level associated with improved control [47]. Glycemic control strongly influences the decline in β cell function in type 1 diabetes. The DCCT identified a "virtuous circle" whereby residual insulin secretion resulted in better glucose control with less hypoglycemia and slower progression to vascular complications; better glucose control, in turn, prolonged β cell function [47]. The natural course of β cell destruction is heterogeneous in the decline of C-peptide. Some patients lose β cells completely soon after the onset of diabetes whereas others retain minute residual β cells over a long period [103]. In healthy subjects, C-peptide response increases with age between 19 and 78 years [104, 105]. Age is a determinant of residual insulin production at diagnosis [105, 106]. Older patients have much higher stimulated C-peptide levels at the time of diagnosis of diabetes [107], whereas younger patients have a more rapid decline in C-peptide. As shown in the DCCT and other studies [108], there is clear evidence of clinical benefit from preserved β cell function in patients with type 1 diabetes: less retinopathy, less neuropathy, and less hypoglycemia with intensive insulin therapy [47, 48]. Potential direct effects of C-peptide include prevention and amelioration of diabetic nephropathy, and neuropathy, decreasing microalbuminuria, improved survival of renal allograft, activation of eNOS with microvasodilatory effect, and activation of transcription factor [107, 109]. The reason that β cells should persist and enhance insulin secretion is not clear.

WHOLE PANCREAS TRANSPLANTATION

Pancreas transplantation has a dramatic effect in curing the problems of hypoglycemia but with the risk of perioperative morbidity. Whole pancreas transplantation is chosen principally for the patient with end-stage renal failure who had or plans to have a kidney transplant. Procedures for whole pancreas transplantation include simultaneous pancreas-kidney transplantation (SPK), pancreas after kidney transplantation (PAK), and pancreas transplant alone (PTA). The procedures have been reviewed by Larsen [110] and by Meloche [111]. The most commonly performed is SPK with enteric drainage and systemic venous outflow. Among whole pancreas transplant procedures, patient survival at 1 and 5 years is 95 and 85%, respectively. Graft survival for whole pancreas recipients is 90, 70, and 45% at 1, 5, and 10 years after transplantation, respectively [112, 113].

Contraindications to transplantation of any type include active malignancy or infection, psychiatric disease which could decompensate after a large surgery, and subjects' inability or unwillingness to take immunosuppressant medications regularly. In successful pancreas transplantation, hypoglycemia-induced glucagon secretion and hepatic glucose production are normalized [114]. Common complications and technical failures include thrombosis, bleeding, leak (6.5%), auto-rejection (12%), and CMV infection (10%) [112, 113].

EFFECT OF SPK TRANSPLANTATION ON DIABETIC RETINOPATHY

Advances in the technical aspects of SPK transplantation which have gradually improved patient and graft survival have lead to investigations of the effect of pancreas transplantation (with normalization or near-normalization of HbA1c) on secondary diabetic complications. The results of investigations of the effect of euglycemia on retinopathy following pancreas transplant can be broadly subdivided as follows:

1. No benefit [115–124]
2. Limited benefit [125–130]
3. Stable for 5 years [131]
4. Significant benefit in the latest publications [132, 133]

The control group has often been too small to be compared statistically with the treatment group. Treatment groups have been controlled for immunosuppression (failed pancreas graft, kidney transplant) or for functioning graft (nontransplanted exogenous insulin-treated patients with type 1 diabetes). There have been no reports of a beneficial or adverse effect of immunosuppressives on retinal microvasculature. The inclusion of a steroid in an immunosuppressive regimen increases the prevalence of cataracts. Studies on diabetic retinopathy have been difficult to interpret because of the predominance of advanced proliferative retinopathy with high prevalence of scatter laser treatment. Candidates for pancreas and kidney transplants are usually persons who are approaching end-stage renal failure or are undergoing renal dialysis after about 22 years duration of disease. Eyes that worsen develop vitreous hemorrhage, traction retinal detachment, neovascular glaucoma, or have persistent neovascularization requiring more scatter laser treatment. It may not be possible to separate the effects of normoglycemia from those of scatter laser treatment. Normoglycemia has a beneficial effect on reducing the risk of severe NPDR and early proliferative retinopathy [25, 33]. In a study of the influence of glycemic control on the initial response to scatter laser treatment in patients with high-risk PDR, Kotoula et al. [134] reported that good metabolic control (HbA1c < 8%) before, during, and after treatment was associated with greater regression of proliferative retinopathy following scatter laser treatment.

An early report from Ramsay et al. [124] evaluated retinopathy using seven-field stereo fundus photography in 22 subjects (34 eyes) who had undergone successful pancreas transplantation and 16 subjects (28 eyes) with failed pancreas transplantation. In the study group, 10 of 22 patients were blind in one eye, and the average grade of retinopathy pretransplant was P6 (elevated neovascularization). Both groups had been treated with scatter laser (44% eyes, study group; 78% eyes, control group). In the study group, after a follow-up of mean 24 months, 19 eyes (56%) were unchanged and 15 eyes (44%) progressed by two or more grades (5 were laser-treated) compared with the control group in which 13 eyes (46%) were unchanged, 14 (56%) progressed, and 1 (4%) improved. One 16-year-old subject with HbA1c ≥ 16% progressed from mild NPDR to PDR within 6 months of successful pancreas transplantation suggesting accelerated retinopathy associated with profound reduction in HbA1c. Wang et al. [116] compared the progress of diabetic retinopathy in 51 subjects with type 1 diabetes who had undergone successful kidney and pancreas transplantation with 21 subjects who had undergone successful renal transplant alone. In both groups, the prevalence of scatter laser treatment was 72%. An evaluation was made on a side-by-side comparison and grading of seven-field stereo fundus photographs taken weeks before transplantation and again 1 year later. They reported that in the 1 year follow-up, near-normalization of glycemia (mean HbA1c 6.4%) associated with successful transplantation did not accelerate retinopathy nor has a beneficial effect on the progression of advanced retinopathy. Pearce et al. [131] found that more than 90% of 17 subjects (33 eyes) examined 5 years after successful SPK had stable retinopathy defined as absence of need for laser treatment. In this series, 25 of 33 eyes had received scatter laser treatment prior to transplantation. Marchetti et al. [127] reported a series of 28 PTA (without controls) followed for 6–24 months with clinical and photographic documentation in which diabetic retinopathy improved in 58.8%, stabilized in 35.3%, and worsened in 5.9%. Improvement was defined as regression to a lower retinopathy grade in the nonproliferative group and a significant reduction of retinal lesions in the proliferative and laser-treated group. Chow et al. [115] examined 46 patients after successful SPK and 8 patients with failed SPK (functioning kidney). Sixty-eight of 82 eyes (83%) had undergone scatter or focal laser treatment. Seventy-five percent of eyes in each group remained stable. In the SPK group, 14% improved and 10% progressed from no diabetic retinopathy to NPDR (2 eyes), or progressed from NPDR to active proliferation (4 eyes), or progressed from inactive to active proliferation (12 eyes). Six eyes developed macular edema. Koznarova et al. [125] compared 43 normoglycemic SPK patients with 45 control patients, who either failed SPK or kidney transplants. On the basis of examinations before and 1 year posttransplantation, the rates of improvement, stabilization, and deterioration in the normoglycemic group were 21.3, 61.7, and 17.0%, respectively and, in the control group, 6.1, 48.8, and 45.1%, respectively. Almost 80% of eyes in each group had been treated with laser. Giannarelli et al. [132] followed 48 patients for median 17 months after successful SPK and a control group of 43 patients with type 1 diabetes. In the NPDR SPK group (12 patients), 42% improved, 25% did not change, and 33% progressed by one grade. In the laser-treated/proliferative retinopathy (LT/PDR) SPK group (36 patients), 97% did not change and one patient (3%) worsened. In the LT/PDR group, the number of improved/stabilized patients was significantly higher in the transplanted group than in the control group. Giannarelli et al. [133] studied the course of retinopathy in 30 PTA recipients and in 35 nontransplanted matched type 1 diabetic patients for 30 months. In both groups, the prevalence of laser treatment and/or proliferative retinopathy was 67%. In the NPDR PTA group, 50% of patients improved by one grade and 50% showed no change. In the LT/PDR PTA group, stabilization was observed in 86% of cases, whereas worsening of retinopathy occurred in 14% of patients. In the NPDR control group, retinopathy improved in 20% of patients, remained unchanged in 10%, and worsened in 70%. In the LT/PDR control group, retinopathy did not change in 43% and deteriorated in 57%. Macular edema in four PTA patients at baseline resolved at follow-up. Despite a relatively short follow-up, successful PTA positively affected the course of retinopathy. However, one PTA and one control patient developed neovascular glaucoma. In both of Giannarelli's studies [130, 131], examinations were masked. Fundus photographs were graded using the EURODIAB Study classification of severity of retinopathy.

Studies on the course of diabetic retinopathy following whole pancreas transplantation have not been designed to investigate the transient early worsening of retinopathy that was found within 12 months of baseline in the DCCT and in the earlier studies [31]. Such occurrences might be expected in the presence of similar important risk factors such as higher level of HbA1c before treatment and its greater reduction with treatment during the first 6 months after transplantation (odds ratio 1.6 for each percent point decrease). Longer duration of diabetes and more severe retinopathy are risk factors when worsening is measured by the development of soft exudates or IRMA, but not when defined as progression of three or more steps on the retinopathy scale. One can also question whether or not the time to recovery from worsening will follow the 50% recovery rate within 6–12 months found in the DCCT and whether or not subsequent progression will be found in 50% of the patients. Large decrease in HbA1c is a risk factor for the development of clinically important outcomes such as severe nonproliferative retinopathy, proliferative retinopathy, and clinically significant macular edema. In the DCCT, the long-term benefits greatly outweighed the risk of early worsening.

ISLET CELL TRANSPLANTATION

Successful islet cell transplantation in experimental animals in 1972 [135] introduced new concepts to avoid complications of whole pancreas graft related to exocrine pancreas and vascular supply and to allow investigation of pretransplant procedures to reduce immunogenicity. The technique of islet cell transplantation improved substantially in 2000 with the development of a glucocorticoid-free immunosuppressive protocol which quickly resulted in exogenous insulin independence with no apparent diabetogenic or toxic effects and improved graft survival [136]. This treatment became known as the Edmonton Protocol. In the Edmonton series, insulin independence was achieved with a transplanted islet cell mass of >9,000 IE/kg of recipient body weight delivered in two or three infusions from 2 to 4 donors [112, 137]. Markmann et al. [138] confirmed the efficacy of the Edmonton immunosuppressive protocol in a series of nine transplanted patients of whom seven achieved insulin independence following either one infusion (five patients) or two infusions (two patients). An international multicenter trial confirmed previous experience with the Edmonton Protocol at single centers [139]. Graft dysfunction requiring renewal of insulin therapy has been observed with longer follow-up. Persistent islet function even without insulin independence provides both protection from severe hypoglycemia and improved levels of glycated hemoglobin [139]. Glucagon secretion in response to hypoglycemia does not improve with islet cell transplantation as described with whole pancreas transplantation [96]. Islet cell transplantation is a minor and safe procedure, performed with local anesthesia via radiologic control of percutaneous cannulation of a portal vein. Islet cell transplantation (predominantly allotransplantation) has an initial graft survival (insulin-free) rate of 44% and C-peptide-producing (but not insulin-free) rate of 80%. Complications include portal vein thrombosis (5%), bleeding (14%), emergency exploratory laparotomy (8%), liver steatosis (23%), and mouth ulcers (77%). Patient survival ≥5 years after transplant is 90% [112, 113].

ADVERSE EFFECTS OF CHRONIC IMMUNOSUPPRESSION

Islet cell transplantation requires long-term, calcineurin-based immunosuppressives with the risk of developing nephrotoxicity, infection, and malignancy. New risks introduced by immunosuppression were reviewed by Larsen [110]. The majority of islet transplant programs use a combination of Sirolimus and Tacrolimus (Edmonton Protocol) and Mycophenolate Mofetil maintenance immunosuppressives. Usually one or more induction immunosuppressive agents are used at the time of the first islet infusion antibody induction therapy [111].

EFFECT OF ISLET CELL TRANSPLANTATION ON RETINOPATHY

As yet, it is unclear what clinical effect an islet transplant will have on secondary diabetic complications. Lee et al. [140], who clinically examined eight patients at least 1 year following islet cell transplantation (>5,000 IE/kg recipient body weight), observed improvement in one patient and no clinical progression of retinopathy in the others. Ryan et al. [141], who presented results of a 5-year follow-up of clinical islet cell transplantation, briefly reported that of 47 completed patients, four required laser treatment or vitrectomy within 5 months of transplantation. A comparison of the efficacy of intensive medical therapy (intensive insulin therapy, and when indicated angiotensin blockade, control of lipids and blood pressure to recommended level) with islet cell transplantation on the progression of microvascular disease, conducted as a prospective nonrandomized crossover cohort study, showed significant differences on the outcomes of metabolic control and retinopathy [142, 143]. Progression was defined as the need for laser treatment or one step or more worsening on the International Disease Severity Scale [144]. Seven-field stereo fundus photography was performed at baseline and annual examinations. Multiple (1–4) islet infusions (total 10,000 IE/kg recipient body weight) were required to induce and sustain insulin independence. Islets were isolated from pancreas of adult heart-beating cadaver organ donors in the Ike Barber Human Islet cell transplantation Laboratory at Vancouver Hospital [145]. Sixty-four percent (16 of 25 patients) remain insulin-independent at mean 36 months follow-up. Glucose control assessed by 3 monthly HbA1c measurements over follow-up 34 ± 17 months (range 6–67 months) improved significantly from mean 8.1 ± 1.2% at entry to mean 7.5 ± 0.9% during medical therapy, and significantly from mean 7.0 ± 0.7% at the time of the first islet cell transplant to mean 6.7 ± 0.7% posttransplant. At baseline, there were 44 eyes with nonproliferative retinopathy and 41 eyes with proliferative retinopathy of which 39 eyes had been treated with laser. The follow-up interval was sufficiently long to assess the outcome of progression of retinopathy. Progression occurred significantly more often in all subjects in the medical group (10/82 eyes, 12.2%) than after islet cell transplantation (0/51 eyes, 0%). Considering only subjects who had received transplants, progression occurred in 6/51 eyes while on medical treatment and 0/51 eyes posttransplant. Endpoint determination was not masked, but the decision to treat subjects with laser was made by retina specialists independently of the study team. This is the first pancreas or islet transplant study to include a medical control group treated to current standards.

Pancreas islet cell transplantation is an evolving minor procedure for the physiologic delivery of insulin with the aim of normalization or near-normalization of HbA1c. It achieves quality of life benefits which accrue for reasons of insulin independence and the promise of a lower prevalence of complications at a later date, as well as a longer lifespan.

REFERENCES

1. EURODIAB ACE Study Group. Variation and trends in incidence of childhood diabetes in Europe. Lancet. 2000;355(9207):873–6.
2. Onkamo P, Vaananen S, Karvonen M, Tuomilehto J. Worldwide increase in incidence of Type I diabetes—the analysis of the data on published incidence trends. Diabetologia. 1999;42(12):1395–403.
3. Klein R, Klein BEK. Vision disorders in diabetes. In: Diabetes in America. 2nd ed. Washington: National Diabetes Data Group. National Institute of Diabetes and Digestive and Kidney Diseases; National Institutes of Health; 1995. p. 293–338.
4. Cusick M, Meleth AD, Agron E, et al. Associations of mortality and diabetes complications in patients with type 1 and type 2 diabetes: early treatment diabetic retinopathy study report no. 27. Diabetes Care. 2005;28(3):617–25.
5. The Diabetes Control and Complications Trial (DCCT). Design and methodologic considerations for the feasibility phase. The DCCT Research Group. Diabetes. 1986;35(5):530–45.

6. Writing Team for the Diabetes Control and Complications Trial/Epidemiology of Diabetes Interventions and Complications Research Group. Effect of intensive therapy on the microvascular complications of type 1 diabetes mellitus. JAMA. 2002;287(19):2563–9.
7. Klein R, Klein BE, Moss SE, DeMets DL, Kaufman I, Voss PS. Prevalence of diabetes mellitus in southern Wisconsin. Am J Epidemiol. 1984;119(1):54–61.
8. Klein R, Klein BE, Moss SE, Cruickshanks KJ. Association of ocular disease and mortality in a diabetic population. Arch Ophthalmol. 1999;117(11):1487–95.
9. Klein R, Klein BE, Moss SE, Cruickshanks KJ. The Wisconsin Epidemiologic Study of diabetic retinopathy. XIV. Ten-year incidence and progression of diabetic retinopathy. Arch Ophthalmol. 1994;112(9):1217–28.
10. Klein R, Knudtson MD, Lee KE, Gangnon R, Klein BE. The Wisconsin Epidemiologic Study of Diabetic Retinopathy XXIII: the twenty-five-year incidence of macular edema in persons with type 1 diabetes. Ophthalmology. 2009;116(3):497–503.
11. Klein R, Klein BE, Moss SE, Davis MD, DeMets DL. The Wisconsin Epidemiologic Study of Diabetic Retinopathy. IX. Four-year incidence and progression of diabetic retinopathy when age at diagnosis is less than 30 years. Arch Ophthalmol. 1989;107(2):237–43.
12. Klein R, Klein BE, Moss SE, Cruickshanks KJ. The Wisconsin Epidemiologic Study of Diabetic Retinopathy: XVII. The 14-year incidence and progression of diabetic retinopathy and associated risk factors in type 1 diabetes. Ophthalmology. 1998;105(10):1801–15.
13. Klein R, Klein BE, Moss SE, Davis MD, DeMets DL. The Wisconsin epidemiologic study of diabetic retinopathy. II. Prevalence and risk of diabetic retinopathy when age at diagnosis is less than 30 years. Arch Ophthalmol. 1984;102(4):520–6.
14. Klein R, Knudtson MD, Lee KE, Gangnon R, Klein BE. The Wisconsin Epidemiologic Study of Diabetic Retinopathy: XXII the twenty-five-year progression of retinopathy in persons with type 1 diabetes. Ophthalmology. 2008;115(11):1859–68.
15. Early Treatment Diabetic Retinopathy Study Research Group. Grading diabetic retinopathy from stereoscopic color fundus photographs—an extension of the modified Airlie House classification. ETDRS report number 10. Ophthalmology. 1991;98(5 Suppl):786–806.
16. Early Treatment Diabetic Retinopathy Study Research Group. Fundus photographic risk factors for progression of diabetic retinopathy. ETDRS report number 12. Ophthalmology. 1991;98(5 Suppl):823–33.
17. Klein R, Klein BE, Moss SE, Davis MD, DeMets DL. Glycosylated hemoglobin predicts the incidence and progression of diabetic retinopathy. JAMA. 1988;260(19):2864–71.
18. Klein R, Klein BE, Moss SE, Cruickshanks KJ. Relationship of hyperglycemia to the long-term incidence and progression of diabetic retinopathy. Arch Intern Med. 1994;154(19): 2169–78.
19. Klein R. Hyperglycemia and microvascular and macrovascular disease in diabetes. Diabetes Care. 1995;18(2):258–68.
20. Lauritzen T, Frost-Larsen K, Larsen HW, Deckert T. Effect of 1 year of near-normal blood glucose levels on retinopathy in insulin-dependent diabetics. Lancet. 1983;1(8318):200–4.
21. Dahl-Jørgensen K, Hanssen KF, Brinchmann-Hansen O, Barbosa J, Micossi P, Brancato R, et al. What happens to the retina as diabetic control is tightened? Lancet. 1983;1(8325):652–3.
22. Lauritzen T, Frost-Larsen K, Larsen HW, Deckert T. Two-year experience with continuous subcutaneous insulin infusion in relation to retinopathy and neuropathy. Diabetes. 1985;34 Suppl 3:74–9.
23. Dahl-Jorgensen K, Brinchmann-Hansen O, Hanssen KF, et al. Effect of near normoglycaemia for two years on progression of early diabetic retinopathy, nephropathy, and neuropathy: the Oslo study. Br Med J (Clin Res Ed). 1986;293(6556):1195–9.

24. The Kroc Collaborative Study Group. Diabetic retinopathy after two years of intensified insulin treatment. Follow-up of the Kroc Collaborative Study. JAMA. 1988;260(1):37–41.
25. Reichard P, Nilsson BY, Rosenqvist U. The effect of long-term intensified insulin treatment on the development of microvascular complications of diabetes mellitus. N Engl J Med. 1993;329(5):304–9.
26. Wang PH, Lau J, Chalmers TC. Meta-analysis of effects of intensive blood-glucose control on late complications of type I diabetes. Lancet. 1993;341(8856):1306–9.
27. Diabetes Control and Complications Trial. Implementation of treatment protocols in the Diabetes Control and Complications Trial. Diabetes Care. 1995;18(3):361–76.
28. The DCCT Research Group. Diabetes Control and Complications Trial (DCCT): results of feasibility study. Diabetes Care. 1987;10(1):1–19.
29. The Diabetes Control and Complications Trial Research Group. The effect of intensive treatment of diabetes on the development and progression of long-term complications in insulin-dependent diabetes mellitus. N Engl J Med. 1993;329(14):977–86.
30. Epidemiology of Diabetes Interventions and Complications (EDIC). Design, implementation, and preliminary results of a long-term follow-up of the Diabetes Control and Complications Trial cohort. Diabetes Care. 1999;22(1):99–111.
31. Diabetes Control and Complications Trial. Early worsening of diabetic retinopathy in the Diabetes Control and Complications Trial. Arch Ophthalmol. 1998;116(7):874–86.
32. The Diabetes Control and Complications Trial. The effect of intensive diabetes treatment on the progression of diabetic retinopathy in insulin-dependent diabetes mellitus. Arch Ophthalmol. 1995;113(1):36–51.
33. Diabetes Control and Complications Trial Research Group. Progression of retinopathy with intensive versus conventional treatment in the Diabetes Control and Complications Trial. Ophthalmology. 1995;102(4):647–61.
34. Diabetes Control and Complications Trial. The relationship of glycemic exposure (HbA1c) to the risk of development and progression of retinopathy in the diabetes control and complications trial. Diabetes. 1995;44(8):968–83.
35. Diabetes Control and Complications Trial (DCCT). The absence of a glycemic threshold for the development of long-term complications: the perspective of the Diabetes Control and Complications Trial. Diabetes. 1996;45(10):1289–98.
36. Kilpatrick ES, Rigby AS, Atkin SL. The effect of glucose variability on the risk of microvascular complications in type 1 diabetes. Diabetes Care. 2006;29(7):1486–90.
37. McCarter RJ, Hempe JM, Chalew SA. Mean blood glucose and biological variation have greater influence on HbA1c levels than glucose instability: an analysis of data from the Diabetes Control and Complications Trial. Diabetes Care. 2006;29(2):352–5.
38. Kilpatrick ES, Rigby AS, Atkin SL. A1C variability and the risk of microvascular complications in type 1 diabetes: data from the Diabetes Control and Complications Trial. Diabetes Care. 2008;31(11):2198–202.
39. Lachin JM, Genuth S, Nathan DM, Zinman B, Rutledge BN. DCCT/EDIC Research Group. Effect of glycemic exposure on the risk of microvascular complications in the diabetes control and complications trial—revisited. Diabetes. 2008;57(4):995–1001.
40. The DCCT Research Group. Epidemiology of severe hypoglycemia in the diabetes control and complications trial. Am J Med. 1991;90(4):450–9.
41. The DCCT Research Group. Adverse events and their association with treatment regimens in the diabetes control and complications trial. Diabetes Care. 1995;18(11):1415–27.
42. Fanelli CG, Porcellati F, Pampanelli S, Bolli GB. Insulin therapy and hypoglycaemia: the size of the problem. Diabetes Metab Res Rev. 2004;20 Suppl 2:S32–42.

43. Egger M, Davey Smith G, Stettler C, Diem P. Risk of adverse effects of intensified treatment in insulin-dependent diabetes mellitus: a meta-analysis. Diabet Med. 1997;14(11): 919–28.
44. EURODIAB IDDM Complications Study. Microvascular and acute complications in IDDM patients: the EURODIAB IDDM Complications Study. Diabetologia. 1994;37(3):278–85.
45. The Diabetes Control and Complications Trial Research Group. Influence of intensive diabetes treatment on body weight and composition of adults with type 1 diabetes in the Diabetes Control and Complications Trial. Diabetes Care. 2001;24(10):1711–21.
46. The DCCT Research Group. Weight gain associated with intensive therapy in the diabetes control and complications trial. Diabetes Care. 1988;11(7):567–73.
47. The Diabetes Control and Complications Trial Research Group. Effect of intensive therapy on residual beta-cell function in patients with type 1 diabetes in the diabetes control and complications trial. A randomized, controlled trial. Ann Intern Med. 1998;128(7):517–23.
48. Steffes MW, Sibley S, Jackson M, Thomas W. Beta-cell function and the development of diabetes-related complications in the diabetes control and complications trial. Diabetes Care. 2003;26(3):832–6.
49. Klein R, Moss S. A comparison of the study populations in the Diabetes Control and Complications Trial and the Wisconsin Epidemiologic Study of Diabetic Retinopathy. Arch Intern Med. 1995;155(7):745–54.
50. The Diabetes Control and Complications Trial/Epidemiology of Diabetes Interventions and Complications Research Group. Retinopathy and nephropathy in patients with type 1 diabetes four years after a trial of intensive therapy. N Engl J Med. 2000;342(6):381–9.
51. White NH, Sun W, Cleary PA, et al. Prolonged effect of intensive therapy on the risk of retinopathy complications in patients with type 1 diabetes mellitus: 10 years after the Diabetes Control and Complications Trial. Arch Ophthalmol. 2008;126(12):1707–15.
52. Nathan DM, Cleary PA, Backlund JY, et al. Intensive diabetes treatment and cardiovascular disease in patients with type 1 diabetes. N Engl J Med. 2005;353(25):2643–53.
53. The DCCT Research Group. Reliability and validity of a diabetes quality-of-life measure for the diabetes control and complications trial (DCCT). Diabetes Care. 1988;11(9): 725–32.
54. Diabetes Control and Complications Trial Research Group. Influence of intensive diabetes treatment on quality-of-life outcomes in the diabetes control and complications trial. Diabetes Care. 1996;19(3):195–203.
55. Johnson JA, Kotovych M, Ryan EA, Shapiro AM. Reduced fear of hypoglycemia in successful islet cell transplantation. Diabetes Care. 2004;27(2):624–5.
56. Huebschmann AG, Regensteiner JG, Vlassara H, Reusch JE. Diabetes and advanced glycoxidation end products. Diabetes Care. 2006;29(6):1420–32.
57. Goh SY, Cooper ME. Clinical review: the role of advanced glycation end products in progression and complications of diabetes. J Clin Endocrinol Metab. 2008;93(4):1143–52.
58. Monnier VM, Bautista O, Kenny D, et al. Skin collagen glycation, glycoxidation, and crosslinking are lower in subjects with long-term intensive versus conventional therapy of type 1 diabetes: relevance of glycated collagen products versus HbA1c as markers of diabetic complications. DCCT Skin Collagen Ancillary Study Group. Diabetes Control and Complications Trial. Diabetes. 1999;48(4):870–80.
59. Genuth S, Sun W, Cleary P, et al. Glycation and carboxymethyllysine levels in skin collagen predict the risk of future 10-year progression of diabetic retinopathy and nephropathy in the diabetes control and complications trial and epidemiology of diabetes interventions and complications participants with type 1 diabetes. Diabetes. 2005;54(11):3103–11.
60. Cahill GFJ. Metabolic memory. N Engl J Med. 1980;302(7):396–7.

61. Ceriello A, Ihnat MA, Thorpe JE. Clinical review 2: the "metabolic memory": is more than just tight glucose control necessary to prevent diabetic complications? J Clin Endocrinol Metab. 2009;94(2):410–5.
62. Brownlee M. The pathobiology of diabetic complications: a unifying mechanism. Diabetes. 2005;54(6):1615–25.
63. Beckman JS, Koppenol WH. Nitric oxide, superoxide, and peroxynitrite: the good, the bad, and ugly. Am J Physiol. 1996;271(5 Pt 1):C1424–37.
64. Foury F, Hu J, Vanderstraeten S. Mitochondrial DNA mutators. Cell Mol Life Sci. 2004;61(22):2799–811.
65. Piconi L, Ihnat MA, Ceriello A. Oxidative stress in the pathogenesis/treatment of diabetes and its complications. Curr Nutr Food Sci. 2007;3(3):194–9.
66. Goldin A, Beckman JA, Schmidt AM, Creager MA. Advanced glycation end products: sparking the development of diabetic vascular injury. Circulation. 2006;114(6):597–605.
67. Ihnat MA, Thorpe JE, Kamat CD, et al. Reactive oxygen species mediate a cellular "memory" of high glucose stress signalling. Diabetologia. 2007;50(7):1523–31.
68. Frank RN. Metabolic memory in diabetes is true long-term memory. Arch Ophthalmol. 2009;127(3):330–1.
69. Skrivarhaug T, Fosmark DS, Stene LC, et al. Low cumulative incidence of proliferative retinopathy in childhood-onset type 1 diabetes: a 24-year follow-up study. Diabetologia. 2006;49(10):2281–90.
70. Hovind P, Tarnow L, Rossing K, et al. Decreasing incidence of severe diabetic microangiopathy in type 1 diabetes. Diabetes Care. 2003;26(4):1258–64.
71. Backlund LB, Algvere PV, Rosenqvist U. New blindness in diabetes reduced by more than one-third in Stockholm County. Diabet Med. 1997;14(9):732–40.
72. Wong TY, Mwamburi M, Klein R, et al. Rates of progression in diabetic retinopathy during different time periods. Diabetes Care. 2009;32(12):2307–13.
73. Lecaire T, Palta M, Zhang H, Allen C, Klein R, D'Alessio D. Lower-than-expected prevalence and severity of retinopathy in an incident cohort followed during the first 4–14 years of type 1 diabetes: the Wisconsin Diabetes Registry Study. Am J Epidemiol. 2006;164(2):143–50.
74. Knudsen LL, Lervang HH, Lundbye-Christensen S, Gorst-Rasmussen A. The North Jutland County Diabetic Retinopathy Study: population characteristics. Br J Ophthalmol. 2006;90(11):1404–9.
75. Pambianco G, Costacou T, Ellis D, Becker DJ, Klein R, Orchard TJ. The 30-year natural history of type 1 diabetes complications: the Pittsburgh Epidemiology of Diabetes Complications Study experience. Diabetes. 2006;55(5):1463–9.
76. Mauer M, Zinman B, Gardiner R, et al. Renal and retinal effects of enalapril and losartan in type 1 diabetes. N Engl J Med. 2009;361(1):40–51.
77. Heller SR. Minimizing hypoglycemia while maintaining glycemic control in diabetes. Diabetes. 2008;57(12):3177–83.
78. Davis S, Alonso MD. Hypoglycemia as a barrier to glycemic control. J Diabetes Complications. 2004;18(1):60–8.
79. Polonsky WH, Anderson BJ, Lohrer PA, Aponte JE, Jacobson AM, Cole CF. Insulin omission in women with IDDM. Diabetes Care. 1994;17(10):1178–85.
80. Rodin G, Olmsted MP, Rydall AC, et al. Eating disorders in young women with type 1 diabetes mellitus. J Psychosom Res. 2002;53(4):943–9.
81. Morris AD, Boyle DI, McMahon AD, Greene SA, MacDonald TM, Newton RW. Adherence to insulin treatment, glycaemic control, and ketoacidosis in insulin-dependent diabetes mellitus. The DARTS/MEMO Collaboration. Diabetes Audit and

Research in Tayside Scotland. Medicines Monitoring Unit. Lancet. 1997;350(9090): 1505–10.
82. Lee WC, Balu S, Cobden D, Joshi AV, Pashos CL. Prevalence and economic consequences of medication adherence in diabetes: a systematic literature review. Manag Care Interface. 2006;19(7):31–41.
83. UK Hypoglycaemia Study Group. Risk of hypoglycaemia in types 1 and 2 diabetes: effects of treatment modalities and their duration. Diabetologia. 2007;50(6):1140–7.
84. Donnelly LA, Morris AD, Frier BM, et al. Frequency and predictors of hypoglycaemia in Type 1 and insulin-treated Type 2 diabetes: a population-based study. Diabet Med. 2005;22(6):749–55.
85. Leichter S. Is the use of insulin analogues cost-effective? Adv Ther. 2008;25(4):285–99.
86. Hermansen K, Davies M. Does insulin detemir have a role in reducing risk of insulin-associated weight gain? Diabetes Obes Metab. 2007;9(3):209–17.
87. Ashwell SG, Amiel SA, Bilous RW, et al. Improved glycaemic control with insulin glargine plus insulin lispro: a multicentre, randomized, cross-over trial in people with Type 1 diabetes. Diabet Med. 2006;23(3):285–92.
88. Jeitler K, Horvath K, Berghold A, et al. Continuous subcutaneous insulin infusion versus multiple daily insulin injections in patients with diabetes mellitus: systematic review and meta-analysis. Diabetologia. 2008;51(6):941–51.
89. Wentholt IM, Hoekstra JB, Devries JH. Continuous glucose monitors: the long-awaited watch dogs? Diabetes Technol Ther. 2007;9(5):399–409.
90. Chetty VT, Almulla A, Odueyungbo A, Thabane L. The effect of continuous subcutaneous glucose monitoring (CGMS) versus intermittent whole blood finger-stick glucose monitoring (SBGM) on hemoglobin A1c (HBA1c) levels in Type I diabetic patients: a systematic review. Diabetes Res Clin Pract. 2008;81(1):79–87.
91. British Cardiac Society, British Hypertension Society, Diabetes UK, HEART UK, Primary Care Cardiovascular Society, Stroke Association. JBS 2: Joint British Societies' guidelines on prevention of cardiovascular disease in clinical practice. Heart. 2005;91 Suppl 5:v1–52.
92. American Diabetes Association. Standards of medical care in diabetes—2009. Diabetes Care. 2009;32 Suppl 1:S13–61.
93. Canadian Diabetes Association Clinical Practice Guidelines Expert Committee. Canadian Diabetes Association 2008 clinical practice guidelines for the prevention and management of diabetes in Canada. Can J Diabetes. 2008;32:S1–201.
94. Cryer PE. The barrier of hypoglycemia in diabetes. Diabetes. 2008;57(12):3169–76.
95. Bolli G, de Feo P, Compagnucci P, et al. Abnormal glucose counterregulation in insulin-dependent diabetes mellitus. Interaction of anti-insulin antibodies and impaired glucagon and epinephrine secretion. Diabetes. 1983;32(2):134–41.
96. Kendall DM, Teuscher AU, Robertson RP. Defective glucagon secretion during sustained hypoglycemia following successful islet allo- and autotransplantation in humans. Diabetes. 1997;46(1):23–7.
97. Cryer PE, Gerich JE. Glucose counterregulation, hypoglycemia, and intensive insulin therapy in diabetes mellitus. N Engl J Med. 1985;313(4):232–41.
98. Cranston I, Lomas J, Maran A, Macdonald I, Amiel SA. Restoration of hypoglycaemia awareness in patients with long-duration insulin-dependent diabetes. Lancet. 1994; 344(8918):283–7.
99. Lingenfelser T, Buettner U, Martin J, et al. Improvement of impaired counterregulatory hormone response and symptom perception by short-term avoidance of hypoglycemia in IDDM. Diabetes Care. 1995;18(3):321–5.

100. Pampanelli S, Fanelli C, Lalli C, et al. Long-term intensive insulin therapy in IDDM: effects on HbA1c, risk for severe and mild hypoglycaemia, status of counterregulation and awareness of hypoglycaemia. Diabetologia. 1996;39(6):677–86.
101. Fanelli C, Pampanelli S, Epifano L, et al. Long-term recovery from unawareness, deficient counterregulation and lack of cognitive dysfunction during hypoglycaemia, following institution of rational, intensive insulin therapy in IDDM. Diabetologia. 1994;37(12):1265–76.
102. Palmer JP, Fleming GA, Greenbaum CJ, et al. C-peptide is the appropriate outcome measure for type 1 diabetes clinical trials to preserve beta-cell function: report of an ADA workshop, 21–22 October 2001. Diabetes. 2004;53(1):250–64.
103. Nakanishi K, Inoko H. Combination of HLA-A24, -DQA1*03, and -DR9 contributes to acute-onset and early complete beta-cell destruction in type 1 diabetes: longitudinal study of residual beta-cell function. Diabetes. 2006;55(6):1862–8.
104. Gottsater A, Landin-Olsson M, Fernlund P, Gullberg B, Lernmark A, Sundkvist G. Pancreatic beta-cell function evaluated by intravenous glucose and glucagon stimulation. A comparison between insulin and C-peptide to measure insulin secretion. Scand J Clin Lab Invest. 1992;52(7):631–9.
105. Sherry NA, Tsai EB, Herold KC. Natural history of beta-cell function in type 1 diabetes. Diabetes. 2005;54 Suppl 2:S32–9.
106. The DCCT Research Group. Effects of age, duration and treatment of insulin-dependent diabetes mellitus on residual beta-cell function: observations during eligibility testing for the Diabetes Control and Complications Trial (DCCT). J Clin Endocrinol Metab. 1987;65(1):30–6.
107. Palmer JP. C-peptide in the natural history of type 1 diabetes. Diabetes Metab Res Rev. 2009;25(4):325–8.
108. Nakanishi K, Watanabe C. Rate of beta-cell destruction in type 1 diabetes influences the development of diabetic retinopathy: protective effect of residual beta-cell function for more than 10 years. J Clin Endocrinol Metab. 2008;93(12):4759–66.
109. Hills CE, Brunskill NJ. Cellular and physiological effects of C-peptide. Clin Sci (Lond). 2009;116(7):565–74.
110. Larsen JL. Pancreas transplantation: indications and consequences. Endocr Rev. 2009;25(6):919–46.
111. Meloche RM. Transplantation for the treatment of type 1 diabetes. World J Gastroenterol. 2007;13(47):6347–55.
112. Ryan EA, Lakey JR, Paty BW, et al. Successful islet cell transplantation: continued insulin reserve provides long-term glycemic control. Diabetes. 2002;51(7):2148–57.
113. Vrochides D, Paraskevas S, Papanikolaou V. Transplantation for type 1 diabetes mellitus. Whole organ or islets? Hippokratia. 2009;13(1):6–8.
114. Barrou Z, Seaquist ER, Robertson RP. Pancreas transplantation in diabetic humans normalizes hepatic glucose production during hypoglycemia. Diabetes. 1994;43(5):661–6.
115. Chow VC, Pai RP, Chapman JR, et al. Diabetic retinopathy after combined kidney-pancreas transplantation. Clin Transplant. 1999;13(4):356–62.
116. Wang Q, Klein R, Moss SE, et al. The influence of combined kidney-pancreas transplantation on the progression of diabetic retinopathy. A case series. Ophthalmology. 1994;101(6):1071–6.
117. Bandello F, Vigano C, Secchi A, et al. Effect of pancreas transplantation on diabetic retinopathy: a 20-case report. Diabetologia. 1991;34 Suppl 1:S92–4.
118. Scheider A, Meyer-Schwickerath E, Nusser J, Land W, Landgraf R. Diabetic retinopathy and pancreas transplantation: a 3-year follow-up. Diabetologia. 1991;34 Suppl 1:S95–9.

119. Sutherland DE, Dunn DL, Goetz FC, et al. A 10-year experience with 290 pancreas transplants at a single institution. Ann Surg. 1989;210(3):274–85; discussion 285–8.
120. Petersen MR, Vine AK. Progression of diabetic retinopathy after pancreas transplantation. The University of Michigan Pancreas Transplant Evaluation Committee. Ophthalmology. 1990;97(4):496–500; discussion 501–2.
121. Caldara R, Bandello F, Vigano C, et al. Influence of successful pancreaticorenal transplantation on diabetic retinopathy. Transplant Proc. 1994;26(2):490.
122. Munda R, First MR, Kranias G, Alexander JW. Effects of pancreatic transplantation on diabetic complications. Transplant Proc. 1989;21(1 Pt 3):2865–6.
123. Zech JC, Trepsat C, Gain-Gueugnon M, Lefrancois N, Martin X, Dubernard JM. Ophthalmologic follow-up of type I diabetic patients after kidney and pancreas transplantation. Transplant Proc. 1992;24(3):874.
124. Ramsay RC, Goetz FC, Sutherland DE, et al. Progression of diabetic retinopathy after pancreas transplantation for insulin-dependent diabetes mellitus. N Engl J Med. 1988;318(4):208–14.
125. Koznarova R, Saudek F, Sosna T, et al. Beneficial effect of pancreas and kidney transplantation on advanced diabetic retinopathy. Cell Transplant. 2000;9(6):903–8.
126. Sosna T, Saudek F, Dominek Z. Effect of successful combined renal and pancreatic transplantation on diabetic retinopathy. Acta Univ Palacki Olomuc Fac Med. 1998;141:75–7.
127. Marchetti P, Boggi U, Coppelli A, et al. Pancreas transplant alone. Transplant Proc. 2004;36(3):569–70.
128. Konigsrainer A, Miller K, Steurer W, et al. Does pancreas transplantation influence the course of diabetic retinopathy? Diabetologia. 1991;34 Suppl 1:S86–8.
129. Landgraf R, Nusser J, Muller W, et al. Fate of late complications in type I diabetic patients after successful pancreas-kidney transplantation. Diabetes. 1989;38 Suppl 1:33–7.
130. Ulbig M, Kampik A, Thurau S, Landgraf R, Land W. Long-term follow-up of diabetic retinopathy for up to 71 months after combined renal and pancreatic transplantation. Graefes Arch Clin Exp Ophthalmol. 1991;229(3):242–5.
131. Pearce IA, Ilango B, Sells RA, Wong D. Stabilisation of diabetic retinopathy following simultaneous pancreas and kidney transplant. Br J Ophthalmol. 2000;84(7):736–40.
132. Giannarelli R, Coppelli A, Sartini M, et al. Effects of pancreas-kidney transplantation on diabetic retinopathy. Transpl Int. 2005;18(5):619–22.
133. Giannarelli R, Coppelli A, Sartini MS, et al. Pancreas transplant alone has beneficial effects on retinopathy in type 1 diabetic patients. Diabetologia. 2006;49(12):2977–82.
134. Kotoula MG, Koukoulis GN, Zintzaras E, Karabatsas CH, Chatzoulis DZ. Metabolic control of diabetes is associated with an improved response of diabetic retinopathy to panretinal photocoagulation. Diabetes Care. 2005;28(10):2454–7.
135. Reckard CR, Barker CF. Transplantation of isolated pancreatic islets across strong and weak histocompatibility barriers. Transplant Proc. 1973;5(1):761–3.
136. Shapiro AM, Lakey JR, Ryan EA, et al. Islet cell transplantation in seven patients with type 1 diabetes mellitus using a glucocorticoid-free immunosuppressive regimen. N Engl J Med. 2000;343(4):230–8.
137. Ryan EA, Lakey JR, Rajotte RV, et al. Clinical outcomes and insulin secretion after islet cell transplantation with the Edmonton protocol. Diabetes. 2001;50(4):710–9.
138. Markmann JF, Deng S, Huang X, et al. Insulin independence following isolated islet cell transplantation and single islet infusions. Ann Surg. 2003;237(6):741–9; discussion 749–50.
139. Shapiro AM, Ricordi C, Hering BJ, et al. International trial of the Edmonton protocol for islet cell transplantation. N Engl J Med. 2006;355(13):1318–30.

140. Lee TC, Barshes NR, O'Mahony CA, et al. The effect of pancreatic islet cell transplantation on progression of diabetic retinopathy and neuropathy. Transplant Proc. 2005;37(5):2263–5.
141. Ryan EA, Paty BW, Senior PA, et al. Five-year follow-up after clinical islet cell transplantation. Diabetes. 2005;54(7):2060–9.
142. Warnock GL, Thompson DM, Meloche RM, et al. A multi-year analysis of islet cell transplantation compared with intensive medical therapy on progression of complications in type 1 diabetes. Transplantation. 2008;86(12):1762–6.
143. Thompson DM, Begg IS, Harris C, et al. Reduced progression of diabetic retinopathy after islet cell transplantation compared with intensive medical therapy. Transplantation. 2008;85(10):1400–5.
144. Wilkinson CP, Ferris III FL, Klein RE, et al. Proposed international clinical diabetic retinopathy and diabetic macular edema disease severity scales. Ophthalmology. 2003;110(9):1677–82.
145. Warnock GL, Meloche RM, Thompson D, et al. Improved human pancreatic islet isolation for a prospective cohort study of islet cell transplantation vs. best medical therapy in type 1 diabetes mellitus. Arch Surg. 2005;140(8):735–44.

Index

From: *Ophthalmology Research: Visual Dysfunction in Diabetes*
Edited by: J. Tombran-Tink et al. (eds.), DOI 10.1007/978-1-60761-150-9

MIX
Papier aus verantwortungsvollen Quellen
Paper from responsible sources
FSC® C105338

If you have any concerns about our products,
you can contact us on
ProductSafety@springernature.com

In case Publisher is established outside the EU,
the EU authorized representative is:
Springer Nature Customer Service Center GmbH
Europaplatz 3, 69115 Heidelberg, Germany

Printed by Libri Plureos GmbH
in Hamburg, Germany